Introduction to PC Communications

Philip L. Becker

with

Mike Robertson

Mark Chambers

Phil James

Alan C. Elliott

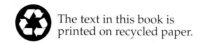 The text in this book is
printed on recycled paper.

Introduction to PC Communications

Copyright ©1992 by Que® Corporation.

All rights reserved. Printed in the United States of America. No part of this book may be used or reproduced in any form or by any means, or stored in a database or retrieval system, without prior written permission of the publisher except in the case of brief quotations embodied in critical articles and reviews. Making copies of any part of this book for any purpose other than your own personal use is a violation of United States copyright laws. For information, address Que Corporation, 11711 N. College Ave., Carmel, IN 46032.

Library of Congress Catalog No.: 91-62313

ISBN 0-88022-747-8

93   5 4 3

Interpretation of the printing code: the rightmost double-digit number is the year of the book's printing; the rightmost single-digit number, the number of the book's printing. For example, a printing code of 92-1 shows that the first printing of the book occurred in 1992.

Screens reproduced in this book were created using Collage Plus from Inner Media, Inc., Hollis, NH.

Publisher: Lloyd J. Short

Associate Publisher: Karen A. Bluestein

Acquisitions Manager: Rick Ranucci

Product Development Manager: Thomas H. Bennett

Managing Editor: Paul Boger

Book Designer: Scott Cook

Production Team: Claudia Bell, Brad Chinn, Brook Farling, Sandy Grieshop, Audra Hershman, Betty Kish, Phil Kitchel, Bob LaRoche, Laurie Lee, Diana Moore, Anne Owen, Juli Pavey, Cindy L. Phipps, Tad Ringo, Linda Seifert, John Sleeva, Kevin Spear, Bruce Steed, Suzanne Tully, Lisa A. Wilson, Phil Worthington

Product Director
Walter R. Bruce III

Production Editor
Lori A. Lyons

Editors
Sara Allaei
Sandra Blackthorn
Kelly Currie
Beth Hoger
Heidi Weas Muller

Technical Editor
Phil James

Composed in Garamond and Macmillan
by Que Corporation

Phil Becker's career in computer communications was launched in 1966 when he first used a modem and a Teletype to call a computer from his apartment. Phil received his electrical engineering degree from Vanderbilt University and then moved to Colorado where he worked at Martin Marietta on computer systems for the Apollo Moon program and the Viking Mars Lander program. While Phil's specialties in those years were operating system and language compiler design, his fascination for computer communications caused him to be frequently involved in large synchronous and asynchronous computer communications projects. When microcomputers appeared, he used them to design numerous communications projects including the communications system for the hundreds of mirror units (heliostats) in the Sandia and Barstow solar generating projects.

In 1983 Phil left Martin Marietta and formed Becker Systems, a computer communications consulting house. During this period of his career he designed communications front ends and X.25 switches for mainframe computers. Since 1985 Phil has been president of eSoft, Inc., a company that manufactures and sells his TBBS bulletin board software. He is also a popular speaker on the subject of computer communications, and his articles on computer communications and the computer industry have been widely published.

Mike Robertson is the creator and original system operator of the Datastorm Technical Support Forum on CompuServe. His favorite pastimes include writing short stories and scripts for his video-artist wife, playing bass, and writing music for local jazz and reggae groups in Columbia, Missouri.

Mark Chambers has been a PC software technical writer and bulletin board system operator for over five years. He and his wife Anne live in Columbia, Missouri, with their daughter, Erin.

Phil James is manager of Product Development at DATASTORM TECHNOLOGIES in Columbia, Missouri. Phil also is a free-lance writer, editor, and performance artist.

Alan C. Elliott, M.A.S., is assistant director of Academic Computing Services at the University of Texas Southwestern Medical Center in Dallas. He is the author of several books, including *Using Norton Utilities, Introduction to Microcomputing with Applications, A Daily Dose of the American Dream,* and *PC Programming Techniques.* He is coauthor of the 1985 and 1988 editions of the *Directory of Statistical Microcomputing Software.* His programming credits include Kwikstat (statistical data analysis) and PC-CAI (a computer-assisted instruction language), published by TexaSoft. His articles have appeared in professional and popular periodicals, including *PCWeek, Collegiate Microcomputer Journal,* and *Communications in Statistics.*

TRADEMARK
ACKNOWLEDGMENTS

Que Corporation has made every effort to supply trademark information about company names, products, and services mentioned in this book. Trademarks indicated below were derived from various sources. Que Corporation cannot attest to the accuracy of this information.

3Com is a registered trademark and EtherNet is a trademark of 3Com Corporation. Amiga 1200RS is a registered trademark of Commodore-Amiga, Inc. ANSI is a registered trademark of American National Standards Institute. Apple and Macintosh are registered trademarks of Apple Computer, Inc. ARCNET is a registered trademark of Datapoint Corporation. AT&T is a registered trademark and UNIX is a trademark of American Telephone & Telegraph Company. The Brooklyn Bridge is a trademark of Fifth Generation Systems, Inc. Carbon Copy Plus, Virex, and Virex-PC are trademarks of Microcom Software Division. cc:Mail is a trademark of ccMail, Inc. CompuServe is a registered trademark of CompuServe Incorporated. Crosstalk for Windows and Remote 2 are trademarks of DCA/Crosstalk Communications, Inc. DaVinci and eMail are trademarks of DaVinci Systems. DEC, DECnet, and (CP/M) are registered trademarks of Digital Equipment Corporation. DESQview is a trademark of Quarterdeck Office Systems. Freeware is a trademark of Andrew Fluegelman. GEnie is a service mark of General Electric Corporation. Hayes, Smartcom, and Smartmodem are registered trademarks of Hayes Microcomputer Products, Inc. Higgins is a trademark of Enable Software. IBM, Personal Computer AT, and PS/2 are registered trademarks and Token-Ring Bridge Program is a trademark of International Business Machines Corporation. IRMA Windowlink is a trademark of Digital Communications Associates Inc. LapLink is a trademark of Traveling Software, Inc. Lotus and 1-2-3 are registered trademarks of Lotus Development Corporation. MCI and MCI Mail are registered service marks of MCI Communications Corporation. Microsoft is a registered trademark of Microsoft Corporation. NetWare and Novell are registered trademarks of Novell, Inc. Network File System (NFS) is a trademark of Sun Microsystems, Inc. PROCOMM Lite and PROCOMM PLUS are registered trademarks of DATASTORM TECHNOLOGIES, INC. pcANYWHERE IV is a trademark of Dynamic Microprocessor Associates. Prodigy is a registered trademark of Prodigy Services Corporation. TBBS (The Bread Board System) is a trademark of eSoft, Inc. Telenet is a registered trademark of GTE Telenet Communications Corporation. The Norton Utilities is a trademark of Symantec Corporation. The Source is a service mark of Source Telecomputing Corporation, a subsidiary of The Reader's Digest Association, Inc. Tymnet is a registered trademark of Tymnet, Inc. VSAFE is a trademark of Central Point Software, Inc. VIRUSCAN is a copyright of McAfee Associates. WordPerfect and WordPerfect Office are registered trademarks of WordPerfect Corporation. WordStar is a registered trademark of WordStar International Incorporated. Zenith is a registered trademark of Zenith Electronics Corporation. ZMODEM-90 is a trademark of OMEN Technology, Inc. Trademarks of other products mentioned in this book are held by the companies producing them.

Contents at a Glance

TABLE OF CONTENTS ▽

I   Learning PC Communications Fundamentals

When I began my career in computers in 1966, I found that the one area which even the most capable computer people feared was computer communications. Because I was fascinated by computer communications, I learned about this area early in my career, and in that process I discovered that it wasn't as difficult as it was portrayed to be at all. The problem, it turns out, is that no one ever really quite tells you how computer communications work. Often those who use computer communications everyday don't know themselves how they work; they have just discovered enough about communications through trial and error to get by.

Today, over 25 years later, I see millions of people facing that same hurdle to understanding computer communications that I once faced. I have created this book because of my desire to put in one place as much information as I can about how computer communications really works in terms that you can understand. My goal with this book is to give you a fundamental understanding of computer communications of all types and also to acquaint you with the techniques and capabilities of this unique aspect of computing.

This book will enable you to gain access to the world of computer communications in a way that was previously possible only if you spent years surrounded by experts in the field. This book will not only lead you to understand communications if you don't know anything about them, but it will also stay with you as a reference work even if you become a programmer who works in the field of computer communications. Finding out how computer communications works is difficult at every level. I hope you find this book will clear away much of the mystery about computer communications and enable you to understand why I find it to be the most exciting area of computing today.

Introduction

From their inception computers have held out the promise of world-wide instant communications. Science fiction novels predicted home delivery of newspapers, automated accounting of households, and most tantalizing of all—universal access to any information you may want. Computers were supposed to make this access so easy that we could take it for granted.

Today technology is finally making many of these visions real. From your home or office you can do research that once was impossible, or prohibitively expensive. You can converse with people from all walks of life and use this personal networking to find out where to get the information you need. Rapid response to changing situations now is possible from anywhere in the world.

In a world in which information is power, you need to know how to gain access to that information. PC communications (also known as tele-computing) can put you in touch with information that previously was available only to a small number of experienced people. All that is required is your PC, a modem, and the knowledge of how to use them.

PC communications is at once exciting and frightening—exciting because of the potential of sharing information over a wide area, and frightening because it seems that so much needs to be learned and so few opportunities to find the needed information are available.

Introduction to PC Communications clears away the mystery that surrounds modems and serial data connections. This book teaches you what you need to know to understand and effectively use this rapidly growing area of computing. You no longer will fear using PC communications because you will be able to make it work for you.

What Is PC Communications?

PC communications is any method by which one computer can connect remotely to another and exchange information. This covers a very large variety of technologies. At one extreme are Wide Area Networks (WANs), using dedicated satellites and fiber-optic connections. At the other is a simple cable connecting a laptop computer to a desktop PC to transfer a file. This book focuses on that area of PC communications which offers you the most return for the least investment—dial-up modem and direct connect serial communications. This book also discusses alternate ways to connect PCs together to share files and printers as well as how to connect PCs to mainframe and minicomputers to share information.

Dial-up PC communications uses the same telephone equipment you use to make normal voice calls. This is why it also is called *telecomputing*—an abbreviation for "computing through the telephone." With the addition of a modem, a computer is able to use the telephone system to make and receive data calls. This enables any two computers to connect and exchange information anywhere there is a telephone.

Any information that may be stored on a computer may be exchanged using PC communications. Data files of any type may be sent from one computer to another. Central sites may do electronic publishing of documents to many remote sites, and central databases may be accessed remotely.

PC communications gives you the ability to connect any two computers as needed (and only when needed) to transfer information and, therefore, is an ideal method for sharing information among a very large group of computer users. This is especially true if the users are widely dispersed, or if they travel from place to place. Because the normal telephone system is used to make the connection between computers, mobile computer users (such as salesmen or reporters) can always connect with an office computer using only a laptop computer and any telephone. Because the connections are only present and being paid for when they are in use, PC communications also is the most economical method of connecting computers.

Why Is PC Communications So Confusing?

Although PC communications is simple in overall concept, it immediately becomes very confusing when you begin to use it. You seem to be confronted with a nearly endless list of obscure new terms you must learn before you can even begin, which can seem overwhelming and unordered.

The reason for this is twofold. First, communications is a rapidly evolving area of computing, requiring you to keep up with many changes and innovations. Second, communications is not something that was developed by a single manufacturer, so everyone has to agree on how to do each new step before there is a single way to do it. As a result, PC communications today is an example of design by committee—a market-driven democratic committee at that.

As an example, the following names are used for exactly the same piece of communications hardware: serial port, com port, SIO card, asynchronous communications adapter, communications interface, and serial interface. You probably have heard others as well. And the serial port is the oldest and most fundamental piece of hardware involved in serial communications.

Serial communications began several decades ago with stock market ticker tapes and news wire service Teletype machines. Serial communications was adapted to computers because it allowed use of existing remote printing equipment. However, computers quickly needed much more capability than the original system provided. Each computer manufacturer built its own version of an enhanced way to do this job. Each was proprietary and none would connect to another.

Because a nearly infinite number of ways exist to connect wires, choose operating voltages, arrange data formats, and so on, serial communications quickly became a "Tower of Babel" and has never fully recovered from this. Parts of the technology you must learn only because no universal agreement has been reached on which of several methods everyone should use.

Examples of this in hardware occur in the cable and connector you must use. If you have two pieces of equipment to connect that don't use the same connector or cable, you have to understand enough about what is being connected to be able to forge the proper connection. This is eliminated to some degree by purchasing all of your equipment from one company, but sooner or later you will need to solve a connection problem. *Introduction to PC Communications* will be a valuable resource when this happens.

In the software is a set of "communications parameters" that you also need to set properly. Here again, most of these have no real technical merits or demerits. They simply represent different methods of doing the same thing. Just as the cable and connectors must match on both ends of a connection, these parameters must be set the same way or things will not work properly. This book helps you understand what these settings mean so that you can intelligently troubleshoot any problems they are causing you.

As you read this book and learn about communications, remember that communications really is very simple at its core. You are trying to make a phone call from one computer to another or connect two computers with wires and send data between them. You are trying to send the same data you type from your keyboard and see on your display through the wires from one computer to another. The confusion arises from the many different terms and from the fact that if you don't match up everything on both ends, it won't work.

At one time, a standard floppy disk had over 50 incompatible formats. Time has made that problem largely disappear. We can hope that as PC communications becomes more widely used, the same type of agreements will occur and you no longer will need to know many of the details to use the technology. Until then, you need a book like this one to sort out all of your options. You also need to learn how to recognize when you have a simple problem which is the result of not having all of the settings set the same on both ends.

What Tasks Use PC Communications?

PC communications can be used to solve almost any data access or data transportation problem. Some examples of the types of problems that PC communications may be used to solve are as follows:

- Local and wide area document access (electronic publishing)
- Sharing a printer among more than one PC
- Financial data acquisition and distribution
- Remote order entry and catalog presentation
- Customer support services
- Wide area database access
- Electronic mail
- Message based or real-time conferencing
- Opinion surveys or polls
- Gathering data files from remote sites into a central office

With the proper software and hardware, PC communications can solve an amazing variety of problems. These tasks, however, all tend to fall into one or more of the following categories:

- Electronic mail and messaging

- File transfer

- On-line research

- On-line transaction processing

- Entertainment

Electronic Mail and Messaging

Electronic mail enables you to send computer messages to another person using PC communications. Today you can send a computer message directly or through data networks to nearly anywhere in the world. This electronic version of normal paper mail is used in many applications when information needs to be sent from one computer to another. Electronic mail has even spawned commercial services such as MCI Mail, which provides only electronic mail.

Messaging is different from electronic mail because it is a group activity. Electronic mail is person to person in its nature, in the same way that normal paper mail is. You may send copies of a message to several people, but it still is a private event. Messaging, on the other hand, is a public activity because each message may be read and responded to by any member of the group.

At first this may seem to be a minor difference, but it totally changes the nature of the medium. Message-based conferences allow lively discussions on many topics by a large group of people. The public nature of a message conference also allows questions to be posed to a group without knowing who will have the answer. This creates an ideal forum for discussion among members of a group working on a problem, while removing the need for them all to be in the same place at the same time. An automatic written record also is created for reference later.

A message conference also creates a truly social environment in an electronic medium. People get to know each other and feel the same bonding that normally requires face-to-face meetings. This makes message conferences a unique medium which solves problems that can be solved by no other technology, allowing both social and professional contact to be maintained electronically.

Message conferences can be quite valuable when you need to do on-line research. You can ask an electronic community for help in beginning your search. Often you can find sources of information you would not locate any other way. You can almost always find information more quickly by using this electronic form of professional and social networking.

File Transfer

File transfer is the act of moving a data file from one computer to another. It can be an extension of electronic mail by attaching files to a message. It also is the foundation for a wide variety of "data library" applications. In these applications, one or more groups of data files are made available at a central site. A method is usually made available to search these files, or abstracts of these files, to locate the needed information. After the required files are located, they then can be transferred to your computer.

In file transfer you want to know that the files you move from one computer to another arrive without any data changed. This requirement has caused several file transfer protocols to be developed. These protocols are rules which the underlying software uses to ensure that any errors are detected and any bad data blocks are retransmitted until the file arrives intact. These protocols are discussed in detail in Chapter 3.

File transfer is at the heart of many PC communications tasks. It may be used to solve nearly any problem in which data files are created on one computer and need to be made available on another computer.

On-line Research

PC communications has made research much easier than it used to be. Incredible amounts of information are available on-line. Many commercial research services exist to aid on-line research. In addition, many local services are provided by government, libraries, and even private citizens that enable you to find information in electronic form.

You can find an on-line database on nearly any topic today, and you can use the electronic community as a starting point to help locate that database as well. We truly are approaching the day when information on any subject is available on-line electronically. In Part II of this book, you learn how to use both public and private services to do such research.

On-Line Transaction Processing

Tasks such as purchasing products on-line or updating a database of sales orders or inventory are known collectively as transaction processing. These applications differ from file transfer, research, and messaging because they directly alter a central database.

On-line transaction processing traditionally has required large central computers and has been very complicated to implement. This no longer is true. Today software is available that enables a PC to provide direct solutions to many multiuser database transaction problems. In Chapters 13 and 14 of this book, you learn how this use to type of software to solve your remote transaction processing problems.

Entertainment

Not all PC communications applications are strictly business. Throughout the history of computers, entertainment and games have been a major theme (although often an underground one). Today PC communications enables multiuser interactive games to be developed at reasonably low cost. This has spawned many very sophisticated entertainment systems. Although some charge for their use, others are provided by private citizens as part of their own hobbies.

PC communications today commonly allows multiple players to play as a team in a multiline environment. Game playing with computers may be derided by some, but it often is some of the most sophisticated computer use in existence. Games also provide a "safe," fun environment for new users to learn about the capabilities of PC communications technology.

Who Should Read This Book?

If you are new to PC communications, *Introduction to PC Communications* is for you. This book introduces you to PC communications and teaches you the fundamentals you need to know to use this technology quickly. This book also will stay with you as you grow in your knowledge of PC communications and be your guide each step of the way.

If you are an experienced user of PC communications, this book is also for you. It collects into a single volume detailed explanations of every area of modem-based PC communications as well as several related topics. Much of this information has been very difficult to find before and usually was presented in a very technical manner when you did find it. This book presents the information in a way that is easy to use and easy to understand.

From the beginner, through the power user, to the programmer—if you want to understand and learn to use PC communications, *Introduction to PC Communications* is for you.

What Is in This Book?

This book is organized so that you can learn about PC communications from the simplest material through the most complex. This book is organized so that you may proceed section by section and end up with a thorough understanding of not only what PC communications can do, but how it is done. This book explains each area of PC communications in detail, building on your knowledge at each step.

This book is organized so that each section also is a stand-alone reference on a single aspect of PC communications. Because this book collects a great deal of information that has never been presented in a single place before, it will remain a valuable reference as you grow in your understanding and use of PC communications.

The overall book is organized into four parts as follows:

Part I: Learning PC Communications Fundamentals

Part II: Using PC Communications

Part III: Examining PC Communications

Part IV: Using Modems

Even the most advanced user of PC communications will learn something new from each section. However, you may skip directly to a section if you need immediate information on a particular topic.

Part I: Learning PC Communications Fundamentals

Chapter 1, "Understanding PC Communications," enables you to find out how much you already know about PC communications. It explains terms such as stop bits, parity, asynchronous data, and others—terms you must understand to be able to use serial communications software and hardware. This chapter also gives you some history to put PC communications in context.

Chapter 2, "Understanding Hardware Requirements," explains what extra hardware you need to use PC communications. It also removes the mythology from cable construction, serial port configuration, and the RS-232 standard. In this chapter you learn the basics of modem setup and operation as well as how to connect the pieces together and make them work.

Chapter 3, "Understanding Software Requirements," discusses the software used in PC communications. The discussion moves from the simplest terminal software, through single-user host mode software, all the way to multiuser BBS software. You learn what each type of software does, and what applications it may be used for. You also learn the meaning of terms such as terminal emulation, upload, download, scripting, and others.

Chapter 4, "Understanding Shareware and Freeware," explores one of the most exciting aspects of on-line communications—low-cost and free software. This chapter gives you the history of shareware and also explains exactly what shareware is and where to obtain it.

Chapter 5, "Understanding Computer Viruses," explains one of the less savory aspects of public access on-line computing. In this chapter you learn the truth and the myths associated with computer viruses. This chapter explains what viruses are, how they work, and most importantly, how you can protect your computer from them. Fear of viruses runs high, and the knowledge in this chapter enables you to replace fear with understanding.

Part II: Using PC Communications

Chapter 6, "Examining PC to PC Communications," introduces you to the most important PC communications software—the terminal program. This chapter also surveys the many communications options available to share files and printers between PCs such as printer buffers and hardware-free LANs. It also introduces you to what is available in the world of on-line computing.

Chapter 7, "Using PCs on a Network," introduces you to the power of local area networks (LANs). You see how a LAN enables you to share files, printers, modems, and programs, and learn how to use these capabilities.

Chapter 8, "Using Electronic Mail," explains this most important part of PC communications. You learn the difference between electronic mail, bulletin board message systems, and conference systems, as well as how to use them.

Chapter 9, "Using Commercial Information Services," introduces you to the world of commercial services. Services such as CompuServe, Prodigy, MCI Mail, The Well, and others enable you to purchase access to large amounts of information directly from your PC. This chapter explains how these services operate and what you can find there.

Chapter 10, "Using CompuServe," explains in-depth how to use the oldest and largest on-line data service.

Chapter 11, "Using Prodigy," explains how to use the newest large information service in depth.

Chapter 12, "Using GEnie," explains how to use this popular commercial service in detail.

Chapter 13, "Using Electronic Bulletin Board Systems," introduces you to the world of public BBS systems. You learn how to locate BBS systems near you and how to use them effectively.

Chapter 14, "Becoming a BBS SYSOP," takes you to the next step and shows you what is required to set up and operate a BBS of your own. You learn what you need to consider and how to select hardware and software for your application. You also find tips on how to make your BBS a profitable business.

Chapter 15, "Examining Store and Forward Networks," introduces you to another exciting aspect of computer communications—store and forward message networks. You learn how these networks link hundreds of thousands of people internationally. This chapter describes how these networks operate with a focus on the largest network of them all—Internet. This chapter also shows you how to become a part of these networks to send and receive international messages directly from your PC.

Part III: Examining PC Communications

Chapter 16, "Understanding Terminal Emulators," introduces you to the concepts involved in terminal emulation. You learn what terminal emulation is, when you need it, and what forms it can take. Finally, you understand what it takes to get the screen displays and keyboard inputs you want to work remotely.

Chapter 17, "Understanding PC to Mainframe Connections," explains in-depth how mainframe data networks operate, and how you can connect a PC to them. You learn what SNA, DECnet, and TCP/IP networks are and how they work. You also learn more than one way to hook a PC to each of these networks.

Chapter 18, "Understanding Network Communications," explains in detail how LAN asynchronous servers operate. You learn about communications program interfaces such as NASI, NCSI, and LANACS, among others. You see what the purpose of these programs is and how to use them.

Chapter 19, "Programming PC Communications," is a complete tutorial for those who want to write software that performs serial communications. You learn how UARTS work and how to program interrupts. A complete serial driver is developed and explained as a working programming example.

Part IV: Using Modems

Chapter 20, "Learning Modem Fundamentals," explains what a modem is and how it operates. You are given the information you need to select, install, and troubleshoot this most important part of your communications system.

Chapter 21, "Understanding Advanced Modems," builds on the basics you learned in Chapter 20 and explains the newer advanced, high-speed and error-correcting modems. You learn about the "alphabet soup" of modem specifications (for example, V.32, V.32bis, V.42bis, MNP, and so on). You also learn how to tell when you need these features and when you don't, as well as what impact their use has on your other hardware and software requirements.

Chapter 22, "Modem Troubleshooting Guide," is a step-by-step procedure to help you get your modem and terminal program software working. If you have a problem with your modem installation, this guide will help you pinpoint the cause and fix it.

Appendixes

Appendix A, "Glossary of Terms," provides a comprehensive glossary of communications terminology so that you can understand the buzzwords you encounter in PC communications.

Appendix B, "ASCII and Extended IBM Codes," provides a quick reference to the codes used for display and keyboard input.

Appendix C, "The AT Command Set," is an explanation of the basic Hayes compatible modem commands.

Appendix D, "Cable Diagrams," shows you how to quickly connect computers and modems.

Appendix E, "Using PROCOMM Lite," teaches you how to use PROCOMM Lite, a compact version of the best-selling telecommunications program, PROCOMM PLUS (from DATASTORM TECHNOLOGIES, Inc.). PROCOMM Lite is included free with this book.

Part I

Learning PC Communications Fundamentals

Includes

Understanding PC Communications

Understanding Hardware Requirements

Understanding Software Requirements

Understanding Shareware and Freeware

Understanding Computer Viruses

Understanding PC Communications

P C communications is the most confusing part of modern computing. More than 20 million computers are equipped with modems, (which connects a computer to the telephone line), and the number is growing rapidly. Although nearly 2.5 million people use PC communications regularly, only a few of these users probably understand why or how it all works.

This lack of knowledge doesn't have to be so. Communication among PCs is inherently simple in concept. In fact, PC communications is the logical extension of the first electronic communications method ever invented—the telegraph. The confusion arises, however, because after 100 years of development and growth, many options, buzz words, and slight variations on the theme have evolved. In addition, getting a straightforward explanation is difficult because so few people—even if they know *how* to make things work—understand *why* things are the way they are. To complicate matters further, much of the material written about PC communications is wrong because the authors don't fully understand the subject.

The history of PC communications precedes electronic computers by more than 50 years. This history (which actually begins with the telegraph) has been filled with many twists and turns of innovation and refinement and has strongly influenced current technology. In addition, getting industry agreement on the many small details of the changes and enhancements that have occurred has been difficult. The result is many nearly equivalent ways to do the same job, and you are forced to know enough about all these minute details to make things work properly.

Agreements by different companies on how to implement technical details are called *standards*. Standards come in two forms: codified and de facto. A *codified* standard is one that has been formally adopted and published by one of the recognized industry standards organizations. A *de facto* standard evolves when a large number of hardware or software manufacturers agree informally to do a certain thing the same way—usually close to the way a popular program or computer operates. Often no official written document exists for a de facto standard.

A standard has no meaning if it is not widely adopted. Many formal standards are not used because people in the industry feel that the standards don't provide the needed solutions. A de facto standard frequently has more force than a codified standard. The IBM PC plug-in card bus, for example, is a de facto standard. It became so useful that it was codified later in a reasonably formal fashion but never has become a formal standard. PC users' lives have been made much easier, however, by the fact that many different brands of computers use the same interface.

In communications, many standards exist. Some standards are powerful because they have been widely adopted, and other standards are not. Examples are the formal, codified RS-232 standard and the informal, de facto Hayes-compatible modem standard (Chapter 2 discusses these standards in more detail). Specific standards are discussed in this book where they apply.

Using the technology is much easier if you have an understanding of how it works. This book attempts to clear the confusion. As you will see, tackling the topic of PC communications is not that difficult, and if you understand only this chapter, you will know more about PC communications than 90 percent of all the people who use it.

Storing Bits and Bytes

The purpose of PC communications is to send information from one computer to another. To understand that process, you first must understand how a computer stores information.

Computers are *binary* (meaning two states) devices. They perform so many amazing tasks that you may have trouble believing computers really know only two things: on and off. Computers originally were made up of switches (or electronic switches known as relays). A switch only has two positions—

on or off. The ability to distinguish between these two conditions is the smallest possible amount of information and is called a data *bit*.

Today's modern electronic computers use millions of electronic switches but they still must reduce everything they do to one of two conditions—on or off. The numbers 1 and 0 commonly are assigned to the on and off states. A data bit therefore can have a value of either 1 or 0.

To store or manipulate a useful amount of information (such as a text file or program data), computers group several bits together into a larger unit of information known as a byte. A byte is made up of 8 bits and can express a number from 0 to 255 (2^8–1). Computer memory is specified in bytes, kilobytes, and megabytes.

In the normal metric notation system, kilobyte would mean 1,000 bytes. Computers, however, use powers of 2 quite naturally and use decimal numbers only with a great deal of work. Therefore, a *kilobyte* is defined as 1,024 bytes because 1,024 is the exact power of 2 that is closest to 1,000. Similarly a *megabyte* is 1,024 kilobytes. This method of calculation explains why 64K of memory has 65,535 bytes rather than 64,000 bytes.

A computer can express a number directly in one or more bytes, but users often need to store text information. A computer handles this task by assigning each letter or character a number from 0 to 255 (the range of numbers possible in a byte). No technical reason decrees that any particular number should be assigned to any particular letter. Thus, you theoretically can devise many *code sets* to express text in a computer. In fact, several different code sets have been used by different older computers. Thankfully, however, the PC computer world has agreed on a standard here—at least for the alphabet and many of the punctuation characters. This standard code set is known as the *ASCII character set*.

The other character code set you may encounter is EBCDIC for Extended Binary Coded Decimal Interchange Code. This character set is widely used by IBM in its mainframe environment. Thus, you may have to work with this code set if you move text files between a PC and an IBM mainframe. Almost all other alternative code sets exist only in small areas that you will encounter rarely. In a PC environment, you use only the ASCII code set.

ASCII, which stands for American Standard Code for Information Interchange, standardizes only the codes used for 128 characters, leaving an "extra" 128 possible numbers in an 8-bit byte. As a result, those always inventive computer programmers have two options. They can save a bit and store characters in only 7 bits, or they can use the "extra" bit for other purposes. As you may guess, programmers have done both from time to time.

One common use for this 8th bit is to create "extended ASCII" character sets. The only problem is that the extended characters aren't ASCII at all, but apply only to a given brand of computer. One extended character set that is in wide use, however, is the IBM Graphics character set. In fact, this set is used so frequently that many computers and terminals other than IBM and compatibles are beginning to implement it. The IBM Graphics character set is an important character set for PC communications.

The ASCII character set is a codified standard, and the IBM Graphics character set is a de facto standard. This combination of a published standard with a de facto extension can cause confusion. And, in Chapter 3, you learn that yet another standard often is combined with ASCII to enable cursor positioning, reverse video, and color control. This codified standard is known as *ANSI*.

The combined extensions (IBM Graphics characters and ANSI control sequences) are referred to in the same breath as *ANSI Graphics*. But already existing is a codified ANSI Graphics standard that is different from the IBM standard and doesn't include the ANSI control sequences. What is almost always meant by ANSI Graphics is the combination of the ANSI control sequences and the IBM Graphics character set, even though this terminology technically is incorrect.

Now that you know how a computer stores information in bits and bytes, you are ready to learn how one computer can send that information to another computer.

Examining Serial and Parallel Communications

Before data can be sent between computers, the data must be placed on one or more wires. Each wire can express a single bit of data by showing one of two voltages. Thus, each wire represents a single bit at any time. Of course deciding which voltage should represent a 1 and which a 0 (which is positive or negative, 5 volts or 10 volts, and so on) is another area in which endless variation is possible.

Before a computer can send several bits down the wires, some method has to be devised to let the computer know when one bit is done and another is being sent. The simplest method is *parallel communications*. It works by using one wire per bit in a byte and an extra wire as a strobe to signal when the data is changed. In figure 1.1, you can see this sort of a connection shown schematically.

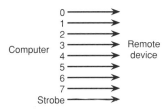

Fig. 1.1. An example of a simple 8-bit parallel circuit.

To send a byte of information on this 9-wire parallel connection, the computer places each bit of the byte on a separate wire and then sets the strobe wire to 1 for a specified period of time. The strobe then is set back to 0, the next byte is placed on the wires, and the process is repeated.

This procedure is called *parallel data transmission* because all the bits of a byte are sent at the same time (in parallel). The example shows an 8-bit parallel interface. If you use more wires, you can send any number of bits at the same time. The more wires, the faster the data is sent at the same strobe repetition rate.

Parallel interfaces also are used *inside* your computer for sending information among the various components. In fact, the parallel interface determines whether your computer is 8-bit, 16-bit, or 32-bit. The number of bits refers to the number of data bits contained in the parallel interface between your computer and its memory. This figure is also the number of bits the computer can transfer to or from memory in a single access.

Your PC's parallel printer interface operates in much the same way. The printer interface has a few more wires to show things like paper out and so on, but otherwise it functions the same way as the example in figure 1.1 when sending characters to the printer.

Notice that the effect of the strobe in this interface is to divide the data on any single wire by time. The strobe marks when in time the "next" data bit is present on the wire. One bit follows another in order on a single wire, and the strobe lets the computer know when a new bit has been sent.

A parallel interface works well when the distance is short. Over long distances, however, running all the wires and supplying the electronics to control them properly becomes expensive and cumbersome.

In an effort to reduce the number of wires, you could use a single data bit along with the strobe. This design would be slower because only one bit could be sent per strobe period, but you would need only two circuits— one for data and one for the strobe.

If you could eliminate the strobe, you could get the data transmission down to a single circuit. In fact, if you were to make it a rule that the strobe is always rigidly periodic (like a clock ticking), you wouldn't have to send the strobe at all. The receiving end would know to look at the wire every "tick" of the strobe. This method works, and the assumed clock rate determines the speed at which data can be transmitted.

Data bits, therefore, may be sent down a single wire circuit one after another sequentially. This method is *serial data transmission* and is used in PC communications. The speed at which data is sent down the wire is *bits per second*, which is abbreviated as *bps*.

The speed of serial transmission—the clock rate—also is known as the *baud rate*, after Jean-Maurice-Emile Baudot. In 1874 Baudot received a patent on a telegraph coding method that used a five-unit combination of current on and off signals of equal duration. This code represented a significant advance over Morse code, which was the first widely used serial data transmission code.

In 1894 Baudot also invented a mechanical system that could transmit and receive text by using his code method. This system was the foundation of modern serial communications and formed the basis for stock market ticker tapes and later the Teletype machine for news wire services.

When the data is being sent down a serial wire, the baud rate and the speed in bps are always the same. With the advent of newer advanced modems, the bps rate and baud rate of a modem's audio signal are not necessarily the same. For this reason, using only the term bps to specify serial data speed is becoming more common. Just remember that on a wire, baud rate and bps are always the same; inside a modem, they may not be. (Refer to Chapter 2 for modem basics, and Chapters 21 and 22 for more advanced information about modems.)

Baud rate technically is defined as the number of signal changes per second. In high-speed modems, several bits are sent per signal change, so the baud rate and the bps rate are different. A standard 2400-bps modem, for example, transmits data at 600 baud on its audio signal. The speed on the wire between the modem and your computer, however, is both 2400 baud and 2400 bps because each bit is represented by one bit per signal time (see Chapter 21 for more discussion on modems).

Now that you have learned about serial data transmission and clock speed, you are ready to put all your information together and learn how PC communications information is sent on a serial data link.

Comparing Asynchronous with Synchronous Communication

Serial communication is possible because each "tick" of the clock represents a bit and each group of bits is a character. The receiving end can synchronize a clock with the predetermined baud rate and decode the bits as they come by. If this synchronization is all that is done, you have *synchronous* serial communications.

Synchronous communication is efficient because only data bits are sent. No overhead is wasted. If you think about it, however, you can see that this system is also a rat race. The sender must send something every single bit forever! This structure is fine if a great deal of data is being sent, but what if you want to start and stop the transmission? With synchronous communications, you cannot stop, so you must have an agreed-on idle sequence to fill up the time when you have no data to send. Therefore, protocols and other complications are required for making synchronous communications work.

Synchronous communication is frequently used for communicating with a mainframe computer at high speeds. You find synchronous communication in remote terminal emulators that use such things as IBM 3270 protocols. Synchronous communications require different serial port hardware than asynchronous communications do. In PC to PC communications, synchronous communication is not used.

If you add one bit to indicate when a character is starting and another bit to indicate that the character has ended, you can send single groups of bits (a character) any time you want. This method is *called asynchronous communications* and is used for most PC modem communications. This method is called asynchronous because the implied clocks can start and stop between characters and thus the data can be sent asynchronously.

The added bits that frame each data character in asynchronous communications are called the *start bit* and the *stop bit*. The start bit is always a 0, and the stop bit is always a 1, which guarantees that the start bit always results in a change of voltage on the line. The start bit always is detected as the start of a character regardless of the data value of the character itself.

Your new knowledge of synchronous and asynchronous serial communications and start and stop bits gives you the fundamentals necessary to understand the communications parameter terms discussed next.

Using Parity

As you learned in the preceding section, asynchronous serial data is sent in a packet that is "framed" by a start bit and one or more stop bits. This frame may be of any size but is generally the length of one data character.

Because the ASCII character set requires only 7 bits to express the 128 defined codes, only 7 bits need to be sent to transmit text information. Sending only 7 bits represents a 10 percent savings in the amount of data sent, which translates to a 10 percent increase in effective speed. (7 data bits plus start and stop bits require 9 bit times compared with 10 bit times for 8 data bits plus start and stop bits. 1 bit thus represents 10 percent of the total time needed to transmit a character.)

Computer programmers, however, weren't content to let things alone. They saw 7 as an odd number and imagined that a bit was "left over."

Naturally they looked for some way to use that extra bit. First, it was used for trying to detect errors in transmission. Such errors may be caused by signal distortion or noise and are quite common when serial data is sent over great distances.

In an effort to detect errors, programmers experimented with a mathematical concept called *parity*. Parity means that the extra bit is set to either a 1 or a 0 based on the number of 1s in the other 7 data bits. *Even parity* means that the parity bit is set to guarantee that the 8 bits always include an even number of 1s; *odd parity* means that the parity bit is set to guarantee an odd number of 1s.

Parity is the total count of bits in a character that are set to 1. The letter A in ASCII, for example, is expressed as the decimal number 65. In binary, this number is expressed as 1000001 and thus has 2 bits set to 1. Without any additional bits, the parity of this number is even. If the parity setting is even, the added parity bit is a 0. The letter C is the number 67 in ASCII. In binary, this number is 1000011, which has 3 bits set to 1. In this case, the added parity bit is set to a 1 to ensure even parity with four 1s in the transmitted data.

Using parity, a receiver can verify quickly that an even or odd number of bits is set to 1 in every data packet. If the parity is wrong, an error has occurred. This method seems like an easy way to detect errors. But, when studying the electrical characteristics of serial communications, mathematicians discovered that parity doesn't detect well at all the kinds of errors that occur in serial transmission. In fact, many errors may occur that parity *never* can detect. Thus, using parity in serial communications doesn't do much good.

By the time this fault was discovered, however, mainframe computers had begun to use serial communications quite heavily. The use even reached the point where the American National Standards Institute (ANSI) issued a standard (ANS X3.16) indicating that text was to be communicated using 7 data bits and even parity. This large installed base continued to use this format because it worked for them, and coordinating a change was not realistically possible.

When personal computers began to use serial communications, however, programmers saw the parity bit as an extra bit again. Because binary data is stored in 8-bit bytes and because parity does almost no good in detecting errors, PC software quickly adopted the use of all 8 bits for data transmission and dropped the use of parity.

The result of this evolution is that you must know the number of data bits and whether parity is used in order to set the proper format for the computer with which you want to connect. If your computer is not set the same as the system you are calling, the data will not be transmitted properly.

Some communications programs enable you to set the parity bit to *mark parity* or *space parity*. Mark and space are synonyms for 1 and 0 and indicate that the parity bit is to be forced always to a 1 or always to a 0, regardless of the value of the data being sent. These settings usually are used to "fudge" a 7-bit data setting into looking like an 8-bit data setting in which the high-order bit is always a 0 or always a 1.

The final setting you may need to understand is whether 1, 1.5, or 2 stop bits are required. This setting isn't critical any longer, and the only difference you may find in practice is a slight difference in effective speed. The original purpose for the various stop bit settings was to allow time for mechanical devices such as Teletype machines to return the print hammer to the basket. With modern electronic devices, even mechanical printers no longer require this extra time.

In asynchronous transmission, data bits are sent from the least significant to the most significant bit. If a parity bit is used, it is considered the most significant bit. The start bit is always a 0, and the stop bit is always a 1.

Between characters, the line is left at the voltage of a stop bit. For this reason, a start bit always means a change in voltage, regardless of the data value, and indicates when the receiver should start timing data bits again. Because the stop bit is always a 1, the receiver can verify that it properly timed the character.

The timing diagram of a single character with 7 data bits, even parity, and 1 stop bit looks like the one shown in figure 1.2.

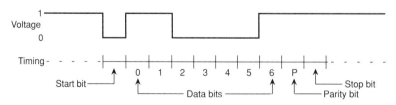

Fig. 1.2. Timing of a serial character during transmission.

Figure 1.2 shows the way a serial data signal is divided by time. Each marked interval on the timing line represents one bit time. (The baud rate determines the actual length of a bit time.) The character timing begins when the signal voltage changes from the value that represents a 1 to the value that represents a 0. This begins the start bit, which also begins the character timing.

The seven data bits are numbered 0 through 6. (Numbering commonly begins with 0 instead of 1 for simplicity when working in binary numbers.) Notice that the data bits are sent in reverse of the order you may think, with the least significant bit (units bit) sent first. The standard specifies this method, and as long as everyone agrees, this method will work—the receiver will assemble the bits back in the proper order.

After the data bits have been sent, the parity bit is sent. The voltage always returns to a 1 during the stop bit time interval. In figure 1.2, you notice that the line "idles" or waits between characters at a voltage of 1. This guarantees that the next start bit will always make a voltage transition from 1 to 0. This is how the receiver always can detect the start bit and begin timing each character.

The data shown in figure 1.2 is the ASCII letter C, which has a binary value of 1000011 and thus requires a parity bit of 1 to give even parity. This order of bits in an asynchronous packet is defined by the formal standard ANS X3.16. This standard defines both the "7-bit environment," which it indicates should have even parity, and the "8-bit environment," which it indicates should have no parity. Thus ANS X3.16 indicates that an asynchronous character packet always should have a total of 10 bits when the start and stop bits are counted because 1 stop bit is specified.

Because asynchronous data always is framed by start and stop bits, one other function is possible. This function is known as a *break* signal. The break signal occurs when the line is switched to a start bit value and left there for more than two character times. The receiver can detect this condition as different from a normal data character because the stop bit is absent. This signal is named "break" because in the original days of mechanical teleprinters, the signal was generated by a switch breaking the connection for a specified length of time. This signal is used as a signal that is separate from any data pattern to interrupt output and perform control functions. The break signal commonly is used in PC to Mainframe communications, but rarely is used in PC to PC communications.

Chapter Summary

In this chapter, you learned the fundamentals of how serial data communications operates. You usually can ignore these details except for knowing how to set the baud rate and data format in your communications program. When problems arise, however, understanding what is going on can help you avoid becoming lost and overwhelmed as you track down the causes of your troubles.

Understanding Hardware Requirements

Everything about computers falls into one of two categories: hardware or software. *Hardware* consists of everything you can see or touch. Examples of computer hardware are keyboards, integrated circuits, disk drives, and cables. *Software* consists of computer instructions and programs. You cannot see or feel software, but computers cannot do anything without it. Examples of computer software are DOS and your word processing program.

Serial communication requires specialized hardware and software. Because so many possible combinations of communications options are available, you may end up needing to configure any part of the communications software or hardware to solve problems that arise. The next chapter covers the software portion of communications. In this chapter, you learn about the hardware you need in order to use serial communications. Installing and configuring this hardware can be a source of frustration if you don't understand how the hardware works. If a switch is set wrong or a cable is wired incorrectly, nothing will work properly.

NOTE A middle ground between software and hardware is called *firmware*. You may encounter this term in PC communications, so you need to understand it. Normally software is captured electrically on recording media (such as a floppy disk or magnetic tape) or in computer memory. You can change software without changing any hardware—by rewriting a disk or altering memory. Firmware is software that is captured permanently in an integrated circuit (called a ROM or PROM) so that you cannot change the software without replacing the hardware that holds it. The line between software and firmware is not in itself firm. The important thing for you to know is that firmware is just software that is more difficult to change.

All communications hardware falls into one of the following categories:

- The communications port that connects the computer to a serial data circuit

- The modem (*mo*dulator/*dem*odulator) that connects a serial data circuit to an audio circuit (such as a telephone line)

- The cables that connect these pieces

In this chapter, you learn how each piece of hardware used in PC communications works. You also learn how to connect and configure the hardware properly, and you become familiar with the terminology involved.

Understanding Communications Ports

The most important hardware element in serial communications is the *serial port*. It is known by several names (IBM calls it the asynchronous communications adapter), but its function is identical in all cases.

As discussed in Chapter 1, a computer operates internally by moving data in parallel one or more bytes at a time. The serial port converts internal computer data bytes into a serial data stream to send by accepting a byte from the computer and then shifting it out one bit at a time to the serial data line. The serial port also receives a serial data stream one bit at a time and combines these bits back into bytes, which are then transmitted into the computer.

In addition to the serial to parallel data conversion, a serial port is responsible for presenting the serial circuitry with the proper voltages and signal timings to control its operation as well as send and receive a serial data stream.

The serial port must communicate with the computer as well as with the serial link. The serial port must accept both commands and data from the computer, and it must transmit both status and data back to the computer. For these communications, the serial port uses a special addressing space on the computer which is reserved for input/output (*I/O*) ports (as opposed to memory addresses). By reading or writing to the assigned I/O port addresses (often called just *port addresses*), the computer can control the serial port, read its status, and read and write data bytes to and from the serial port hardware.

In addition to transmitting data and status, the serial port has a way to indicate to the computer that it needs attention. This type of signal "interrupts" the computer from whatever it may have been doing to tell it the serial port has information the computer should look at. This enables the computer to pay attention to the serial port only when data actually is present so that the computer can spend the rest of its time on other tasks. This signal is called the *serial port interrupt*.

In an IBM PC, the port addresses are defined as COM1 and COM2, and you can connect and operate up to two serial ports at the same time. On computers that use the original IBM plug-in card bus, the addition of two more ports at defined addresses COM3 and COM4 has become a de facto standard. These addresses, however, are not the same on Microchannel systems as they are on PC bus systems.

Other computer brands have different rules for where the communications port addresses occur, and you have to investigate the manual for your computer if it is not an IBM compatible. All serial ports operate in the same way, however, using port addresses and an interrupt.

The important thing to know is that no two communications ports may occupy the same address. Two serial ports set to the same base address produce a *port address conflict*. The computer cannot talk to the ports if such a conflict occurs. The solution is to change the base address of one of the serial ports.

The "guts" of a serial port is a device called a *Universal Asynchronous Receiver/Transmitter (UART)*. The UART converts the data bytes to and from a serial data stream. Before the invention of the UART, a serial communications port required a large number of components—as much as an entire rack of electronics. The UART has a place in history as the first commercially successful large-function integrated circuit. The UART has made serial interfaces inexpensive.

On the IBM PC expansion card bus, I/O port addresses are expressed as three- or four-digit hexadecimal numbers. COM1 always uses addresses 0×3f8 through 0×3ff, and COM2 always uses addresses 0×2f8 through 0×2ff. Because an IBM standard serial port always uses eight consecutive addresses, referring to the first address is sufficient. All software knows that a serial port also uses the next seven consecutive addresses as well. This single first address is often referred to as the *port base address*. On the PC bus (also known as the Industry Standard Architecture [ISA] bus), COM3 uses a base address of 0×3e8 and COM4 uses a base address of 0×2e8.

On the IBM Microchannel bus, COM1 and COM2 use the same base address that they do on the PC bus because those two addresses truly are standardized. COM3 becomes 0×3220, and COM4 becomes 0×3228, on a Microchannel computer, however, because IBM never standardized the base addresses of any serial ports beyond COM1 and COM2.

If you use many serial ports (on the multiuser bulletin board systems discussed in Chapter 14, for example), you probably will have to deal directly with the port base addresses rather than the COM1 and COM2 definitions. The reason is that only COM1 and COM2 have a fixed base address in all cases.

Normally a serial port ends in a connector to which a cable is attached. A serial port also can be integrated into an internal modem card. In this case, all the electronics and connections are on a single card, but the electronic parts are still the same as they are for separate components.

Serial ports from two computers may be connected directly by cable if the systems are not far apart (if, for example, they are on desks in the same room). To do serial communications over great distances, however, a second piece of hardware called a *modem* is required. The next section describes what a modem is and how it extends the distance over which serial communications is possible.

Exploring Modem Basics

One of the most exciting parts of PC communications is that it enables you to connect computers by using normal telephone lines. For this connection to work, the communications have to be converted into sound patterns that a telephone can transmit. The device that completes this conversion is the *modem*. Modem is an acronym built from the words *mo*dulator and *dem*odulator, which describe the functions of a modem.

A modem modulates a sound wave with information that the modem later can demodulate to reconstruct serial data patterns. Modulation takes many forms. The simplest form of modulation consists of turning a tone on and off. When a tone is on, for example, it represents a high voltage on the serial line, and when the tone is off, it represents a low voltage. A modem built on this principle turns the tone on and off to match the serial data stream that is sent to the modem. Another modem may "listen" to the phone line and set the output serial line high when hearing the tone and low when no tone is heard.

This simple modulation scheme (turning a tone on and off) works for slow-speed serial transmission. The turning of the tone on and off is the modulation of the tone in this case. A much better method, however, is to use two tones and switch between them. If one tone means on and the other means off, then you truly are modulating the tone by changing its pitch based on the serial data stream. This scheme is known as *frequency shift keying* (FSK) and is used by 300-bps modems.

The tone emitted also is known as the *carrier* because it carries the information over the phone line. Because a tone always is present in the FSK modulation scheme, the modem can tell whether it is still properly connected. When the carrier tone is detected, the modem feeds it to the computer, using the RS-232 carrier detect (CD) signal and thus lets the computer know that the telephone connection is still good.

Modem operation is discussed in much more technical detail in Part IV. For now you need to understand only that the function of a modem is to enable the serial data signals to be converted to sound, which can be sent on a telephone line and to decode that sound back into serial data signals on the receiving end. Many processes, not just FSK, are available for this job.

The FSK method is limited to about 450 bps because of the frequency range that a telephone is guaranteed to pass. Newer high-speed modems (which support speeds of 14,400 bps or faster) use exotic modulation methods, which are discussed in Chapter 21. Aside from the speed of transmission, however, the purpose is always the same—to make it appear that the wires entering the modems on each end are connected to each other when the link actually passes only sound.

Using Direct versus Acoustic Connection

Early modems were large devices that connected to the telephone through a speaker and microphone in a special cradle. The telephone handset was placed in this cradle after the telephone number was dialed. This type of connection is known as *acoustic coupling* because sound is used to connect the modem to the telephone.

Acoustic coupling is not used much today because of two changes in technology and regulations. The first change was the modular telephone jack, which enables you to connect and disconnect telephone devices easily without the need for tools or special knowledge. The modular jack also has standardized how such connections are made. Today in the United States nearly all telephones are connected with the RJ-11 style of telephone jack.

The second change was the deregulation of the telephone company. Before the deregulation, no piece of electrical equipment could be connected to a telephone line except through a special interface box that only the telephone company could make or sell. This restriction made the direct connection of any electrical device to the telephone difficult and expensive. An acoustic connection was allowed because it doesn't connect electrically to the telephone line. Now that the telephone industry has been deregulated, any company can make direct-connection equipment as long as the company obtains FCC registration (or DOT registration in Canada). Before it can be registered, the device has to pass specific electrical tests designed to ensure that the device will not harm the telephone lines.

A direct connection to the telephone line eliminates the distortion that can occur with the playing of tones through a speaker and back into the microphone of a telephone. As a result, much higher speeds are possible with fewer noise errors. Therefore, most modems now use direct connection to the telephone line. Acoustic couplers are still available for use in those cases in which a direct connection isn't possible.

Comparing External and Internal Modems

A modem can be made to fit entirely inside your computer. In addition, the serial port can be built on the same circuit card as the modem, and that circuit card can be built to plug directly into your computer's expansion bus. A modem built in this fashion as a single piece for a specific type of computer is an *internal modem*. An internal modem works only on the type of computer for which it was built and usually draws its power directly from the computer into which it is plugged.

On the other hand, an *external modem* connects to the computer through an RS-232 cable (described in this chapter's section "Understanding the RS-232C Standard"). You can use this type of modem with any computer that has an RS-232-compatible serial port. An external modem also needs its own power connection separate from the computer.

Because fewer parts, no power supply, and no case are involved, an internal modem is usually less expensive than an external version of the same modem. You may not enjoy this price advantage, however, if the internal modem is built for a single computer brand. In that case, the number of modems of one type sold is small enough that the price becomes higher than that of an external modem. This pricing effect is especially true of internal modems for laptop computers.

Another advantage of an internal modem is that you eliminate extra cables and boxes, reducing clutter and resulting in a tidier installation.

Electrically, internal and external modems are identical. Choosing the right type for you is a decision based on cost and flexibility—not performance. Internal modems often cost less and always give a more integrated installation. External modems give you more flexibility; you easily can move one modem among multiple computers of different brands. In addition, external modems supply status lights, which can be quite useful in following the operation of PC communications. Internal modems rarely have such status lights because the modems are mounted inside a computer case. See Chapters 20 and 21 for more discussion about modems.

Understanding the AT Command Set

As discussed in a previous section, the primary function of a modem is to convert digital data to and from sound waves. But a modem also must perform other functions. The modem, for example, must be capable of answering incoming telephone calls and dialing outgoing telephone calls. In addition, you may need to change certain settings of the modem in order to "tune" it to varying telephone line conditions.

Originally these control functions and settings were accomplished through extra port connections to the modem or through several switch or jumper settings. The controlling computer used this extra port for dialing and modem control. Each modem was designed with its own interface, and each was different from the others.

With the advent of microcomputer technology, putting a computer inside a modem for the sole purpose of controlling modem operation became possible. The computer in the modem could be made "smart" enough to take its commands through the data stream sent to the modem. But because this process was still fairly arbitrary (any command could be linked to any character sequence), different manufacturers made modems with many different command structures. This disparity caused a great deal of trouble for programmers. They had to modify their software constantly just to keep up with different modem brands.

One of the largest problems that occurs when you mix data and commands on the same channel is that the computer in the modem must know when commands are being sent and when data is being sent (because data may be anything and thus may look like commands). Early intelligent modems dealt with this dilemma in a variety of ways, none of which solved all the problems. When Dennis Hayes developed the Hayes Smartmodem 300, he incorporated a method that solved this problem quite effectively. The Hayes method was to define a special sequence of both characters and idle time, which is now commonly known as the +++ sequence. If the modem sees a period of no data, followed by the characters +++, followed by another period of no data, then the modem knows to switch from data to command mode. The requirement for idle time before and after the character sequence enables these characters to be part of a normal data stream without being interpreted as a command.

As the Hayes SmartModem 300 became a dominant force in the marketplace, other modem manufacturers began to implement identical command structures in their modems. Thus arose the concept of a Hayes-compatible modem. This compatibility occurs through the use of the *AT command set*, the command set used in the original Hayes modems.

This command set is known as the AT command set because each of its commands begins with the letters AT to get the modem control computer's attention. This de facto standardization on the Hayes command set has made possible the correct operation of software with many different makes of modems.

The basic Hayes AT command set is detailed in table 2.1. A modem must recognize this command set in order to be considered Hayes compatible. (See Chapter 21 for detailed information on the AT command set.)

Table 2.1
The Hayes AT Command Set

Command	Function
AT	Prefixes all other commands (by itself, results in an OK response indicating that the modem is accepting commands)
A/	Reissues the last command given
A	Manually answers an incoming call
C0	Disables transmit carrier
C1	Enables transmit carrier
DP	Dials, using pulse method (number follows)
DT	Dials, using DTMF tone method (number follows)
E0	Disables echo of command input
E1	Enables echo of command input
F0	Sets half-duplex mode
F1	Sets full-duplex mode
H0	Places phone on hook (hangs up)
H1	Takes phone off hook (goes busy)
I	Returns modem product code (manufacturer code)
I1	Returns internal ROM checksum (manufacturer code)
I2	Performs modem internal memory test
L1	Sets speaker to low volume

continues

Table 2.1 *(continued)*

Command	Function
L2	Sets speaker to medium volume
L3	Sets speaker to high volume
M0	Turns speaker off
M1	Turns speaker on when dialing and off when carrier detected
M2	Turns speaker off always
O	Returns on-line if connected
Q0	Returns modem command result codes
Q1	Does not return modem command result codes (acknowledgment of command operation or failure)
Sn=x	Sets S register to value (controls several internal timing and modem features)
Sn?	Displays current value of an S register
V0	Returns numeric result codes
V1	Returns text result codes
Xn	Sets result code completeness level to n (higher values enable call progress monitoring, baud rate detection, and other extended result codes)
Z	Resets the modem internally and restores last stored values for all parameters
+++	Enters command mode if currently in data mode

Most modems today use an extended AT command set. *Extended* means that one or more commands unique to that brand of modem have been added to the widely agreed-on AT commands. Because these extended commands are not the same in different brands of modems, things are becoming unique again.

As new features are added, the same extended functions in different modems commonly use different commands. Therefore, much of the value of Hayes-compatible standardization is being lost. As a result, you frequently need to learn about your modem in enough detail to design or use *initialization strings*, which are composed of AT commands to program the proper settings. Because your modem may have one or more unique features, it may well require a unique initialization string to be programmed for proper operation—even if the modem is Hayes compatible.

Chapters 20 and 21 discuss more about modems, but for now just remember that your modem is controlled by an internal computer. Many advanced modems have scores of control and configuration options so that you can use the modems in a variety of ways. Because the control of these options has not been standardized yet, you may have to learn more than you want to know about your modem in order to set it up properly. In time, the control of these new options most likely will become standardized as well. This cycle of innovation (with its attendant chaos) followed by standardization is one that undoubtedly will be repeated several times in the growth of PC communications.

Examining the RS-232 Standard

External modems and serial ports require cables to connect them. Before you can learn about cable construction, however, you must learn about the RS-232 standard, the formal standard that describes the serial port connection to a modem or other serial device.

In Chapter 1, you learned that serial communications can send data on a single circuit. This question then arises: "Why does my serial cable plug have so many pins?" The reason is that in addition to the data circuits, several control signals are necessary.

In the early days of computer communications, all these signals were implemented by each manufacturer as required by its specific device. Different companies used different voltages or currents to make the connections, making the connection of one computer interface to another company's equipment impossible.

To bring some order to this chaos, the RS-232 standard was developed and adopted. The development of this standard was an attempt to bring together all the manufacturers of communications equipment and get as much agreement as possible. In order to be as inclusive as possible, the standard defined many optional items that often are not needed, indicating how they were to be implemented if they were used.

The RS-232 standard originally was adopted by the Electronic Industries Association (EIA) in October 1969. At that time, connector technology was undergoing a significant revolution. Cable technology also was being affected by new plastic compounds and manufacturing innovations, such as ribbon cables that today's users take for granted. For this reason, the RS-232 standard intentionally did not mandate a connector type but left this item open. As a result, many different types of connectors have been used for RS-232 serial ports.

To address connector standardization, the ISO (International Organization for Standardization) adopted the ISO 2110 standard in 1972. This standard specifies the familiar DB-25 connector that is in wide use today. Thus, the DB-25 connector itself became widely used because of the ISO 2110 standard and is not part of the RS-232C standard. An RS-232 interface may use any connector and still be standard.

The major victory of the RS-232 standard has been to define the voltages and circuits that all serial interfaces use. To accommodate a wide range of existing equipment and to provide flexibility in implementation, many voltages are supported. The standard defines a binary 1 (or mark condition) as a voltage between –3 and –25 volts. A binary 0 (or space condition) is defined as any voltage between +3 and +25 volts. Any voltage between –3 and +3 volts is considered indeterminate or undefined. In addition, maximum and minimum signal switching speeds are defined to enable filtering for noise resistance. These specifications establish an upper limit to cable length and speed of about 50 feet at 20,000 bps. Slower speeds may have longer cable lengths, and higher speeds must have shorter cable lengths. The practical upper limit of speed for an RS-232 interface is about 150,000 bps on a cable of 5 feet or less.

RS-232 was adopted quickly and is the reason that today you can connect most serial hardware to most computers regardless of who manufactured them. The RS-232 standard went through two major revisions (RS-232B and finally RS-232C) before arriving at its current state. These revisions corrected problems encountered by early users of the standard.

Because a standard does no good if most manufacturers don't use it, successful standards address only what everyone can agree on. The RS-232 standard had to satisfy a wide variety of both technical and political requirements to be successful, and its final form reflects the emotional nature of the job it had to accomplish. You should not blame the RS-232 committee for what it left out or failed to require. Instead, you should salute the committee for finding *some* agreement and beginning the long march toward the day when you will not have to know this level of detail to use serial communications. Someday, serial communications always will be done the same way by every computer equipment maker.

You now know the history of the RS-232 standard. Because this standard is such a central feature of all communications cables you will deal with, this chapter also looks at what the standard defines. But first you must learn how the "ends" of the serial connection are defined and the terminology used. Remember, the material isn't hard; it's just more detail than you normally need to know about most parts of your computer.

Understanding DTE, DCE, and Signal Definitions

Serial communications circuits go in only one direction. Only one end can send and the other end receive on a single circuit. Some method had to be used to define the names and directions of each standard circuit. The RS-232 standard defined one end of the connection as DTE and the other as DCE.

DTE stands for data terminal equipment, and DCE stands for data communications equipment. When the RS-232 standard was developed, serial communications were used primarily for hooking terminals to computers through modems. Thus, the terminal was DTE and the modem was DCE. The CCITT (Consultative Committee on International Telephone and Telegraph) standards committee, when drafting compatible international standards (such as X.24, which incorporates the RS-232 definitions), referred to DCE as Data Circuit-Terminating Equipment, so you may encounter this name as well. Just one more thing to make something simple look hard.

Because the computer user usually sits at the computer terminal, the RS-232 signal directions were established relative to the terminal (DTE). (At least the RS-232 committee did remember for a moment that people are the reason all this standardization is necessary!) With this orientation, a circuit named Transmit Data means that the terminal (DTE) sends data on this line, and the modem (DCE) receives data on this line. The important thing to remember is that you must plug DTE-configured cables into DCE-configured cables, or the signals cannot cross-connect properly. This connection is explained in more detail later when cable fabrication is examined in the section "Constructing Cables."

Cable requirements cause most of the problems encountered during the connection of communications hardware. The RS-232 standard defines 25 total signals to enable you to connect and control nearly any type of modem and computer. This large number of signals was defined to support a wide range of communications devices—some extremely specialized and sophisticated. Rarely are all 25 signals required. In fact, only 9 (or fewer) signals are implemented in most PC serial interfaces.

Table 2.2 defines all 25 RS-232 circuits. The signals are discussed one by one in the following paragraphs. In this table, the pin numbers used are those for an ISO 2110 standard DB-25 connector. Those pin numbers also are used for the following discussion of the 9 signals that commonly are implemented in PC serial interfaces. The signal names are always the same because they are defined by the RS-232 standard, but the abbreviations used for them do vary a bit.

Table 2.2
RS-232 Signal Definitions

ISO Pin #	RS-232 Name	DTE/ DCE	EIA Code	CCITT Code	RS-232 Function
1	FG		AA	101	Frame (protective) ground
2	TD	DCE	BA	103	Transmitted data
3	RD	DTE	BB	104	Received data
4	RTS	DCE	CA	105	Request to send
5	CTS	DTE	CB	106	Clear to send
6	DSR	DTE	CC	107	Data set ready
7	SG		AB	102	Signal ground
8	CD	DTE	CF	109	Received line signal detect
9		DTE			Reserved for testing (+ voltage)
10		DTE			Reserved for testing (– voltage)

ISO Pin #	RS-232 Name	DTE/ DCE	EIA Code	CCITT Code	RS-232 Function
11		DTE	SA		Supervisory transmit data
12	SCD	DTE	SCF	122	Secondary received line signal detect
13	SCTS	DTE	SCB	121	Secondary clear to send
14	STD	DCE	SBA	118	Secondary transmitted data
15	TC	DTE	DB	114	Transmitter clock
16	SRD	DTE	SBB	119	Secondary received data
17	RC	DTE	DD	115	Receiver clock
18					Unassigned
19	SRTS	DCE	SCA	120	Secondary request to send
20	DTR	DCE	CD	108.2	Data terminal ready
21	SQ	DTE	CG	110	Signal quality detect
22	RI	DTE	CE	125	Ring indicator
23		DCE	CH	111	Data rate selector (direction optional)
		DTE	CI	112	EIA/CCITT code changes with direction
24	ETC	DCE	DA	113	External transmit clock
25		DCE			Unassigned

Control Signals: + = on
                  – = off

Data Signals: 0 (space) = +3v to +25v
               1 (mark) = –3v to –25v

Pin 1 is a safety ground (FG). This pin isn't related to data transmission at all but is like the third wire in your AC wall socket, which protects you against shock hazards. Often this pin is not even connected.

Pin 2 is the transmitted data (TD) circuit. The terminal (DTE) uses this wire to send data to the modem (DCE).

Pin 3 is the received data (RD) circuit. The modem (DCE) uses this wire to send data to the terminal (DTE).

Pin 7 is the signal ground (SG). Because each active circuit is a single wire, its voltage must be referenced against the signal ground. Therefore, pin 7 must always be connected from one end to the other; otherwise, nothing can work.

The minimum number of connections required to have data flow in both directions is three wires: pin 2 (transmitted data), pin 3 (received data), and pin 7 (signal ground). If these three wires are connected properly, you can send data in both directions between the terminal (DTE) and the modem (DCE). Figure 2.1 shows a cable with this minimum wiring. (For this and all subsequent figures, arrows indicate direction of data flow and control-signal flow.)

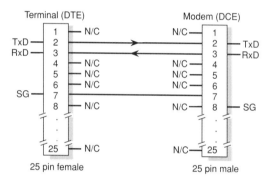

Fig. 2.1. *Minimum wiring for a 25-pin RS-232 serial data cable.*

The connectors in figure 2.1 (and elsewhere in this book) are shown as Male and Female. The *male* connector has pins showing while the *female* connector has sockets or holes showing. These connectors often are referred to by their pin count and as *Socket* (female) or *Plug* (male) type, resulting in the term DB-25S (female 25-pin D connector) or DB-25P (male 25-pin D connector). In this book, the terms male and female will be used along with the pin count to describe these connectors, but you may encounter references using the DB-25 or DB-9 notation. Both notations refer to the same type of connectors.

Understanding Handshake Signals

You have learned that you can send serial data in both directions, using only three wires, so you may be wondering why the other signals listed in table 2.2 are necessary. The reason is that serial communications often take place over switched connections (such as a dial-up telephone). One end of the connection also may need to ask the other end to pause momentarily while the first end processes the information it has received.

These signals generally are called *handshake* signals because they enable the computer and equipment involved to meet (shake hands) and establish agreements on how to handle various conditions. The handshake signals implemented on most PC serial interfaces are divided into two categories: *flow control* and *modem control*.

Examining Flow Control Signals

The two flow control signals include pins 4 and 5. Pin 4 is the request to send (RTS) signal. This signal is sent from the terminal (DTE) to the modem (DCE) to request that the terminal be allowed to send data. Pin 5, the clear to send (CTS) signal, is the handshake response to the RTS from the modem (DCE) to the terminal (DTE). CTS tells the terminal (DTE) that it is allowed to send data.

These two signals (RTS/CTS) are used for controlling data flow. They enable one end of the connection to ask the other end to quit sending data if it cannot be processed currently. These two signals often are referred to as *hardware flow control* because they are actual wires (and thus are hardware rather than software). For an illustration of this flow control, refer to figure 2.2, which shows a cable with data connections and also RTS and CTS flow control signal connections.

When the terminal wants to send data to the modem, the terminal asserts RTS. The modem then responds by asserting CTS if data can be accepted at this time. If the modem cannot accept the data, the modem removes CTS, and the terminal stops sending until CTS returns. This design enables the modem to control the flow of data from the terminal as needed. The RS-232 committee envisioned this type of operation when the committee named these signals.

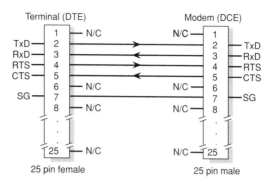

Fig. 2.2. Wiring for a 25-pin RS-232 serial data cable that enables hardware (RTS/CTS) flow control.

When the RS-232 standard originally was created, many modems operated in only one direction at a time. They had to be switched from one direction to the other before information could be sent. For this reason, the flow control signal pair is named *request to send* (RTS) and *clear to send* (CTS). When the terminal asserted "request to send," the modem electronically switched directions so that the data could flow from the terminal. This process took time, and so the modem asserted CTS to indicate that the link had "turned around" and was "clear to send" data. When the terminal was done sending, it removed RTS, and the modem shifted back from the send state to the receive state, ready to receive data. The modem then removed CTS to indicate that it was in receive mode. The RS-232 signal naming was based on this type of operation.

Today, however, you often want to send information in both directions at the same time (full duplex). If the modem is connected, or if no modem is present because the circuit is directly connected with wires, then the circuit unquestionably can carry data in either direction at any time. This design is the kind of circuit most often used in PC communications, and so a variation of RTS/CTS flow control is used.

This variation uses RTS not as a "request to send" but as a reverse "clear to send" to the modem. Thus, flow control is independently available in both directions without using all the wires the RS-232 standard designed for full-duplex control. PC serial interfaces don't implement most of the RS-232 signals, so some method had to be devised to enable bidirectional flow control. As usual, those inventive souls who make hardware and software for use with PCs found a way to get the job done—even when they didn't have everything the standard says they need.

In this PC variation of RTS/CTS flow control, the terminal asserts RTS if it can *receive* data from the modem and removes RTS to stop the modem from sending data. The modem asserts CTS if it can receive data from the terminal—without regard to whether the terminal has RTS asserted. Thus, with this variation, the two signals are completely independent and could be renamed "modem clear to send" (RTS) and "terminal clear to send" (CTS). If you think of them in these terms, you can understand more easily how they are used in PC communications to control the flow of serial data in both directions.

Examining Modem Control Signals

The final four control signals implemented on a PC serial interface exist primarily to control the operation of a modem. This control and monitoring are needed because modems connect to the telephone lines. Thus, modems aren't always connected to other computers, and the process of making a connection involves dialing and signal matching, which has several steps and takes time.

Pin 6 is the *data set ready* (DSR) signal. This signal is sent from the modem (DCE) to the terminal (DTE) to indicate that the modem has made a connection and is ready to operate.

Pin 8, the *carrier detect* (CD) signal, is sent by the modem (DCE) to the terminal (DTE) and indicates that the modem is receiving a valid signal from another modem—a stable signal that meets the signal quality requirements necessary for the modem's operation.

Pin 20 is the *data terminal ready* (DTR) signal. This signal is sent from the terminal (DTE) to the modem (DCE) to indicate that the terminal is ready for the modem to accept incoming connections. If this signal is turned off by the terminal (DTE), the modem (DCE) is supposed to disconnect any current connection (it may, for example, hang up the phone) and refuse any new connections. Often the DTR signal is dropped briefly and then asserted again, which tells the modem to hang up the line.

Pin 22, the *ring indicator* (RI) signal, is sent from the modem (DCE) to the terminal (DTE) and is set on to indicate that the telephone is ringing. This signal goes on and off roughly in synchronization with the audible ringing sound and so is a pulsing signal.

During the discussion in the following paragraphs, you can refer to figure 2.3, which shows a diagram of a cable that connects a computer terminal (DTE) to a modem (DCE) with all the signals normally implemented in a PC serial interface.

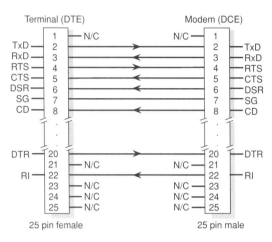

Fig. 2.3. Wiring for a 25-pin RS-232 serial data cable that enables hardware (RTS/CTS) flow control and full modem control.

The two most important signals of these four modem control signals are carrier detect (CD) and data terminal ready (DTR). These two signals are often all that is used for controlling a serial interface. DTR controls the modem by allowing or disallowing connections. DTR also can be used for ending a connection. CD indicates that a connection exists and is used for determining when the other end has disconnected.

The data set ready (DSR) signal originally was used for indicating when a modem that shared a telephone line was connected to that line. Today DSR is used as a status signal in a variety of ways, but almost all of them are some variation on the theme that the modem is ready to be used.

The ring indicator (RI) is the least often used signal of this group. Most modems today handle the function of answering the telephone when it rings internally. If RI is used, however, its function is to alert the computer that the telephone is ringing. Using the other control signals to answer the phone and make the connection is then up to the computer.

You now have learned the theory behind the serial signals that are implemented and used most commonly on a PC. You can learn more detail about the handshake and control signals in Chapter 19, where programming of the serial port is discussed. You have learned enough here, however, to understand the signals' functions and why things don't work if they aren't hooked up properly.

Constructing Cables

Now that you are somewhat familiar with the RS-232 serial interface, you can look at the cables used for communications. One of the most commonly used cables is the one shown in figure 2.3 in the preceding section. This cable connects a 25-pin computer port to a 25-pin modem. The DTE (computer terminal) end of the cable is a female connector, and the DCE (modem) end of the cable is a male connector. Because the cable is connecting a DTE connector to a DCE connector, all wiring is "straight through," connecting the same pin numbers on both ends.

Not all computers use the ISO-2110 25-pin connectors, which complicates cable selection. Because only 9 of the 25 RS-232 signals are usually implemented on a PC, IBM has maintained the practice on some of its serial ports of using a 9-pin connector. Figure 2.4 shows how that connector is wired on an IBM PC serial port. This connector is the terminal (DTE) configuration and is a male connector.

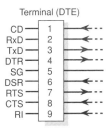

Fig. 2.4. Pin configuration for the IBM 9-pin serial port connector, which is configured as DTE and is a male connector.

Connecting this port to a 25-pin modem requires a cable with a 9-pin connector on one end and a 25-pin connector on the other. Because you still are connecting a DTE port to a DCE connector, you need to build a cable that logically connects the signals "straight through" as the cable in figure 2.3 does. This requirement results in the cable wiring shown in figure 2.5.

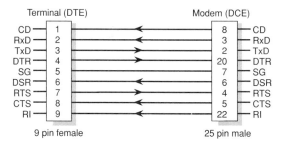

Fig. 2.5. IBM 9-pin port to 25-pin modem cable wiring.

Other cable design considerations are involved when you need to connect two computers without using modems. The next section examines that type of connection, which is referred to as a *direct connection*. It also is sometimes called a *hard-wired connection* because only wires connect the two serial ports.

Understanding Direct Connections

As you learned earlier in this chapter, the RS-232 standard expects that you are connecting a terminal (DTE) to a modem (DCE). At times, however, you may want to connect two terminals (or computers) directly—without any modems at all. In this case, you must construct a cable that fools each terminal into thinking that it is talking to and controlling a modem. Such cables are known as *null modem cables*.

Because you are engaging in trickery, no single null modem cable design works for all cases. Two designs are used most commonly, depending on whether you want each end to control itself or to control the other end. These designs are known as the *full-handshake* and the *loopback* null modems. Figure 2.6 shows the configuration for the full-handshake version, and figure 2.7 shows the configuration for the loopback design.

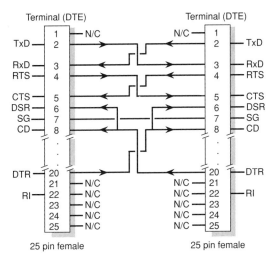

Fig. 2.6. *Wiring for a 25-pin RS-232 full-handshake null modem cable that enables hardware (RTS/CTS) flow control and simulation of modem control.*

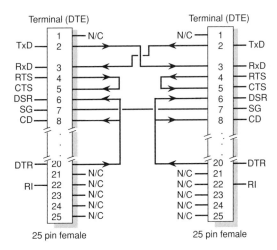

Fig. 2.7. *Wiring for a 25-pin RS-232 loopback null modem cable that does not enable RTS/CTS flow control or simulation of modem control.*

The full-handshake null modem simulates end-to-end modem control. Because the DTR of one terminal furnishes the CD and DSR signals for the other terminal, the normal end-to-end control sequences, such as using DTR to signal a disconnection, operate. Also, CTS from one end furnishes RTS for the other, so this arrangement enables full hardware RTS/CTS flow control to be used.

The full-handshake null modem is the most commonly used null modem cable. Because the DTR signal from one end is used to supply the CD and DSR signal of the other end, a complete modem connection sequence may be emulated. Also, lowering DTR on one end causes the other end to believe that the modem's carrier went away, so modem disconnect sequences also can be simulated.

A full-handshake null modem requires that the terminal on each end supply a proper DTR signal, however, so that the opposite end in turn will receive a proper CD and DSR signal. This modem also assumes that both ends will supply a proper CTS signal so that the opposite end will receive a proper RTS signal for flow control.

In practice, null modem cables are used to connect a variety of devices and not all of them implement all of these signals. Two common variations of the null modem cable may be used when one or both ends do not implement all of these signals.

The loopback null modem cable uses each end's own signals to simulate the other end's signals. In this case, flow control is not possible (because the other end cannot observe the RTS signal), and neither is link disconnection through the use of DTR. This type of null modem is used when the software on one end does not respond as the software on the other end expects. With a loopback circuit, each end sees the modem as responding to its own control. When the modem lowers DTR, it sees CD and DSR go away, indicating that the "modem" has disconnected.

One variation on this theme is a hybrid in which you require full-handshake flow control but loopback modem control. This design results in a null modem cable like the one shown in figure 2.8.

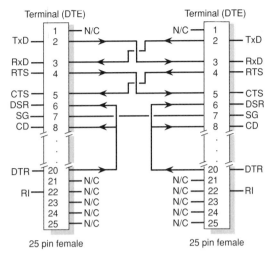

Fig. 2.8. *Wiring for a 25-pin RS-232 hybrid null modem cable that enables hardware (RTS/CTS) flow control but not simulation of modem control.*

If you are connecting computers that do not use the ISO-2110 25-pin connectors, you should concentrate on the signal names. Figure 2.9 shows a full-handshake null modem cable for two IBM computers using the IBM DB-9 connectors on their serial ports. Note that the signal names that are connected are identical to figure 2.6, but the pin numbers are not. This design results in the same electrical signal connection as in figure 2.6 and therefore performs the same function.

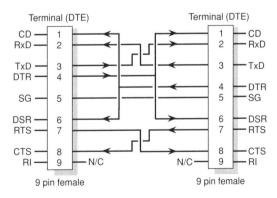

Fig. 2.9. *IBM 9-pin full-handshake null modem cable.*

When purchasing or constructing null modem cables, you must keep in mind that the purpose is to trick the two computers involved into thinking that a modem is present. At the same time, you must understand a bit about how the software on each end expects the modem to work and what portion of the flow and modem control the software implements. You will encounter still other hybrids of the null modem adapter, but if you understand how the modem and flow control signals operate, you should be able to understand what type of null modem cable your application requires.

Although all of these interconnections have been discussed as being cables, they also are available in the form of small adapter blocks. These blocks have a connector at each end and a solid plastic case between them. The wiring described is contained inside this plastic case. You probably should standardize your installation with one type of cable (usually a male to female cable with all pins wired straight through) and use adapter blocks to make the interconnections. This gives you the maximum flexibility when you are faced with varying interconnection tasks.

Adapter blocks that are wired straight through, but have either two female or two male connectors, are sometimes called *sex change adapters* or *gender benders*. They are used to solve problems when you need to connect two cables that have the proper wiring, but are both the same "sex" and therefore will not plug into each other. Adapter blocks also are available that adapt 9 pin to 25 pin connectors.

If you purchase prefabricated null modem adapter cables or null modem adapter blocks, you should be sure you obtain a wiring diagram. As indicated previously, three different types of null modem adapter are available, and if you don't have the proper adapter for your application, it will not work correctly.

Troubleshooting Cable Problems

By far the most common reason for problems with serial communications is improper cable connections. Diagnosing and correcting these problems can seem an overwhelming task because the most common symptom for any problem is total failure of the serial circuit. But you can use some simple yet effective techniques to pinpoint and solve such problems.

If the circuit is partially operating, you often can refer to the discussion of cable operation and RS-232 covered earlier in this chapter to determine what portion of the circuit is failing. This process can lead you to know whether you need a cable with more wires connected or of a different type. Often, however, an incorrect cable gives you no symptoms to decipher. The cable just doesn't work.

In this case, the problem you face is that you have no visibility into what is happening in the RS-232 cable. Knowing what the various signals *should* be doing is all well and good, but if you cannot see them, that knowledge doesn't help much. The solution to this problem is an inexpensive RS-232 read-out device, which you plug into the circuit. This device has indicator lights for the RS-232 signals so that you can see what is happening in your cable. A read-out device is available for as little as $5 from many computer hardware stores.

To troubleshoot a cable, you plug the read-out block into the line at the DTE end. Because signals are named relative to the DTE end of a connection, this ensures that the signal names on the lights are correct if the cable you are examining is anything other than a straight-through cable. Check first that the data lines are wired correctly. To do this, press a key on your computer keyboard and see whether the TD (transmit data) light on your indicator block flashes. If it does not, then you know you have a problem with that signal. The most common problem is the failure to use a null modem when you need to, but you also may have a broken wire.

If the data lights work properly, but the connection still doesn't function, the problem must lie in one or more of the control signals. By using your software, you can check most of these signals relatively easily. If you are using a null modem connection, see if raising DTR on one end of the cable results in CD and DSR appearing on the other (assuming your cable is a full-handshake null modem). If you are using one of the alternate null modem cables, check the drawings and cable discussion earlier in this chapter and verify that each source signal (output from the RS-232 connector) causes the proper input signals to which the source should be connected to operate.

Another troubleshooting device that can be handy is called a breakout box. In addition to indicator lights like the read-out block, an RS-232 breakout box has a series of jumper wires and switches that enable you to build any cable connections you need "on the fly" by moving jumpers and switches. When you have an unusual application, a breakout box enables you to try various cable types easily until you determine which type of cable you require. Breakout boxes are fairly expensive, however, and only make sense if you must build and debug many RS-232 connections.

Remember that more than one set of connections exist which are called a null modem. If you purchase a commercial null modem, verify that it is wired the way your application requires. Your software probably gives you some idea of what is required. If not, you will have to study the RS-232 theory presented in this chapter and determine which type of null modem you need for your application.

After you determine which connections are wrong, you can try to repair the cable yourself (if you are adept at using a soldering iron) or purchase the proper cable now that you have determined what type you need.

Chapter Summary

In this chapter, you learned about the hardware that is particular to PC communications. You learned about serial ports and modems and the cables that connect them. In addition, you learned about the serial interface on your PC. This chapter's discussion on cable operations and RS-232 can help if you need to troubleshoot problems with serial cables or connections. This chapter also can help if you need to connect two dissimilar computer ports or if you encounter RS-232 ports that have unusual connectors. Using the information in this chapter and your computer manual, you can construct or specify the proper cable for the signals you need to connect.

3

Understanding Software Requirements

After you connect the proper hardware to your PC, you need to add the appropriate software before you can use PC communications. Although a wide variety of PC communications software is available, the types all fall into one of the following categories:

- *Terminal emulation programs*. These programs enable your computer to call another computer and communicate interactively with it.

- *Remote control programs*. These programs enable your computer to control (or be controlled by) another computer through the telephone line.

- *Host programs*. These programs enable your computer to act as a remote host for other computers. The type of host can range from a single user file transfer host to a multiuser information center with electronic messaging and database capability (bulletin board system software).

Some software combines more than one of these functions into a single integrated package. A particular software package, however, usually focuses on one of these categories and adds some functions of the others.

This chapter examines each of these communications software categories. You are introduced to each type of software and learn what types of applications are best suited for which software. You also learn the meaning of such terms as terminal emulation, file upload and download, file transfer protocols, and scripting—as well as how to use them.

Using Terminal Emulation Software

The first category of communications software—terminal emulation programs—is by far the most common PC communications application. In fact, it is the *only* type of communications software with which most users of PC communications are personally acquainted. Examples of popular programs in this category are PROCOMM PLUS from DATASTORM TECHNOLOGIES, Crosstalk from DCA, and Smartcom from Hayes Microcomputer Products. PROCOMM Lite, a scaled-down version of PROCOMM PLUS, is included with this book. Refer to Appendix E for instructions on using the program.

The primary function of a terminal emulation program is to enable you to type on your computer's keyboard and have the information sent to the serial port, and to display any characters received from the serial port on your computer's screen. In other words, through terminal emulation software, two personal computers can connect to each other, and the people at each computer may type to each other and send files back and forth. This function was originally the domain of dedicated computer data terminals, which give this type of software its name.

Early data terminals contained no computers. A *data terminal* was a collection of electronics that performed only the single function of a keyboard and display. With the use of terminal emulation software, your computer can emulate these data terminal functions. The power of your computer, however, enables additional functionality beyond simple terminal emulation. Those functions also are examined later in this chapter.

The modern terminal program has made the use of computers and modems popular today. This software has become so flexible and capable that many communications problems can be solved by robust terminal program software alone.

Configuring the Serial Port

In Chapter 1, you learned that serial communications involves several attributes such as number of data bits, parity, and so on. You also learned that these attributes must be set properly in order for serial communications to succeed. To help you with this setup, all communications software includes a configuration section.

Each program is designed slightly differently, but figure 3.1 shows a good example—the serial port configuration screen for the popular terminal program PROCOMM PLUS. This configuration screen is where you tell your software about the hardware to which it is connected. You can see that PROCOMM PLUS enables you to set all the parameters discussed in Chapter 1—baud rate, parity, data bits, stop bits, and port.

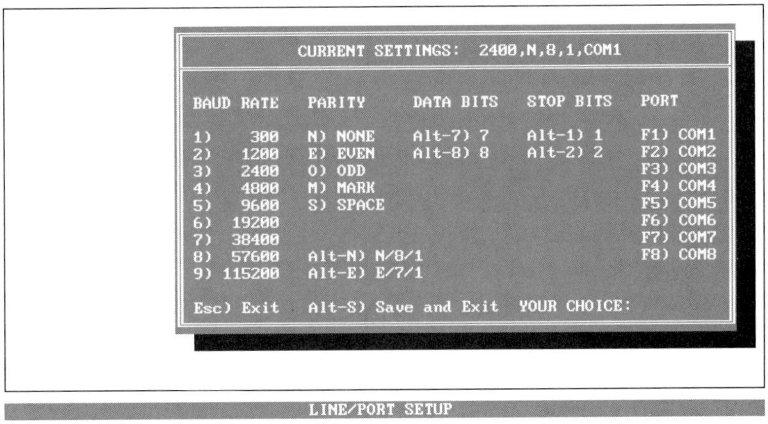

```
        CURRENT SETTINGS:   2400,N,8,1,COM1


 BAUD RATE    PARITY       DATA BITS     STOP BITS     PORT

 1)     300   N) NONE      Alt-7) 7      Alt-1) 1      F1) COM1
 2)    1200   E) EVEN      Alt-8) 8      Alt-2) 2      F2) COM2
 3)    2400   O) ODD                                   F3) COM3
 4)    4800   M) MARK                                  F4) COM4
 5)    9600   S) SPACE                                 F5) COM5
 6)   19200                                            F6) COM6
 7)   38400                                            F7) COM7
 8)   57600   Alt-N) N/8/1                             F8) COM8
 9)  115200   Alt-E) E/7/1

 Esc) Exit    Alt-S) Save and Exit   YOUR CHOICE:
```

LINE/PORT SETUP

Fig. 3.1. The serial port configuration screen for PROCOMM PLUS.

In addition to the specific serial port settings, you must tell your software where to find your serial port. You provide this information by giving the port base address—COM1 or COM2. PROCOMM PLUS also enables you to use nonstandard IBM ports, as described in Chapter 1. To do so, you must define the port's base address and interrupt in the menu shown in figure 3.2. PROCOMM PLUS then assigns the port to the mnemonic COM1 through COM8 so that you can select that port in the main configuration screen. Remember that only COM1 and COM2 have standard definitions. Any number higher than COM2 is used only as a label for which you must define the port address and interrupt.

After you tell your software where the serial port is located, you must define the serial data stream format you want to use. With this setting, you indicate the appropriate number of data bits, parity presence and type, and number of stop bits. Because the data format is determined by the other computer with which you are going to connect, the format often varies for different connections. Many terminal emulation programs help you deal with this assortment by enabling you to associate a particular serial data format with each telephone number in a dialing directory. Figure 3.3, for example, shows the PROCOMM PLUS Dialing Directory screen. Notice that you can specify each serial port parameter for each telephone number you may want to dial.

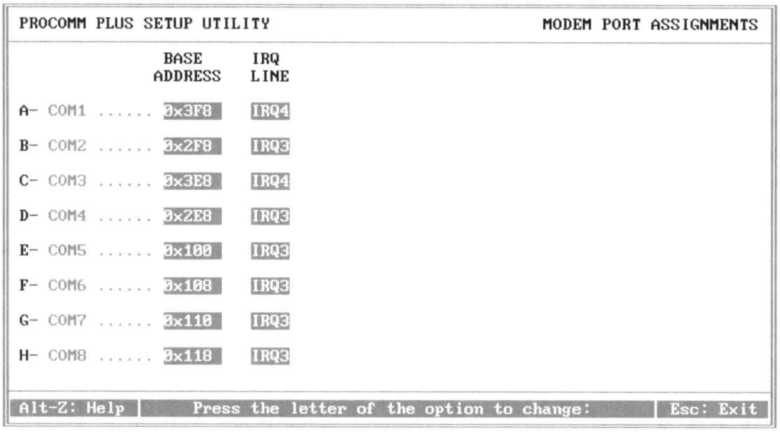

```
PROCOMM PLUS SETUP UTILITY                      MODEM PORT ASSIGNMENTS

                    BASE        IRQ
                    ADDRESS     LINE

A- COM1 ......  0x3F8     IRQ4

B- COM2 ......  0x2F8     IRQ3

C- COM3 ......  0x3E8     IRQ4

D- COM4 ......  0x2E8     IRQ3

E- COM5 ......  0x100     IRQ3

F- COM6 ......  0x108     IRQ3

G- COM7 ......  0x110     IRQ3

H- COM8 ......  0x118     IRQ3


Alt-Z: Help  |     Press the letter of the option to change:     | Esc: Exit
```

Fig. 3.2. Defining base addresses and interrupts for nonstandard IBM ports.

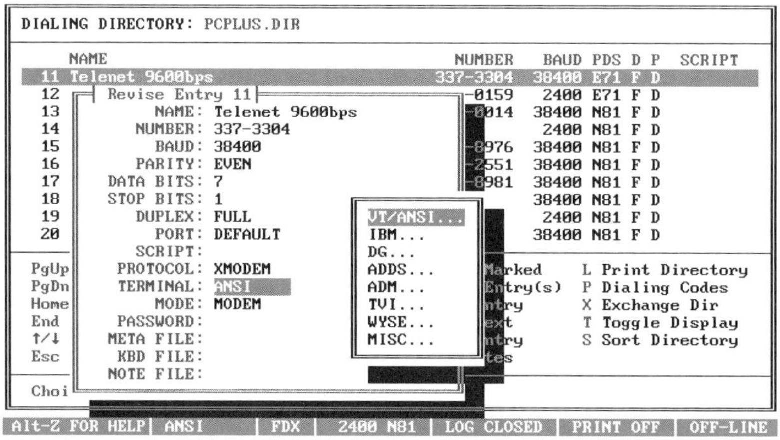

```
DIALING DIRECTORY: PCPLUS.DIR

      NAME                              NUMBER   BAUD PDS D P   SCRIPT
11  Telenet 9600bps                   337-3304  38400 E71 F D
12  ┌─ Revise Entry 11 ──────────────┐ -0159    2400 E71 F D
13  │         NAME: Telenet 9600bps  │ -0014   38400 N81 F D
14  │       NUMBER: 337-3304         │          2400 N81 F D
15  │         BAUD: 38400            │ -8976   38400 N81 F D
16  │       PARITY: EVEN             │ -2551   38400 N81 F D
17  │    DATA BITS: 7                │ -8981   38400 N81 F D
18  │    STOP BITS: 1                │         38400 N81 F D
19  │       DUPLEX: FULL        ┌─ VT/ANSI... ─┐  2400 N81 F D
20  │         PORT: DEFAULT     │ IBM...       │ 38400 N81 F D
    │       SCRIPT:            │ DG...        │
PgUp│     PROTOCOL: XMODEM      │ ADDS...      │ arked   L Print Directory
PgDn│     TERMINAL: ANSI        │ ADM...       │ ntry(s) P Dialing Codes
Home│         MODE: MODEM       │ TVI...       │ ntry    X Exchange Dir
End │     PASSWORD:            │ WYSE...      │ ext     T Toggle Display
↑/↓ │    META FILE:            │ MISC...      │ ntry    S Sort Directory
Esc │     KBD FILE:            └──────────────┘ tes
    │    NOTE FILE:
Choi└

Alt-Z FOR HELP| ANSI    |  FDX  | 2400 N81 | LOG CLOSED | PRINT OFF | OFF-LINE
```

Fig. 3.3. Specifying the serial data stream format for a particular connection with PROCOMM PLUS.

Understanding How Terminal Emulation Works

To understand fully the operation of terminal emulation software, you first must look at the hardware device that this type of program emulates—the data terminal. As mentioned briefly at the beginning of the chapter, a data terminal is a keyboard and display connected to a serial port. When you press a key on the data terminal's keyboard, the corresponding ASCII character is sent to the serial port, and any character received from the serial port is displayed on the data terminal's screen.

A program that performs this emulation often is referred to as a *terminal program*. The simplest possible terminal program would send every key you pressed to the serial port and display every received character on your computer's screen. Modern terminal programs, however, go beyond this minimal functionality to perform terminal emulation of specific, widely used brands of data terminals.

True terminal emulation therefore means that the software attempts to make your computer operate as identically as possible to the particular brand of terminal being emulated. For that reason, you must select a particular make and model of data terminal—for example, a VT-100 (a terminal made by Digital Equipment Corporation) or a Televideo 950 (made by Televideo Corporation). Figure 3.4 shows the terminal emulation selection options of the program Smartcom Exec.

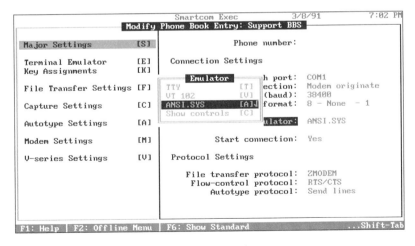

Fig. 3.4. Choosing the type of terminal to emulate.

When you select a particular terminal model to emulate, your software maps your computer's keyboard as closely as possible to the keyboard of that terminal model, enabling you to use any special function keys the terminal may offer. The software also makes your computer's display respond to the special character sequences the selected model of data terminal uses to control such actions as cursor positioning, dim or bright display, underline, reverse video, and possibly even color.

The simplest terminal to emulate is a Teletype (often abbreviated as TTY). A Teletype was a printing device that had no display to control (characters were always one color and one brightness and could not be reversed) and a keyboard with only the ASCII characters and control codes (no function keys). To emulate a TTY terminal, therefore, software merely has to send letters, numbers, punctuation, and control characters to the serial port while displaying received characters to your computer's screen without any interpretation.

The only real emulation done by a TTY emulator is to properly handle the limited control functions such as carriage return and line feed. These functions are triggered by specific single control characters and are primarily for print head and paper movement. On a CRT screen they translate to cursor movement (because there is no print head or paper). As an example, a carriage return on a mechanical printer is functionally the same as moving the cursor to the first column of the line on a CRT screen. A line feed (feed the paper forward one print line) is emulated on a CRT by lowering the cursor to the next line on the screen and scrolling the screen if the cursor is already on the bottom line. A top of page function is emulated by clearing the CRT screen and moving the cursor to the home position. Because these control functions are the most basic possible (and are always required by any terminal) they are each assigned a specific single control character from the 32 codes the ASCII character set reserves for control functions.

In most cases, however, you want more capability to control the display than TTY emulation provides. You want to enable the remote computer to clear your screen, position the cursor at will, and display characters with attributes such as intensity, blink, reverse video, or color. You also may want to send function keys, arrow keys, delete keys, insert keys, and other keys to the remote computer. For such capability to be possible, both your terminal program and the remote computer (the other computer with which you are communicating) must agree on how each extended function is to be done—that is, which character sequences will represent each of these enhancements of the TTY protocol.

If your terminal program and the remote computer know about a particular model of data terminal that supports the functions you require, you can obtain those functions by emulating that data terminal with your software. Because many models of data terminals have been popular through the years, you normally are given a menu of terminal emulations from which to choose. The same terminal to emulate must be chosen on both your terminal program and the remote computer. The two most widely used terminal emulations today are the VT-100 and the ANSI terminal emulation.

The most important thing to remember is that you must match your terminal emulation to the type required by the computer you are calling, or your video display does not properly show the screens as they are intended. Often, an incorrect emulation results in garbage characters showing on your screen. Also, an incorrect terminal emulation means that you cannot use your computer's function keys to control the remote computer correctly.

Understanding File Transfers

Terminal emulation software enables your computer to perform as a data terminal and interact with remote computers. But you are using a complete computer and not just a data terminal, so your terminal program has the option of providing many other functions.

One of the most common "extras" such software offers is the capacity to move a disk file from the remote computer to your computer's local disk and vice versa. This process is known as *file transfer* (some of these buzzwords actually make sense).

In a discussion of file transfer, two new terms come into play: *upload* and *download*. To understand these terms, you should think of the computer you are calling as being on top of a hill. From this mental picture, you can see that you must send a file up to the remote computer (upload) and that the remote computer must send a file down to you (download). Thus, the term upload means that you send a file to the remote computer, and download means that you receive a file from the remote computer.

The simplest way to transfer a file from one computer to another is to have the sending computer simply print the entire file to the receiving computer just as if it were intended for display on-screen. You can tell the receiving computer's software, however, that it should capture the characters being sent and write them to disk rather than (or in addition to) the screen. This simple form of file transfer is an *ASCII download* or *ASCII upload* depending on the direction. It also is sometimes known as a *screen capture* because you are capturing the data written on your computer's screen to a disk file.

You see in the ASCII file transfer the beginnings of the structure of any type of file transfer. The software must set down some rules so that it knows when to start capturing the data stream to disk and when to stop. These rules may be as simple as "start when I press a function key and stop when I press another function key," but because computers are literal devices, they must have a fully specified, formal set of rules for everything they do. A formal set of rules often is called a *protocol* (especially in the diplomatic corps). In PC communications, therefore, any set of rules for transferring a file from one computer to another is a *file transfer protocol*.

The ASCII file transfer protocol has some limitations. First, you have no good way to tell the program exactly when to start or stop the transfer. Also, this protocol transfers only text files, and often you may want to transfer binary data files (such as programs or work files). Finally, if an error occurs, you have no way to detect or repair it, and the received file thus may not match the original.

The XMODEM Protocol

In 1978, Ward Christensen invented the XMODEM file transfer protocol to solve these problems. This protocol has since become the most widely used file transfer protocol in PC communications. XMODEM can detect and correct errors that may occur during the transfer, and it exactly defines the start and finish, enabling you to synchronize the file transfer process properly.

Christensen originally developed the XMODEM protocol to record data onto tape from his computer. He had a computer that was not compatible in its disk system with any other computer of its day, so he wrote a program that implemented his XMODEM protocol, which he designed to be simple to implement on any computer. By recording on tape the XMODEM tones, he was able to transfer data from one computer to another.

Because Christensen also invented the computer bulletin board software concept (discussed in Chapters 13 and 14) that popularized PC communications and used his XMODEM protocol there, it has become the most widely used file transfer protocol in the world today.

In order to correct errors in transmission, a protocol must be able to detect that an error has occurred. In order to perform this detection, all error-resistant protocols must add information to the data that is predictable on the receiving end, and which also is highly likely to be affected by any errors. In order to accomplish this, the data is broken into *packets* (sometimes called blocks) and the sending end performs a mathematical calculation on the data to generate a *signature* number. This number then is sent with the data block. The receiving end performs the same mathematical formula on the data it received and compares the results with the signature number it received. If these two numbers don't match, then a transmission error has occurred in either the data or the signature number itself for that data block. The receiver then knows that the data is bad and can request that the data be sent again.

The original XMODEM protocol used a simple checksum mathematical formula for error detection. This plan was not an optimal choice for use on a communications link, so XMODEM-CRC was developed. XMODEM-CRC uses a different and superior CRC (cyclic redundancy check) error-detection scheme. Because of this change in XMODEM after it gained wide popularity, you may have two XMODEM choices on your protocol menu. If you have the option, you always should use XMODEM-CRC because its superior error detection is a better guarantee of file integrity during the transfer.

Although XMODEM is an effective protocol for file transfers, it has limitations. Foremost among these limitations is that XMODEM requires an 8-bit serial data path in order to operate. You may remember from Chapter 1 that many mainframe computers use 7-bit serial data paths; XMODEM cannot be used with them. In response to this problem, Bill Catchings and Frank da Cruz of Columbia University developed KERMIT, another file transfer protocol.

The KERMIT Protocol

KERMIT was designed to operate over any kind of serial data connection. Thus, KERMIT restricts itself to a small subset of the ASCII character set and avoids any use of control characters.

KERMIT uses a method called *quoting* to handle binary characters. Quoting converts a single nonprintable binary character into two printable characters that can be decoded by the receiver back to the original single binary character. This method enables such data to be sent down any serial link, but the process is slower because it takes extra characters. To combat this expansion of data and regain some of the lost speed, KERMIT also incorporates a primitive data compression routine that can express repeated characters as three characters.

Remember that the purpose of the KERMIT protocol is to support file transfer over any serial data link and through any mainframe hardware in existence. KERMIT performs that job well.

Columbia University released the specifications of the KERMIT protocol into the public domain and also makes available fully operational terminal programs that incorporate the KERMIT protocol for most of the computers in the world. The only charge for this software is for the tapes or disks required to copy it. As a result, KERMIT is as close to a file transfer *lingua franca* as any file transfer protocol ever invented.

NOTE

Although KERMIT has achieved its goal of near universal implementation, it has paid a price for this. In its efforts to be usable in nearly every type of environment, KERMIT has sacrificed performance. Newer versions of KERMIT are adding variations such as SuperKermit and Long Block Kermit, which overcome most of its early limitations. You cannot expect KERMIT to give you the fastest file transfers, however, and often it will be very slow. KERMIT's primary advantage is that it will connect almost any type of computer to any other type of computer and guarantee the data gets through correctly, which it does very well.

You can use KERMIT to transfer files between different models of computers over nearly any serial data link in the world. By comparing KERMIT and XMODEM, you can begin to see why such a proliferation of file transfer protocols has evolved. Each protocol addresses a particular set of problems that result from serial data transfer. These problems include error correction, data size of the serial link, propagation delays in long distance or satellite links, high-speed data links, and even ease of use.

As modem technology has evolved, modems have become much faster and more capable than they used to be. The newest modems are capable of serial data speeds more than 100 times faster than the original 300 bps of the Bell 103 modems. Because XMODEM waits for an acknowledgment between each file block and because its blocks are small, XMODEM is not efficient at high speeds. In addition, XMODEM has no formal method for sending the file's name, exact size, and date; the protocol sends only the file data. Many ad-hoc extensions of XMODEM exist to handle these problems. The most widely used of these variants are MODEM7 and TELINK. MODEM7 sends only the file names and enables you to send several files in a single transfer. TELINK adds the exact size and file date to the file name. Both MODEM7 and TELINK, however, just add a header to the XMODEM transfer and have the identical performance loss on high-speed data links.

Many developers experimented with extensions to XMODEM to solve these problems. One of the most successful and widely adopted of these extensions was the YMODEM protocol.

The YMODEM Protocol

In the early 1980s, Chuck Forsberg of Omen Technology, Inc., entered the protocol design field for PCs with his YMODEM protocol. He was trying to devise a simple extension to XMODEM that would solve both the file attribute problem and the high-speed performance problem. His first efforts resulted in the YMODEM protocol and its variant YMODEM-g.

YMODEM uses data blocks that are eight times as large as XMODEM. Thus YMODEM is roughly eight times as efficient on very high-speed links, because the majority of the time wasted by XMODEM is between blocks. Other than adding a file name and attribute header and enlarging the blocks, YMODEM is essentially the same protocol as XMODEM.

YMODEM-g is a variant of the YMODEM protocol and is designed for a specific set of high-speed modems. Today V.32 and V.32bis modems are the standard for high-speed transfers. Only a few years ago, however, this type of modem was prohibitively expensive. As a result, modem makers offered a number of proprietary high-speed modems at a much lower cost. These modems all had a common shortcoming: they became inefficient unless data moved in only one direction. But these modems also all guaranteed that the data they sent was error free. To provide optimum speed in this setting, YMODEM-g streams data blocks in one direction only and assumes that everything is received correctly. If an error does occur, the receiving end sends an error signal, and the entire file transfer is aborted.

Because of its design, YMODEM-g is the fastest possible error-detecting file transfer protocol. You can use it, however, only if the serial data link is guaranteed to be error free.

The ZMODEM Protocol

By 1987, file transfer protocols had become quite a "Tower of Babel." Many specialized protocols that solved specific problems had come into wide use. When you want to transfer a file, knowing what protocol to use can be difficult. A protocol may give superior performance in one set of circumstances but be very inefficient in another. Chuck Forsberg created the ZMODEM protocol in an attempt to make a protocol that incorporated all the benefits of previous protocols but was smart enough to automatically choose the proper methods to use for each serial link.

Early protocols such as XMODEM became popular quickly because they were easy to implement. With the number of optional modes among which it must automatically select and handle, a protocol such as ZMODEM is extremely complex. For this reason, ZMODEM has been slow to be adopted. ZMODEM does, however, solve most of the user interface problems faced by file transfer protocols, making it one of the easiest protocols to use. It also maintains high efficiency over nearly all types of data circuits and can recognize and use automatically a 7-bit data circuit for mainframe transfers. Therefore, you can expect to see ZMODEM become the protocol of choice over time, slowed only by the fact that implementing it correctly requires a large software effort.

The ZMODEM protocol incorporates many of the benefits of the XMODEM, YMODEM, and KERMIT protocols (although its rules are not the same). Thus, ZMODEM can be efficient with all types of modems and can operate over 7-bit data links. ZMODEM has two levels of implementation, with the most complete being ZMODEM-90. The public domain ZMODEM lacks the capability to do data compression or operate over 7-bit data links.

ZMODEM makes a number of decisions automatically, which enables it to be as efficient as the data link allows. ZMODEM even has the capacity to overlap sending and acknowledging data blocks, which often is referred to as *sliding windows* in other file transfer protocols. With this feature, ZMODEM can maintain top speed even over satellite links where a significant time delay occurs between the sending of a character by one end and the receipt by the other.

ZMODEM also shares with KERMIT the capacity to avoid the use of any control characters and to use only text characters for its transfer. The protocol thus can work over packet-switched networks or other data links that reserve certain characters for their exclusive use.

Finally, ZMODEM also enables you to implement a file restart feature. This feature is helpful if you are disconnected when you have partially transferred a file. With most other protocols, the time you spent before disconnection is lost. With ZMODEM you can restart the file transfer where it left off and transfer only the remaining portion.

As you can see, ZMODEM doesn't offer anything that some other protocol hasn't offered before. The real benefit of ZMODEM is that it integrates all these capabilities into a single protocol and makes its decisions automatically without requiring you to understand which ones are needed at which times. It also has a smooth user interface where you can make all file requests in a single command to the sending end's software. Other protocols can require you to type commands to both your software and the remote computer to initiate a file transfer.

ZMODEM has several modes because different communications links require different modes for optimal operation. High-speed, half-duplex modems require streaming (ACK-less) transmission, for example, to prevent data flow in both directions simultaneously for the fastest operation. Packet-switched networks require sliding windows modes to compensate for the delays without overrunning the network as a streaming protocol would. Long distance circuits that go through satellites (such as overseas calls) may have delays on the order of seconds. Although streaming protocols can solve the delay problems, a sliding windows protocol provides faster error detection and repair if errors are likely. Both streaming and sliding windows protocols remove the slowdowns caused by end-to-end delays.

Because of this need for different protocol operation under different circumstances, ZMODEM allows all of the modes to occur. Normally it will pick the proper mode automatically. If ZMODEM fails to determine the correct mode itself, you can set your terminal program to pick the correct mode manually, but you rarely will ever need to do this.

Choosing a Protocol

With so many protocols available, you may be confused as to which is the best protocol for each application. So many protocols exist because each addresses a different set of concerns. These concerns are based primarily on the many different types (and characteristics) of data communications links. No one protocol is best for all situations.

Your choice will be narrowed a bit because you can choose only from those protocols that are supported by both your computer software and the software on the computer to which you are connected. You also must choose from those protocols that support the characteristics of the data link you are using such as 7- or 8-bit data, and avoiding any restricted characters. No such restrictions usually are on directly dialed modem to modem data links. But data links via packet switched networks or data PBX connections may allow only certain types of data.

If after narrowing your choices you still have more than one choice, then efficiency considerations and ease of use will be your only remaining reasons for picking one protocol over another. The efficiency of a file transfer is affected by the nature of the data link you are using.

If you are using an asymmetrical or half-duplex modem that guarantees error-free data transmission (such as one of the proprietary 9600 bps modems discussed in Chapter 21), then you must use a streaming protocol for maximum data link efficiency.

A *streaming protocol* sends data in only one direction and does not generate responses for good records. It only responds when a record is received that has transmission errors. A streaming protocol may not be used unless the modems (or other data link) support full end-to-end hardware flow control (RTS/CTS) or are capable of running at their rated speed 100 percent of the time with no slowdowns. This is because a streaming protocol will overrun packet networks or other data links that don't provide flow control because it never stops sending data under any circumstances. Examples of streaming protocols are YMODEM-g, SEAlink Overdrive, and ZMODEM.

For data links that can send data in both directions at the same time without slowing below their rated speeds (such as V.32 9600 bps or V.32bis 14,400 bps modems), a *sliding windows protocol* is as good as a streaming protocol for efficiency (if the window size is larger than the end-to-end link delays). A sliding windows protocol also can be used to limit the maximum unacknowledged data sent, and thus will adjust to prevent overrunning packet networks (unlike a streaming protocol).

A sliding windows protocol still acknowledges receipt of each record whether it is received with or without errors. However, a sliding windows protocol will continue to send a certain number of records ahead of the last acknowledgment it has received. The number of records it will send ahead is called the window size and determines the amount of link delay the protocol will cover up without losing any speed. A streaming protocol, therefore, is really just a special case of a windowed protocol in which the window size is infinite and the lack of a negative acknowledgment (NAK) is considered a positive acknowledgment (ACK).

Like a streaming protocol, a sliding windows protocol removes the effect of link delays on throughput. Unlike a streaming protocol, a sliding windows protocol retains control over the maximum amount of data it can send without acknowledgment that the data has traveled all the way through the link. You therefore can adjust the window size not only to compensate for delays, but also to avoid loading the data link buffers (if any) with more than they can hold. This prevents data overrun in the link even if the link does not support end-to-end hardware flow control (RTS/CTS). A sliding windows protocol therefore is usually the optimal choice for a packet-switched network.

One other factor that can influence the choice of protocol is the expected error rate of the data link. If the data link you are using is likely to have a high error rate, you will want to choose a protocol with smaller record sizes so that less time is wasted retransmitting data to correct the errors. A sliding windows protocol with bad record insertion is the best in these cases, and SuperKermit is the only common protocol that has this feature. If the error rate is moderate, a protocol like SEAlink or ZMODEM, which dumps all "sent ahead" blocks on an error, is acceptable. On a link with a moderate error rate, you will want to use these protocols in sliding windows mode with smaller record sizes to control the amount of retransmitted data on an error. If you are using a data link that guarantees no errors (such as MNP modems or V.42 modems), then larger records generally are better because there is no retransmission overhead, and larger records produce fewer transactions in a protocol which is acknowledging receipt of each record. Also, the ratio of data characters to inserted protocol characters is better on larger records.

If the error rate is not character based, but is due to link reliability (in other words, you expect the link to disconnect frequently), then a protocol that can restart a transmission where it left off is a good choice. ZMODEM currently is the only widely available protocol that offers this option.

If, after these considerations, you still have more than one choice of protocols, ease of use is the determining factor. ZMODEM usually wins here because it will automatically start the download on the receiving end without operator intervention and automatically adjust record size for the prevailing error rate. If you are looking simply for raw speed, however, YMODEM-g is always the fastest protocol. It requires an 8-bit transparent (no reserved characters) error-free link with end-to-end hardware (RTS/CTS) flow control, however, to operate. Any errors will cause YMODEM-g to fail unrecoverably.

Understanding Scripting

Because your terminal program is operating on a computer, the program can do many other things to make your use of PC communications smoother and easier. One of the most important labor savers many terminal programs provide is *scripting*. Scripting is a process by which you can create a "script" that tells the terminal program how to perform certain functions automatically when you can predict what will happen.

Scripting is useful for automating any sequence of events you perform frequently. You always may go through the same steps, for example, to log on to a particular remote computer system. Rather than type the same answers in response to the same questions every time, you can write a script that tells your terminal program how to perform that sequence. Then, when you log on to that system, you can tell the terminal program to use that script and you are saved the effort of typing all the answers again.

The scripting capability of many terminal programs has become so powerful that an entire scripting language has been developed. This language enables you to write simple programs in scripts that make decisions based on the characters received. With these powerful scripting languages, automating entire transactions has become possible—from dialing the phone through logging into the remote system, transferring files, and logging off.

Terminal program scripts can be helpful. You can use them for small portions of any transaction that you do frequently. You also can write intricate scripts that completely hide the details of an entire communications session—making PC communications much easier to use. The details of creating scripts are different for each terminal program. You need to study the manual for your terminal program in order to learn how to write scripts for it. If you use PC communications routinely, this effort will be rewarded with less time and effort expended on each transaction.

As an example of how scripts actually work, you will examine a sample QMODEM script to log on to a BBS. The dialing directory is set to invoke this script after the number is dialed and the modems connect. This script is designed to answer all of the logon questions until you get to the top menu. Then it should return control to your keyboard. The system the following script will log onto is a PCBoard BBS.

```
TurnON    8_BIT
TurnOFF   LINEFEED
TurnON    XON/XOFF
TurnON    SCROLL
TurnOFF   PRINT
TurnOFF   ECHO
TurnOFF   SPLIT
;
TimeOut   30      ; Set Waitfor timeout to 30 seconds
Waitfor   "want ANSI Graphics (Enter)=no?"
Delay     100
Send      "n^M"
```

```
Waitfor   "What is your first name?"
Delay     100
Send      "Firstname Lastname^M"     ; Your logon name
          would go here
Waitfor   "Password (Dots will echo)?"
Delay     100
Send      "Password^M" ; Your password would go here!
Waitfor   "Since 'Last Read' (Enter)=yes?"
Delay     100
Send      "^M"
Waitfor   "Press (Enter) to continue?"
Delay     100
Send      "^M"
Waitfor   "Command?"
Delay     100
Exit      ; End of Script
```

Notice that the first thing the script does is to configure the terminal parameters properly for this BBS. Then the script waits up to 30 seconds for the opening question to be printed about whether you want graphics. You have chosen to have the script always answer that question with a NO response. The script delays a bit after it sees the text before sending the N followed by a return so that the PCBoard software will not miss the response. *Note:* The sequence ^M indicates the control character return that simulates your pressing the Enter key on your keyboard.

This script continues to answer all of the logon questions, including entering your name and password and requesting a scan for all new messages since your last logon. Finally the script waits for the prompt at the top menu and exits back to your keyboard.

This is a very simple script. Scripts can become very lengthy, and not all terminal program script languages are as simple as this one. But the effort you expend to learn how to script your terminal software will pay off when you can automate most of your common PC communications tasks.

Using Remote Control Software

Terminal emulation software enables two personal computers to connect to each other. The people at each computer may type to each other and send files back and forth. A logical extension of this connection is to control one computer from the other. In many cases, you want to run programs on the remote computer and want your computer's keyboard and display to operate as though it is the remote computer's keyboard and display.

Remote control software uses PC communications to accomplish this complete control of another personal computer. Examples of this type of software are CloseUp from Norton Lambert, PC Anywhere from DMA, and Carbon Copy from MicroCom.

Remote control software enables you to connect to another computer and run nearly any software that is on it just as though you were at that computer's keyboard and display. The software also enables you to observe on your computer the normal operations of another user on the remote computer.

Remote control software has been used quite successfully by PC software customer service representatives to troubleshoot problems a customer may have in using his or her software. This type of remote control over a modem connection enables the technical person to observe a problem firsthand without having to travel to the customer's site. Problem resolution speeds up tremendously without the cost of a trip. The cost savings this use of remote control software generates has made it a rapidly growing area of PC communications.

With a few simple enhancements, you also can use remote control software to call an unattended remote computer and run programs on it. If the remote control software has added the file transfer capability of the terminal programs discussed in the preceding section, you even can create data remotely and transfer it to your computer! Depending on how often this job needs to be done, and by how many people, remote control software can be a cost-effective way to obtain a tremendous amount of remote access functionality.

This capability can be used to solve many data problems. If you are at home and you have data at the office in a database, for example, you can call in, extract the reports or records you need, and download them to your home computer where you can print them out. Remote control software also allows remote debugging of applications. A customer support technician can see and operate a customer's computer without actually going to the customer's site. This greatly reduces the costs involved and leverages talented people's time.

When remote control software adds a caller logon and security check capability, simple single-user remote host operation is possible. A computer may be left operating unattended (no one watching or operating it) and still be used remotely. A caller to such a system can run any program on that computer from any site that has a modem and telephone.

This type of remote host operation has a slightly different focus from the remote host options of the terminal programs discussed previously. Terminal programs in host mode are focused primarily on transferring files to and from the remote computer. When remote control programs are used in host mode, the focus is on running programs on the remote computer.

This difference in focus results in radically different requirements for the host software—making remote control programs a different type of software even though they share many of the same functions of terminal programs.

With this comparison of terminal programs and remote control programs in host mode, you can see that using a computer as a host for other computers to call is a large application category. Within this category fall many quite different functions that require different types of software to perform. One of the most interesting of these areas—because of the variety of data movement and presentation applications it can handle—is the computer bulletin board.

Using Bulletin Board Software

Bulletin board system (BBS) software is a category of PC communications that is concerned entirely with remote host access to a computer. BBS software enables messages and files to be exchanged through a central PC for a variety of purposes. Examples of software in this category are PCBoard from Clark Development, TBBS (The Bread Board System) from eSoft, Inc., Wildcat! from Mustang Software, and The Major BBS from Gallacticomm.

The category of software known today as BBS software began in 1978 when Ward Christensen decided to place the contents of his computer club's cork bulletin board on a computer for remote access by the club's members. His original focus was on the messages, which were written on 3-by-5-inch cards and tacked to the club's cork board. He patterned the way the software presented these messages after the way you look at such a cork board. From a distance you first see the large headings on the cards. You then pick a few cards that interest you and look at them closely, reading their contents.

The response to this computer bulletin board was immediate and intense. Club members were quick to see that this structure also could support the cataloging and transferring of files as well as messages. The system's organization made browsing a large amount of information and finding what you needed an easy process. With file transfer capability, you then could retrieve what you had located. The message system structure also promoted interaction in an electronic forum that was much like what occurred in a club meeting.

Today's BBS software has taken these concepts to high levels of sophistication. Unlike terminal programs in host mode or remote control software in remote host mode, BBS software creates an entire on-line data service. A BBS shifts the focus to any of several areas, and you often can tailor BBS software to present information exactly as you want it. Applications for which BBS software is appropriate follow:

- Data file gathering into a central site

- Data library maintenance for remote lookup and retrieval

- Electronic message centers for remote access

- Public message forums

BBS software comes in many forms, from simple single-user types to multiuser types capable of creating enormous systems. Because BBS software covers such a wide range of applications, the programs available approach things from quite different positions. One of the most fundamental technical differences between a BBS and the other host software discussed in this chapter is that BBS software is capable of providing multiuser access. In other words, a BBS enables many other computers to call your central PC and use it to send or receive messages and files at the same time.

The more sophisticated BBS software also allows a variety of on-line functions such as interactive conferencing and even on-line games. Most people first experience BBS software as a caller (which is described in Chapter 13). As a result, they often form the opinion of what BBS software is and can be used for from this experience. BBS software is being used to increase productivity in business, however, by providing instant access to information at remote locations. BBS software also is being used to create entire businesses based on selling on-line BBS services to the public.

At the time this book is being written, about 30,000 public access BBSs are in existence. About 100,000 business BBSs also currently are being used to solve corporate connectivity problems. It took nearly 10 years to reach this level of BBS use, but the number of BBS installations is expected to double annually for the next few years as this technology moves from the "early adopters" into the mainstream.

The growing popularity of BBS software is based on its unique ability to address the category of connectivity problems which I have termed "loose connectivity." These are situations in which dial-up access is required in order to collect or disseminate information. This may include the need for store and forward networks of either a temporary or permanent nature. Usually a "floating component" of people exist who either move from place to place or who need access over a wide and unpredictable area on demand. BBS applications can be categorized into the following three general areas:

- Data collection

- Data retrieval and distribution

- E-mail and message conferencing

An example of the first category are franchise businesses collecting daily store operations data at the central office to properly manage the flow of inventory and resources. Among businesses that use a BBS for this purpose are Honey Baked Hams, Arby's Roast Beef, and San Francisco French Bread. These tend to be routine data collection exercises. In some cases they feed the collected files into a LAN system; in others they feed into a mainframe or minicomputer system.

Examples of the second category are publishing and library functions. Some businesses that use a BBS for this purpose are the U.S. Dept. of Commerce, the U.S. Navy, Scripps Howard/TV Data, and the San Francisco Public Library. In these functions, documents or reports are stored at a central site and are published electronically. Users electronically can request and obtain reports or documents on demand from the central site. Again, in some cases these documents and reports may originate on a PC LAN, or on a mainframe or minicomputer. These documents may in some cases also arise in a stand-alone PC environment. In all cases they must be placed on the BBS in a timely fashion for publishing.

The final category is the most easily understood. E-mail has made itself a part of our lives since its appearance on LANs. As MCI mail, CompuServe, and other widely used commercial data services demonstrate, however, the need exists for handling the "loose connectivity" problem here as well, and BBSs do this job better than any other software available. Companies that use a BBS to solve message-based problems are Microsoft, Borland, and the U.S. Army (which uses a commercial BBS in the field to handle the logistics of moving men and machines).

Today these categories tend to merge and blur in many applications. Because data collection and dissemination is the heart of any company, problems always need to be solved. Whenever communications can be improved, productivity improves as well. As businesses see the benefits of the added communications LANs have made possible, they begin to look to the next level of problems. These usually lie in the area of loose connectivity, and a BBS is often the preferred solution.

Meanwhile, BBS software use continues to grow in the public arena as well. We are seeing the emergence of BBS systems that generate well over a million dollars a year in income for their owners. These systems supply the same types of on-line services and games to their subscribers as the commercial on-line services described in Chapter 9. They offer these services, however, at much lower prices. In fact, commercial BBS systems introduced the flat-rate pricing structure that now has been picked up by GEnie and Prodigy. This is an indication of the growing importance of these grass roots systems (often literally "mom and pop" operations) which BBS software has made possible.

Today BBS software is evolving into a complete full-service central computer facility. The use of BBS software for such functions as questionnaires, credit card verification, and database transaction processing is becoming common. In many ways, a BBS now emulates the large central computer service capabilities that traditionally have been provided by mainframe and minicomputers.

Chapter Summary

In this chapter, you learned the types of PC communications software that are available. You learned that this software ranges from simple terminal programs to sophisticated multiuser remote host applications. You now should be able to evaluate PC communications software with a better understanding of the purpose and function each type of software provides. You also learned how to configure software to match the serial port hardware your computer has as well as the other computer to which you are connecting.

In the next chapter, you learn how you can use BBSs that others have created. This approach is a good way to learn the capabilities of BBS in addition to learning how to use the resources these systems provide. In Chapter 14, you learn how to evaluate and choose BBS for an application of your own. You also learn what is involved in setting up and operating a BBS service.

4

Understanding Shareware and Freeware

From their invention in 1975 until the early 1980s, microcomputers formed a utopian environment for their users. This relatively small community of early microcomputer users all felt they were part of computing history; a spirit of cooperation existed because everyone in the field felt they were part of what was happening. PC communications, although in their infancy, served to bond this community together.

Most of the software this group developed for the early Apple, TRS-80, and CP/M microcomputers was distributed freely and placed in the public domain. Through modems and bulletin boards, this software was circulated widely, and a grass roots distribution system developed.

By the early 1980s, the microcomputer revolution had created a user base large enough to support commercial software. The opportunity clearly was present for a programmer to make money writing microcomputer software. Most programmers who wrote microcomputer software did so as a sideline, however, and didn't want to make it their primary business. Most programmers continued to release their efforts into the public domain, and only a few struggled to form truly professional companies to sell their software.

Into this arena came Andrew Fluegelman. In 1982, Fluegelman wrote a terminal program for the newly released IBM PC called PC-TALK. Fluegelman was the first to see that the natural processes of trading public domain software via modems had built a software distribution channel. From this insight he developed the idea of copyrighting his software (as if it were commercial), distributing it through the on-line channel at no charge, and then asking for a contribution from anyone who used this software and found it of value. This new marketing approach counted on the feeling of community that existed in the computer world to cause those who used the software to make a contribution. At the same time, this approach removed the expense of advertising and distribution, which meant the cost of the software could be much lower.

Andrew Fluegelman coined the name *Freeware* for his approach. By the time he released his landmark terminal program PC-TALK III, he had refined his new marketing approach to his satisfaction. Figure 4.1 shows the opening banner from the PC-TALK III program, which has the original user-supported software notice.

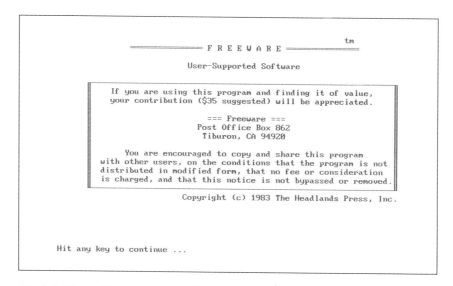

```
                                                          tm
              ========= F R E E W A R E =========
                    User-Supported Software

        If you are using this program and finding it of value,
        your contribution ($35 suggested) will be appreciated.

                         === Freeware ===
                        Post Office Box 862
                        Tiburon, CA 94920

            You are encouraged to copy and share this program
        with other users, on the conditions that the program is not
        distributed in modified form, that no fee or consideration
        is charged, and that this notice is not bypassed or removed.

                Copyright (c) 1983 The Headlands Press, Inc.




    Hit any key to continue ...
```

Fig 4.1. The notification screen from PC-TALK, the first shareware program (then called freeware).

PC-TALK III was a tremendously successful program. Its user interface set several new standards that were so widely adopted, they still are used today. Examples are the use of the Page Down and Page Up keys to initiate file transfers, the use of the Alt-X key to exit, the Alt-P key to set communications parameters, and the Alt-D key to invoke a dialing directory. The dialing directory operation itself still feels remarkably like that used in the most modern terminal software.

A terminal program was a natural first candidate for this revolutionary software distribution method. The concept that you could sell software by giving it away seemed odd at first, but soon many others wanted to give it a try. Fluegelman had trademarked the name Freeware, however, and therefore he was the only one who could use it. After many on-line discussions, the term *shareware* was adopted by the microcomputing community as the general term for his new software marketing method. As you learn later in this chapter, Fluegelman's original term freeware has taken on a new meaning.

Andrew Fluegelman died in the late 1980s, and to honor his pioneering in the computer industry, the Andrew Fluegelman award was created. This award is given each year to someone who has contributed to the advancement of the microcomputer industry. Ironically, of the first five Fluegelman awards given, only one went to a shareware author—Tom Jennings for the development of FidoNet.

The shareware industry that Andrew Fluegelman's vision began is healthy and growing. In this chapter you learn what modern shareware is all about. You also learn the difference between shareware, public domain software, and modern-day freeware. You learn how to evaluate shareware programs and how they are packaged. You also learn about ten programs that are shareware classics, where to obtain shareware, and what you need to do if you want to become a shareware author.

What Is Shareware?

Unlike a commercial program, shareware is not neatly packaged or heavily advertised. Shareware usually is downloaded from a local bulletin board or received through the mail in *archived* form, in which the program and its documentation are compressed into a single, smaller archive file. After you receive the archive file, you can decompress the file so that it expands into its original format. Figure 4.2 shows a typical file archived in the popular ZIP format. To run the program, you use the PKUNZIP.EXE utility to separate the individual files. You learn more about PKUNZIP later in this chapter in the section "Ten Classic Shareware and Freeware Programs." You can find more information on how to receive archived files from a bulletin board in Chapter 15.

```
PKUNZIP (R)    FAST!    Extract Utility    Version 1.1    03-15-90
Copr. 1989-1990 PKWARE Inc. All Rights Reserved. PKUNZIP/h for help
PKUNZIP Reg. U.S. Pat. and Tm. Off.

Searching ZIP: PC-FLOW.ZIP

 Length  Method   Size  Ratio   Date    Time   CRC-32  Attr  Name
 ------  ------   ----  -----   ----    ----   ------  ----  ----
  16512  Implode  1240   93%  07-10-86  02:48  3fc5f5cd --w  DEMO1.PIC
  16392  Implode  1243   93%  01-01-80  00:17  bb0fbbcc --w  DEMO2.PIC
  16512  Implode  1677   90%  01-01-80  01:16  5a112c6d --w  DEMO3.PIC
  16512  Implode  1867   89%  01-01-80  00:43  b91e00b2 --w  DEMO4.PIC
  16512  Implode  1452   92%  01-01-80  00:40  fb37c82e --w  DEMO5.PIC
  16392  Implode   662   96%  01-01-80  01:06  391c5839 --w  FLOWSIDE.PIC
   4011  Implode  1606   60%  07-27-87  00:05  426a5afb --w  HELP.DOC
    267  Shrunk    184   32%  07-14-86  03:20  d7f6ef55 --w  M_FLOW.MSC
   8519  Implode  3045   65%  05-17-88  02:16  4a456863 --w  PC-FLOW.DOC
  46848  Implode 27636   42%  07-17-86  01:09  137eb77a --w  PC-FLOW.EXE
     17  Stored     17    0%  07-14-86  02:59  4ed28688 --w  PCFLOW.BAT
     38  Stored     38    0%  07-14-86  03:00  35528624 --w  PCFLOWM1.BAT
     38  Stored     38    0%  07-14-86  03:00  dfd45b46 --w  PCFLOWM2.BAT
 ------          ------  ---                                 ------
 158570          40705   75%                                   13

E:\WORK>
```

Fig. 4.2. A file archived in the popular ZIP format.

If you purchase a shareware program, you register your copy by paying the program's author directly. In return, the author sometimes sends a bound manual or a supplemental program not included in the unregistered version. Some shareware programs have extra features that cannot be used unless you are registered. If you buy such a program—commonly labelled *crippled*—the author sends you a special file that adds your name and serial number to the program and enables the extra features. Normally, however, a shareware program is fully functional and documented when you receive it. After all, the author wants you to be able to evaluate the program.

Because shareware is distributed independently by its author, the cost to you usually is a mere fraction of a similar commercial package. Shareware fees typically range from 5 to 150 dollars, depending on the program's power, uniqueness, and versatility. The author receives all the profit, eliminating the need for a manufacturer or distributor. In addition, most shareware programs are copyrighted by the author, who retains full control over the product.

Of course, a shareware program differs from its commercial competitor in another important way: you can try a shareware program before you buy it. Because most shareware is not crippled, you can do the following:

- Fully test the program to make sure that it performs as promised.

- Evaluate the program's convenience and user friendliness.

- Judge the quality of the documentation.

You may find, for example, that one shareware database program supports a mouse and that another program does not; or you may find that a communications program does not offer an acceptable number of features. With shareware, you can choose between similar packages without risking any money.

Because a prospective buyer pays no money up front, the author relies on the buyer's honesty and conscience to make a profit. Usually, a shareware program can be evaluated only for a month or two; after that time, you must register to continue using the shareware.

Support for a shareware program almost always is provided by the author—usually through a voice telephone call, through messages on CompuServe or a bulletin board, or by mail. Shareware programs often are improved or updated to remain competitive with commercial products; therefore, most authors welcome suggestions from users of their shareware.

For these reasons, shareware has become increasingly popular. Two other classifications of noncommercial software, however, deserve attention—freeware and public domain software.

Freeware and Public Domain Programs

Like shareware, freeware and public domain software are not written for a software developer or sold in stores. Freeware and public domain software are downloaded from bulletin boards and information services or ordered through the mail, usually in archived format.

Freeware and public domain programs, however, are completely free to the public. The author expects no profit and will not support freeware or public domain software. If a program is popular enough, the author may update the software from time to time. Updates on freeware and public domain software, however, are rare. For most of these programs, only a single version exists.

In the case of modern *freeware*, the author retains control of the program by obtaining a copyright; however, the software is released to the public without a request for payment. The author may not want to support a program or may feel that enough copies would not sell as shareware. Regardless of the reason, the author normally does not release the original source code used to create the program. Programs written by a college's computer science students or faculty often are distributed as freeware.

In the case of *public domain software*, the author has released the program to the public domain—meaning that the program is not copyrighted and may be included in other software packages. Typically, public domain programs are simple and specific. The programs often have been written as quick projects by computer science students or professionals. An engineer, for example, may write a BASIC program that calculates the volume of a given container. Such a program is not worth trying to sell. The program could be used by other engineers, however, and could eventually be added to a larger software package. Public domain software often is distributed in program and source code form, making adaptation by other programmers easy.

Freeware and public domain programs often lack the polish, complete documentation, and impressive features of commercial software or shareware. Such programs, however, still can be useful additions to your software library. A freeware program can provide the basic features you're looking for or help familiarize you with a particular type of program—an introductory database or spreadsheet, for example. If you are learning a programming language or would like to speed the development of a software package you're writing, experiment with public domain programs; you may be pleasantly surprised.

How To Evaluate Shareware

Shareware authors allow you to use and evaluate a program before you decide whether to register. This time period begins when you decompress the archived file. If you still are using the program after the evaluation time has elapsed, you are obligated to erase the program and supplementary files from your disk. Even if you do not register the program, however, you still may distribute the **original archive** to other users.

If the software package is complex (as in the case of a computer-aided drafting program) or requires substantial data entry (as in the case of a bookkeeping system), several weeks may not be enough time for you to decide whether to register. How do you evaluate shareware? What do you look for in a computer program?

The evaluation process begins immediately after you decompress a shareware program, at which time you should follow these steps:

1. *Read or print any READ.ME file.* Typically, you will find a READ.ME, a README.1ST, or a similarly named file in the archive. This file usually contains important information on the program.

2. *Scan the decompressed program for viruses.* If you have a virus-checking program, take a few seconds to make sure that the new program will not damage your PC's data with a virus. One of the best virus scanning programs available today is a shareware program itself—VIRUSCAN, from McAfee Associates, which is available on most bulletin boards. You learn more about VIRUSCAN later in this chapter in the section "Ten popular Shareware and Freeware Classics," and computer viruses are discussed in the next chapter.

Checking every new program for viruses before you run it is a good idea!

To display a file on-screen, use the DOS TYPE command followed by the name of the text file you want to display. To display the file READ.ME, for example, enter the command *TYPE READ.ME* and press Enter. You can pause the display by pressing Ctrl-S; you can restart the display by pressing Ctrl-Q. Vernon Buerg's excellent LIST shareware utility displays one page at a time and enables you to scroll back and forth. (More details about the LIST program can be found later in this chapter in the section "Ten Popular Shareware and Freeware Classics.")

If you want to print a file, first check to see that your printer is on-line. Next, use the DOS PRINT command followed by the name of the text file you want to print. To print the file READ.ME, for example, enter the command *PRINT READ.ME* and press Enter.

Now you are ready to perform an important step that many computer users ignore: read or print the instruction manual. The file containing the manual usually has a DOC or TXT extension. (Many programs contain more than one instruction file; however, one file should be designated as the manual.) Taking time to read the manual is essential to using and evaluating the program. You also should rate the manual itself. Is the manual easy to read? Do the instructions seem complete and comprehensive? Does the manual provide examples? Does the manual provide the author's address or telephone number and directions for how to get help with the program?

If you are planning to use the program heavily or if the program is complex, print the manual if possible. Printing the manual will save you from having to read instructions repeatedly on-screen.

After you have read the program's manual, evaluate the shareware using the following guidelines:

- *Use the program as often as possible.* During the evaluation period, take every opportunity to use the program in your daily tasks. You also can discover a program's strengths and weaknesses by experimenting with its options. Possible options include color, mouse support, and memory-resident operation. Also check to see whether the new shareware is compatible with other programs you currently use.

- *Check the program's revision history.* Most shareware programs are accompanied by a list of past versions. This list may be a separate text file or may be included in the program's manual. A program's revision history can reveal quite a bit about the quality and upgrades that you can expect. Has the program repeatedly been modified to fix bugs? Have new features been added regularly?

- *Compare the program to a commercial equivalent.* Comparing the features and cost of a shareware package with its commercial equivalents is wise—especially if you aren't familiar with the application. If you familiarize yourself with the features offered by the commercial programs, you will be better prepared to look for those same features when evaluating similar shareware programs. If you're evaluating a shareware word processing program, for example, jot down its functions; you can compare the list with features offered by commercial packages such as Word or WordPerfect. Although the commercial programs usually feature many more options and commands, you have to determine how many functions you actually will use—and whether those functions are worth additional hundreds of dollars. In many cases, you will find that a shareware equivalent performs all but the most powerful tasks at a fraction of the cost of commercial packages.

- *Ask other users for their opinions.* If you received the shareware archive from a bulletin board, post a public message and ask others whether they have used the program. If they have not used the program, ask whether they have used a similar product. Shareware authors depend on word-of-mouth advertising; you too can benefit from other users' input. If you're evaluating a shareware program that will be used by your fellow employees, you should elicit their opinions. (Remember that most shareware programs require a site license if a program is to be used on more than one PC. Although a site license typically costs more, such a license enables you to make copies for other office employees.)

- *Call the author for more information.* Check to see whether the program has a support bulletin board or a voice telephone number. Calling the author is a good way to check on future support. You want to know that if you register the program the author will provide assistance upon request. In addition, the author often can answer questions during your evaluation or provide information on future upgrades and features.

- *Research the application.* Books are an important source of information on shareware and commercial programs. Check local bookstores and libraries for tutorials on the application and user's guides for similar commercial programs. Shareware books and magazines usually list features of popular packages as well as the author's evaluation of their performance. Magazines devoted to shareware often include comparisons between specific shareware and commercial programs.

- *Consider the cost of the program.* Although shareware typically is much less expensive than commercial equivalents, the price you pay to register is not always the last. If a commercial equivalent in the same price range does not exist and the shareware program seems to fit your present and future needs, then you probably have found the right program. If, however, the commercial program costs only twenty-five dollars more and offers more data storage or speed, paying for the extra features may be wise. (Again, the important question to ask is whether you will use the extra features.) On the other hand, more shareware packages are offering registered users inexpensive upgrades to the basic program. A shareware author, for example, may offer extra fonts for a word processor or a set of preconfigured templates for a spreadsheet program. Check the program's documentation to see whether the program can be upgraded in the future.

As the evaluation period draws to a close, sum up your impressions. Ask yourself the following important questions:

- Was the program easy to use?

- Did the program offer the features you use most often?

- Did the program perform as promised—without errors?

- If necessary, can the program be expanded?

If your answers are positive, the program is valuable and worth registering.

Remember that for some computer tasks, you can choose from more than one shareware package. CompuServe, for example, offers many shareware word processing programs—each with its own special features and options. Because no risk is involved, you can pick and choose among programs before registering your favorite.

Sources of Shareware

One attraction of shareware is its easy availability. If you are interested in finding shareware, try the following sources:

- *Bulletin boards.* At least one or two public bulletin boards probably exist in your local calling area. (Most larger towns and cities have many bulletin boards.) If you have a modem, these host systems—usually run by computer hobbyists—are a treasure trove of programs. Some boards are geared toward particular computers and offer software especially for IBM, Apple, Atari, or Amiga computers; other boards specialize in certain types of software, such as ham radio, graphics, or games. Most bulletin boards do not charge for downloading and are open 24 hours a day. Larger services have several incoming data lines and normally charge by the hour or offer yearly subscriptions. Along with shareware, you're likely to find public message bases where you can ask technical questions and seek advice.

When you first log on to a bulletin board, check to see whether the board offers a master list of available programs that you can download. An archived master list usually contains a complete description of every file, which you can review at your leisure or print for later reference. Figure 4.3 shows part of a typical BBS download file listing, which contains a short description of each file and its size. If you need the telephone numbers for bulletin boards in your area, call a local computer store, which probably will have several numbers on hand.

```
ASEASY30.ZIP    119984  11-01-89  A full Lotus 123 spreadsheet clone
BACKMAIL.ZIP    123047  04-14-90  A PD office electronic mail system!
    If you've got a need for electronic mail over direct connection or
    modem, this is a great application ... runs in background, too!
BANKER.ZIP       20261  11-20-89  A personal accountant program
BUSCARDS.ZIP     41407  12-05-89  Create & print business cards on the PC
CDISK430.ZIP    148224  03-12-90  Catdisk 4.3 disk cataloging program
CHECKING.ZIP    213290  03-16-91  Good checkbook manager, prints checks
CHEX.ZIP         84620  02-28-90  Simple-to-use checkbook manager...
CLUBEZ.ZIP      118226  06-01-91  Management tool for any type of club
CNC.ZIP          35784  02-28-90  Excellent encryption program
CRITIC2.ZIP      36499  04-03-90  Writing analyzer, checks your style!
CTDESK73.ZIP    113272  03-12-90  Nice PD desktop publisher!
    This publisher accepts graphics, other fonts.  Register for full use.
DESKTEAM.ZIP     29148  11-01-89  Organizational utilities set
DPLUS332.ZIP    201172  12-14-89  Good user-friendly shareware database
EDNACOOK.ZIP    110046  12-14-89  A quite comprehensive recipie system
EMPLMNGT.ZIP    119808  03-01-91  Employee Management system for business
EZCASE16.ZIP    287830  03-01-91  Computer-aided Engineering package
EZLBLR10.ZIP     42353  03-25-91  Powerful label maker for businesses
FASTBUCK.ZIP    239533  03-12-90  A great financial managment program.
FE426PRG.ZIP    200704  11-01-89  File Express! Good database (SYSOP)
FE426SUP.ZIP    189440  11-01-89  Supplemental files for File Express!

E:\WORK>
```

Fig. 4.3. *A typical BBS download file listing.*

- *Information services.* Information services such as CompuServe, Prodigy, and GEnie offer thousands of shareware, freeware, and public domain programs. These services usually charge according to how long you are connected; therefore, capturing (or *logging*) the file listing to disk for future reference is wise. Information services typically are the first to receive new shareware programs and the latest updates for existing programs. Like bulletin boards, information services offer message conferences for almost every special interest, enabling you to ask questions and receive technical support. Finally, most shareware authors maintain memberships in large information services so that you can communicate with authors through electronic mail.

- *User groups.* You can find user groups wherever you find PCs. Typically, a user group offers its software library to members who can check out all types of programs to use and evaluate. In addition, group members are a great source of support, offering hands-on software experience and the chance to exchange shareware programs. Many user groups also arrange volume discounts on hardware and software for their members. Call a local computer store and ask for more information on PC user groups in your area.

- *Mail-order companies.* As shareware's popularity grows, more companies are offering programs on disk through the mail—

usually for a nominal price. Although perhaps the most expensive method of buying shareware, ordering through the mail certainly is convenient. Most advertisements describe each program and its requirements, and many mail-order companies print their own shareware catalogs. To find a shareware distributor, check popular computer magazines such as *BYTE, PC Magazine,* and *Computer Shopper.*

• *Computer stores.* PCs often are sold with shareware, freeware, and public domain programs already loaded on their hard drives. Your local retailer, therefore, may be able to provide you with a disk full of programs—for free.

• *PC-SIG.* For a long time, PC-SIG has been the leading vendor of quality shareware on floppy disk. In its catalog, PC-SIG offers complete descriptions of thousands of programs in every software category. PC-SIG also releases a CD-ROM library, which contains the entire current shareware collection on a single CD. For more information about PC-SIG, call the customer service line at (800) 245-6717.

Ten Popular Shareware and Freeware Classics

As shareware has evolved, certain programs have become classics. If you frequent a PC user's group, a bulletin board, or an information service, you hear the names of such programs repeated frequently. Perhaps one program was the first to offer a timesaving new interface; another program may have introduced a new application; yet another program may have featured graphics and sound never found before in a shareware program.

The following ten programs have changed the way many PC owners use their machines. These programs cover the entire range of applications— from communications and graphics to data retrieval and entertainment. These classics ably demonstrate that shareware and freeware can be as good as—if not better than—commercial software. You should be able to find any of these programs easily through one of the shareware sources mentioned in the preceding section. Many of these programs constantly are updated with new features, and all offer excellent documentation and support.

LIST

LIST is a good example of a shareware program that takes up where DOS left off. LIST is a replacement for the DOS TYPE command, which has been dubbed user-hostile by many PC owners. LIST displays a full screen of ASCII text without requiring you to press Ctrl-S to pause the display. The cursor-movement keys enable you to scroll through the file in both directions, and you can return quickly to the beginning or end of the file with a single key. LIST supports a mouse and can search for a particular text string. If your PC is equipped with an EGA or VGA display, LIST even can show 43 or 50 lines of text on-screen. Basic DOS functions also are available within LIST; you can copy a file, delete a file, or run a program. After you have used LIST, you will have a hard time living without it.

CompuShow

CompuShow is the premier shareware graphics viewer for the PC. This program displays Apple Macintosh MacPaint pictures and pictures in the CompuServe GIF and RLE formats. CompuShow supports all common picture resolutions and a wide range of graphics hardware, enabling PCs equipped with monochrome monitors to view color pictures in black and white. Pictures can be displayed individually or viewed in a PC graphics slide show complete with its own script language. As GIF pictures become more popular, CompuShow will continue to grow in support and features.

PC-Write

PC-Write was one of the first shareware programs on the market. The program has grown from a basic text editor into one of the best shareware word processors available today. PC-Write provides most features needed by a small business but still is simple enough for home or student use. PC-Write can be run from a single floppy, enabling owners of older PC- and XT-class machines to create headers and footers, multiple columns, and indexes.

ProComm

ProComm is the PC terminal program preferred by thousands of users across the United States and Europe. Known for its intuitive user interface, this easily mastered program pioneered many features that are standard on today's communications software. As shown in figure 4.4, the ProComm command structure emphasizes easily memorized Alt key sequences, enabling a user to perform functions (such as downloading and dialing) without having to use a cumbersome menu system.

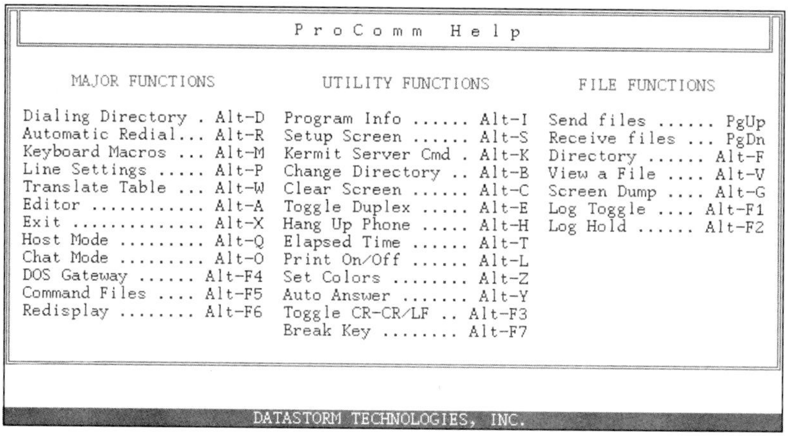

```
                        P r o C o m m   H e l p

       MAJOR FUNCTIONS          UTILITY FUNCTIONS         FILE FUNCTIONS

   Dialing Directory . Alt-D  Program Info ...... Alt-I  Send files ...... PgUp
   Automatic Redial... Alt-R  Setup Screen ...... Alt-S  Receive files ... PgDn
   Keyboard Macros ... Alt-M  Kermit Server Cmd . Alt-K  Directory ...... Alt-F
   Line Settings ..... Alt-P  Change Directory .. Alt-B  View a File .... Alt-V
   Translate Table ... Alt-W  Clear Screen ...... Alt-C  Screen Dump .... Alt-G
   Editor ........... Alt-A   Toggle Duplex ..... Alt-E  Log Toggle .... Alt-F1
   Exit ............. Alt-X   Hang Up Phone ..... Alt-H  Log Hold ...... Alt-F2
   Host Mode ........ Alt-Q   Elapsed Time ...... Alt-T
   Chat Mode ........ Alt-O   Print On/Off ...... Alt-L
   DOS Gateway ...... Alt-F4  Set Colors ........ Alt-Z
   Command Files .... Alt-F5  Auto Answer ....... Alt-Y
   Redisplay ........ Alt-F6  Toggle CR-CR/LF .. Alt-F3
                              Break Key ........ Alt-F7


                     DATASTORM TECHNOLOGIES, INC.
```

Fig. 4.4. The ProComm Help screen.

ProComm is noteworthy for another reason as well. ProComm proved so successful in the shareware market that a commercial product, PROCOMM PLUS, was developed. PROCOMM PLUS added features, terminal emulations, and file transfer protocols to the original shareware program. Wisely, the developers of ProComm still support the original shareware version for those users who don't want the extra power (and cost) of the commercial version. ProComm still is distributed through user groups and bulletin boards. PROCOMM Lite often is licensed for distribution by modem manufacturers, who package PROCOMM Lite with their hardware.

PKZIP and PKUNZIP

PKZIP and PKUNZIP probably are the most popular shareware products ever written. As mentioned earlier in this chapter, the PKZIP utility can compress an entire directory of files into one convenient archive. The archive is considerably smaller than the total size of the original files—depending on the type of files being compressed. The archived file, which usually is denoted with the extension ZIP, can be uploaded easily to a bulletin board. This process, then, requires only one file transfer and greatly reduces the necessary transfer time. When another user receives the archived file, it can be decompressed into separate files with PKUNZIP. At that point, the program is ready to run.

The PK utilities are not valuable just to BBS users. Many people simply use PKZIP to store information that they don't need often but need to keep handy. A spreadsheet, for example, may contain files for all twelve months of last year as well as the current year. To avoid cluttering the directory with extra files, you can compress the files with PKZIP into LASTYEAR.ZIP and then delete the old files, saving disk space and headaches. If you suddenly need the data for last January, you can use PKUNZIP to decompress and extract just that month. The PK programs have proven so valuable that the technology they're based on has been integrated into many commercial packages to provide invisible compression of data files—proving that commercial software developers also know the value of good shareware.

VIRUSCAN

The recent threat of the computer virus has inspired a completely new shareware application—the virus scanner. A virus scanner searches any (or all) of the files on your PC's hard drive for known viruses. VIRUSCAN, from McAfee Associates, is the most popular shareware virus detector. VIRUSCAN also is one of the most often updated shareware programs, because new viruses are being discovered constantly. A new version of VIRUSCAN usually is released every month or two—complete with a detailed listing of all detectable virus strains.

If VIRUSCAN finds that a virus has infected a file on your disk, the file's name is displayed. You can choose to erase the file (and the virus). VIRUSCAN, however, does not remove a virus from a program; VIRUSCAN can delete the entire file only. Other commercial and shareware packages can be used to scrub the infected file clean without deleting the entire file. Chapter 5 discusses computer viruses in depth.

DSZ

The shareware program DSZ is well known to most bulletin board users as the premier implementation of the ZMODEM protocol—probably the most popular BBS file-transfer protocol in use today. DSZ's ZMODEM is fast and stable. ZMODEM enables automatic file downloading and batch transfers (in which multiple files are sent at the same time). ZMODEM even features point-of-interrupt restart, which means that an interrupted file transfer can be restarted from the point where the connection between the user and the BBS was lost. If you are downloading a large file from another PC and your neighborhood loses power for a few minutes, for example, you can reconnect and resume the download instead of starting over again. Much like the PK utilities, DSZ has been incorporated into nearly every PC communications program; however, registering enables you to access special features in DSZ that aren't found in other programs. DSZ also provides other protocols, including YMODEM-g for error-correcting modems.

QEdit

Programmers and PC hobbyists often make use of a text editor, which has a purpose similar to the DOS EDLIN commands. The editor, however, includes many more features and usually is easier to use. QEdit was one of the first full-featured text editors. QEdit offered a WordStar command set and was highly configurable, enabling customization by users. Today's QEdit boasts many additional features, including a user-friendly pull-down menu and user-configurable commands and macros. Figure 4.5 shows the standard QEdit help screen, which registered users can customize.

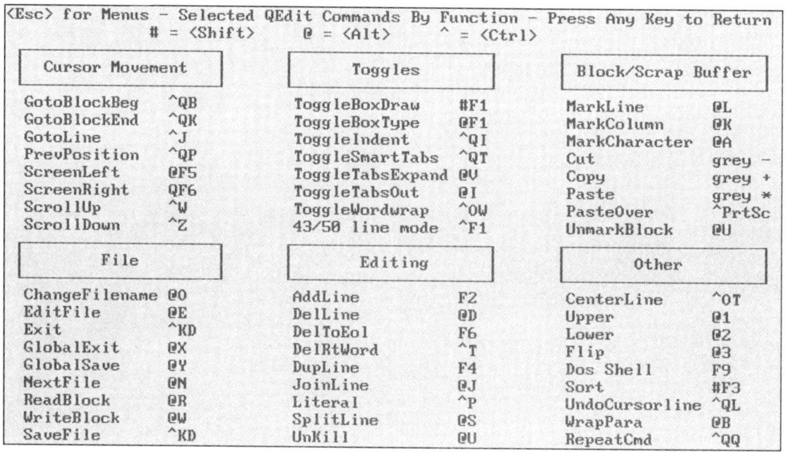

```
<Esc> for Menus - Selected QEdit Commands By Function - Press Any Key to Return
              # = <Shift>      @ = <Alt>      ^ = <Ctrl>

   ┌─── Cursor Movement ───┐   ┌─────── Toggles ───────┐   ┌── Block/Scrap Buffer ──┐

    GotoBlockBeg    ^QB       ToggleBoxDraw    #F1        MarkLine         @L
    GotoBlockEnd    ^QK       ToggleBoxType    @F1        MarkColumn       @K
    GotoLine        ^J        ToggleIndent     ^QI        MarkCharacter    @A
    PrevPosition    ^QP       ToggleSmartTabs  ^QT        Cut              grey -
    ScreenLeft      @F5       ToggleTabsExpand @V         Copy             grey +
    ScreenRight     @F6       ToggleTabsOut    @I         Paste            grey *
    ScrollUp        ^W        ToggleWordwrap   ^OW        PasteOver        ^PrtSc
    ScrollDown      ^Z        43/50 line mode  ^F1        UnmarkBlock      @U

   ┌───────── File ────────┐   ┌─────── Editing ───────┐   ┌──────── Other ─────────┐

    ChangeFilename  @O        AddLine          F2         CenterLine       ^OT
    EditFile        @E        DelLine          @D         Upper            @1
    Exit            ^KD       DelToEol         F6         Lower            @2
    GlobalExit      @X        DelRtWord        ^T         Flip             @3
    GlobalSave      @Y        DupLine          F4         Dos Shell        F9
    NextFile        @N        JoinLine         @J         Sort             #F3
    ReadBlock       @R        Literal          ^P         UndoCursorline   ^QL
    WriteBlock      @W        SplitLine        @S         WrapPara         @B
    SaveFile        ^KD       UnKill           @U         RepeatCmd        ^QQ
```

Fig. 4.5. *The QEdit help screen.*

A memory-resident version of QEdit also exists; this program need be loaded only once, when you turn on your PC, yet it is available instantly at any time until you turn off the computer. You can call the memory-resident version with a hot-key sequence—even within other programs. QEdit is a fine example of shareware's value and is recommended highly for programmers and PC hobbyists.

FRACTINT

FRACTINT stands out as a classic freeware program. In fact, the authors clearly state their contribution policy in the documentation: "Don't want money. Got money. Want admiration." If you enjoy displaying graphics on your PC, you certainly will admire FRACTINT. The program uses configurable mathematical formulas to produce spectacular, highly detailed fractals. You can choose from dozens of different patterns, and FRACTINT supports most PC monochrome, color, and high-resolution color graphic adapters. Figure 4.6 shows the large selection of basic fractal formulas available with FRACTINT.

```
                    FRACTINT  Version 15.1

                 Select a Fractal Type

barnsleyj1      barnsleyj2      barnsleyj3      barnsleym1      barnsleym2
barnsleym3      bif+sinpi       bif=sinpi       biflambda       bifurcation
cmplxmarks.jul  cmplxmarksmand  complexbasin    complexnewton   diffusion
fn(z*z)         fn*fn           fn*z+z          fn+fn           formula
gingerbreadman  henon           ifs             ifs3d           julfn+exp
julfn+zsqrd     julia           julia4          julibrot        julzpower
julzzpwr        kamtorus        kamtorus3d      lambda          lambdafn
lorenz          lorenz3d        lsystem         magnet1j        magnet1m
magnet2.j       magnet2m        mandel          mandel4         mandelfn
mandellambda    manfn+exp       manfn+zsqrd     manowar         manowar.j
manzpower       manzzpwr        marks.julia     marksmandel     newtbasin
newton          pickover        plasma          popcorn         popcorn.jul
rossler3d       sierpinski      spider          sqr(1/fn)       sqr(fn)
test            tetrate         unity

          Use the cursor keys or type a value to make a selection
        Press ENTER for highlighted choice, ESCAPE to back out, or F1 for help
```

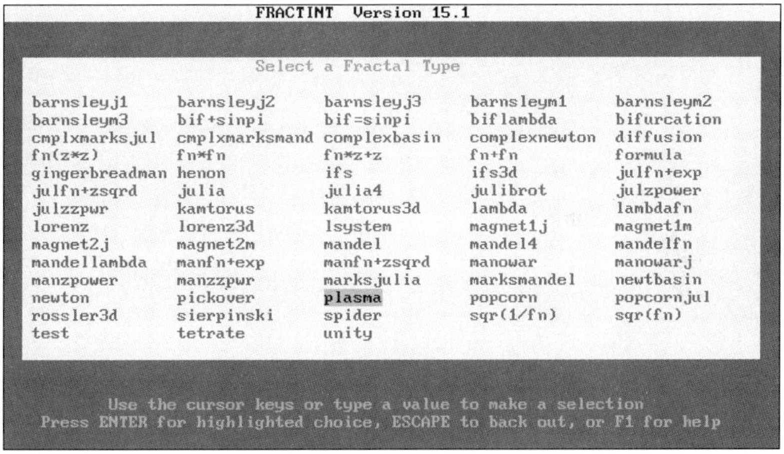

Fig. 4.6. Basic fractal formulas available with FRACTINT.

After you find a fractal design you like, you can save it to disk or zoom in on a portion of the picture. Pictures also can be animated with color cycling or wrapped around a solid shape (try creating a fractal planet, complete with mountains and oceans). FRACTINT designs can even be pasted on top of a

GIF format picture. The documentation is superb, and a convenient on-line help feature is built in. FRACTINT embodies the principle of freeware by displaying a scrolling list of users who have contributed new features and video drivers. Like many shareware and freeware packages, the four primary authors list their CompuServe numbers on the title screen, enabling you to send mail and ask questions about the program. A Windows version of FRACTINT also exists.

Captain Comic

Another exceptional freeware program is Captain Comic, an arcade adventure game with outstanding animated EGA color graphics, sound effects, and music. In the game, you play the hero, Captain Comic, who must travel across a dangerous planet to recover stolen treasures. You battle a large variety of creatures and pick up items and treasures as you go. Captain Comic was one of the first freeware game programs to feature commercial-quality EGA graphics and is still fun to play today. In fact, the game's popularity has prompted a sequel—Captain Comic II. Figure 4.7 shows a typical Captain Comic screen.

Fig. 4.7. *A Captain Comic game screen.*

Quality: Shareware versus Commercial Products

As mentioned earlier, most shareware programs don't offer all the features and power of a commercial package. Because most shareware programs are written by a single author, information and features can be limited. The amount of time that can be spent perfecting and testing a package is limited as well. A shareware author also may not be an accomplished programmer, making program development a time-consuming process.

A PC software developer, on the other hand, usually has many skilled, professional programmers. Powerful, complex programs can be written or updated relatively quickly. Commercial programs tend to offer many more features, more attractive graphics, sound, and expensive packaging and documentation.

These advantages seem to make a commercial package a buyer's clear choice. As you have seen, however, shareware has proven to be an attractive alternative to commercial software, and more programs are released as shareware every year. Of course, much of the secret of shareware's success can be found in its low registration cost—but who would register a poorly written program?

The question, then, is one of quality: how do commercial software and shareware programs compare? What elements of a shareware program indicate real value, and what can you expect when you register a shareware product instead of buying a program off the shelf? The answers to these questions are discussed in the following sections.

Packaging and Documentation

One element of a commercial program that cannot be matched by a shareware competitor is fancy packaging. Shareware almost always is distributed in archived format with documentation on disk—no expensive box, no quick reference card, no extras. The shareware principle holds that the program itself is the most valuable part and that the expensive package does nothing more than catch the customer's eye and add to the price. You should know, however, that many shareware authors send registered users a bound manual. The essential documentation for a shareware program, therefore, often is as well presented as commercial software documentation.

As with any software, the quality of shareware documentation varies. Because the shareware author usually writes the documentation, the instructions benefit from the author's first-hand experience with the program; however, documentation can suffer because of the author's poor writing skills. Often, a shareware author completely ignores the manual and relies on help screens within the program and a simple READ.ME file.

In the documentation category, the commercial product (which usually is written by a professional technical writer) has the advantage; the software developer can afford larger, better-illustrated manuals.

Technical Support

The importance of technical support is growing rapidly throughout the computer software industry. Today, most people recognize that no program or documentation—no matter how well written—can provide all the answers for every user. A well-trained source of technical support is required. Software packages that do not provide support will not last long in today's marketplace.

Opinions about the quality of shareware's technical support vary widely. Although commercial developers often boast a large technical support staff, some shareware users hold that no person can support a program better than its author. The author knows every command and option, whereas a commercial technician may have to provide support for many different commercial programs. In fact, if a programming error turns up, shareware authors have been known to provide the user with a patch over the phone or through the mail. Some shareware programs have been so successful that the author has hired additional staff to support the product.

An author must make the commitment to support his or her shareware program. Whether through the mail, by phone, or on a bulletin board, he or she must be available to answer questions. The author must establish the proper relationship with a user calling for help. As you can imagine, some registered shareware users have been disappointed by a lack of support— perhaps the author couldn't be reached or seemed to ignore them. As mentioned earlier, you should check a shareware product's technical support as you evaluate the program.

A recent development in PC technical support is the arrival of the independent support company. Unlike the traditional technical support department, an independent service isn't a subsidiary of a software developer.

Instead, an independent company provides help for an entire range of programs and applications. The company usually charges you by the minute or offers a 900 number. A user can call an independent support company for assistance in formatting a DOS disk, backing up a hard drive, or using a mouse in a Windows program.

The independent support company can be a valuable source of aid for shareware users—especially those users who are new to an application. If you never have used a database or communications package, the documentation included with a shareware program may not be enough to familiarize you with the basic terms and procedures involved in logging on to a bulletin board. Such general questions can be answered easily by an independent support company—usually in one call.

Revisions and Testing

Commercial software and shareware are governed by the same principle of revisions and updates: if the product is successful, you can expect continued improvements and bug fixes. Commercial software and shareware differ widely, however, in the processes of revising and testing a new program.

When a software package has to be revised, shareware has a clear advantage over commercial software: a software developer takes longer to design and implement a revision of a popular program. User requests must be evaluated, competing packages must be reviewed, and extensive planning must take place before the programming staff begins work. Most commercial software is advertised heavily—a process that adds time and expense. Technical support personnel must be trained to deal with the new version. In addition, owners of a commercial program typically pay an upgrade fee to receive the new version.

A shareware author faces none of these constraints. Program changes can be made in a fraction of the time required by a software developer, and because the author usually supports the program, user suggestions readily are available. Shareware authors can find ideas for new program features in trade magazines and bulletin board discussions. Releasing a new shareware revision requires little or no advertising; when the program is ready for distribution, the author simply uploads the new archive (usually denoted by a revision number in the file name) to a few large national bulletin boards, and the users take care of the rest. The author also does not need to familiarize a technical support staff with the changes made to the program.

For these reasons, a popular shareware program is revised much more often than a commercial program. Because a shareware user registers only once, upgrading to the newest version of a shareware program usually is free.

The slow pace of commercial software development pays off in another area, however; one strength of a commercial program is its extensive testing before release. The testing of a new or newly revised program often takes many months and sometimes involves hundreds of users inside and outside the company.

Most software developers conduct separate alpha and beta stages of program testing. During the *alpha* phase, copies of the new program are distributed internally, and company employees are encouraged to use the alpha program as often as possible. This internal testing almost always uncovers many bugs, which are fixed on the spot. Additionally, the program is modified to correct minor design flaws or to include last-minute features.

After the program is marked *stable* (meaning that the software exhibits no obvious problems and the design has been finalized), the program passes into the all-important *beta* phase. During the beta phase, copies of the software and its documentation are sent to a select group of external users who begin using the program daily. The opinions and program bugs reported by these users are extremely valuable to the developer.

Testing is an area in which shareware is sadly lacking. Although exceptions exist, most shareware programs receive little or no testing by independent users. The author uses a new program for awhile and may enlist a few dedicated PC owners to help with testing—but this process is equivalent only to the alpha phase. A shareware author rarely beta tests a program; therefore, potential problems may not be uncovered until the archive has been distributed. The lack of testing is yet another reason to fully evaluate a shareware program before you register. Only repeated daily use can uncover bugs or weaknesses in a program and its documentation.

After you have registered a shareware program, make sure you obtain the latest version as it is released. You can obtain the latest version through the author or a support BBS. New versions of popular shareware programs can appear on BBSs very quickly after being released to the public. The latest version of McAfee's VIRUSCAN, for example, often appears on larger bulletin boards and CompuServe within a few hours of release on the company's support BBS.

Writing Your Own Shareware

If you have computer programming experience, shareware may be the perfect way to distribute your best original programs. As long as you copyright your work, releasing shareware involves almost as little risk on your part as on the part of the user—and you collect all the profit. Shareware programs also display your programming skills to a prospective employer. Well-written shareware packages even can launch themselves into the commercial market—as proven by the ProComm communications program. If you would rather not distribute your programs for a profit but still would like to share your work with others, consider releasing your programs as public domain software.

The following programming guidelines apply especially to shareware:

- *Build an attractive user interface.* Because of the growing importance of color and sound in software, your shareware program has to look good. Almost any computer application benefits from mouse support, and the Windows drop-down menu system seems the most successful with shareware users today. Because many PCs are equipped with EGA and VGA displays, color can help define menu boundaries, highlight available commands, and alert users to possible problems. (You also should make the colors easy to configure.) If your program supports EGA and VGA displays, make sure that your program can take advantage of the 43- and 50-line text modes. Remember that your shareware program will be evaluated against commercial software and other shareware contenders; therefore, your program must stand out. Figure 4.8 shows the simple, powerful drop-down menu system used in QEdit.

- *Design the command structure carefully.* A uniform, easy-to-understand set of commands makes your program accessible to the novice and the expert. If you currently are using a similar commercial program, consider how its menu system can be improved— what works and where does the menu fall short? Is the user always aware of the choices available at each level or section of the program? Designing the command structure can be a difficult task for any software author, because you know how the program works; therefore, get another opinion. Ask several users (with different PC-skill levels) to test your prototype. Also make the command structure as bulletproof as possible. Warn the user, for example, when he or she could lose data with the wrong command, and prompt for confirmation when deleting or overwriting files.

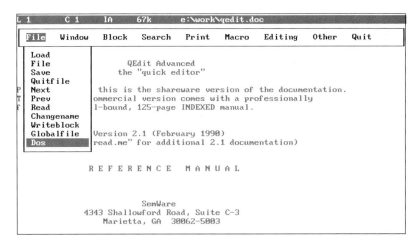

L 1   C 1   IA   67k   e:\work\qedit.doc

```
File   Window   Block   Search   Print   Macro   Editing   Other   Quit

  Load
  File                QEdit Advanced
  Save              the "quick editor"
  Quitfile
P Next        this is the shareware version of the documentation.
T Prev      ommercial version comes with a professionally
f Read      l-bound, 125-page INDEXED manual.
  Changename
  Writeblock
  Globalfile  Version 2.1 (February 1990)
  Dos         read.me" for additional 2.1 documentation)


              R E F E R E N C E    M A N U A L


                      SemWare
            4343 Shallowford Road, Suite C-3
                Marietta, GA  30062-5003
```

Fig. 4.8. QEdit's popular drop-down menu system.

- *Decide whether the program will be crippled.* This major decision should be made early in the development of a program. Crippled software usually takes one of three forms: a *date gate* (the program works only within a certain time period), a *crippled save function* (data can be loaded and modified but never saved) or a *banner pause* (a registration screen halts the program for several seconds before or during execution). Whether you cripple the program depends largely on your faith in human nature; the majority of shareware programs are not crippled in any way. Your program can search for a key file, however, to enable extra features or to display the name of a registered user. Another nice touch is to thank users (as they exit) for registering.

- *Good documentation is a must.* As the author of a shareware program, which would you rather receive: a registration check or an irate call demanding support? These outcomes represent the difference between a program accompanied by a concise, complete, well-written manual and a program supplied with a simple READ.ME file and help screen. Like the program itself, the manual must accommodate all users—regardless of their skill levels.

  A long and complex manual should include a table of contents and an index. Remember to include the maximum, minimum, and default values for input fields. Also include a troubleshooting section and an explanation of possible error messages. On-line help is another important part of your shareware package. On-line help should be available from everywhere within the program and should guide the user toward the proper help sections. If a user calls on-line help from the printing menu, for example, the program should display text on how to print a document or set up a printer.

- *Add features that users will notice.* Certain options and features have become standard equipment in most commercial software; your shareware program should have as many standard features as possible. If your program lists data, for example, can that data be sorted? Does your program enable a user to shell to DOS? Does an "undo" feature enable a user to recover from a mistake? Many programs benefit from expandable features, such as a module of extra questions for a trivia game or a communications program that supports external file-transfer protocols. Applications that demand repetitive work should offer a "learn mode" or macro support to enable automation. Advanced users appreciate a variety of command line switches to speed execution; novice users enjoy a "demonstration" mode that leads them through the program. At the very least, your program should display instructions on how to register and where to go for technical support.

- *Thoroughly test your program.* As mentioned earlier, nothing irritates a potential shareware buyer more than evaluating a program that locks up or returns to DOS with a cryptic error code. A shareware program must be tested fully before you consider a public release. Alpha and beta testing take time; however, no substitute exists for a person constantly using each menu command, entering unexpected data at an input prompt, deleting a required file, or taking a printer off line to uncover problems in your program code. Encourage your testers to recommend changes or additions that make your program easier to use and more attractive to a wider variety of users. If your program processes data and offers macros or automation, set up an unattended marathon stress test—you may be surprised at the results.

If the same questions continue to be asked by your testing team, decide whether program changes are required or whether your manual needs fine-tuning. Test your methods of technical support as well. Can the average user reach you easily? Can you handle a user's problem courteously and quickly over the telephone or through electronic mail? Every problem you locate and correct—no matter how minor—will save you headaches after your program is released.

An extremely simple program may not require many of the extras mentioned in the preceding list. If your shareware project will compete with even one commercial software program, the time and trouble you invest in making your program as attractive, powerful, and user-friendly as possible may make the difference between success and failure.

When you are ready to release your software, the remaining steps are easy. Archive the following components into a single file: your program, any required supplemental files, the manual, a short READ.ME file, and a separate registration form. If you use PKZIP, move the files to a separate, empty subdirectory, type the command *PKZIP filename *.** (in which *filename* is the name of your shareware archive with the extension ZIP) and press Enter. Remember that users can distinguish the newest version more easily if you include the revision number in the archive name. Finally, upload your program to several local bulletin boards and a dozen or so multiline bulletin boards in major cities across the country. If you are not a member of a widely used on-line service such as CompuServe or GEnie, joining is a good idea. You then can upload your shareware package to a wide audience, and you can receive comments and questions through electronic mail.

The Future of Shareware

The future possibilities for shareware seem endless. As you have seen, a shareware program's cost is very attractive, and the best shareware packages now are as recognizable to PC owners as any commercial product. With the arrival of dedicated shareware magazines and books, programs now are being reviewed and documented professionally for the software community. Such focus draws the attention of shareware authors to the necessity of quality documentation, technical support, and extensive testing.

Because shareware depends largely on electronic bulletin boards for its distribution and advertising, the continued growth of the BBS community will provide a stronger and more widespread network for the distribution of archived shareware. Many bulletin boards boast shareware libraries that fill several hundred megabytes of hard disk space. Other boards are tied together in networks that enable messages and files to be sent from coast to coast or from country to country. Users of these bulletin boards can request a shareware archive from another network BBS; the network transfers the file from system to system until the file reaches its destination.

In addition, shareware distribution through CD-ROM continues to accelerate. Titles such as the PC-SIG Library contain thousands of shareware archive files—all of which are available to a PC equipped with an inexpensive CD-ROM drive. Owning a CD-ROM shareware library is similar to opening a bottomless treasure chest. You can evaluate several different programs for each application without the hassle of calling a bulletin board or ordering programs through the mail.

The quality of the average shareware program also has improved. Today's programs are more powerful (offering more features than the public domain software of the past), more user-friendly, and more professional. As more PC hobbyists and professional programmers become interested in the possibilities of shareware, the competition will get tougher; however, you can expect the quality of shareware to continue to rise as well.

Chapter Summary

Shareware and PC communications are a marriage made in heaven. Many large PC software companies owe their existence to the fact that the shareware channel reduced their start-up costs and enabled them to establish their software without millions of dollars of up front marketing costs.

In the documentation of his historic first shareware program, Andrew Fluegelman wrote "Thank you to the entire PC community for your many words of encouragement and financial support. PC-TALK and Freeware started as an experiment. It seems to have worked beyond what anyone suspected. I hope ... that we continue to find ways to share our adventures in computing."

Clearly, his dreams were realized.

5

Understanding Computer Viruses

T his chapter explains the computer virus, detailing what a virus is, how it infects your PC, and the problems it can cause. You also learn the various methods of protecting your PC and how to remove a virus from an infected file. Finally, this chapter discusses several different virus detection and protection programs, both commercial and shareware.

What Is a Virus?

The computer *virus* is a program, much like any other program you may run on your PC. The difference is one of intent: unlike a program you use to create a document or process data, a virus has been written specifically to damage your PC's files, disrupt your PC's operation, or—in the worst case—cripple your PC by causing lockups or formatting your hard drive.

Viruses vary in the amount of damage they cause and the methods they use. Although some are harmless, others may interfere with your PC only at random times. All viruses, however, share the following two common characteristics:

- *Stealth*. A virus can operate effectively only if it's hidden; after the virus is identified, it can be treated. Viruses therefore generally infect files and disks invisibly, while you perform common functions like executing a program, formatting a disk, or copying a file. Only after the virus has infected your PC does it begin to cause trouble.

- *Replication*. Just like a biological virus that attacks the human body, a computer virus grows by "replicating" itself among the files on your disk. Many virus strains, for example, can be introduced in a single file, but they spread to many different programs on your disk while you use your PC. In addition, the "smarter" strains of viruses can copy themselves onto the boot sectors of floppy disks, enabling them to easily infect an entire office or user group as disks are shared between PCs. Other viruses even copy themselves into your PC's memory.

The earliest destructive programs were called *Trojan Horses* because they pretended to be genuine disk utilities, games, or applications. A Trojan Horse usually was accompanied by a READ.ME or documentation file that promised the program would "clean" your hard disk or "organize" your programs. Instead, the program simply erased some or all of the PC's files, sometimes formatting the entire disk or destroying the disk's FAT table. The author of such a program usually couldn't resist displaying a jab at the unfortunate user, so a Trojan Horse was hard to miss. These programs were crude and didn't require much skill to write. They worked, however, and they still make rare appearances today.

The first Trojan Horses, however, were not viruses. They did not copy themselves into memory or replicate themselves onto other programs and disks; they simply damaged or destroyed the hard drive. Only recently have programs surfaced that appear to work as promised, but in fact infect your PC with a virus.

The next development in destructive programming was the software *worm*, which introduced the idea of stealth. If a worm is introduced into a PC or network, it silently attaches itself to a program and wreaks havoc. The worm differs from a virus, however, in that it doesn't replicate itself; a worm resides only within the original system. Worms recently have become more intelligent, "moving" themselves from program to program to avoid detection. Worms are most common in large information systems and networks.

The first true virus program surfaced in July of 1987 at Jerusalem's Hebrew University. Commonly known as the "Friday the 13th" or "Israeli" virus, this strain infects program files, destroying each program you run on any Friday the 13th. Additionally, the Friday the 13th virus slows down your PC and draws a distinctive "black box" at the left edge of the screen.

Although much has been written about the dangers of computer viruses, they are not as common as you may believe—most PC owners have never seen one. Many PC experts commonly believe that bulletin boards and shareware sources are sure carriers of every virus strain ever created. This is a myth! A PC owner has as much to fear from commercial software and disks received from a user's group as programs downloaded from a bulletin board.

As you acquire software from any outside source, you also increase your risk of infection. Newer, more "intelligent" viruses constantly are being written. According to the estimates compiled by the National Computer Security Association in Washington D.C., a new viral strain currently is being created every 48 hours! For these reasons, you should keep a vigilant eye on *all* of the software you receive and the day-to-day behavior of your PC and its programs.

Sources of Viruses

PC users universally ask "Why would someone write a virus? Why not just create a useful program?" The answer is as old as human nature itself: most viruses seem to be written as jokes, though others act as instruments of revenge.

The most common virus author seems to be the computer programmer with a twisted sense of humor. Many viruses have little or no effect on a computer, indicating that they may have been written as a programming experiment. The Hong Kong virus strain, for example, infects the disk's boot sector but seems to cause no significant damage to a PC at all (except, of course, for replicating itself onto other disks).

Other viruses have been written especially to irritate, but not to destroy; an example of such a practical joke is the Fu Manchu virus, which constantly monitors the keyboard of an infected PC for the reboot sequence Ctrl-Alt-Del. When these keys are pressed, the virus displays the message THE WORLD WILL HEAR FROM ME AGAIN, but does not cause any additional damage.

Many viruses, however, are written or spread deliberately to cause damage throughout a company, usually by an employee (or ex-employee) who holds a grudge. An irate programmer could seriously damage a software developer by introducing a virus into the company's computer network— or into a commercial software product. Such a case is rare, but most developers now double-check their master disks for viruses before sending them to the disk duplicator.

The computer software *pirate* is another likely author of a virus. Pirates are known for removing the copy protection from commercial games. For example, a game may require that you enter a word from a certain page and paragraph of the manual before play begins; the pirate removes this documentation "check" and then makes copies of the program. Software pirates have considerable programming skills and an interest in illegal software. They easily can include viruses inside altered software, and many instances of infected pirated programs have been reported. If you're interested in keeping your system virus-free, avoid *any* altered software.

Major Virus Types

The following major virus types currently exist:

- *Viruses that reside in programs.* These viruses infect your PC by "attaching" themselves to another executable file; this can be a program (ending in COM or EXE), a DOS SYS file (ending in SYS), or a program overlay (usually ending in OVL or BIN). The virus replaces the first two- or three-thousand bytes of the target program with a copy of itself, which is loaded back into memory each time you run the host program. This explains why you cannot destroy a virus simply by turning off your PC. Examples of this type of virus include the DataCrime series (which reformats your hard drive on October 12 of any year), Plastique (which prevents pro-grams from running, erases files, and plays music through your PC's speaker), and Keypress (which randomly repeats keystrokes). This is the most common method of virus infection.

- *Viruses that reside in the boot sector.* These viruses infect a system by copying themselves into the protected boot sector found on all floppy and hard disks. Your PC reads the information contained in the boot sector each time you turn it on, so boot viruses constantly are reloaded into memory, ready to infect floppy disks and spread the virus. Examples of this type of virus include Ping Pong (which displays a bouncing ball on your screen, forcing you to reboot), Stoned (which can damage your files while it displays a message advocating the legalization of marijuana) and Disk Killer (which formats a hard drive after a certain number of disks have been infected). Boot viruses are much rarer than file viruses and can be harder to locate. Another characteristic of most boot viruses is the "relocation" of the original boot sector or File Allocation Table to another area on your hard drive, making recovery of data particularly difficult.

- *Viruses that are created in programs.* As mentioned earlier, the Trojan Horse program unfortunately is alive and well. Unlike the older programs that immediately destroyed data, however, today's Trojan Horses actually load viruses. Typically, these viruses then lie dormant for some period of time before attacking the system, and instead of "heckling" the user, the program usually gives no hint that it has infected your PC until well after the damage has been done. Examples of this type of virus include Strain 403 (which changes all occurrences of the letter O to the numeral zero on your screen), Burger (which destroys other programs by overwriting them with a copy of itself), and Leprosy (which destroys every program in a certain directory). Trojan Horse viruses are the rarest of all viruses.

As these examples prove, viruses do varying amounts of damage. Some are imperceptible, slowing down your PC slightly or adding bytes to the ends of files. Others are nuisances, playing music on your PC's speaker or displaying odd graphics at random times. Finally, of course, a rapidly growing number of viruses attempt to cause as much damage and loss of data as possible.

Viruses like the DataCrime series are particularly lethal on a particular date (perhaps the author's date of birth or a historical date of some significance). These "time bombs" may not be evident immediately, saving their destructive surprise for later.

Methods of Infection

The danger of your PC being infected by a virus increases if you work in an office, trade shareware or public domain programs with others, or download programs from a computer bulletin board. As your exposure to new programs grows, so does your risk of infection.

To protect your PC and its software against the threat of viruses, you need to understand how a virus is transmitted from computer to computer and from file to file. The author of a virus depends on the PC user to help spread the infection, so to avoid becoming a carrier you must be aware of the possible dangers of the ordinary disk swap or software installation.

Specifically, viruses can infect your PC via the following methods:

- *Copying or running a program from an infected floppy disk.* Some viruses "attach" themselves to programs already on your hard drive as you copy or run programs from a floppy disk. When you run the newly infected program, the virus replicates itself onto other programs. If the virus is one of the strains that "installs" itself into memory or onto the boot sector of a disk, any floppy disk used on the infected PC probably will be infected.

- *Installing commercial or shareware software from a floppy disk.* Although most software developers thoroughly test their programs for viruses before manufacturing disks, instances of tampering have occurred as the master disks are created. A single pass through an infected machine can place a virus on the manufactured disks. More often, however, a virus has been placed deliberately on the master disks by an irate employee. This brings up an important point: **no software is guaranteed safe from a virus, including the commercial software you purchase at a store!** To protect your PC, monitor the addition of commercial programs to your hard drive as closely as any other files.

- *Running programs from a network.* Many PC users believe that viruses cannot be transmitted through a local area network (or LAN), reasoning that the program (and therefore the virus) never actually resides on their PC. By running an infected network program, however, you load the virus into your PC's memory—from there the virus can infect programs located on your hard drive. Because the office environment is especially susceptible to virus attack, a responsible LAN manager should periodically check the files on the system for viruses.

- *Running programs downloaded from a bulletin board.* A program you receive from a bulletin board service (BBS) also can be infected (even if the program was stored in an archived file). Typically, any user of a BBS can upload a file for distribution; for this reason, most system operators (or SysOps) monitor files received from users to make sure that each archive is virus-free. Many bulletin boards also can perform virus checking automatically as a part of the upload process. If you receive an infected program from a bulletin board, notify the SysOp of that system immediately.

Symptoms of a Virus Infection

Computer viruses differ as much in their visible symptoms as biological viruses that attack the human body. Because most viruses cause damage to files and disks, you may have difficulty distinguishing between the symptoms of a virus and the symptoms of a hardware failure. You therefore should use a virus detection program first if you suddenly encounter problems with your hard drive, floppy drive, keyboard, or display. Conversely, if your PC exhibits a possible symptom, it does not indicate definitely that a virus is at work. As any long-time PC owner can tell you, many other reasons exist for a disk error or keyboard lockup. A good virus detection program is required to verify that your system is free of infection.

The following symptoms commonly are exhibited by viruses:

- *Disk errors* (reported by the DOS CHKDSK program or a program like Norton Disk Doctor). A disk's File Allocation Table (or FAT) is a favorite virus target. The FAT acts as a "road map" for DOS, indicating which areas of a disk contain the data that comprises each file. Many viruses simply scramble or erase the FAT, while others relocate it to another area of the disk. Other viruses copy themselves on specific areas of a floppy or hard disk, destroying the information previously stored there and marking the location as a bad sector or sectors.

   If the DOS CHKDSK command reports a rapidly growing number of bad sectors, lost clusters, or cross-linked files (especially after you repeatedly use the CHKDSK /F option or a program like Norton's Disk Doctor to fix them), you may be seeing the effects of a virus.

- *Unexpected formatting of floppy or hard disks.* A symptom this serious is hard to ignore, especially if your PC displays a message while formatting that obviously didn't come from DOS or one of your programs. Remember, some virus strains—like DataCrime— only reformat a hard disk on certain days of the year, so many months can pass before the virus is programmed to strike.

- *Changing program and file sizes.* Most viruses infect executable programs by copying themselves onto the program, or even overwriting the original program entirely. Typically, this changes the file's size, and the DOS DIR command will show the change. An example of this is the Green Friday virus, which adds a little over

1000 bytes to each infected file. Most PC owners do not observe the size of a program from day to day, however, so this may not be evident immediately, and the virus may continue to infect other files. In fact, many recent viruses are "intelligent" enough to mask these size changes, so a directory listing looks the same. Antivirus programs can monitor a program for these changes to detect the spread of infection; they record the original size of a file and compare it to the current size, or they calculate a *checksum* signature of the original file and compare it to the current signature.

- *Loss of system memory* (reported by the DOS CHKDSK program). Most viruses install themselves as "memory resident," meaning that they occupy a certain amount of your PC's RAM. Older viruses did not attempt to mask or hide this loss of free RAM. The Frere Jacques virus, for example, installs itself in memory if you run an infected program, taking up approximately 2000 bytes. If you use the DOS CHKDSK command, it will display 2000 bytes less of free RAM. Many newer viruses can fool DOS into ignoring this memory loss, so the CHKDSK command reports your full complement of available memory.

- *Unusual time/date stamping*. The author of a virus often leaves clues that a file is infected. A common clue is the alteration of an infected file's creation time and date. In the case of the Hundred Years virus, 100 years are added to the creation date, while the Lisbon virus alters the seconds field in the creation time to 62 seconds. These changes may not be obvious immediately, but they are helpful if you recognize them.

- *System "locks up" frequently*. Several viruses are written especially to lock your PC, forcing you to reboot by pressing the reset button. If the virus is stored in the boot sector, this will likely happen every time you turn on your PC. If the virus uses the program infection method (like the JOJO strain), the computer may lock up only when you run certain programs. If a program does lock your PC, monitor that program closely in the future.

- *Unexplained messages or graphics*. As mentioned earlier, some Trojan Horse viruses mimic DOS messages or pose as disk utilities while they destroy your files or hard disk. Some viruses display a message clearly indicating that your PC has been infected. Between December 24 and January 1 of any year, for example, the Christmas virus displays a full-screen picture of a Christmas tree, while the HELP ME virus displays the message HELP ME whenever you press

F1. Other viruses seem written with an accent on comedy; the Red Cross virus displays an ambulance moving back and forth along the bottom of your PC's screen. These screen displays probably are the only easily recognized, absolute indicators of viral infection.

- *Loss of keyboard or mouse control.* Your PC's keyboard and mouse also are virus targets. A short time after the Keypress virus is loaded into memory, it randomly repeats keystrokes. Fumble is quite subtle, substituting the character to the right of the key you actually pressed. The EDV virus simply locks the keyboard, forcing you to reboot.

- *System slows down.* This is a hard symptom to uncover, especially if you use a multitasking environment like Microsoft Windows or Quarterdeck's DESQview (which can run more than one program at once, but slower than normal). A virus like Friday the 13th actually "steals" processing time away from the programs you're currently running. This system slowdown often is discovered as a PC owner runs a program that displays the real-time performance of the system, such as Norton's SI program.

- *Printer, monitor, or ports act strangely.* Finally, some viruses act upon your PC's peripherals: your printer, monitor, and serial and parallel ports. These viruses typically remove characters or otherwise garble data sent to your printer or modem. Examples are the Mistake virus (which adds random characters to data sent to the printer), the Mixer series (which garbles data transmitted through an infected PC's serial and parallel ports) and ITAVIR (which sends characters to all available input/output ports on your PC, causing distortion or flickering on your monitor and problems with serial and parallel ports).

Antivirus Software

As mentioned earlier, viruses depend on stealth to infect a PC. After the virus has attached itself successfully to a file or copied itself onto the boot sector of a disk, it may begin destroying data or disrupting your PC's operation immediately, or it can wait until a predetermined time.

How, then, can you prevent a virus from infecting your computer? Or, if it's already infected, how can you detect the virus and remove it? Can the damage to your programs and data files be repaired?

The rest of this chapter covers commercial and shareware antivirus software: detecting a virus attack in progress or an already infected file, protecting against viruses, how antivirus software destroys the virus, and methods of file and disk recovery.

Detection Software

One technique used by antivirus software is to detect viruses after infection. These packages are designed to detect a virus after it already has infected your PC. An antivirus *detection* program takes advantage of the viral symptoms and signatures mentioned earlier by scanning a desired disk, directory, or file for traces of infection. The program may search for one or all of the following: memory-resident virus code, unusual or suspicious data in a disk's boot sector, virus code copied or appended onto a program, damaged or relocated FAT, and an altered disk partition table. Additionally, many detection programs can create *validation* files, which contain a checksum image of each program in a directory; later, the program can compare the original validation file with the current checksum to see if the program has been altered by a virus. If a problem is detected, the antivirus software attempts to match the possible virus with one it has been programmed to recognize and reports its findings.

Detection programs vary, however, in their repair capabilities. The popular VIRUSCAN shareware program (distributed by McAfee Associates), for example, will delete infected files if you want, but doesn't repair them. The commercial Anti-Virus package from Central Point Software, however, can remove most strains of viral code from infected files without damaging them.

The next sections examine the virus detection programs included in three antivirus packages—two are commercial offerings and one is a popular (and considerably less expensive) shareware package available on most bulletin boards.

Microcom's VPCSCAN/VIRx

The commercial program VPCSCAN (included in the VIREX-PC package) performs three functions: it detects viruses residing in your PC's memory and disks, removes any virus it finds, and attempts to restore infected programs and files to their original condition. VPCSCAN first searches RAM for memory-resident viruses. If a virus is found, you should immediately turn off your PC completely and reboot from an uninfected floppy (ensuring that the virus cannot replicate itself from memory after it's been purged from your disks). This detection program also checks your PCs boot record and partition table.

If a virus is found and VPCSCAN "recognizes" it, the program may be able to repair the file; this depends on whether the virus has completely overwritten the file, overwritten part of it, or simply added the viral code to the beginning or end of the file. If the virus has overwritten part or all of a program, that file is destroyed permanently and reconstruction is impossible. The only option left is to restore the original file from a backup. If the virus has appended itself or prefixed itself to the file, chances are good that VPCSCAN can remove just the infection.

VPCSCAN also offers two other options: you can choose to delete the file completely, or you can leave the file unchanged until later. If an infected file is deleted, the program "cleans" the erased file from your disk (making sure that it cannot reinfect your PC by being unerased accidentally with another utility).

Microcom distributes a freeware version of VPCSCAN called VIRx through bulletin boards and user groups; it contains only the detection functions available in VPCSCAN. Although you cannot use VIRx to repair a file, it can erase an infected file.

Figure 5.1 shows a typical completion screen displayed by VIRx after it has successfully checked a hard drive for viruses.

```
VIRx Version 1.4 - VIRx is a trademark of Microcom Systems, Inc.
   Copyright (C) 1990-91 by Ross M. Greenberg. All Rights Reserved.
Scanning: C:\
C:\ is clean


63 directories examined.  336 files examined.
0 files infected. 0 viruses removed, 0 files deleted
Boot Record was not infected.
Memory check shows 0 viruses found

C:\>
```

Fig. 5.1. *VIRx discovers no virus for drive C.*

McAfee's VIRUSCAN

VIRUSCAN is a shareware command-line virus detection program. Like VPCSCAN, it identifies viruses on disk and currently loaded in RAM. Unlike VPCSCAN, however, VIRUSCAN does not check the boot record or partition table—another program in the McAfee package, VSHIELD, performs this task.

If VIRUSCAN locates a virus, the program enables you to delete the infected file. McAfee Associates offers a separate repair utility, CLEAN-UP, that's similar to the repair function included in VPSCAN. The same restrictions apply to CLEAN-UP; if the virus has overwritten part or all of the original file, repair is very difficult or impossible.

VIRUSCAN can be registered separately for those who don't require additional functions or a memory-resident antivirus program. Each of the McAfee programs is updated constantly with new virus data (usually once every one or two months); registered users can download the latest version direct from the company's bulletin board. As discussed in Chapter 4, the shareware principle enables you to try McAfee's VIRUSCAN risk-free. If you find the program valuable, the authors ask you to pay a registration fee.

Another VIRUSCAN feature is the optional creation of validation checksum codes. These ten-byte codes are added by VIRUSCAN to each file you select. Thereafter, each time you use VIRUSCAN or the memory-resident VSHIELD, these codes compare the current checksum against the code attached to the file. If the codes are not the same, then either program will alert you—the file probably has been altered by a virus (unless it's been modified by a legitimate program like the PKLITE compression system). Figure 5.2 shows the VIRUSCAN command help screen.

Central Point's Anti-Virus and BOOTSAFE

Anti-Virus, a commercial package from Central Point Software (CPAV), features more than either of the other detection programs mentioned so far. The detection program, CPAV, uses a Windows-like environment, fully supports a mouse, and provides the user with icon displays of files as they're being scanned. This adds up to a very professional, user-friendly program, substituting "point-and-shoot" principles for the stark command-line interface of VIRUSCAN and VPSCAN. Figure 5.3 shows CPAV as it checks files for viruses.

```
C:\>scan
SCAN 7.2V77 Copyright 1989-91 by McAfee Associates.  (408) 988-3832

To scan entire disk(s), just specify the disk(s) you want to scan.
Examples:
          SCAN C:
          SCAN C: D:
          SCAN A:

To scan a single directory, specify the directory.
Examples:
          SCAN \newstuff
          SCAN C:\unknown\things

To scan a single file, specify the file.
Examples:
          SCAN prog.com
          SCAN C:\there\unknown.exe

To remove infected files, use the /d option.
Example:
          SCAN C:\here /d


C:\>
```

Fig. 5.2. Instructions and examples displayed for VIRUSCAN.

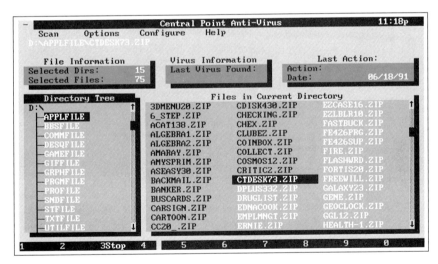

Fig. 5.3. Central Point's Anti-Virus program during the virus detection process.

Anti-Virus offers all of the features mentioned previously for VIRUSCAN and VPCSCAN, including optional repair of infected files (where possible) and the creation of checksum files. These options are fully configurable. If your hard disk is low on free space, for example, you can choose not to create checksum files or an "activity log," which contains entries summarizing the last 400 actions taken by Anti-Virus. Another unique Anti-Virus feature is the ability to display each virus recognized by the program in a convenient pull-down menu. If you select a particular virus strain by highlighting it and pressing Enter (or pressing your left mouse button), Anti-Virus displays a short list of symptoms and possible damage the virus can cause. The virus list can be updated manually (by entering the identifying virus "signatures" into a template) or automatically (by downloading the latest signatures in data-file format from the Central Point bulletin board). Figure 5.4 shows the manual-entry CPAV virus template, ready to receive the hexadecimal codes of a new virus signature.

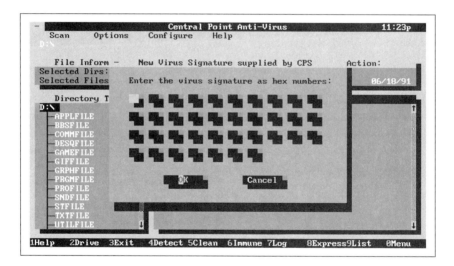

Fig. 5.4. *This screen enables users to enter new virus signatures into Anti-Virus.*

Central Point also includes a separate detection program, BOOTSAFE, for viruses that reside in a disk's boot sector or partition table. BOOTSAFE usually is installed in your AUTOEXEC.BAT and can be used on any bootable disk. When executed, BOOTSAFE checks the current boot sector and the current partition table against the original information it has stored on the disk. If a change has occurred, BOOTSAFE will display a warning. You can update the information BOOTSAFE uses for the comparison at any time (in the rare case that you change the partition table yourself), and this partition information can be saved to a floppy so that you can restore the table if a virus destroys or relocates it.

Pros and Cons of Virus Detection Software

Antivirus programs using the detection method have a few definite advantages over their protection counterparts. Because antivirus programs don't operate in conjunction with other programs, they cause no memory conflicts; your other programs have access to all available RAM; and detection programs don't slow down the operation of your PC.

Antivirus detection programs have one fundamental problem: typically, they're used only *after* a virus has damaged files or has infected other computers. Unless virus detection software is used quite often—perhaps once a day—your PC may be infected with a damaging file or boot sector virus that can spread quickly throughout an office or over a network. Even worse, you possibly may encounter a Trojan Horse. As discussed earlier, these viruses are particularly destructive, and often damage data in the space of a few minutes after you first run them.

How, then, can a PC be protected *before* running potentially infected programs? The next section discusses the second variety of antivirus software: the memory-resident virus protection program.

Protection Software

Another technique used by antivirus software is to protect systems from infection. These packages act as sophisticated "filters," preventing a virus from infecting your PC's memory or disks. Protection antivirus programs usually are loaded memory-resident when you first boot your PC, and they run continuously in the background while you work. Protection programs monitor all activity that takes place in your PC's RAM and examines each write operation to your floppy and hard disks in order to intercept the virus during its attack, before it has a chance to replicate itself onto a file or boot sector. If a program attempts a suspicious disk operation (like initiating a low-level format or attempting to delete files), tampers with the PC's BIOS, loads itself memory-resident, or attempts to alter a disk's boot sector or partition table, the antivirus program displays a message warning of possible virus activity and prompts you for confirmation on whether or not you want the activity to continue. For this reason, protection programs are an especially valuable defense against Trojan Horse viruses because they intercept the low-level software commands favored by these viruses. A viral detection program, however, cannot prevent immediate damage or infection.

Protection programs generally don't offer options to clean or delete infected files. They usually reveal only that a file is attempting a suspicious action that could indicate a possible infection, leaving it up to you to take action. Because a protection program is memory-resident, it must remain small, or it will take up too much RAM for other programs to run. For this reason, antivirus protection software usually is part of a complete system, like the VIREX-PC program offered as a part of Microcom's antivirus package.

Although several commercial and shareware antivirus programs have been released in the last few years, only a few stand out as exceptional because of features and reputation. The following sections use these well-known software packages to demonstrate the common strengths and weaknesses of antivirus software.

McAfee's VSHIELD

The memory-resident protection software offered by McAfee Associates is called VSHIELD; as with VIRUSCAN, you can register it separately. VSHIELD specifically monitors the requests that DOS makes to run a program. When DOS begins to load an executable file, for example, VSHIELD automatically scans the program code for viruses. If the file does indeed carry a virus, VSHIELD will interrupt the program load request, making it impossible for a virus to load itself into memory.

VSHIELD also can optionally check a program's validation code. If you have enabled this feature, VSHIELD automatically calculates the current checksum of a program and compares it to a validation code created previously with VIRUSCAN; if the two don't exactly match, VSHIELD once again will halt the loading process.

VSHIELD is less sophisticated than other memory-resident protection programs, with benefits and problems resulting from that simplicity. On the positive side, VSHIELD takes considerably less processing time and less memory than other protection software. Because it scans for viruses only when a program is loaded, it doesn't have to monitor each and every disk access made by your other programs. VSHIELD also avoids annoying warning messages caused by applications with nonstandard disk and memory routines. Problems occur, however, if a virus is not recognized by VSHIELD—the virus is loaded successfully into memory, where it can safely replicate itself and continue infecting other files.

Central Point's VWATCH and VSAFE

Central Point Software includes two separate memory-resident protection programs with its antivirus system—VWATCH and VSAFE.

VWATCH is essentially the same as McAfee's VSHIELD; it scans programs being loaded for viral infection, preventing them from being executed. Additionally, VWATCH also monitors the information being read during a disk access for suspicious code. This program should be used when your PC is low on available RAM because it takes almost a third less memory to run than VSAFE.

VSAFE is a different program altogether and is fully configurable, enabling you to select exactly what disk and memory operations you will and will not allow a program to attempt. You can configure VSAFE as General Write Protect (no disk writes—a requirement if the program already may be infected), Resident (alerts you if a program attempts to load itself into memory), Protect Executable Files (warns you if a program is being modified), and HD Low-level Format (alerts you of attempts to reformat your hard drive). You also can load VSAFE in extended memory so that you can save conventional RAM for other programs. Figure 5.5 shows the pop-up options window available within VSAFE, in which you can selectively enable or disable each protection level.

Microcom's VIREX-PC

Microcom's entry in the memory-resident category is the program VIREX-PC, which operates much like VSAFE. VIREX-PC also monitors attempts by any program to format disks or load itself into memory, enabling you to cancel them if you want. VIREX-PC also offers the unique feature of detecting "illegal" attempts to write to a file without using standard DOS calls. You definitely should keep an eye on any program that tries to sidestep the normal DOS disk access methods.

Much like VIRUSCAN, VIREX-PC can calculate a checksum for files and the disk's boot sector. After a file is "registered," its checksum code is stored in a separate file, and any change in the program's checksum will be detected. If you choose, VIREX-PC can prevent the execution of unregistered programs, making it ideal for a security-conscious network manager.

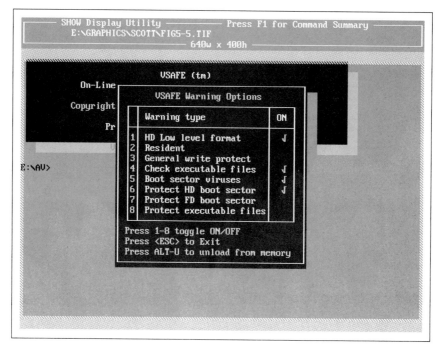

Fig. 5.5. *Selecting the protection level for VSAFE.*

Pros and Cons of Virus Protection Software

Virus protection programs offer numerous advantages. As mentioned earlier, the most important advantage is the opportunity to stop a virus before it can damage your system. Protection software can be loaded as part of your AUTOEXEC.BAT file (so you don't need to remember to run it), and it actually can enable you to test a program for a virus by running the suspect program without damaging your system.

The down side to protection programs is the system resources they demand. Your PC must supply a memory-resident antivirus program with RAM and processing time, which means you have less room to run other programs and your PC may run noticeably slower (depending on processor type and speed). Memory-resident programs also are infamous for interfering with other software. Conflicts often arise, for example, between memory-resident protection programs and communications software. Additionally, most protection programs flag a legitimate program trying to modify a disk boot sector, FAT, or memory location. You therefore may have to confirm the operation of regular programs more often than you intercept viruses.

The worst news is that viruses are getting smarter and more intelligent with each new strain. Viruses recently have appeared, for example, that remove memory-resident software before they infect a PC (effectively removing a protection program before it can do its job). Other viruses use different tactics on virus detection software; some are encoded, making them harder to find, and others now are infecting data files as well as executable programs.

This constant escalation of virus and antivirus has left many PC users asking for a new protection scheme: the "self-protecting" program, which actually fights virus infection by itself. A self-protecting file would not require a memory-resident program to monitor for virus infection. Instead, the file would internally check itself for virus code each time you use it and would alert you if it had been altered in any way. The additional protection code must not interfere with the proper function of the program, however, and it shouldn't add very much to the size of the original file.

Immunizing Files

Central Point's Anti-Virus program offers an example of this new antivirus technique, which it calls *immunization*. If you immunize an executable program, it automatically will inform you if it's been modified. If you want, the file then will attempt to repair itself. The CPAV program is not required for an immunized program (so you can easily swap immunized software), and the process only adds a little less than a thousand bytes to the original file.

After you successfully immunize a file, it checks itself each time you run it against an internal code (which provides the equivalent of a checksum). A slight delay occurs when loading an immunized file (again, this depends on the processor type and speed), but execution is no slower than most memory-resident protection programs.

Immunization also can be reversed, in case you're running low on disk space. Immunized files are restored to their original size, exactly as they were originally. Less than a thousand bytes may not sound like very much, but CPAV enables you to quickly immunize all program files on an entire drive, so those few bytes accumulate very rapidly.

Immunization certainly seems to solve many of the problems of detection and protection antivirus software. Immunization uses little or no system resources, constantly monitors changes to programs, and even repairs infected files. Even immunization, however, cannot guard against a virus of the Trojan Horse variety. Such a virus probably would not affect an immunized file at all; instead, it would act immediately, perhaps destroying the partition table, infecting the disk's boot sector, loading itself into memory, or (in the worst case) completely reformatting the disk.

Although immunization cannot provide the total antivirus protection that many PC users are looking for, it's a most innovative step in the right direction, and this new technique soon may become commonplace in the shareware and commercial marketplace.

Ultimate Protection

You may ask, "What's the ultimate protection? Will a combination of programs ensure that I'm never the victim of a virus attack?" The answer is yes—and no. No matter how good antivirus programs are, they cannot help you if you don't use them regularly. They also cannot guard against a virus that they don't recognize.

The ultimate antivirus protection for your PC is your own common sense. Select the antivirus program (or programs) that you feel most comfortable using, and use them at least once a week. If you download a program from a bulletin board or you purchase a commercial program, check it for viruses before you install or run it. If your PC has extra disk space available, use the checksum validation feature offered by most antivirus packages to monitor changes that are made to your executable programs. If you purchase Central Point's Anti-Virus package, use the immunization feature to make your programs harder for a virus to infect.

Even after you buy or register an antivirus package, you must continue to update it with the latest virus data to continue to detect viruses effectively. New viruses now are appearing weekly, and an out-of-date virus detection or protection program is probably worse than no protection at all. If you don't update your antivirus software regularly, you may think that your files and disks are safe, although they actually are being destroyed by a new virus strain.

Another common sense rule: back up your files regularly. As mentioned earlier, many viruses destroy infected files, so you have no hope of repair no matter what antivirus program you use. The only method of recovering these files is to restore them from a previous backup. A regular disk backup is the best data insurance available today.

Ensure that a virus will have difficulty replicating itself from computer to computer: write-protect floppy disks that you don't use often, and mark essential programs as Read-Only on your hard drive. If you're trading disks with a friend or at the office, take a few seconds to check that disk for viruses. If you work with a computer network, detecting a virus is even more important—before it infects every connected PC.

If you do discover a virus on your disk, *do not copy it or rename it—use common sense and erase it immediately*. If possible, attempt to repair any damaged files. As an added safety measure, try to trace the infected file back to its source. If you received the file from a BBS, inform the board's system operator of a possible infection. If you brought an infected file with you from work or from a user group, inform everyone involved of a possible virus infection. As you can imagine, "no one likes a virus carrier," but a silent virus carrier is much, much worse.

Finally, don't panic. Viruses are not widespread, and your PC very well may never be exposed to one. You should play it safe, however. By following these guidelines, you minimize your risk of viral infection and prevent an infected file from spreading to other PCs.

Chapter Summary

In this chapter you were introduced to computer viruses, the methods they use to infect a PC, the damage they can do, and possible symptoms of infection. This chapter also covered the latest in antivirus software—programs that can identify existing viruses and prevent a virus infection, and even help your programs and files protect themselves from viral damage.

Part II

Using PC Communications

Includes

Examining PC to PC Communications

Using PCs on a Network

Using Electronic Mail

Using Commercial Information Services

Using CompuServe

Using Prodigy

Using GEnie

Using Electronic Bulletin Board Systems

Becoming a BBS SYSOP

Examining Store and Forward Networks

6

Examining PC to PC Communications

Not many years ago, computers were seen as a menacing force. Books such as *1984* and *Brave New World* envisioned a society in which large computers controlled the entire population, suppressing individuality. Many science fiction books and movies portrayed computers as a totalitarian force inherently antithetical to a free society.

Contrary to these predictions, however, personal computers actually have created a wonderful new world of information. Now, with a personal computer and the right software, you can create, file, and manage by yourself more information than the large computer systems that inspired such fear not so long ago. You also can connect your personal computer to any other computer in the world and communicate this information when and where you want.

This capability of personal computers to communicate easily with other computers has changed the power dynamic of computing from centralized to decentralized. The fact that so much information can be shared so easily outside of government or corporate control has prevented the nightmare scenarios of computer oppression from coming true. When you learn how to use the communications capabilities of your personal computer, you also gain access to information of all types.

Nothing changes the world as much as new channels of communication becoming available. Communications is an enabling technology, and each advance in that technology solves a new group of problems. PC communications today are revolutionizing the world of computing as much as the invention of the personal computer did 15 years ago.

This chapter introduces you to the most important PC communications software—the terminal program. This chapter also examines the many communications options available to share files and printers between PCs, such as printer buffers and hardware-free LANs. It also introduces you to what is available in the world of on-line computing.

Defining PC Communications

Whenever two or more people are using computers, you can be certain that they eventually will want to move information from one computer to the other. How easily they are able to accomplish such data transfer dictates the kinds of problems they can solve. Any method that enables them to move information from one computer to the other can be called *PC communications*.

Users may want to move information for a reason as simple as printing a report on a printer attached to another computer. The reason also may be as complex as wanting several people in a group to work simultaneously on a report. In the latter case, the report file must be available to all the people working on it at all times. These two examples represent the extremes of PC communication; most communications applications fall in between these two extremes.

The simplest form of PC communications is the physical exchange of a data recording, such as a floppy disk or tape cartridge. This type of data transfer has become known as a *Sneaker Net* because the information is carried physically from one computer to the other (presumably by people wearing sneakers to ease the impact of walking).

The concept of a Sneaker Net is simple, but it forms a model for you to understand all forms of computer communications. Moving floppy disks between computers in the same room (or the same building) is a *Local Area Sneaker Net*. Putting a floppy disk in an overnight express envelope and shipping it to another city is a *Wide Area Sneaker Net*. This simple data transfer concept embodies all of the functionality of sophisticated LAN, WAN, and modem technology.

In the end, all forms of computer communications must accomplish the same task as the simple act of transferring data on floppy disks—they just make it much easier and quicker to accomplish. Even distributed processing and complex data networks have their parallels in Sneaker Net forms, although they would be so tedious and slow that no one would ever seriously consider carrying them out. At their heart, however, all PC communications are as simple in functional purpose as moving files on a floppy disk—they just require many steps that make them seem complicated.

Understanding Communications Links

The first step in automating computer communications beyond the Sneaker Net is to find a way to connect the computers electrically. This electrical connection is called the *physical link* because it is composed of specific physical components (wires, connectors, and modems, for example). Many different types of electrical connections are possible between computers, but each type represents the physical link that data must pass through to move from one computer to the other.

You may think that the communications problem is solved when a physical link is established between the two computers. For data to be transferred successfully between the computers, however, rules must be established for sending file names, specifying file sizes, correcting transmission errors, and so on. As you learned in Chapter 3, such sets of rules are called *protocols*. When a set of protocols is established that enables you to move a file across a physical link and software is in place to implement these protocols, then a *virtual link* has been established and the file transfer can occur.

The most obvious property of any physical link is its speed. In general, the greater the distance between the two computers, the more difficult (and expensive) it becomes to obtain high-speed physical links. The speed of the physical link connecting the two computers affects the tasks you realistically can carry out over the link. Slower links make transferring very large data files less practical, for example, because they may take hours to transfer. Faster data links enable you to couple the functions of the two computers more closely. When a link becomes fast enough, you can obtain the full functionality of the remote computer from your computer's keyboard in applications such as a Local Area Network (LAN).

In practice, physical communications links between computers are divided into two categories: serial communications links, which have lower speeds but can be made over normal telephone circuits; and high speed links, which require special hardware adapters and cables but are capable of extremely high speeds.

Serial Communications Links

The simplest and least expensive connection between two computers is the serial communications link. In Chapter 1, you learned the concepts of serial communications, and in Chapters 2 and 3 you learned about the hardware and software used to make serial communications possible.

The most basic form of serial communications is two computers in the same room connected by a wire (a null modem cable). This type of serial link often is used to move files from a laptop computer to a desktop PC when the two computers do not have a common floppy disk format. Examples of products using this type of communications are LapLink (Traveling Software, Inc.) and Brooklyn Bridge (Fifth Generation Systems, Inc.). Figure 6.1 shows the screen of a LapLink program in action.

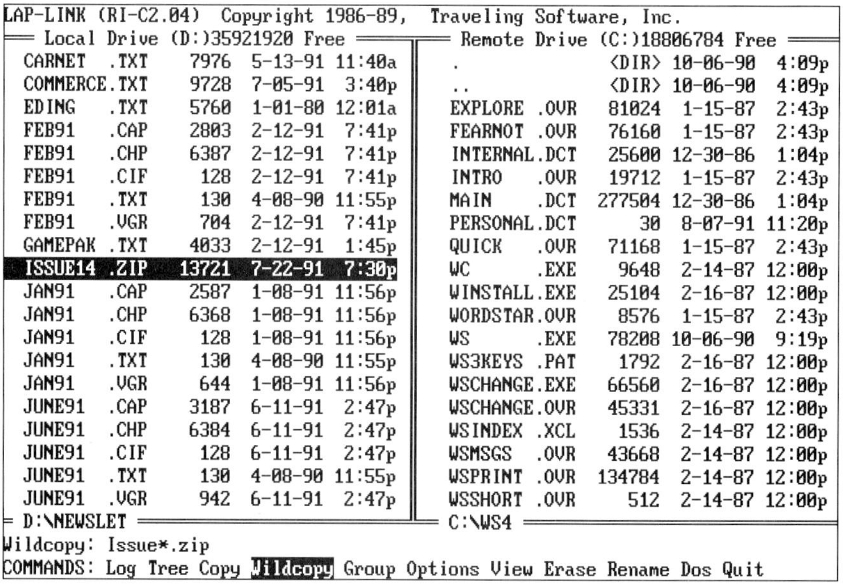

```
LAP-LINK (RI-C2.04) Copyright 1986-89,  Traveling Software, Inc.
= Local Drive (D:)35921920 Free ====   == Remote Drive (C:)18806784 Free ==
   CARNET   .TXT    7976  5-13-91 11:40a    .              <DIR> 10-06-90   4:09p
   COMMERCE.TXT     9728  7-05-91  3:40p    ..             <DIR> 10-06-90   4:09p
   EDING    .TXT    5760  1-01-80 12:01a    EXPLORE  .OVR  81024  1-15-87   2:43p
   FEB91    .CAP    2803  2-12-91  7:41p    FEARNOT  .OVR  76160  1-15-87   2:43p
   FEB91    .CHP    6387  2-12-91  7:41p    INTERNAL.DCT   25600 12-30-86   1:04p
   FEB91    .CIF     128  2-12-91  7:41p    INTRO    .OVR  19712  1-15-87   2:43p
   FEB91    .TXT     130  4-08-90 11:55p    MAIN     .DCT 277504 12-30-86   1:04p
   FEB91    .VGR     704  2-12-91  7:41p    PERSONAL.DCT      30  8-07-91  11:20p
   GAMEPAK  .TXT    4033  2-12-91  1:45p    QUICK    .OVR  71168  1-15-87   2:43p
   ISSUE14 .ZIP    13721  7-22-91  7:30p    WC       .EXE   9648  2-14-87  12:00p
   JAN91    .CAP    2587  1-08-91 11:56p    WINSTALL.EXE   25104  2-16-87  12:00p
   JAN91    .CHP    6368  1-08-91 11:56p    WORDSTAR.OVR    8576  1-15-87   2:43p
   JAN91    .CIF     128  1-08-91 11:56p    WS       .EXE  78208 10-06-90   9:19p
   JAN91    .TXT     130  4-08-90 11:55p    WS3KEYS  .PAT   1792  2-16-87  12:00p
   JAN91    .VGR     644  1-08-91 11:56p    WSCHANGE.EXE   66560  2-16-87  12:00p
   JUNE91   .CAP    3187  6-11-91  2:47p    WSCHANGE.OVR   45331  2-16-87  12:00p
   JUNE91   .CHP    6384  6-11-91  2:47p    WSINDEX .XCL    1536  2-14-87  12:00p
   JUNE91   .CIF     128  6-11-91  2:47p    WSMSGS   .OVR  43668  2-14-87  12:00p
   JUNE91   .TXT     130  4-08-90 11:55p    WSPRINT  .OVR 134784  2-14-87  12:00p
   JUNE91   .VGR     942  6-11-91  2:47p    WSSHORT  .OVR    512  2-14-87  12:00p
= D:\NEWSLET =====                      == C:\WS4 ==
Wildcopy: Issue*.zip
COMMANDS: Log Tree Copy Wildcopy Group Options View Erase Rename Dos Quit
```

Fig. 6.1. The LapLink Control screen.

Programs such as LapLink make transferring files over a serial link easy. To move a file from one computer to the other, you simply move the cursor to highlight the file you want to send and press C to begin the transfer. Options for copying entire directories or groups of file names also are available.

Locally connected serial communications (such as Laplink) on an IBM PC may reach speeds up to 115,200 bps, or 11,520 characters per second. To put this speed in perspective, a 360K floppy disk transfers data at a speed of 250,000 bps, or 21,250 characters per second (the slight difference in the ratio of cps to bps for the two is due to a difference in the way data is formatted for transmission). Therefore, a short, serial hard-wired connection is about half the speed of a floppy disk. At this speed, serial connections may span a distance of only a few feet.

Similar connections also can be made using parallel printer ports in many cases. Where possible, these connections allow even faster speeds. They are still limited to a few feet in length, however, and are not easily extended like serial connections.

Modem Serial Links

Serial communications can be extended from a few feet to any distance by adding modems to the link (see Chapters 20 and 21 for details on modems). Modems enable the serial link to be connected through a telephone circuit instead of a cable.

Adding modems to the physical link affects the link in two ways. First, modems cannot transport serial data as rapidly as a wire can—they currently are limited to an upper speed in the range of 19,200 bps (1,920 characters per second), just one-sixth the speed of a floppy disk. Less expensive modems transmit data even more slowly. Second, because modems have to dial and answer the telephone, you have to add a separate method for controlling this process of establishing and ending the physical link.

After a connection has been made between two modems, the serial modem connection acts the same to the two computers as the single cable did (except for the slower speed). The two computers now can communicate from nearly anywhere in the world, however, because the modems connect through telephone lines. In fact, the advent of cellular telephone technology even enables this type of link to be established without telephone wires.

In addition to establishing international connections, modems enable you to establish the link on demand without any special wiring or preparation for the connection. The only requirement is two people who want to connect their computers; the only planning required is obtaining the telephone number to dial and arranging a time to make the connection. Simply by dialing the telephone number, the other computer's modem is connected, and a link is established between the computers.

The operation of PC communications between two computers connected with modems is different from the operation of computers connected by a wire primarily because both computers are located physically far apart. The computer you are dialing is *remote* from your location, and some coordination is required to make sure it is ready to answer the telephone when you call it. This difference gives rise to the different categories of software for PC communications that you learned about in Chapter 3. By far the most popular type of software for use with modem links is the terminal program, which is covered in detail later in this chapter.

High-Speed Communications Links

For relatively short distances (up to a few thousand feet), several types of physical connections enable high-speed data transfer. Examples of this type of communications link are EtherNet, Token Ring, and ARCnet. These connections usually are made with coaxial cable or twisted pair special wiring; they transmit data at speeds from 8 to 40 times the speed of a floppy disk. These high-speed physical links between computers require special adapter cards along with special cabling.

Because of the very high speed of these physical computer connections, these links can form tightly coupled *local area networks* (LANs). The special LAN software enables you to treat another computer's resources (such as printers or disks) as though they are a part of your own computer. Chapter 7 describes how you can use LAN technology to communicate information between computers.

Although the high speed of these communications links provides many features not available any other way, they are different from serial connections (hard-wired or modem) in one important aspect: they must be wired in a permanent, planned way. This requirement makes installing and using a LAN quite different from installing and using a modem. Modems and serial links are more appropriate for temporary connections or connections that must go through telephones; LAN links offer more integration of the communications into your regular computer usage in settings with fixed computer locations.

Sharing Printers

Moving data files from one computer to another is the primary focus of PC communications, but sharing printers is another driving force behind communications. Specialty printers (such as laser printers) are relatively expensive. They also tend to be used by any one person in spurts followed by long periods of no use. The need for several computers to share a single printer has driven the development of several specialized forms of PC communications.

The simplest printer-sharing device is a *printer switch*. A simple printer switch box, however, can be used only when a small number of computers are located very close to the printer they share. When the distance between all the computers and the printer is more than 25 feet, some form of computer communications is required.

In Chapter 7, you learn how printers can be shared in a LAN setting. A LAN setting is certainly the easiest printer-sharing setup to use, but quite expensive and complex to install if you need to share printers only. Because sharing printers is such a common need, specialty PC communications products are available for just this task. The most commonly used are devices that attach to a serial or parallel port so that you don't have to add any special expansion adapters, thereby greatly reducing your cost.

Intelligent Printer Buffers

Intelligent printer buffers enable printer sharing among a large number of PCs. The *printer buffer* itself is hardware that attaches directly to the printer(s) being shared and connects to the PCs with a serial cable. Intelligent printer buffers are also sometimes called *data switches*, but you don't press any physical buttons to do the switching.

An intelligent printer buffer has a computer and software contained in it that enable it to switch the printer automatically. The PCs don't need any special software because the buffer appears as a regular serial printer to them. Some intelligent buffers also offer a parallel interface to the PC. Although a parallel interface provides much higher speed, it is limited in length to about 25 feet. A serial interface enables a computer to use a printer buffer that is located a few hundred feet away. Figure 6.2 illustrates the connection of an intelligent printer buffer.

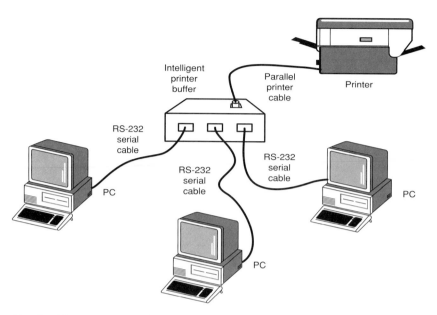

Fig. 6.2. *The connection of an intelligent printer buffer.*

When a PC outputs to what it thinks is a normal printer, the buffer accepts the data into its local memory and then sends the print image output to the printer. The buffer arbitrates the case when more than one computer prints at the same time by queueing each incoming print job and outputting it to the printer in order. Printer buffers with up to 8 megabytes of RAM are available so that several print tasks can be stored before the buffer fills (and slows down the attached PCs to the speed of the printer itself).

Printer buffers solve the printer-sharing problem very inexpensively and are an example of PC communications in a dedicated application. Thinking small is often the best way to solve a problem, and printer buffers are such an example. If you only have a printer-sharing problem, you don't help yourself by purchasing a complex solution that complicates your life for no good reason.

Hardware-Free LANs

A *hardware-free LAN* is a product that enables you to connect two or three PCs with serial cables, load special software in each PC, and share files and printers among the computers. Examples of this type of product are DeskLink, Printer LAN, and MasterLink.

These products are called hardware free because they use no additional hardware beyond the standard serial or parallel ports your computer already has.

A hardware-free LAN uses special cables and software to connect two or three PCs using standard serial ports (see fig. 6.3). One of the PCs is connected to the shared printer in the normal way (through a printer port and cable). Special software is installed on each computer, and the hardware-free LAN is in place.

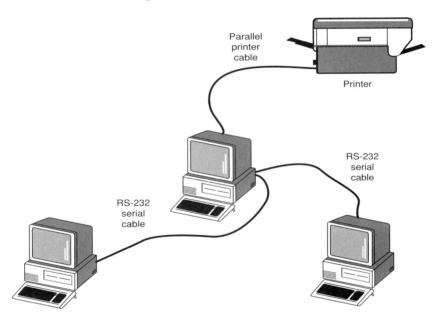

Fig. 6.3. *Connecting a hardware-free LAN.*

Hardware-free LANs (sometimes called *serial LANs*) usually share files as well as printers, but the communications link is very slow; therefore, hardware-free LANs are not usable as LANs in the typical sense. They do provide a very easy method of sharing printers and moving files between two or three computers. They also are well integrated into DOS, so you don't have to learn many special commands or new procedures in order to share the printer or move a file from one computer to the other.

A hardware-free LAN is an inexpensive way to connect two or three computers together and provides an easy method of sharing files and printers. It provides a high degree of functionality at a very low price. A hardware-free LAN is not a true local area network, however, as described in Chapter 7.

The slow speed of a hardware-free LAN means that it really is useful only as a well-integrated software data switch for sharing files and printers among a maximum of three computers. If you understand these limitations, a good hardware-free LAN is inexpensive ($50 to $100 per PC) and performs its sharing tasks well.

Using Terminal Programs

The most commonly used software for PC communications over serial and modem links is the terminal program. A *terminal program* enables you to control your modem, send and receive data over serial links, and transfer data or program files between computers. In the following chapters, you learn about such applications as electronic mail, commercial information services, computer bulletin board systems, and more. You use a terminal program on your PC to access all these systems. This section describes how you can use a terminal program and serial communications.

Assume for a moment that you are in a business in which you work regularly on a personal computer at your desk. On that computer, you develop reports and proposals that you then use in your daily business. Further assume that you have just flown to a distant city to present a proposal to an important client. You have taken a laptop computer with you, and before you left, you transferred the information you needed from your desk PC to the laptop computer. You may have done this transfer by copying it to a floppy disk, which you then read onto your laptop's hard drive, or you may have used a local serial connection and a product such as LapLink to make the transfer.

When you arrive in your hotel room, however, you find that you have forgotten one very important file. Because your laptop has a modem, you call the office and have them bring up the terminal program on your computer and prepare it to answer the phone when you call. You then connect your laptop's modem to the hotel telephone, use the terminal program and modem on your laptop to call your computer at the office, and with the help of the person there, you transfer the missing file to your laptop in time for your presentation.

This scenario demonstrates the benefits and some of the problems with using modem PC to PC communications. The benefit is obvious: your meeting will be a success because you have the information that you couldn't have received in time any other way. The drawback is that you had to explain to someone in your office what you wanted to do, and they had to be present at that end to press the proper keys when you placed the telephone call so that the file transfer could be performed.

This type of operation is known as *attended operation*; a person must be present to *attend* to the other end of the transfer in order for it to work. If no one had been available to run your computer or terminal program, you would have been stuck—even though all the hardware and software were in place. Thus, a useful enhancement to the basic terminal program is *unattended* or *automatic operation*.

In a terminal program, unattended operation is known as *host mode*. If the terminal program you use at the office supports host mode operation, you can set it up in that mode before you leave. Then when you arrive at the hotel and notice the missing file, you can immediately call your computer, which is ready for you at the office. You can call it in the middle of the night when no one is there because you can control the entire process from your hotel room. In effect, host mode on a terminal program turns your office computer into an electronic answering machine.

After you have a host mode terminal program on your office computer, other applications become possible. For example, suppose your presentation is so successful that you get the order on the spot. But you have to answer some unexpected questions about what parts are in stock and when they will be available. You know that your computer at the office has this information in a database. You can use the terminal program and modem in your laptop to call the computer at your office (which is waiting for you in a host mode communications program), run the database program, look up the required information, and be ready with the figures, all during your lunch break. You then are ready to close the deal in one day because of your PC to PC communications capability.

Communicating with On-Line Databases

In the scenario described in the preceding section, the information about parts and availability was on your personal computer at the office. In some cases, however, this type of information is on a mainframe computer at your company. If the mainframe computer has dial-up access capability, you still can use your terminal program and the modem in your laptop to call the mainframe directly and retrieve the information (see Chapter 17 to learn how to connect PCs to a mainframe).

To communicate from your laptop in the hotel room to the mainframe computer at your office, you use the same terminal program that you use for PC to PC communications. You configure your terminal program to emulate one of the data terminals your mainframe uses (see Chapter 16 for more information on terminal emulation). You then use the terminal program and the modem to place a call to the mainframe computer, log on in the same manner as from a data terminal in the office, and run the programs directly—from anywhere in the world!

In Chapter 3, you also learned about advanced PC host programs such as bulletin board software. This type of software enables you to set up a central information system in your home or business that can place a very large number of files and other computerized information on-line. You even can have such personal computer systems answer multiple telephone lines at the same time and perform on-line functions that for many years only mainframes and minicomputers could handle. Using such a large multiuser communications system from your laptop does not appear to you (as the caller) to be much different from using the remote host mode terminal program described in the preceding section. You still use your terminal program and modem to place the call and operate the remote computer. Chapter 14 provides more information on how to set up and use BBS software.

Many commercial on-line database services are also available. These services offer dial-up access to large databases of specialized information for a fee. You still use your terminal program and modem to access these services, covered in Chapters 9 through 12. The specific details of operating each service's remote program to locate and then transfer to your computer the information you want are different for each service, but the process of communicating really is the same in each case.

Using On-Line Entertainment Services

So far the uses for PC communications discussed in this chapter have been strictly business. One of the fastest growing areas of PC communications, however, is on-line entertainment. On-line entertainment services primarily feature interactive on-line games, socially based message areas, entertainment databases, and interactive on-line chat areas.

Entire social worlds have developed on-line in commercial on-line services (described in Chapters 9 through 12) and public access bulletin boards (described in Chapter 13). People meet and chat about nearly every subject imaginable. On-line marriages have even taken place (but no honeymoons; the technology is not that good yet!).

One of the more interesting aspects of on-line entertainment is multiplayer on-line games. These games—available on commercial services and on many multiuser bulletin board systems—are becoming increasingly sophisticated. Several games now are available that permit interactive play among several callers at the same time. In addition to the multiplayer action games, the old favorites of chess and variations on Dungeons and Dragons adventures also are popular.

Many entertainment services feature on-line sports databases. Some systems have daily information for fantasy football and baseball leagues that are quite popular. One commercial service features an auto racing conference giving up-to-date results and on-line interviews with drivers and mechanics after the races.

Chapter Summary

PC to PC communications take many forms, from simple printer and file sharing to world-wide information and message exchange. The same technology that enables you to send files between two personal computers enables you to access large information systems anywhere in the world. You can subscribe to and access with your personal computer and modem commercial information systems as well as public-access bulletin board systems.

All that you need to access this world of information is a computer, a modem, and terminal program software. The following chapters examine each type of PC communications in detail.

Using PCs on a Network

I f you work in an office in which more than just a few people use PCs, chances are you have a Local Area Network, or LAN, installed. LANs enable many users to share programs, data files, and devices such as printers, which provide flexibility in communications, resource sharing, and security. Although "sneaker-net," or walking floppy disks from desk to desk, still may be cost effective in very small offices, the use of LANs is growing rapidly.

Although this book is concerned primarily with asynchronous communications via the serial port on a stand-alone PC, you should understand some of the concepts of Local Area Networking. After introducing the basics of LANs and the various services they provide, this chapter discusses some of the ways LANs can be used to share asynchronous communications among network users by means of "modem pooling" and asynchronous communications servers.

Defining a LAN

Basically, a *LAN* consists of two or more PCs connected to each other so that they can share files. Each PC is still autonomous, using its own processor and its own RAM. This makes a LAN very different from a multiuser system such as UNIX, in which terminals or PCs emulating terminals are connected to a central computer that does all the real work. On a LAN, you may run a program that doesn't reside on your own disk drive but is loaded into your local PC's memory. Your local PC's CPU then does all the processing. Figure 7.1 illustrates the difference between a LAN and a conventional multiuser system.

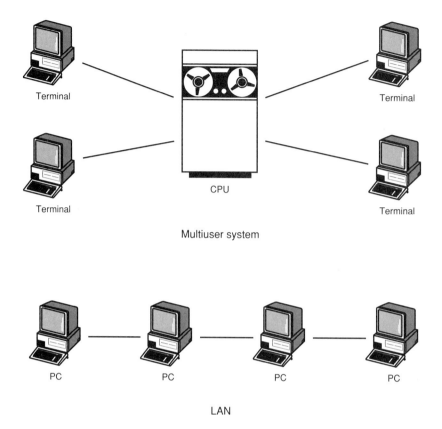

Fig. 7.1. LAN versus multiuser computers.

LANs typically offer several benefits beyond simple file sharing. These include *printer sharing*, in which each user on the network may access a single printer that is attached to the file server or to a specialized print server; *electronic mail (E-mail)* between network users; *bridges* to other LANs; and shared *gateways* to other computers or wide area networks.

PCs on a LAN may be connected in various configurations. On most LANs, one PC usually is designated as a *file server*, which acts as a central clearing house for files to be shared among users. The other PCs may have only a single floppy drive for booting up and connecting to the LAN, or even no drive at all in the case of "diskless workstations." The PCs normally are connected to each other with cabling that requires a special adapter in each machine, although several low-end networking systems utilize ordinary serial cable hooked directly to serial ports. These so-called *RS-232* LANs have much lower throughput and less flexibility than their specialized and more costly big brothers. In the RS-232 system, data transmission rates are measured in bits per second; data transmission rates along coaxial cable on a LAN typically are measured in megabits per second. If all you need to do is move files from machine to machine or share printers, however, RS-232 LANs can provide a cost-effective alternative for the small office.

The physical pattern made by the connected PCs is known as the *LAN topology*. The PCs may be connected in a line, for example, which is known as *bus topology*. You also can have a central server machine that acts as a hub, much as in a conventional multiuser computer system; this is known as *star topology*. More common than simple star topology is a sort of "series of stars" configuration, in which several stations are connected to a special intelligent Media Access Unit (MAU) or hub; these clusters of PCs then may be strung together to create larger networks. Which topology a LAN uses generally is not as important as the hardware and LAN operating system chosen, but different systems use different topologies. An IBM Token Ring system always runs on a "series of stars" topology, for example, and EtherNet LANs generally use a bus or a branching bus topology. Figure 7.2 shows various LAN topologies.

Physical Requirements of a LAN

A LAN requires that all of the PCs be connected by the same kind of cabling and that each PC uses the same kind of LAN adapter card (also known as a *Network Interface Card* or *NIC*). The cabling options include coaxial cable, twisted pair (similar to ordinary telephone wire), and fiber optics. Some recent LANs even use radio transmission rather than physical cabling; this may be used in situations in which retrofitting an entire building with long runs of wire is difficult.

Coaxial cable is the most widely used LAN cabling today and normally is a requirement on larger networks because of its bandwidth and relative immunity to noise. Various systems now are available that enable the cheaper and more convenient twisted pair wiring to imitate some of the electrical properties of coaxial cable by means of special concentrators.

The *file server* on a LAN provides storage space for data and programs that may be accessed by network users. A file server on a LAN should be a fairly powerful PC such as a 386 or 486 with a large, fast hard disk and possibly a sophisticated disk caching system for added access speed. On most LANs

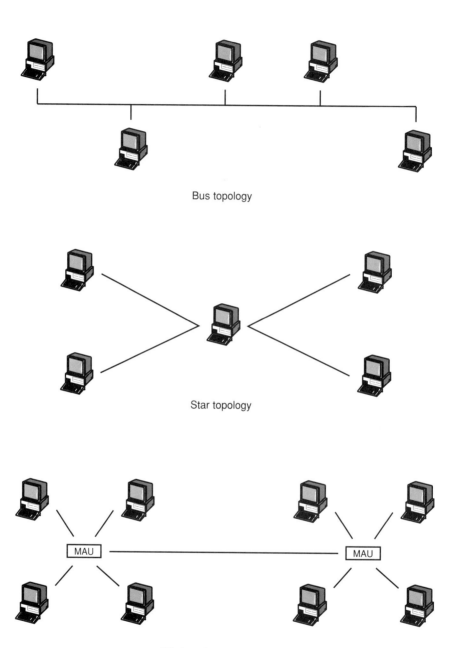

Bus topology

Star topology

"Series of stars" topology

Fig. 7.2. *LAN topologies.*

the file server is a dedicated machine, but on some LANs the file server also may act as a regular workstation. A workstation on the LAN can be anything from an old PC or XT to the latest EISA 486, and it may or may not have its own hard drive. Even if the file server is connected to a network print server and a network Asynchronous Communications Server, it still may have its own printer and modem. LANs therefore enable you to share some resources and use other resources locally.

Broadband versus Baseband

In Chapter 2, you learned how a series of ones and zeros can be represented by changes in wave frequencies on an analog telephone line. Digital data is modulated into analog signals by a modem attached to the PC's serial port. A *broadband* network works much the same way: data is modulated into analog signals, generally in the radio frequency range, which are transmitted over cabling to other network nodes such as workstations or file servers.

In *baseband* signalling, however, the data stays in digital form as it is transmitted over the LAN cabling. No modem is required. Baseband networks are cheaper, simpler, and therefore more common in most business settings than broadband systems. EtherNet, ARCNET, Token Ring, and AT&T StarGroup all are baseband systems. Broadband systems, which include IBM's PC Network, are more common in educational settings that require communications over larger distances; often a broadband LAN spans several buildings. Several baseband networks can be connected via a long broadband "backbone."

Network Cards and Network Protocols

The *network adapter card* is the hardware device that enables a workstation or other type of network node (such as a server) to communicate with the network as a whole. The network adapter card must exchange data with the PC's data bus as well as transmit and receive data through the network cabling. Because data moves very quickly over the network, overrunning the processor in any individual PC is possible. The network card therefore buffers all incoming data. The network card also must put data from the PC into the correct format for transmission over the LAN, and it must strip incoming data of network framing. In this sense, the network card acts somewhat like an RS-232 serial port that adds framing before transmitting and removes framing from received data.

The network card also is similar to an ordinary serial card in that data is transmitted a bit at a time. Many differences exist, however: the packetizing of data is much more complex, data is transferred at a much faster rate, and the network adapter card may use DMA (Direct Memory Access) channels for getting data to and from the PC's data bus rather than, or in addition to, an IRQ line and I/O port. A network adapter card contains firmware (executable code stored in ROM) that it uses to do its job, whereas a regular RS-232 serial port simply provides a UART chip. (See Chapter 2 for discussion on IRQ lines, I/O ports, firmware, and UART.)

Installing a network adapter card is simply a matter of inserting the card into an empty slot in the PC and then attaching the LAN cabling to the card. Several problems, however, may need to be addressed before that node is actually up and running on the network. The most common stumbling block is that the LAN card may use the same IRQ line as some other adapter card in the PC, such as a serial port. The IRQ setting on one of the two conflicting devices must be changed by means of jumpers. Changing the IRQ line on the network adapter card often requires changing some setting in the network software, much as changing the IRQ setting on an RS-232 port requires changing the settings in a communications program (see Chapter 2 for discussion on serial ports). The network adapter also possibly may use the same I/O port address as some other device, or may use extended memory in the PC that already is being utilized by a VGA video card. All these conflicts must be resolved before the PC can communicate on the LAN. IBM's PS/2 computers use Microchannel Architecture, which enables you to resolve these conflicts in software via the system configuration program and the Adapter Descriptor File (ADF).

Several different kinds of Network Interface Cards are available. Obviously, different types of cabling require different physical interfaces on the cards, but you should be aware of some other equally important factors when choosing network adapters. Sixteen-bit cards, for example, may give better throughput than eight-bit cards. The most important consideration in choosing network cards, however, is what network protocol you are planning to use: *EtherNet* or *token passing*.

EtherNet

EtherNet originally was developed by Xerox for bus-topology networks. In 1980, a subcommittee of the Institute of Electrical and Electronics Engineers (the IEEE) met and developed the 802.3 standard, which now is popularly known as the EtherNet standard, even though it deviates in some ways from Xerox's original specifications. The 802.3 document specifies the cabling that is to be used, the maximum length of the cabling, the format for data packets transmitted over the LAN, and, because several people are using the LAN at once, methods for avoiding collision of the data packets. The specifics of the cabling are beyond the scope of this book, so this section instead looks closely at how data is transmitted from station to station on the bus of an 802.3 network.

Suppose that you want to look at a directory listing of files on the file server. This is a little different than doing a "DIR" on a local drive, in which DOS writes the information to your own screen. If the file server simply reports its files to its own screen, that doesn't do you any good. The information somehow must be transmitted along the coaxial cable to your particular PC, just as your request to look at a directory listing has to be passed to the file server. When the information is sent to you, it is put into packets similar to the packets of a file transfer protocol (see Chapter 3 for discussion on file transfer protocols). An EtherNet data packet begins with a special eight byte preamble used for synchronizing, then two fields used for addressing. One of these fields contains the destination address for the packet—in this case the EtherNet address of your machine (each EtherNet card has a unique address). The other address field contains the address of the data's source—in this case the file server. Next is a field that indicates what type of data is being transmitted, and then the data itself. The 802.3 standard specifies that the data field of each packet may contain between 46 and 1500 bytes. Last is a frame-check sequence to ensure data integrity; this is very similar to the CRC byte in each block of a file transfer protocol such as YMODEM (the YMODEM protocol is discussed in Chapter 3). Figure 7.3 shows an EtherNet data packet.

Preamble	Destination address	Source address	Length	Data	Frame check

Fig 7.3. An EtherNet data packet.

Suppose that you want to send a message to a user at the same time he is trying to send a message to you. Obviously, you have the danger of collision and data corruption. To avoid this, EtherNet uses a method called *Carrier Sense Multiple Access with Collision Detection*, or *CSMA/CD*. When you try to send your message, your network card generates a special carrier-sense signal and then "listens" for a carrier-sense signal from the other users. If no other carrier-sense signal is present, the data is sent. Because of timing considerations on all but the smallest networks, however, collisions still are possible. The 802.3 EtherNet standard requires that both users "listen" for collisions while transmitting data. Collisions are announced throughout the network by means of a special jam signal. If your network card detects a collision, it must wait a random amount of time before retransmitting its data. Each workstation waits a different random amount of time after a collision is detected to help prevent repeat collisions. All of this traffic control can degrade network performance on EtherNet LANs with many users.

Token Passing

The IEEE approved two more standard protocol specifications for LANs—802.4 and 802.5. 802.3 establishes standards for a token passing bus architecture, and 802.4 describes a token passing ring. *Token passing* is a method of data transmission that is much more orderly than EtherNet. In a token passing network that conforms with these IEEE standards, a special data entity called a *token* is passed constantly from node to node on the network. The token is like a blank envelope. When a particular station wants to send a message to another node, it puts a message in the envelope, addresses it, and makes sure to include a return address. The token then is passed off to the next node and the next until it reaches its final destination. The recipient reads the data, sets an acknowledgment bit (stamps the envelope "received") and sends it on around the circle to its originator. The originator sees the acknowledgment bit and replaces the message with a brand new "clean" token. If the acknowledgment bit has not been set, the originator knows that the destination address was not currently on the network, but still sends out a new blank token. The 802.5 standard also defines a system for prioritizing messages, making the system much more complex.

This orderly token passing scheme makes EtherNet seem like pure anarchy, but token passing also has its drawbacks. As a token passing network expands in the number of stations, it slows down because the token must travel all the way around the circular list of nodes. When the network is started, each node must be made aware of the address of the node ahead of it in the token ring. In addition, dynamic reinitialization is necessary each time a new station joins the LAN.

Token passing bus-topology LANs, as defined by the 802.4 standard, are not common. ARCNET is a token passing bus network, but it deviates quite a bit from the IEEE specification. Token passing ring LANs include IBM's popular Token-Ring Network. The virtual circle of nodes through which the token is passed should not be confused with the LAN's physical topology: the wiring is never laid out in an actual ring. Figure 7.4 shows the data flow on a token-ring network.

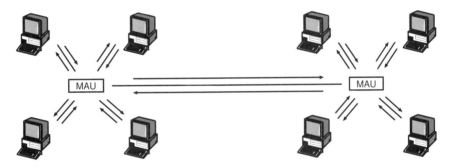

Fig. 7.4. *Data flow on a token-ring network.*

Network Operating Systems

The software that makes a LAN work, that provides the necessary platform for LAN applications and the structure for shared use of resources such as file servers, is called the *Network Operating System*. This operating system normally "sits on top of" a lower-level operating system such as DOS or OS/2 and replaces some basic functions of the lower-level system with its own network-specific services. To describe and compare the various Network Operating Systems available today (such as 3Com, Banyon Vines, IBM PC LAN, IBM LAN Manager, AT&T StarGroup, and so on) is beyond the scope of this book. Instead, the next sections look at how some LAN applications (programs that are written specifically to utilize the services of a Network Operating System) work, first from a technical point of view and then from the perspective of an actual user.

LAN Applications and Transport Services

A workstation on a LAN may run its own software locally just like a stand-alone PC. A workstation also can access information and run programs stored on a file server. Special programs also are designed specifically for LAN use. These programs may include special security features to lock out LAN users who don't need access, they may allow for simultaneous reading or writing of data, or they may provide the ability to communicate across the LAN to other users or other running programs. Suppose that you enter the words *Send AMY Hello* at your local keyboard. The message `Hello` magically pops up on Amy's screen at the other end of the building. Obviously, this is something that an ordinary DOS program cannot do. Specialized LAN programs use standardized "transport services" that are built into the LAN Operating System, the ROM on a network adapter card, or a special device driver or TSR to transport information from one PC on the network to another. Some LAN Operating Systems are able to support several different transport services. A transport service may be thought of as a "black box": programmers writing a specialized LAN application do not necessarily have to understand exactly how the particular protocol works; they simply follows the rules for asking the protocol to provide the required functionality. In other words, the protocol provides an interface for the applications programmer.

NETBIOS

NETBIOS is one of the most popular interfaces for communications services on a LAN. NETBIOS was developed originally for the Sytek adapter cards on IBM PC Networks. On a network that supports NETBIOS, the network adapter card or a software NETBIOS emulator contains a table of names. These are all the names that may be used to "talk to" this particular workstation or to a device attached to it (such as a printer). An application may add names to the name table, which initially contains only the unique permanent name for the particular network device or card. An application may send a message to any particular name on the network, or it may send a "broadcast datagram" that's intended for all nodes. Messages are issued and read by placing data in a specially constructed Network Control Block (NCB) and then issuing a software interrupt for a particular function such as sending a message to a particular point on the network. NETBIOS-compatible applications also must "listen" for incoming messages.

NETBIOS has become a sort of "lingua franca" for LAN applications. Many modern networks have NETBIOS emulators that enable you to run applications specifically intended for NETBIOS LANs. The programming interface is identical for any of these networks, allowing total portability of applications. What happens "underneath" the NETBIOS interface, however, may vary greatly from LAN to LAN. NETBIOS insulates the programmer from these functional complexities.

Using a LAN

So far this chapter has covered the physical components of a LAN and a little about how data is transmitted over LAN cabling. You have learned about Network Operating Systems and transport services for specific LAN software. Now that you know some of the technical fundamentals, you should be ready to look at an example of a LAN from a user's point of view. This demonstration uses Novell NetWare, one of the most popular Network Operating systems.

Logging onto the Network

Assume that you are a user on a NetWare LAN in a medium-sized office, and that much of your work requires you to communicate regularly with other LAN users. When you first turn on your machine, DOS loads any device drivers listed in your CONFIG.SYS file and then automatically executes your AUTOEXEC.BAT file. CONFIG.SYS and AUTOEXEC.BAT are simple text files that may be edited with any text editor. CONFIG.SYS is used to extend and fine-tune DOS functionality. Like any batch files, AUTOEXEC.BAT tells DOS to execute a series of different commands. The only difference between AUTOEXEC.BAT and other batch files is that AUTOEXEC.BAT runs automatically when you boot your machine. Near the end of your AUTOEXEC.BAT are two commands that are part of the NetWare system: *IPX* and *NET5*. IPX.COM initializes the network adapter card in your machine; NET5.COM loads what is known as the network shell, a network-specific replacement for many DOS services. It is called NET5 because you booted up with MS-DOS Version 5.0. If you had booted up with MS-DOS 4.01, you would have had to run NET4.

After NET5 is loaded, you are ready to log onto the network. The next line in the AUTOEXEC.BAT simply reads F:. One thing that NET5 provides for is immediate access to a network drive. Because you already have drives A:,

C:, D:, and E: (which happens to be a ramdrive) on your local machine, NetWare maps the first network drive as F:. The next, and last, line in the AUTOEXEC.BAT file reads `Login Phil`. (For this example, the name Phil is used.) After this last line is executed, the following prompt appears on-screen:

```
Enter your password:
```

All LANs must provide security in the form of password protection of individuals' accounts. Some of the files that you can access on the file server are files that only you can access, while others are shared with various groups of users. Still other files are utilized by everyone on the LAN. NetWare can be set up so that it requires you to change your password at set intervals, and it can allow or disallow logging in from different physical workstations on the LAN. These options usually are set by the Network Administrator, somebody with a thorough working knowledge of the LAN who is responsible for its configuration and smooth functioning. Obviously the Network Administrator should be somebody that everyone trusts because he or she has access to all files.

After you type in your password, the Network Operating System runs your "login script," a sort of LAN-specific AUTOEXEC.BAT. The Network Administrator originally created copies of a basic login script that everyone uses, and you have customized your script to suit your particular configuration and needs. The following are some of the entries in your login script:

```
ATTACH SERVER2

ATTACH SERVER3

MAP P:=SERVER1\VOL2:

MAP SEARCH1:=G:\

MAP SEARCH2:=D:\BRIEF

FIRE PHASERS

EXIT "CLS"
```

The ATTACH command attaches you to different file servers. When you ran NET5, you were attached to SERVER1, but your login script hooks you up to SERVER2 and SERVER3 as well. You now can access data and programs on all three servers.

The MAP command enables you to name particular network drives alphabetically. For example, you have decided to call the second partition (VOL2:) of the hard drive on SERVER1 your P: drive. When you type _P:_ at the DOS prompt, you will be taken to that drive. You can rename network drives at any time using the MAP command. These alphabetic names for different drives on the file server(s) are specific to your network account; somebody else on the LAN may choose to call the second partition of the drive on SERVER1 Q: instead of P:.

The MAP SEARCH command is very similar to the DOS PATH command; it adds the specified directories into the search path stored in your local PC's environment.

The FIRE PHASERS command shows that even designers of Network Operating Systems can have some fun. This is a rather peculiar sound effect that can be used to indicate that the logon process is complete.

The EXIT command followed by a quoted string enables you to specify a command to run when the login script has ended and control has returned to DOS. In this case, you simply are clearing the screen.

When the login script has finished running, you are back at DOS. Nothing looks any different than it would have if you hadn't logged onto the network, but you now have many network drives available to you. You can type _P:_ at the DOS prompt, for example, and then look at the files or run programs located on Volume 2 of SERVER1.

Suppose that you want to see what time other employees log onto their systems in the mornings. You can type the command *USERLIST* at any workstation to get this information, unless the Network Administrator has restricted access.

TIP

Security Considerations

Because so much sharing of information occurs on the LAN, a Network Operating System must provide security. A user may have different access rights on different drives and directories because NetWare provides a great deal of flexibility in the assignment of security levels. If you go to your "home" directory on the network (which the Network Administrator set up as \PHIL on the second partition of SERVER2) and type in the command *RIGHTS*, the following appears on your screen:

```
SERVER1\VOL2:PHIL
```

Your Effective Rights for this directory are [SRWCEMFA]

```
        You have Supervisor Rights to Directory.     (S)

        * May Read from File.                         (R)

        * May Write to File.                          (W)

        May Create Subdirectories and Files.          (C)

        May Erase Directory.                          (E)

        May Modify Directory.                         (M)

        May Scan for Files.                           (F)

        May Change Access Control.                    (A)
      * Has no effect on directory.
```

Entries in Directory May Inherit [SRWCEMFA] rights.

You have ALL RIGHTS to Directory Entry.

On the other hand, if you go to your boss's home directory and type *RIGHTS*, you are presented with the following message:

```
SERVER1\VOL2:BOSS
```

Your Effective Rights for this directory are [   ]:

You have NO RIGHTS to this directory area.

You can't even see what files he has, much less examine them. If you do a DIR on his home directory, you get the following message:

```
Volume in drive I is VOL2

Volume Serial Number is 5555-5555

Directory of I:\BOSS

File not found.
```

You know for a fact that your boss stores many files in his home directory, but they are hidden from you. Because he has what is known as *Parental* or *Supervisory* permission in his own home directory, he could grant you various rights if he wanted to. He may want to let you search for and view files, for example, but not revise them or write new ones. In that case, he would type _GRANT R S to PHIL_ at the command line, meaning "grant read and search permission to Phil in this directory." Besides the command line interface, NetWare provides a variety of full-screen, menu-driven utilities for performing these same routine administrative tasks.

Particular rights on the network may be assigned not only to individuals, but also to particular groups of users. The Network Administrator may set up a group called SALES that is able to access all the files that relate to sales activities, or a group called R&D that has full access to any source code on the LAN. An individual user can be a member of any number of groups. This concept of groups is a great time saver for the Network Administrator, who can change the privilege levels of an entire set of LAN users at once.

A network obviously saves the company from having to buy each user an absolutely gigantic hard drive. If that were the only convenience provided by LANs, however, they would not have achieved anywhere near their current popularity in business and institutional settings. The next sections look at some of the software that really makes a LAN useful.

E-Mail

One of the first things you probably do after logging onto the LAN is to check your E-mail (electronic mail) for messages from other network users. The program you will be using here for demonstration purposes is the WordPerfect Office Mail program. This is one of the better examples of LAN mail programs.

When you type _ML_ at the DOS prompt, the screen in figure 7.5 appears.

The top window, which is called the IN BOX, shows messages that you have received and haven't deleted yet. You delete most messages as soon as you read them, but others you keep around in your IN BOX for rereading at a later time. If a dot appears to the left of a message, it means you haven't read it. In this case, the screen shows one unread message from JACKIE. You can use the arrow keys to move the highlight bar down to that message and press Enter. Figure 7.6 shows the next screen that appears.

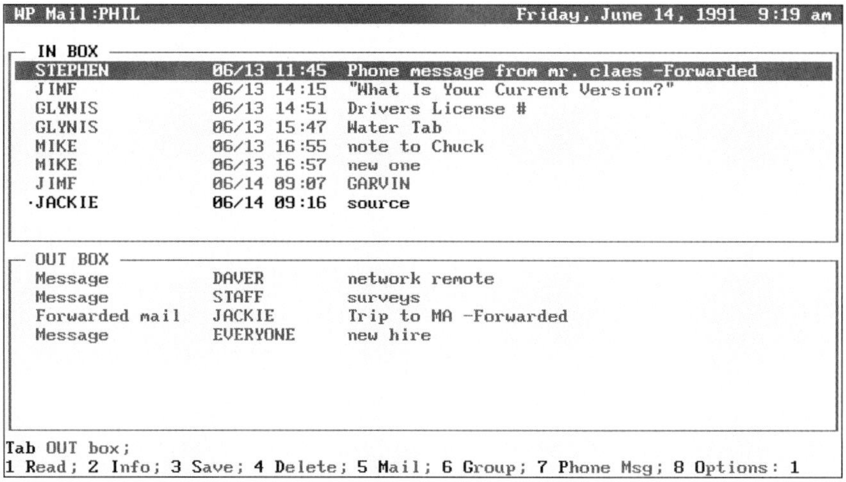

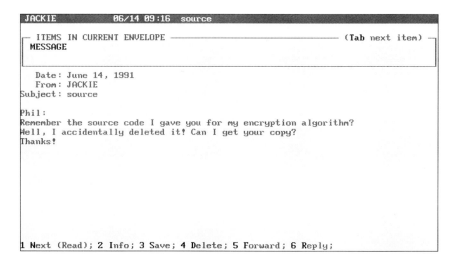

Fig. 7.5. *Main WordPerfect Office Mail screen.*

Fig. 7.6. *A message from a user.*

The bottom of the screen shows various options you can choose. You decide to reply to JACKIE's message by pressing R. The program responds with a blank area for you to enter your reply. You can edit as you write or even import pre-existing text or a WordPerfect file as your message. The program also enables you to attach any sort of file to the message, such as the source code JACKIE wanted. When you complete your message, press F9 to send it. You have an option set so that just seconds after you send the message, a notice will pop up on Jackie's screen informing her she has new mail from PHIL. NetWare enables an individual user, however, to prevent these messages from popping up by using the CASTOFF command. If Jackie is in the middle of compiling some code, she may not want to be interrupted by notifications of new mail.

When you go back to your original WordPerfect Office Mail screen, you see that a new entry is in the bottom window, the OUT BOX, that corresponds to the message you just sent JACKIE (see fig. 7.7). You can read the message again if you want, you can delete it from this OUT BOX, or you can even delete it from JACKIE's IN BOX by pressing Del and then choosing the option From All Mailboxes.

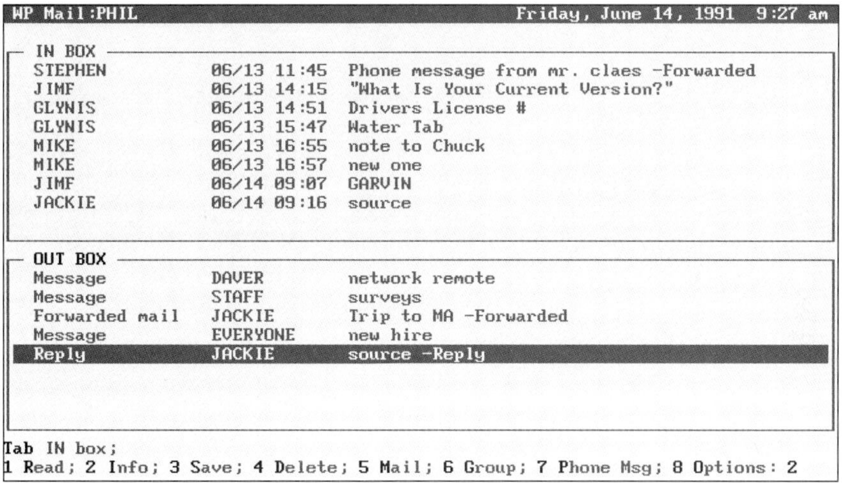

```
WP Mail:PHIL                               Friday, June 14, 1991  9:27 am
┌─ IN BOX ──────────────────────────────────────────────────────────────
  STEPHEN              06/13 11:45   Phone message from mr. claes -Forwarded
  JIMF                 06/13 14:15   "What Is Your Current Version?"
  GLYNIS               06/13 14:51   Drivers License #
  GLYNIS               06/13 15:47   Water Tab
  MIKE                 06/13 16:55   note to Chuck
  MIKE                 06/13 16:57   new one
  JIMF                 06/14 09:07   GARVIN
  JACKIE               06/14 09:16   source

┌─ OUT BOX ─────────────────────────────────────────────────────────────
  Message              DAVER         network remote
  Message              STAFF         surveys
  Forwarded mail       JACKIE        Trip to MA -Forwarded
  Message              EVERYONE      new hire
  Reply                JACKIE        source -Reply



Tab IN box;
1 Read; 2 Info; 3 Save; 4 Delete; 5 Mail; 6 Group; 7 Phone Msg; 8 Options: 2
```

Fig. 7.7. *Updated opening screen.*

If you check the message's status (the Info option), the plus sign tells you that Jackie already has read the message. Go ahead and exit the mail program, and seconds after you're back at the DOS prompt you receive the following message across the bottom of your screen (see fig. 7.8):

Fig. 7.8. *A message from a user.*

LAN E-mail programs usually enable the Network Administrator to assign users to particular mail groups. These groups do not have to coincide with the LAN security groups discussed earlier. If you want to send a message to a mail group called PROGRAMMERS, the message would go to JACKIE and to all her programming cohorts. WordPerfect Office Mail also enables you to set up individual mail groups that are not available to other users. You may want to send out party invitations, for example, to a personal mail group called FRIENDS.

Software such as this uses the network's transport services to communicate across the LAN. The transport service used by Novell is called *IPX*, although NetWare also comes with a NETBIOS emulator that may be loaded to increase the range of usable software. LAN E-mail programs, including newer versions of WordPerfect Office Mail, enable users to communicate with each other on a single network and send messages to and receive messages from users on other networks via special gateways with attached modems. This will be discussed more in depth in the section "Internetworking," later in this chapter, and also is discussed in Chapter 13.

Network Database Programs

Suppose that you need to look up a particular individual in a large database that resides on one of the file servers. One of the common uses of a LAN is to share database files among users. This ensures that all of the information is centralized and that workers aren't wasting valuable disk space with redundant data. With a LAN, the Network Administrator also can choose who has access to the database files. Centralizing also ensures that data stays correct and prevents the "drift" that occurs between separate copies of the same database.

As you run the database program, you know that at least ten other people currently are accessing the same program and the same database file. Database programs written especially for LANs allow this nonexclusive use of files; at the same time they prevent "data collisions" when two users try to update the same record at once. This normally is done through what is known as "record locking." Suppose that you pull up the record in figure 7.9.

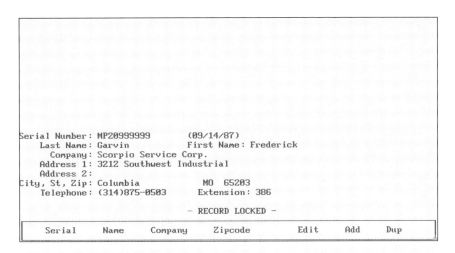

Fig. 7.9. A locked record in a LAN database.

The words `Record Locked` at the bottom of the screen in this Clipper program indicate that another user already is working on the same record. Until the user has left this record, you cannot save any changes to it. Besides this simple locking, a database program may include other methods of dealing with simultaneous use of a particular record. Some applications exist in which users may see each other's changes as they are being made, allowing for a collaborative editing of the record in question.

A surge of interest recently has developed in what are known as *database servers*, machines on a LAN that are dedicated to doing some of the database processing. Instead of a local workstation receiving an entire database file and then using its own CPU to search for a particular record, for example, the workstation simply requests that one record from the specialized server. This eliminates a great deal of network traffic and the delays associated with sending and receiving large files. Many database servers respond to a standardized interface known as *SQL*, or *Structured Query Language*. Database servers must be very powerful PCs with plenty of RAM.

Print Services

In the course of a typical workday, you may have to write many letters. Unfortunately, the only printer attached to your PC is an old, slow dot matrix that nobody else wanted. Fortunately, however, a slick, new PostScript laser printer is attached to each file server, and the LAN operating system enables users to share these resources. Suppose that you have a batch file that issues the following NetWare-specific command:

```
CAPTURE /S=SERVER1 /L=1 /Q=QUEUE1
```

This command tells your system to redirect any output normally intended for the printer port LPT1 (/L=1) to the print queue QUEUE1 (/Q=QUEUE1) on SERVER1 (/S=SERVER1). After this command is run, anything you direct to your local printer port will go instead to the laser printer on SERVER1. Because somebody else's work already may be printing when you send your letter to the network printer, a print queue acts as a "waiting room" for print jobs that need to be serviced. Your PC, therefore, is not tied up waiting for the printer to become available, and you can go on to other chores. The NetWare program PCONSOLE can be used to show the list of jobs waiting to be serviced on any particular print queue (see fig. 7.10).

You can determine from this list that you won't have to wait long before your letter actually gets to the printer. If you change your mind and decide you don't want to print the letter, you also can delete it from the queue.

Special network versions of word processors and other document-oriented programs enable you to specify a network queue instead of a local parallel or serial port as the destination for data to be printed. This eliminates the need for the CAPTURE command and enables you to choose on the fly whether to print to a local or a shared network printer. Microsoft's Windows 3.0 environment also enables you to specify network print queues so that any true Windows program can access these services.

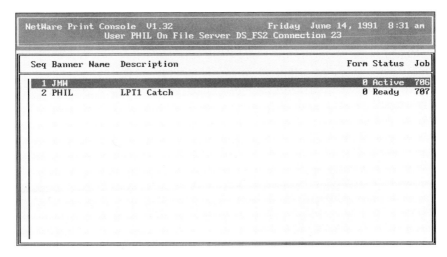

Fig. 7.10. *A list of queued print jobs.*

Other Network Programs

As you go through your workday, you use many other applications specifically designed for the LAN environment. You occasionally may use a network accounting package that enables several people to access the data at once. A special integrated *groupware* program even enables certain LAN users to share ideas and work on the same textual material, adding notes to each other about necessary revisions. Software for internetworking is another large category that deserves its own section.

Internetworking

LANs can be tied together and linked to other types of networks in various ways. This commonly is known as *internetworking*. Various connection services enable users to access programs and data and communicate with each other over greater distances. These services, normally a combination of hardware and software, include various sorts of bridges and gateways.

Bridges

A *local bridge* normally is a combination of hardware and software that interconnects similar LANs via network cabling. A typical use may be to extend a network over several floors of an office building. The bridge is transparent to users; any service available on one LAN is available to users on any bridged LAN.

A *remote bridge* is functionally similar, but it connects two LANs that are too far apart to directly cable together. Instead, two PCs, one on each of the bridged LANs, are connected by a telecommunications link. These PCs must both be running specialized bridge software. Once again, the two networks appear as one to all users except for delays due to the bottleneck of the telecommunications link.

Several LANs often are interconnected via bridges. With Novell's Advanced NetWare, you can connect up to fifteen LANs by local bridges, with one PC acting as a bridge for every four LANs. Remote bridges do not provide quite the same flexibility. Figure 7.11 shows four LANs connected by a single local bridge and two LANs connected by a remote bridge.

Local and remote bridges require special protocols for disseminating information across connected LANs. These protocols, sometimes referred to as *service protocols*, add routing information to all data so that it arrives at the intended destination, even across geographically separated networks. Some common service protocols are XNS, APPC, and parts of the TCP/IP protocol suite. In addition, many bridges use *name services*, which are services built into the software that provide standardized formats for addressing users on bridged LANs. Using the XEROX Clearinghouse name service that is common on 3Com networks, for example, you may send a message to jackie@r&d@building2. XEROX Clearinghouse requires that all addressing be in the format Name@Group@Domain. A list of names, groups, and domains already must be known to the network system as a whole and, in this case, associated with the unique EtherNet addresses of particular nodes.

Routers are similar to bridges, only less intelligent. Routers simply route data to a point on another network without providing any translation of differing protocols. If you think of transmitted data as passing through several network layers, each of which adds routing, error checking, or other control information to the data itself, routers and bridges operate on different layers.

Brouters are hybrid devices combining the features of bridges and routers, operating on several layers simultaneously. Like bridges, routers and brouters may be local or remote.

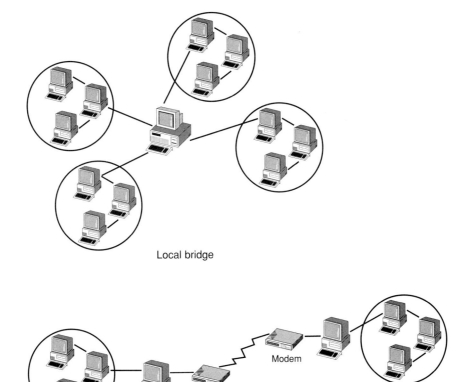

Fig. 7.11. Local and remote bridges.

Gateways

Bridges connect similar LANs. *Gateways*, however, connect LANs to dissimilar networks. Gateways also must connect users on a LAN to a mainframe system, providing the necessary terminal emulation and communications protocol (see Chapter 17 for further discussion on mainframes).

Gateways include software for converting protocols so that unlike networks can "talk to each other." A TCP/IP gateway, for example, enables a LAN workstation to call out and connect to one of the large UNIX-based networks.

Perhaps the most common type of LAN gateway is known as a *3270 gateway*, sometimes also called an *SNA gateway* (see Chapter 17 for more detailed discussion of the 3270 gateway). In a local 3270 gateway, one of the PCs on the LAN is connected directly by coaxial cable to a 3274 cluster controller, which is the front-end processor for the host IBM mainframe. The specially designated "gateway PC" is running 3278 terminal emulation software, which may be accessed by any workstation on the LAN. A remote 3270 gateway works a little differently. In this scenario, the gateway PC is equipped with a special board and software so that it actually emulates a cluster controller. The gateway PC uses a synchronous modem to link with the front-end processor at the mainframe end. Any PC on the LAN may access the mainframe by going through the special gateway PC (see fig. 7.12). The gateway PC also may be available as a regular workstation, or it may be dedicated solely to the task of providing access to the mainframe.

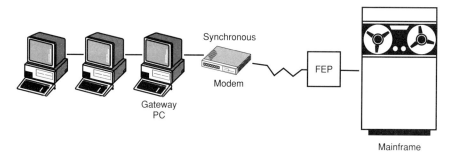

Fig. 7.12. *A remote 3270 gateway.*

Gateways for accessing X.25 networks are becoming more common. X.25 is a standard, much like RS-232, that specifies the interface between DTE and DCE devices, but in X.25 it is assumed that the DCE is connected directly to a packet-switched data network rather than the ordinary switched telephone network (see Chapter 2 for discussion on DTE and DCE devices). Because X.25 requires that data be assembled into special packets for transfer over the network, an interface board in the X.25 gateway acts as a *PAD* (packet assembler/disassembler) available to all users on the LAN.

E-mail gateways enable you to communicate by electronic mail with individuals or groups on other networks. In a system in which the local E-mail software is well integrated with the gateway, sending a message to a remote site is no more complex than sending a message to a local LAN user. The E-mail program may contain a directory of remote sites; sometimes an addressing scheme similar to the XEROX Clearinghouse (mentioned under the section "Bridges") is used. The gateway, usually a dedicated PC, may store several messages intended for a particular destination and then automatically place a modem call to the remote site. Some E-mail gateway systems such as cc:Mail can link sites together into a private store and forward network; many also can link into the large commercial mail services such as MCI Mail, or into systems like IBM PROFS or DISOSS. (See Chapter 15 for detailed discussion of store and forward networks.)

Several different protocols and addressing standards currently are used by E-mail gateways—as if instead of one United States Postal Service, many smaller companies are delivering mail, each with its own requirements for the shape and size of envelopes, each with its own stamps and postal routes. To enhance and broaden the use of E-mail, the CCITT has come up with the X.400 protocols for message handling on E-mail systems. Novell also has developed a de facto standard called MHS, or Message Handling System, which offers a predefined interface for store and forward networked E-mail programs.

Fax gateways for LANs are an important recent development. With a fax gateway, LAN users can share a phone line to route electronic documents to remote fax machines. Suppose that you are in the middle of a WordPerfect letter and you decide it's timely enough that you want to send it by fax rather than by U.S. mail. You press a hotkey and a terminate-and-stay resident program pops up asking you where you want to send your document. Perhaps you have a directory of sites you normally communicate with by fax. After you choose a destination site or phone number, your letter is translated by the fax software into the proper fax format and queued to be sent out. As soon as the phone line is free, the call is placed and the data is transferred directly to the remote fax machine. Some of the large public networks such as CompuServe and AT&T Mail also provide fax gateways. On CompuServe, for example, an ordinary mail message may be faxed simply by specifying a fax address on the TO: line rather than the ID number of another CompuServe user. (See Chapter 10 for detailed discussion of CompuServe.)

Asynchronous Communications for LAN Users

After a brief introduction to local area networking, you now come full circle back to the main thrust of this book—asynchronous communications—by looking at how communications programs are specially implemented for LANs.

Asynchronous Communications Servers

An *Asynchronous Communications Server,* or *ACS*, really is a sort of gateway. An ACS enables users on the LAN to access shared modems. Providing a modem to each desk in the office can be costly, so an Asynchronous Communications Server provides for what is often referred to as *modem pooling*. An Asynchronous Communications Server can be attached to direct serial lines as well as modems.

An Asynchronous Communications Server most often is a PC dedicated to the task of providing asynchronous services. In this sense, it is like a file server on the LAN. The ACS always is running software that makes its serial ports accessible to network users. The ACS may have only one accessible port or it may have eight or even sixteen. Special multiport serial boards allow for this increased number of available ports. A network often has more than one Asynchronous Communications Server. Figure 7.13 shows a typical setup with an ACS that provides several modems and a direct connection to a minicomputer.

An ACS does not need to be a dedicated computer, however. With some systems, a workstation on the LAN does "double duty" by means of a terminate-and-stay resident program that provides access to the machine's serial ports.

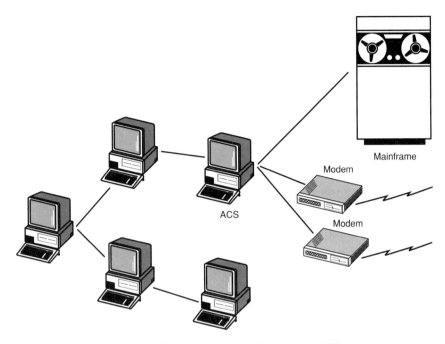

Fig. **7.13.** *An Asynchronous Communications Server on a LAN.*

Components of an ACS

Ordinary communications programs such as Qmodem or ProComm send characters directly to the serial port of the computer on which they are running. These programs have no way of communicating with the modems or ports on the ACS. "Network versions" of communications programs therefore must be designed to work together with a specific ACS. But how do they manage to get the data from the computer you are using to a specific computer across the network? To understand this, you need to look at the various software and hardware elements that comprise an ACS.

An ACS is comprised of three elements: the server hardware (a PC and perhaps a multiport board), the server software (the ACS program), and a special module that runs on any LAN node wanting access to the server. This special module acts as a sort of "black box" and provides an Applications Program Interface, or API (see Chapter 18 for detailed discussion on APIs). A communications program, rather than communicating directly with a local serial port, must send messages to this black box, which then are forwarded over the network to the ACS program itself. These messages involve activities such as transmitting and receiving characters, setting baud rate and parity, and so on. The communications program does not need to know how the black box manages to convey the data over the network, nor does it need to know how the ACS program manipulates its hardware. The program only needs to know how to "talk to" the black box and to understand any messages that may be returned. In a sense, then, a network communications program may be simpler than a stand-alone package: it does not have to involve itself with the intricacies of UARTs and hardware interrupts. The methods for communicating with the ACS black box module, however, may be complex or poorly documented. Figure 7.14 shows a diagram of the logical components of an ACS.

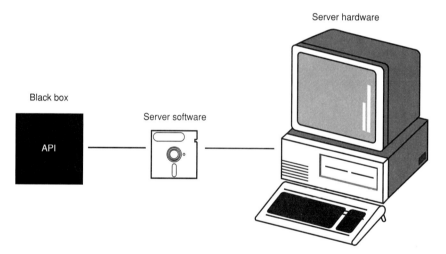

Fig. 7.14. *Logical components of an ACS.*

If every ACS manufacturer developed a unique method for communicating with the local workstation module, as many network versions of communications programs as there are brands of ACS would exist. Fortunately, several universally accepted protocols have been developed. These protocols are known as *APIs*, or *Applications Program Interfaces*. All of the common APIs use software interrupts. A communications program calls the API with a specific value that refers to a provided function such as sending a character. Any data, such as the character to be sent, may be loaded in a CPU register or may be contained in a buffer at a particular memory location. Similarly, any returns from the interrupt call (such as an error message) may be found in a particular CPU register or in a defined buffer. Chapter 18 covers software interrupts more thoroughly.

You should remember that the API used by the ACS must match the API used by the communications program. This can be a significant factor in choosing software. These days, both the ACS and network communications program commonly are configurable to work with a variety of APIs.

APIs for LAN Communications Programs

This next section covers several of the more popular APIs for interfacing between network communications programs and ACS software.

INT14

The most common API is known as *Interrupt 14h*, or simply *INT14*. This API has been around since the beginning of IBM PC history; it describes the interface to the PC BIOS communications routines. You normally can use various functions of software interrupt 14h (14 hexadecimal) to send characters to and receive characters from your PC's COM ports, as well as set baud rates and line settings. This method is slow, however, so it is used rarely in stand-alone PC communications programs, which instead write directly to the UART. The workstation module that comes with an INT14 ACS replaces the normal INT14 BIOS code with its own code. When a software interrupt 14h is issued, the CPU starts executing not BIOS code, but the new code in RAM that now is pointed to by the interrupt vector table. This replacement code does not communicate with the local UART but instead sends characters to and receives characters from the ACS, which may be physically located anywhere on the network—even miles away if a remote bridge exists. The ACS in turn sends the characters to or receives the characters from its own local serial ports.

INT14 is used by many software-only ACS. INT14 also is included as an option in several communications packages normally intended for use on stand-alone PCs. In addition, the Banyon Vines network now has built-in INT14 support. The advantages of INT14 are its simplicity and its ubiquity. A disadvantage is that the API does not offer any functions beyond the basic ones of sending and receiving characters and changing line settings. The API cannot query for available ports, for example, or check the status of

RS-232 lines. A slight variation of INT14 known as the *EBIOS* interface is used by the IBM LANACS program as an alternative API and by AT&T StarGroup's *Asynchronous Gateway Server* and several other ACS. This interface is based on the PS/2 BIOS INT14 capability.

ACSI

ACSI is an API developed by IBM for its LANACS program, an ACS that runs on the IBM PC Network system. ACSI has been adopted by a number of other ACS vendors for their products and commonly is supported by network communications programs.

ACSI is very full featured, but writing a communications program that talks to this black box is more complex than writing a program that talks directly to the PC's UART. This is because all messages must be sent to the ACSI API via NETBIOS. In other words, another protocol layer exists for the applications programmer to deal with, which also is a complex layer. Because of this tight association of ACSI with NETBIOS, an ACS that uses the ACSI API can be developed only for a NETBIOS network or for a network with a completely compatible NETBIOS emulator.

NASI

NASI (NetWare Asynchronous Services Interface) is both the name of an API and the name of the workstation module that is part of the Novell NACS/NASI ACS system. Virtually identical systems are sold by two other vendors as well, both for use on Novell's NetWare: Network Products Corporation markets *NCSI*, and Gateway sells *G/Asynch II*. All three of these products require the same programming interface, which we will simply call NASI, as does another interesting ACS by Network Products Corporation called *NMP*. NMP is a software-only ACS for nondedicated asynchronous servers; it runs on IPX-based operating systems (NetWare) and various NETBIOS networks.

NASI is perhaps the most full-featured API available today. NASI allows querying for available ports, maintaining multiple on-line sessions, baud rates up to 38,400, and status checking of RS-232 lines, including Carrier Detect. A communications program written for the NASI API will have a much more "automatic" feeling than a program written for other APIs. Querying for available ports, in particular, enables the application to report to you a list of modems you currently can use. Programs written for more primitive APIs may require that you remember and type in a command such as *CONNECT PORT1*.

Choosing an ACS

You should ask yourself several questions when choosing an Asynchronous Communications Server for your LAN. First of all, is the ACS designed to run on your Network Operating System? Although some ACS products will run on any NETBIOS LAN, most were developed by companies like IBM, 3Com, and Novell specifically for their own networks, or by independent developers for use under particular Network Operating Systems. I have heard of a Network Administrator who changed his entire LAN system just so it could use a particular ACS he liked!

The second question to ask is if the ACS supports an API that also is supported by the communications program you want to use. You will be spending your time in the communications program, not the ACS software that generally is transparent to you. The best ACS in the world is useless if it works only with the worst communications package.

Only after you determine the ACS's compatibility with various network communications programs should you investigate the "behind the scenes" features of the ACS itself. What baud rates does it support? How may ports? How expensive is it? Two more questions become especially important in the complex arena of network management: how easy is the package to install, and how good is the documentation.

Choosing a LAN Communications Program

Obviously, any LAN communications program you choose must be compatible with an ACS that runs on your network. If your program supports only the Ungermann-Bass API, for example, chances are you will not find an ACS to run with it on a 3Com LAN. But aside from this essential issue of compatibility, choosing a LAN communications program is not very different from choosing a stand-alone package. Ease of use, flexibility, a strong script or macro language, and a good track record—these should be the criteria for choosing one product over another. You also should consider the following additional network-specific issues:

- *Technical support.* This becomes more important with the complexities of setting up a program for shared operation by users with different network access rights and different levels of skill.

- *Installation*. This tends to be the most difficult aspect of using a specialized network program. Make sure that the package you choose is easy to install and configure for your particular ACS. You may think that ease of installation isn't much of an issue because you only have to do it once, but anyone who has worked with LANs for very long knows you probably will have to reinstall the program before too long as the network grows or changes.

- *Licensing*. Licensing for network products varies greatly. Some products are sold for use by a specific number of users, with the option to buy additional licensing as you add users to the system. Other products are licensed according to the number of ports on your ACS. You should determine the licensing requirements for a network product before a major purchase—not only to avoid violating the license, but also to get an idea of potential long-range costs. The initial price of a particular product may seem much lower than that of comparable products until you look at the additional fees for added users.

Using a LAN Communications Package at Work

Most LAN communications programs are produced by developers who already have a stand-alone PC product on the market. The "network version" should not differ in too many respects from the package intended for use on a single machine. The most important difference is likely the requirement that you establish what is known as a *session* with a particular modem or port on the ACS. A session is a reliable point-to-point link with a service on the LAN; before initiating this link, the communications program has "no one to talk to." A session in this special network sense of the word should not be confused with the more common concept of "on-line session," referring to an actual asynchronous communications link with a host. In a network communications program, a session must be established with a port on the ACS before you can dial out and start an "on-line session" with a remote host, BBS, and so on.

Figure 7.15 shows the initial screen of PROCOMM PLUS Network Version in NASI mode. Some of the available services are particular modems, and some are direct connections to in-house computers. The baud rates and line settings may vary for the different ports.

```
┌─────────────────────────────────────┬─────────────────────────────────────┐
│ COMMUNICATIONS SERVICES ON NETWORK   │  PROCOMM PLUS ADD SERVICES MENU     │
│                                      │                                     │
│   GENERAL   SPECIFIC      SERVER     │  ↑/↓ .... Highlight Service         │
│                                      │                                     │
│   BBS         *             *        │  Enter .. Connect Highlighted Service│
│   PORT03      *             *        │                                     │
│   PORT04      *             *        │  PgUp ... Scroll Up One Page        │
│   MT-V32      *             *        │                                     │
│   DT-V32      *             *        │  PgDn ... Scroll Down One Page      │
│   DSI9600     *             *        │                                     │
│   PORT08      *             *        │  Home ... First Service             │
│   DRM_HOST    *             *        │                                     │
│   UNUSED      *             *        │  End .... Last Service              │
│                                      │                                     │
│                                      │  Alt-E .. Expand/Contract Services  │
│                                      │                                     │
│                                      │  Alt-M .. Manual Connect            │
│                                      │                                     │
│                                      │  Alt-X .. Exit PROCOMM PLUS         │
│                                      │                                     │
│                                      │  Alt-Z .. Help                      │
└─────────────────────────────────────┴─────────────────────────────────────┘
```

Fig. 7.15. PROCOMM PLUS Network Version's initial screen.

To establish a session with one of the modems or direct lines, you simply move the highlight bar to the selection you want and press Enter. At that point, a session will be established and using the program will be virtually the same as using the non-network PROCOMM PLUS.

Chapter Summary

In this chapter, you learned some of the technical basics of local area networking. You also have seen some of the specialized services a LAN can provide in the areas of file sharing, printer sharing, E-mail, and internetworking. Finally, you have been introduced to the concept of Asynchronous Communications Servers and how they work together with network communications programs to provide modem pooling on a LAN.

8

Using Electronic Mail

Your perception of electronic mail usually comes from the system on which you first use it heavily. Many users first learn about electronic mail on a local area network (LAN). Others first experience electronic mail in a UNIX or UNIX-variant operating system environment, through a commercial on-line service such as Prodigy or CompuServe, or through computer bulletin board systems.

Each electronic mail system gives its users a very different feeling of what electronic mail is because the focus of the task is quite different in each case. Some systems are national or global in scope; others are limited to a single office area. The message-handling capabilities vary from system to system as well, but all are presented as electronic mail.

Electronic mail has so many faces because it arose through several isolated paths on many computers over many years. The developers and users of LAN-, mainframe-, and minicomputer-based electronic mail systems, as well as of microcomputer dial-up electronic mail systems (BBSs), are only now becoming aware of one another's work in developing and refining electronic mail.

Electronic mail often is combined with other features (such as scheduling, database programs, and so on) and presented as electronic mail. Today, many electronic mail systems are being linked together (as described in Chapter 15), and some users develop the impression that only such linked or internetworked systems are electronic mail.

Because of its many faces, electronic mail has come to mean different things to different people, but it really isn't that complicated. Electronic mail is any method by which two people communicate typed messages addressed to each other, using a computer as a storage and transmission medium. The definition of electronic mail is based on the task itself, not on how the task is accomplished (beyond the use of a computer).

As electronic mail has evolved, it also has been integrated with public or group message conferencing and BBS messaging systems. This integration further clouds the issue of what electronic mail is exactly, but public message systems are fundamentally different from electronic mail systems—even though they use many of the same techniques—because they are intended for group discussions, not for a one-to-one message exchange between two individuals.

Today's electronic message systems offer a variety of messaging structures, many of which are called electronic mail. The correct term for all forms of computer message systems is *computer-mediated communications* (CMC). This term is rarely used outside academic circles, but it is the correct term for what is routinely (and often incorrectly) called electronic mail because it covers all the variant structures, including public conference and BBS message systems. You need not understand the exact terminology, but you should understand the distinctions among the different types of messaging systems. Each messaging system structure is appropriate for solving particular types of problems.

In this chapter, you learn about the history of electronic mail and the several forms electronic mail (and computer messaging in general) takes today. You see how electronic mail is applied in today's computer systems to solve communications problems, and you learn how to use the more commonly encountered forms of electronic mail and computer-mediated conferencing systems.

Understanding the Origins of Computer Messaging

From the earliest days of multiuser computers, the desire to use the computer to pass messages has been evident. The earliest manifestations of this urge were simple on-line message transmission functions that enabled one computer user to type a text string and send it to another user currently logged into the system. Only two persons could participate in this simple one-to-one real-time message exchange. In order to send a message to another user, a user had to be logged on and had to know the terminal number (or ID) of the terminal the other user was on.

The concept of storing a text message on a computer's disk in order to save it until the person it was addressed to logged on appeared early in the development of multiuser software. This simple step of creating a disk database for storing messages until the recipient logged on was the true birth of electronic mail.

The origin of the first such disk-based electronic mail system is not certain, but by the late 1960s, this type of system was in common use on most multiuser computers. The structure of an electronic version of the paper mailbox was fully formed as well. At this stage, electronic mail was still focused on communications between two individuals, but now those communications could be displaced in time because the two people did not have to use the computer at the same time in order to have an electronic discussion.

As electronic mail became more common, extensions based on the way paper memos are circulated in an office were added. Such features as "carbon" copies, return receipts, distribution list mailings, and message forwarding were added to the basic send and receive operations. Because electronic mail was being developed in many isolated computer installations, each system's interface was different, as was the format of the messages each system stored on disk.

Message-Conferencing Systems

In the early 1970s, some visionaries were beginning to see that computer messaging had possibilities beyond the simple electronic replacement of the traditional interoffice paper mail system. Several research facilities and universities experimented with a form of group-oriented message systems they called conferencing. Conferencing was an attempt to expand electronic mail beyond the one-to-one orientation and develop a way to coordinate electronic discussions among a group of people.

In 1971, the U.S. Office of Emergency Preparedness decided to use this emerging technology to help handle the paperwork associated with their governmental tasks. The resulting system was namcd EMISARI and was the first large-scale use of *computer conferencing*—a type of electronic messaging system organized around topics. The designer of EMISARI, Dr. Murray Turoff, carried on the development of his ideas until they became a system called EIES in 1975. EIES in turn inspired Harry Stevens to design a conferencing system called Participate, which became the foundation for the commercial on-line service named The Source (later purchased by CompuServe).

Computer conferencing developed along several divergent lines during the 1970s, and several different people developed systems that took differing approaches to enhancing electronic group discussions. Some of the significant systems inspired by the research of that time that survive today are Caucus (developed by Charles Roth) and PicoSpan (by Marcus Watts), both of which were inspired by research into conferencing at the University of Michigan.

Each message conferencing system has remained true to the basic idea that public discussions should be organized and accessible by topic. The development of computer conferencing as an approach to group electronic discussions remains largely confined to universities and research centers. Because UNIX is the operating system predominantly used in these settings, most computer conferencing systems today are UNIX-based.

Mainframe E-Mail Systems

As the university and research world was developing message conferencing, the pure electronic mail function was being widely adopted by the commercial mainframe and microcomputer worlds. This acceptance in a business environment marked the first successful introduction of electronic mail into the computing mainstream.

Mainframe electronic mail systems such as IBM's Professional Office Systems (PROFS) and Distributed Office Support System (DISOSS) came into widespread use. PROFS was very popular on the older IBM 4300 series computers, and DISOSS was widely used on the 3080 series of IBM computers. It is estimated that nearly 10 million people still send and receive electronic mail through a PROFS or DISOSS system today.

In the minicomputer world, Wang VS systems used Wang Office, Data General supplied CEO for all of its minicomputers, and DEC minicomputers used VMS Mail or All-In-1. These mainframe products not only popularized electronic mail but also pioneered the integration of word processing, scheduling, and database retrieval with the basic electronic mail function.

BBS Message Systems

As it did with most areas of computing, the microcomputer revolution opened up new paths of software development in electronic messaging. In

1978, Ward Christensen developed a microcomputer-based message system called the Computer Bulletin Board System (CBBS). This invention began a totally different path of computer message system development—microcomputer bulletin board systems (BBSs). The microcomputing community was enthralled by the capability of BBS software to communicate messages over modems, and BBS software quickly became a fertile area of group message system development. (See Chapter 13 for detailed discussion of bulletin board systems.)

Bill Blue was the leading pioneer of Apple-based BBS software with ABBS (Apple Bulletin Board System) and PMS (People's Message System). In the TRS-80 computer arena, Bill Abney developed Forum-80 software, and I developed The Bread Board System (TBBS). Each system marked significant points in the development and refinement of BBS message systems.

By the early 1980s, many innovators were experimenting within the BBS software arena, trying out their ideas of how the BBS message system could be improved. Each system carried public message access in a different but related direction, adding greatly to the knowledge base of how people interact in a public electronic setting.

Whereas conferencing systems evolved in near isolation from one another, BBS software was very public; nearly everyone involved knew what everyone else was doing. Several very sophisticated models for public message system operation were attempted in the early days of BBS experimentation. With the large number of BBS users, development was rapid and message system designs evolved quickly. BBS software today has some of the easiest to use and most powerful public access message structures found anywhere.

LAN E-Mail Systems

Another portion of the microcomputer revolution—the local area network—also affected electronic message system development. Because a LAN connects several computers and their users together, electronic mail is a natural LAN application. In the mid-1980s LAN technology reached the point that LAN-based electronic mail software began to appear.

User interaction in a LAN environment is more like the mainframe or minicomputer environment than the dial-up modem-based BBS environment. As a result of this similarity, early LAN-based electronic mail systems followed the mainframe electronic mail system models. A LAN environment, however, is different from a mainframe in one very important way—

users have the full power of a PC at their own workstations, not just a data display terminal as in a mainframe or minicomputer setting. LAN electronic mail software, therefore, soon began to develop a different approach from the other two electronic mail environments (mainframe and BBS).

Because LANs have existed primarily in office environments from the beginning (in contrast to the university or research environment of conferencing and the public enthusiast environment of the BBS), LAN electronic mail software was closely patterned after the established inter-office paper-based mail system. The focus was on extending this model to include functionality appropriate to computer-based systems.

The LAN workstation environment has some of the characteristics of the individual PCs that make up the network. One characteristic is that many workstations have a local private file section in addition to the shared file sections of the file servers. Because of this structure, transferring files from one workstation to another with attached memos explaining their contents was an early focus of LAN electronic mail software.

As LAN usage grew, LAN electronic mail was driven to provide connections between different LANs in widespread locations. Development proceeded on internetworking LAN electronic mail software in the middle and late 1980s. When Novell began including Action Technology's Message Handling Service (MHS) software with its network software, MHS quickly became the most widely used method of internetworking LAN electronic mail. Today almost all LAN-based electronic mail software has a gateway to provide MHS internetworking with other LANs in remote locations around the world.

Understanding Electronic Messaging Today

As the preceding section explained, the history of electronic messaging has resulted in three separate areas of electronic mail software: mainframe electronic mail and conferencing, PC-based BBS electronic mail and public message systems, and LAN electronic mail. Each was developed to serve a very different purpose, and for many years most of the designers of one type of software knew very little about the existence or purpose of the other types.

Today, however, the internetworking of all types of electronic mail and public message systems is occurring rapidly. The interconnection of these types of electronic mail services has exposed the three worlds to each other (see Chapter 15). BBS message systems are being networked and have grown to serve a user base numbering in the millions. The mainframe and minicomputer electronic mail and conferencing systems built the world's largest internetworking system in isolation, but now BBS software is being connected to that system as well. At the same time, BBS software is now being internetworked with LAN electronic mail systems. Almost all of this interconnection has occurred in the last two or three years.

The resulting awareness that the users and designers of the three major areas of electronic mail now have of one another is already beginning to have an effect: the technologies are beginning to merge. This merger already has happened in BBS software (with TBBS) and in LAN-based electronic mail systems (with Brainstorm from Mustang Software), and you can expect to see much more technology merging in the future. Many of the commercial on-line message systems have adopted BBS message system structures in addition to their traditional electronic mail systems.

In the next few years, LAN electronic mail software, BBS message systems, and computer conferencing software will likely continue to adopt more capabilities from one another. Internetworking links among all these systems also will likely become more common, and this will be one of the most rapidly changing and developing areas of computing for the next several years.

Using LAN Electronic Mail

LAN-based electronic mail (E-mail) software is the most refined form of the pure electronic mail structure, likely because LAN E-mail was developed to satisfy requirements in the business world and was always commercial software. The competitive demands of selling software to users who had several products to choose from has resulted in the development of very sophisticated capabilities and user interfaces.

The most widely implemented LAN electronic mail software today is cc:Mail, which is used by more than 20 percent of all LAN electronic mail users. cc:Mail recently was bought by Lotus Development Corporation, and the product is almost certain to change and grow in the near future. Other popular LAN electronic mail systems are DaVinci (by DaVinci Systems), The

Coordinator (by Action Technology), WordPerfect Office (WordPerfect Corp.), and Higgins (the Higgins Group). Microsoft is also making an entry into the LAN electronic mail arena with Microsoft Mail.

Understanding LAN E-Mail Structure

The following sections examine cc:Mail in some detail in order to provide an example of how LAN electronic mail systems operate. This system is the one you will most likely encounter, and its operation is representative of the entire LAN electronic mail category.

The cc:Mail system structure is broken into mail servers called *post offices* and user areas called *mailboxes*. cc:Mail is sold in a variety of sizes and packages, but the basic unit is called the Platform Pack. The Platform Pack includes a post office (mail server) and one mailbox along with a system administration program. Additional user mailboxes are sold in packages of 8, 25, or 100. Other LAN electronic mail packages may bundle their options in different sizes, but the post office/mailbox structure is the most common LAN electronic mail software packaging.

Understanding LAN E-Mail Gateways

In addition to post offices (mail servers) and mailboxes (user interface software), LAN E-mail packages usually provide modules called *gateways* that enable multiple post offices to communicate with each other. Gateways enable the post offices to connect to one another using a variety of transmission methods. The most common gateway connection methods are through a wide-area network (WAN) bridge using TCP/IP or a similar network protocol or through the normal telephone line using a modem.

Gateways also enable a LAN electronic mail system to internetwork with other types of software. A gateway can provide data format and protocol conversions to enable a LAN electronic mail system to exchange mail with such systems as IBM's PROFS and DISOSS, DEC's VMS Mail, and public data services such as MCI Mail and CompuServe.

cc:Mail provides gateway modules that link to a wide variety of electronic mail systems following popular standards such as Novell's Message Handling Service (MHS), TCP/IP's Simple Mail Transfer Protocol (SMTP), and X.400, as well as to direct gateways to several popular specific software packages.

Another type of LAN electronic mail gateway that has gained tremendous popularity is the *fax* gateway. This type of gateway enables you to create a message at your workstation and have the system deliver it automatically by calling a normal fax machine on the telephone. This type of gateway requires the presence of a PC fax card. cc:Mail requires a fax card that meets the DCA/Intel Communicating Application Specification (CAS), such as the Intel Connection Co-Processor or SatisFAXtion board. Fax gateways are very popular with users of LAN E-mail systems but often are an administrative nightmare. The greatest problem is controlling the outgoing fax telephone charges, which can mount rapidly in this environment.

Outbound fax gateways are quite easy to use, but receiving fax on a LAN E-mail gateway is rarely a good idea with today's technology. Incoming fax on computers is far more complicated and much harder to work with than a simple fax machine.

Incoming fax can be stored and manipulated only as a large binary graphics image. You cannot turn a fax file into an ASCII text file or normal message with today's technology. This is because a fax transmission is really a series of dots arranged to print on paper, and not text as computers normally handle it. To translate a fax image file into a text file requires character recognition processing. Today, software for such image processing is not capable of reliably translating a fax image into an E-mail message. Until it is, receiving incoming fax with an E-mail system is not as useful as simply printing incoming fax directly on a fax machine for normal distribution.

Understanding the User Interface

The user interface for most LAN electronic mail software provides at least the following features:

- A TSR message-waiting module

- A TSR pop-up window for reading messages

- The option of importing text files into messages

- The option of attaching binary files to outgoing messages

- The option of using the editor of your choice to prepare messages

- The option of requesting a return receipt

- The capability to keep messages in groups (often called *folders*) by subject

Common additions to these features are a remote access module (enabling you to use the electronic mail system remotely via a modem) and the capability to encrypt message text during transmission or as the message is stored on disk.

For cc:Mail, the TSR portion of the user interface is provided by two programs called NOTIFY and MESSENGER. NOTIFY announces any new mail you receive while on-line. You have the option of just having NOTIFY sound a beep or open a pop-up window describing the message received by subject, sender, date, and time. The MESSENGER program gives you a hot-key entry into a stripped-down version of the complete mail interface, in which you can read, edit, reply to, forward, or store messages. cc:Mail has an interesting additional interface program called SNAPSHOT. This program enables you to capture the screen image of any software you are running and attach it to a message you are sending.

The full cc:Mail user interface is available when you run the program MAIL on your workstation. You supply your password to access the MAIL program, after which you see the screen shown in figure 8.1. As you can see, the user interface at this level is quite simple; you move the highlight bar to the proper selection and press Enter, or you press the single key command to select the desired option.

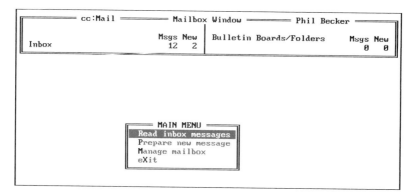

Fig. 8.1. *The cc:Mail user interface main menu.*

Other LAN electronic mail programs have more complex interfaces. For comparison, a DaVinci user interface screen is shown in figure 8.2. As you can see, it provides largely the same functions as the cc:Mail user interface, but they are presented quite differently.

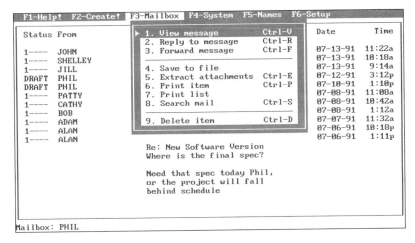

Fig. 8.2. The DaVinci user interface main menu.

Using cc:Mail: an Example

To understand the capabilities offered by LAN electronic mail packages, this section closely examines the cc:Mail program. Because all LAN electronic mail programs are similar in the capabilities they offer, this example will give you a good idea of what LAN electronic mail is all about.

Figure 8.1 showed a sample cc:Mail screen with messages waiting. To read a waiting message, you select the Read Inbox Messages line in the menu and press Enter, or you simply press R. When you make this selection, cc:Mail displays your message in the form shown in figure 8.3.

After you have read the message text, you press Enter to obtain the next cc:Mail menu screen, the Action menu. This screen gives you several options (see fig. 8.4).

You can select the Display Next Message option to move to the next message, or you can select one of the options in the right portion of the menu to take an action with the message (such as move or copy the message; forward it to another person; or print, archive, or delete the message). One common action, used here to illustrate how electronic mail works, is Reply to Message.

When you select this option, you are moved into an editor to enter your reply message text. After composing your reply, you press F10 to send the message. The screen shown in figure 8.5 appears.

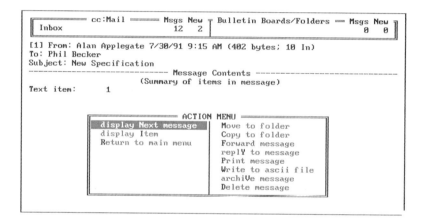

```
[1] From: Alan Applegate 7/30/91 9:15 AM (402 bytes; 10 In)
To: Phil Becker
Subject: New Specification
----------------------------- Message Contents -----------------------------
      Phil,

      I really need that new specification today.  If you can't get me
      the entire thing, at least get me the portion which discribes the
      new compiler commands.  I will be stuck on the documentation after
      today if I don't have that information.  Don't mean to press, but
      it was you who set the deadlines!

      Patiently!
      Alan Applegate
```

Fig. 8.3. *The cc:Mail message display.*

```
┌──────── cc:Mail ═══════ Msgs New ┬ Bulletin Boards/Folders ═══ Msgs New ─┐
│ Inbox                     12   2 │                                 0    0 │
└──────────────────────────────────┴──────────────────────────────────────┘
[1] From: Alan Applegate 7/30/91 9:15 AM (402 bytes; 10 In)
To: Phil Becker
Subject: New Specification
----------------------------- Message Contents -----------------------------
                   (Summary of items in message)
Text item:       1

                   ┌════════════ ACTION MENU ═════════════┐
                   │ display Next message │ Move to folder │
                   │ display Item         │ Copy to folder │
                   │ Return to main menu  │ Forward message│
                   │                      │ replY to message│
                   │                      │ Print message  │
                   │                      │ Write to ascii file│
                   │                      │ archiVe message│
                   │                      │ Delete message │
                   └──────────────────────┴────────────────┘
```

Fig 8.4. *The cc:Mail message Action menu.*

From this menu, you can see that you have the option to attach a file or other text and then display or reedit the message or its subject. Because this message is a reply, it is automatically addressed to the sender of the message you were reading. When you enter a new message, however, you have to supply an address so that the electronic mail system knows where to send the message.

Electronic mail systems vary the most in message addressing. Many systems address a message by name alone, but others add addresses based on where the reader's computer is located. These extended addresses have many forms—some simple, some complex. Chapter 15 explains the concept of wide-area message addressing in detail.

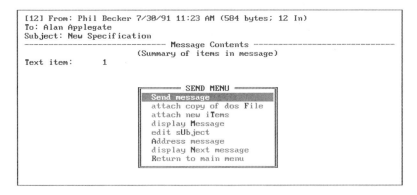

```
[12] From: Phil Becker 7/30/91 11:23 AM (584 bytes; 12 In)
To: Alan Applegate
Subject: New Specification
------------------------------- Message Contents -----------------------------
                       (Summary of items in message)
Text item:       1



                         ====== SEND MENU ======
                        | Send message           |
                        | attach copy of dos File|
                        | attach new iTems       |
                        | display Message        |
                        | edit sUbject           |
                        | Address message        |
                        | display Next message   |
                        | Return to main menu     |
                         ========================
```

Fig. 8.5. *The cc:Mail Send menu.*

You need to learn about addressing on any electronic mail system you use. Normally, the larger the number of users and computers a given electronic mail system links together, the more complicated the addressing becomes.

To continue with the cc:Mail example, if you select the Address Message option from the Send menu, you are presented with the Address menu shown in figure 8.6.

```
From: Phil Becker 7/30/91 11:23 AM (586 bytes; 12 In)
To:
Subject: New Specification
------------------------------- Message Contents -----------------------------
                       (Summary of items in message)
Text item:       1


                    ============ ADDRESS MENU ============
                   | eNd addressing     | Copy to person      |
                   | Address to person  | copy to mailing List |
                   | address to Mailing list | Blind copy to person |
                   | address to board/Folder | set Priority level |
                   | Return to main menu | reQuest receipt     |
                    ======================================
```

Fig. 8.6. *The cc:Mail Address menu.*

This menu shows the many options available for addressing a message. You can send a single message to many people by addressing it to a mailing list containing all their names. You can copy and send the message (as a *carbon copy*) to individuals by name or to a mailing list. You also can request the computer system to send you a return receipt when the message is read by the recipient so that you can be sure the mail was delivered.

Copies normally are listed in the message in the same way a copy list is included on an office memo; each person who receives the message or a copy can tell at a glance who else has received it. If you want to send a copy of the message to someone without informing all the other recipients of the message, the copy is known as a *blind copy*. The only difference between a normal copy and a blind copy is that the person who receives the blind copy is not listed in the copy list on the message for all to see.

The final option of interest here is the Set Priority Level option. This option enables you to indicate how important the message is and how much money the system should spend to send it quickly. A high priority message may be sent immediately, even if it takes an immediate long distance telephone call. A lower priority message waits until the normal times set by the system administrator to be delivered.

Priority is important only if you install a system with more than one post office. The post office function is to store mail and send it to the recipient. When you have only one post office, that task does not require transmission of mail beyond the confines of the LAN itself. If you use more than one post office on LANs that are not located in the same place, however, then the post offices may have to place telephone calls to each other and use modems to transfer the mail. The Priority setting indicates whether an immediate phone call needs to be made, or if the post office is able to wait until a later time when telephone rates are lower to save money. This "non-local" post office function is an example of store and forward electronic mail, which is discussed in more detail in Chapter 15.

LAN electronic mail shows the general model for sophisticated electronic mail software. Electronic mail takes on other characteristics in other environments, however, as the following sections show.

Using BBS Electronic Mail

Electronic mail in a BBS environment (see Chapter 13) is somewhat different from LAN electronic mail for two reasons. First, electronic mail is often a secondary focus in this medium (public message areas tend to be first priority). Second, electronic mail may be networked by store and forward technology (as is the case with LAN electronic mail), but often it is not. As a result, BBS electronic mail is different in some ways from LAN electronic mail.

Local Electronic Mail

All BBS software provides local electronic mail. This type of mail is addressed from one user to another on the same BBS computer and would be equivalent to a LAN electronic mail system that had only a single post office (and therefore no wide area store and forward function). Addressing for this type of BBS electronic mail is almost always by user name or *handle* (an alternate shorthand name used by some BBS systems).

Many BBS local electronic mail systems do not have all the features (such as distribution lists, copies, forwarding, return receipts, and so on) of the LAN electronic mail systems because electronic mail has not been a focus for this software. As BBS software is maturing, however, you now can find most or all of these features in newer BBS software packages.

Internetworked Electronic Mail

Many BBS systems provide internetworked electronic mail. In this case, the message must be addressed with an extended address indicating on which computer in the network the other user is to receive the message. These extended addresses may be quite complex or relatively simple, depending on the type of BBS network in use.

Some BBS software also can be internetworked with LAN or mainframe electronic mail systems. In this case, the extended address you use to send mail on the BBS is a slight variation on the type of addressing the LAN or mainframe system uses. Chapter 15 provides additional information on very large internetworked electronic mail systems.

Most BBS software does not focus on true one-to-one electronic mail when internetworking message areas. More commonly, the software uses shared or distributed public message conferences similar to local public message conferences, except that they are repeated on many BBS systems around the country or around the world.

Despite the size and technical complexity of such installations, these distributed conferences (often called *echomail* or *networked message areas*) are extremely easy to use. All the complexity is dealt with by the BBS software and the system operator.

As a user of a distributed public message conference, you simply use the message area as if it were local. No specific addressing is required to propagate shared conference messages on the network; the BBS software takes care of this automatically. Depending on the BBS software, the conference messages may be organized by topic for easy retrieval, and other message conference options may be available to make the conference easier to use.

Understanding Electronic Mail Privacy

Electronic mail privacy is an important issue today. Employees have objected in court to their employers reading their electronic mail, and the issue of whether or not a BBS SYSOP (system operator) can release electronic mail publicly is being hotly debated.

Currently, you probably should assume that truly private electronic mail does not exist. Assume that anything you place on an electronic system is a written document, such as a memo, that others can read. If you keep this perspective, you can avoid trouble.

On most electronic mail systems, the system operator can read any message entered. You should assume that this is the case on any system you use—even if that system provides message encryption. The SYSOP usually doesn't check most messages because the job is simply too large, so in a sense, you are protected by the sheer volume of messages. At present, however, no legal sanctions exist against a system owner or operator reading any messages that appear on their computer.

In the future, you can expect these issues to be clarified legally. Unlike the telephone company, however, computers are not common carriers, so wiretap laws and communications act disclosure protections do not apply. The courts have upheld that an employer is free to read any electronic mail stored on the company computer no matter how private it is labeled. On some commercial systems, privacy likely can be made legally binding by contract, but even this is still an open issue.

Chapter Summary

From its humble beginnings in the early days of computing, electronic mail has evolved into an exciting medium that is changing the way people do business. The rapid communications, the capability to organize and store messages, and the ease of response make electronic mail a unique communications method. It combines the informality and ease of communicating at your desk with a written record of all transactions.

Electronic mail forms the foundation of all computer-messaging systems. The purest form of electronic mail is found in the LAN electronic mail software, but BBS and commercial on-line services also provide electronic mail capability as well as adding their own unique electronic message conferencing capabilities.

Electronic mail enables a great deal of information to be available in a single location, making the temptation to use electronic mail as an information gathering service strong. You should determine how confidential messages really are in any electronic mail system you use. Never assume that anything you write in an electronic mail system is truly private, as court cases have ruled that information obtained from an electronic mail system can be used against its author.

Public message systems that go far beyond basic person-to-person electronic mail are used to build message-based conferences. These conferences have very different forms based on the background of the software that implements them. BBS software today has the most robust public conferencing software. The rapid growth of this technology, however, will soon result in the integration of the best capabilities of these technologies.

Using Commercial Information Services

C ommercial information services provide access to a world of information, games, software, and social activity. You can purchase access to several services, each with its own focus and character. Access to these services is available from most parts of the United States and, in many cases, from other countries as well.

Many computer software and hardware companies have customer support sections on these services, in which company representatives answer questions and provide user tips on their software. Users of the software also meet to share their experiences and usage tips. These support sections often are some of the best sources of information on using your computer software and hardware effectively.

This chapter gives an overview of these services as well as information on sign-up procedures and pricing.

CompuServe

CompuServe is the oldest commercial information service still active today. It began as a service called MicroNet that offered *bulletin board services*—public message areas based on a particular topic—to personal computer users (similar to those services discussed in Chapter 15). Today CompuServe has grown into a collection of nearly 15,000 services in a single service.

With over 600,000 subscribers, CompuServe is the largest "full service" information utility. No other commercial service offers as many selections or as much information as CompuServe. Special interest groups (SIGs) carry on the original MicroNet tradition and provide software for downloading as well as valuable help and information on using the software. These software libraries are one of CompuServe's greatest strengths. Nearly all serious shareware and public domain software appear on CompuServe. (Shareware software is discussed in Chapter 4.)

In addition to its large collection of SIGs, CompuServe also offers many additional services such as the Dow Jones stock service and OAG airline reservations and information. You can even order transcripts from many television news shows on-line. CompuServe also offers electronic mail (E-mail) that can be sent to other E-mail network users or to fax machines.

CompuServe has one of the largest and most active "CB" chat services available anywhere. This service enables you to type to a large group of people in real time in a sort of on-line CB radio.

In the entertainment arena, CompuServe has pioneered multiuser on-line games. Entries such as MegaWars have been megahits with users for many years.

As you can see from a sample CompuServe logon screen in figure 9.1, just a list of what is added in a single week can be extensive. This screen advises you that you can enter the command GO RATES at any time to check on your current billing and usage charges. The screen also offers 11 selections which print extended news or bulletin files about new features that have been added. You can read these news files now, or enter a command to go to a particular section of the CompuServe system. When measured by the volume of information contained on-line, CompuServe is the largest of the public commercial information services. Chapter 10 discusses how to use CompuServe in detail.

CompuServe is organized as a *text-driven* service. You can call CompuServe with any terminal program and use it through menus or a command-driven interface. CompuServe also offers a program called the CompuServe Information Manager (CIM), which creates a mouse-driven windowed menu environment for accessing CompuServe.

In the text driven mode, you navigate CompuServe by typing commands that direct it to perform specific functions. The output you receive is plain typed text with no graphics or color. This mode enables CompuServe to be used by any data terminal or terminal program. Although this type of interface provides you with the maximum flexibility in selecting where you want to go within CompuServe and what information you want to retreive, some learning is involved. With color monitors and mouse driven programs becoming common, this interface also looks a little old-fashioned. That's why CompuServe developed the *CompuServe Information Manager (CIM)* terminal program.

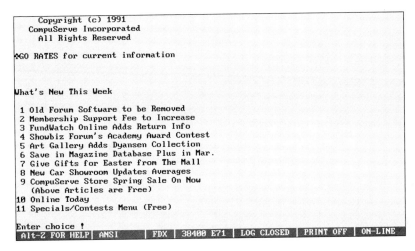

Copyright (c) 1991
CompuServe Incorporated
    All Rights Reserved

GO RATES for current information

What's New This Week

 1 Old Forum Software to be Removed
 2 Membership Support Fee to Increase
 3 FundWatch Online Adds Return Info
 4 Showbiz Forum's Academy Award Contest
 5 Art Gallery Adds Dyansen Collection
 6 Save in Magazine Database Plus in Mar.
 7 Give Gifts for Easter from The Mall
 8 New Car Showroom Updates Averages
 9 CompuServe Store Spring Sale On Now
   (Above Articles are Free)
10 Online Today
11 Specials/Contests Menu (Free)

Enter choice !
Alt-Z FOR HELP | ANSI | FDX | 38400 E71 | LOG CLOSED | PRINT OFF | ON-LINE

Fig. 9.1. CompuServe logon in ASCII text mode.

CIM enables you to navigate CompuServe in a much more user-friendly fashion. With CIM, you can view a list of available files and quickly locate and download the information you want. CIM only works with CompuServe and is available for $24.95 from CompuServe. This price includes a $15 on-line time credit; if you use the service frequently, the program actually costs you only $9.95.

Figure 9.2 shows the same initial logon as figure 9.1, but using CIM. You can see the windowed presentation of the information. By using your mouse and clicking on what you want, you can obtain the same information, but in a more windowed environment. Although CIM is quite easy to learn to use, it can be quite slow. So advanced CompuServe users have developed still other approaches to access CompuServe quickly and easily. Among these are programs such as TAPCIS, which are used to handle message sending and receiving off-line. These programs call CompuServe, quickly download all of your waiting mail to your computer, and then hang up. You read and respond to this mail off-line and then your computer calls CompuServe again and sends back any replies you have automatically. The operation of these programs is described in detail in Chapter 10.

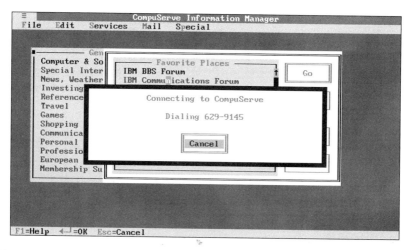

Fig. 9.2. CompuServe logon using CIM.

Prodigy

Prodigy is the most heavily advertised on-line service in history. It is the joint venture of IBM and Sears and is targeted more at ordinary consumers than any other service. With more than 600,000 users, Prodigy has quickly become one of the largest on-line services. It has a teletext rather than interactive text-based structure and includes many graphics and colors. Figure 9.3 shows the Prodigy logon screen.

Teletext services are primarily output based and show text to you much more than they take text from you. Prodigy is not a pure teletext system because it does support E-mail and limited bulletin board message forums. But even when using Prodigy as a message service, its teletext orientation is evident. With an advertisement on every screen and only a few short lines using very large characters, visual presentation clearly is more important to the Prodigy interface than making any large amount of text easy to access. In return for this orientation, you get a visual style that truly is stunning and an ease of navigation and use that is unparalleled.

You cannot access Prodigy with a conventional terminal emulation program. It requires the Prodigy software that must be purchased from the Prodigy service and used with that service only. This custom software can provide graphics and colors that normal text-based systems cannot. The Prodigy program also uses the mouse effectively. Prodigy therefore is the easiest of all commercial systems to use, as well as the most visually appealing. (Prodigy is discussed in depth in Chapter 11.)

Fig. 9.3. *The Prodigy logon screen.*

The focus of any teletext system is one-way information such as news, weather, stock quotes, and so on. Prodigy is strong in these areas and has added an easy on-line shopping service. Prodigy even has teamed up with grocery stores in many areas to provide on-line grocery shopping with delivery to your door. Because Prodigy was developed with novices in mind, new users to telecommunications find Prodigy very easy to use.

In addition to the traditional teletext services, Prodigy offers bulletin boards. Although Prodigy's bulletin board services are no match for services such as CompuServe, these message areas can be informative. Prodigy moderates these message areas by having reviewers authorize each message before it can be posted publicly, so the discussions are not as "free wheeling" as on most bulletin board services. The bulletin boards on Prodigy are strictly message areas; you cannot download or upload files.

Prodigy has a great deal of on-line advertising, with an ad on nearly every screen. If you want more information on a particular ad, click it and a full screen version appears. Because Prodigy is developed as a teletext system, these ads are not as annoying as you may think, and they keep the cost of the system quite low.

Prodigy pioneered the practice of charging a flat rate for commercial information services. For a fee of $12.95 per month, you have unlimited use of every part of the system except E-mail. The E-mail service costs 25 cents per message. (E-mail is discussed in depth in Chapter 8.)

GEnie

GEnie is similar in structure and content to CompuServe but is a smaller service, with about 250,000 users. What CompuServe calls a SIG, GEnie calls a Roundtable, but the combination of message areas and file libraries provides the same content. GEnie has an active on-line games section, a CB-style chat section, an on-line encyclopedia, news, weather, and airline reservations sections. Figure 9.4 shows the GEnie logon screen.

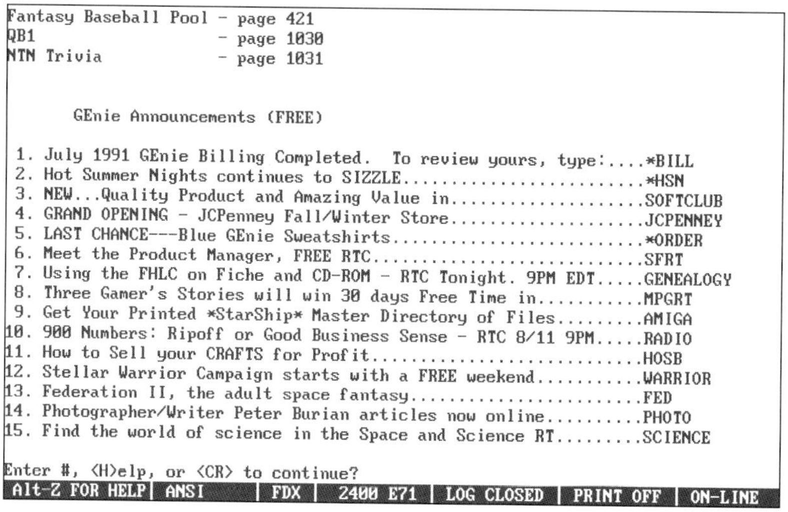

```
Fantasy Baseball Pool - page 421
QB1                     - page 1030
NTN Trivia              - page 1031


        GEnie Announcements (FREE)

 1. July 1991 GEnie Billing Completed.  To review yours, type:....*BILL
 2. Hot Summer Nights continues to SIZZLE........................*HSN
 3. NEW...Quality Product and Amazing Value in..................SOFTCLUB
 4. GRAND OPENING - JCPenney Fall/Winter Store..................JCPENNEY
 5. LAST CHANCE---Blue GEnie Sweatshirts........................*ORDER
 6. Meet the Product Manager, FREE RTC..........................SFRT
 7. Using the FHLC on Fiche and CD-ROM - RTC Tonight. 9PM EDT....GENEALOGY
 8. Three Gamer's Stories will win 30 days Free Time in.........MPGRT
 9. Get Your Printed *StarShip* Master Directory of Files.......AMIGA
10. 900 Numbers: Ripoff or Good Business Sense - RTC 8/11 9PM....RADIO
11. How to Sell your CRAFTS for Profit..........................HOSB
12. Stellar Warrior Campaign starts with a FREE weekend.........WARRIOR
13. Federation II, the adult space fantasy......................FED
14. Photographer/Writer Peter Burian articles now online........PHOTO
15. Find the world of science in the Space and Science RT.......SCIENCE

Enter #, <H>elp, or <CR> to continue?
 Alt-Z FOR HELP| ANSI    |  FDX  |  2400 E71  | LOG CLOSED | PRINT OFF | ON-LINE
```

Fig. 9.4. The GEnie logon screen.

Like CompuServe, GEnie is text-based, so you can access it with any terminal emulator program. Its menu structure is easy to follow and gives GEnie a friendly feeling that is consistent with its somewhat smaller size. In the text-based, full-service arena, GEnie currently is number two and trying hard to draw customers with innovative approaches to pricing and service.

GEnie has taken the lead among full-service utilities by introducing flat-rate pricing. For only $4.95 per month, you can use as many minutes of GEnie's "basic services" as you want. These services are called *Star Services* and are marked clearly with an asterisk (*) on each menu item. Extended services not part of the start services are billed by the minute. With this pricing structure, GEnie is the most economical, full-service system around.

If you need help navigating the GEnie service, you can download Aladdin, a free software package from GEnie. Although Aladdin is not as easy to use as the CompuServe Information Manager, it greatly simplifies many of GEnie's features.

You can set up Aladdin to call the service automatically and go directly to the areas you want. Aladdin also can retrieve messages according to your specifications, saving you a great deal of time—especially on routine activities.

BIX

BIX, a service of McGraw-Hill, is different from other commercial services because its primary focus is conferencing. BIX uses a message conferencing system and offers many technical and entertainment conferences. It also offers file uploads and downloads as well as multiuser chat systems. Figure 9.5 shows the BIX logon screen.

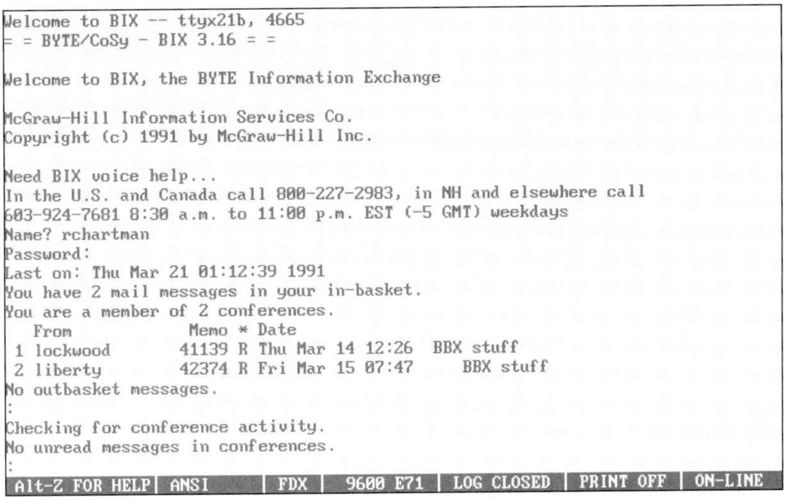

```
Welcome to BIX -- ttyx21b, 4665
= = BYTE/CoSy - BIX 3.16 = =

Welcome to BIX, the BYTE Information Exchange

McGraw-Hill Information Services Co.
Copyright (c) 1991 by McGraw-Hill Inc.

Need BIX voice help...
In the U.S. and Canada call 800-227-2983, in NH and elsewhere call
603-924-7681 8:30 a.m. to 11:00 p.m. EST (-5 GMT) weekdays
Name? rchartman
Password:
Last on: Thu Mar 21 01:12:39 1991
You have 2 mail messages in your in-basket.
You are a member of 2 conferences.
   From         Memo * Date
 1 lockwood      41139 R Thu Mar 14 12:26  BBX stuff
 2 liberty       42374 R Fri Mar 15 07:47     BBX stuff
No outbasket messages.
:
Checking for conference activity.
No unread messages in conferences.
:
Alt-Z FOR HELP| ANSI     | FDX |  9600 E71 | LOG CLOSED | PRINT OFF | ON-LINE
```

Fig. 9.5. The BIX logon screen.

BIX stands for Byte Information Exchange, and this system provides just that—an exchange of information with readers of *Byte* magazine. The system has three major functions: message conferences, file areas (called listings), and real-time chat sections. Many software and hardware vendors have support conferences on BIX, and discussions tend to go into more technical detail here than on other services.

BIX is a gathering place for hardware and software engineers, consultants, and others who are interested in the technical side of the computer industry. BIX users are a close-knit group and have adopted the name BIXEN for themselves. Many computer professionals consider BIX a good place to make contacts and obtain information to further their careers.

Entire conference sections exist on protocol design, programming for most operating systems and in most languages. In these sections you can find the complete specifications documents on such technical topics as how file transfer protocols such as XMODEM and YMODEM operate. This type of technical information is required by programmers to develop communications software. Although such conferences do exist on other services, BIX tends to have more participants who are active at the most technical levels of the industry.

One drawing card for BIX is its relationship to *Byte* magazine. Several *Byte* columnists are active in BIX conferences and are willing to discuss the issues of the day with you. In addition, MicroBytes (a computer industry news service) is updated daily and contains news and features of interest to the computing community. Program listings from past *Byte* magazine articles are available for downloading.

BIX draws a more technical group of users than many other commercial services, making it a good source of technical information. It also has an active real-time chat section similar to the CB services of CompuServe and GEnie, although it is not called a CB section.

BIX uses a UNIX conferencing system called CoSy, which is a command-based system. In many ways this software is reflective of the more technical nature of BIX because it is more technical software. Although a menu-driven "superstructure" has been added to BIX, it is intended for use in its native command-driven mode. This makes learning BIX a bit more complicated, but after you learn the commands and structure you have complete control over the system. With BIX you build a list of the conferences in which you want to participate by using the commands JOIN (to add a conference) and RESIGN (to delete a conference).

The appearance of the system is quite spartan and has a feel much like the UNIX operating system itself. Don't let this simple appearance fool you, however—BIX provides a lot of capability with a very few commands. You soon learn how you quickly can search for and locate information from the large file libraries and message bases. And the fact that BIX is a text-based system means you can access it with any terminal emulation program from any type of computer (see Chapter 16 for discussion on terminal emulators).

Because BIX is a related service of *Byte* Magazine, it offers many of the articles from that magazine on-line—often before they appear in print. Several of the regular authors for the magazine also are active on-line, with some even having their own conferences. *Byte* prides itself on covering all aspects of the computer industry and all operating systems. BIX reflects this wide-ranging technical outlook as well by providing more on-line coverage of computers like the Amiga and Macintosh and operating systems like OS/2 and Unix than most other services.

BIX costs a flat rate of $13 per month. You can call BIX directly at speeds up to 9600 bits per second (bps), but it is long distance outside of New Hampshire. You also can access BIX via BT Tymnet (the same network used for accessing Prodigy in many locations), which has local telephone numbers in many areas of the United States. Tymnet connections, however, have an hourly surcharge.

WELL

No discussion of public on-line services is complete without the WELL. The WELL is an abbreviation for *Whole Earth 'Lectronic Link* and was started in 1985 as an outgrowth of Stuart Brand's "Whole Earth Catalog." The WELL was designed to be an inexpensive on-line meeting place and has become a gathering place for many of the most notable people in the computer industry. Today the WELL is viewed with a near cult-like reverence by many in the on-line world.

The WELL has a definite social and cultural aspect that is quite in keeping with its San Francisco home. Frequent users of the WELL become part of a social group that has much more depth than most on-line services. A feeling exists of being at the source of much of what is happening in the on-line world when you are on the WELL. Users span the spectrum from writers to educators, from artists to programmers, from surviving '60s counter-culture members to businessmen and lawyers.

The WELL operates on a Sequent minicomputer, using a variant of the UNIX operating system. The WELL uses the PicoSpan conferencing system for its message conferences, but the caller also has direct access to many of the UNIX utilities and programs. In fact, regular users of the WELL consider this one of the best parts of the system. A caller can retrieve information on other users at any time and do other utility functions without leaving the thread of the message conference. Learning to operate within the somewhat obscure UNIX environment is considered sort of a "rite of passage" for new users of the WELL.

The WELL also is connected to most large store and forward electronic mail systems such as Internet (see Chapter 15 for a discussion of store and forward mail). This enables any user of the WELL to send electronic mail to any user of a computer in the worldwide UUCP, Internet, or BitNet community. The WELL also carries many of the USENET news groups on-line. Figure 9.6 shows the logon screen for the WELL.

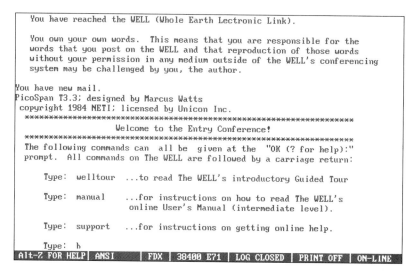

Fig. 9.6. The WELL opening screen.

A subscription to the WELL costs $10.00 per month. In addition, each call has a connect time charge of $2.00/hr at 2400 bps or $6.00/hr at 9600 bps. You also can log onto the WELL via the CompuServe packet network. After you connect to the network, you are prompted for the host name. Enter *WELL* and you connect to the WELL directly. When using the CompuServe network to call the WELL, you have a connect charge of $7.00 per hour.

MCI Mail

MCI Mail is not an information service in the same sense as the other services listed in this chapter. This service focuses on E-mail in all its forms. Through MCI Mail, you can send messages to other MCI Mail subscribers and to fax machines; you even can have messages printed in remote cities and delivered overnight as paper mail.

MCI Mail is the most "connected" commercial on-line service. If you are an MCI Mail subscriber, you can send messages to almost anyone with an electronic mailbox—subscribers to CompuServe; IBM Information Network users (many systems, including PROFS, DISOSS, AS/400, System/36, and System/38, connect to this network); Telex machine users; LAN users with Microsoft Mail or any LAN E-mail system with an MHS gateway; UNIX computer users at an InterNet address; and PC users at a FidoNet address.

MCI Mail has become the predominant on-line service for sending E-mail. Computer companies, journalists, software authors—nearly everyone you need to contact in the computer business has an MCI mailbox. Because this service is strictly for mail, it provides a more comfortable environment for reading mail than some other services.

Many PC productivity packages such as Norton Commander and Lotus Express have added an automatic MCI Mail handler. With many of these programs, you can install a background program on your computer that automatically calls MCI Mail and retrieves any waiting messages or sends responses. This feature can simulate a fully private mailbox on your personal computer for use at your leisure, approaching the feel of paper mail.

MCI Mail has a small annual fee; all other pricing is "by the piece." Like paper mail, you pay only for the messages you send. For heavy MCI Mail users, bulk discount rates are available. MCI Mail does not have telephone connect charges because all connections are made via an 800 number.

MCI Mail enables you to connect to some other services such as the Dow Jones financial service. In these cases, you do have connect time charges added to your bill. MCI also enables companies to create small bulletin boards to handle a group of people. Nevertheless, MCI Mail is primarily for E-mail, and it does this job better than any other service.

Addresses for Services

Addresses for contacting the services discussed in this chapter follow, along with their pricing structure at the time this book was written.

CompuServe
5000 Arlington Center Blvd.
Columbus, OH 43220
(800) 848-8199
Start-up price: $39.95
Monthly minimum: $ 1.50
Price per hour: $12.80

Prodigy
445 Hamilton Ave.
White Plains, NY 10601
(800) 776-3449
Start-up price: $49.95
Monthly rate: $12.95
Price per hour: Free

GEnie
401 N. Washington St.
Rockville, MD 20850
(800) 638-9636
Start-up price: Free
Monthly rate: $4.95
Price per hour: Free Star Services
                                   $18/hour prime time
                                   $6/hour nonprime time

BIX
McGraw Hill, Inc.
1 Phoenix Lane
Peterborough, NH 03458
(800) 227-2983
Start-up price: Free
Monthly rate: $13 (in quarterly increments)
Price per hour: Tymnet surcharge
                                   $6/hour prime time
                                   $3/hour nonprime time
                                   $20/month unlimited nonprime time

The WELL
27 Gate Five Road
Sausalito, CA 94965
(415) 332-4335
Start Up Price: None
Monthly rate: $10.00
Price per hour: $2/hour—2400 bps
                                   $6/hour—9600 bps
                                   $7/hour—CompuServe Packet Network

MCI Mail
1111 19th Street N.W.
Washington, DC 20036
(800) 444-6245

Start-up price:	$35.00	year
Price per message:	$ 0.45	up to 500 characters
	$ 0.75	501 to 2,000 characters
	$ 1.00	2,001 to 7,500 characters
	$ 1.00	each additional 7,500 characters

Chapter Summary

In this chapter, you learned about several commercial on-line services that offer a tremendous resource for information and entertainment. The wide variety of these services means that nearly anything you are interested in is available from one or more of them.

10

Using CompuServe

This chapter acquaints you with one of the richest resources available to users of PCs and modems—the CompuServe Information Service. You discover how to join this service and how to take advantage of some of its most popular features. This chapter covers some of the techniques for navigating a virtual ocean of information and how to keep your costs under control. This chapter also explains how to set your service options; how to use forums—special interest areas where you can download files and leave messages; how to gather information useful to you or your business; how to play games with other users interactively; how to send and receive electronic mail; and even how to reach other on-line services through CompuServe.

The CompuServe Information Service, introduced in 1979, is the public-access branch of CompuServe Incorporated, a leading provider of computer-based information, software, and communications services to business and personal computer owners. Formed in 1969 as a computer time-sharing service, CompuServe Inc. has grown rapidly and now provides services to more than 2,000 corporations and government agencies. H&R Block, Inc., a large financial and tax services corporation, bought CompuServe Inc. in 1980 and has further expanded its markets and services.

The company absorbed The Source, a major competitor, in 1989, dramatically increasing its membership and on-line resources. In 1987, CompuServe went international by signing an agreement with Nissho Iwai Corporation and Fujitsu Limited of Japan to offer a version of the CompuServe

Information Service in that country. It is currently expanding its local access nodes and services in selected European countries. Altogether, CompuServe links you to more than 1,400 databases covering hundreds of subject areas and has more than 800,000 subscribers.

Getting Started

You can log onto CompuServe using any one of the many text-based communications programs available for personal computers. You don't need special or proprietary software, unlike some of the newer, graphics-based on-line services such as Prodigy. The following sections show you how to get signed up and connected to CompuServe for the first time, help you find your way around, get help while on-line, adjust your display options, and log off correctly.

Signing Up

Probably the easiest way to sign up to CompuServe is to call the service's toll-free voice number: 1-800-848-8199. If you have a major credit card, you can sign up immediately; you will receive a special identification number and password that gives you instant access to the service. If you prefer, you can ask for a membership pack that includes the ID and password with a bill for the sign-up fee ($25). You also can get a membership kit through your local computer store.

Many commercial communications packages include a free membership sign-up packet to CompuServe (PROCOMM PLUS, for example), waiving the normal sign-up fee and providing you with a certain amount of charge-free on-line time. The packet contains instructions for connecting to the service, explanations of the major services offered by CompuServe, a list of local access numbers in major cities, and a special identification number and password. When you sign up, you can have CompuServe charge your credit card (MasterCard, VISA, or American Express), or you can have the monthly fee drawn directly from your bank electronically.

Fees

The cost of using the service depends on how much time you spend on-line. You are charged a nominal monthly maintenance fee (currently $2.00), $12.50 per hour of on-line time for basic services, and a small amount ($.35/hour) to reach the service through the CompuServe Packet Net (CPN), an X.25 packet-switched system that links users around the world to computers in two installations in Columbus, Ohio.

Your ID Number and Password

CompuServe, like all bulletin boards and on-line services, must be able to verify who you are when you call in to use the service. This verification is done by presenting you with a user ID prompt followed by a password prompt. CompuServe user IDs are all numbers, usually five digits (starting with 7), then a comma, then up to four more digits (76702,1130, for example). Some European users are given numbers that start with 10 instead of 7. Your ID number identifies you uniquely on CompuServe. You can leave messages in forums or send CompuServe Mail to any user whose ID you know. If you only know another user's first and last name, however, you can find out the ID number by accessing CompuServe's Subscriber Directory, described in the section "Accessing the Subscriber Directory."

When you first sign up, you are given a temporary ID number and password. After you answer the appropriate questions on the initial sign-on screen (see fig. 10.1) during your first session, CompuServe issues you a permanent ID number and a second temporary password that is unique to your new ID number. You then have full access to the system, like any other user. After your billing arrangements have been verified, you receive a letter with your permanent password.

A *password* is a string of characters, usually in the form of an easily remembered phrase (BOAT*SAILOR, for example). Password entry is not case-sensitive. You can enter a password all uppercase letters, all lowercase, or any combination.

CompuServe recommends that you change your password periodically as a security measure. To change your password at any time on CompuServe, all you need to do is type *go password*, and then follow the screen prompts (see fig. 10.2). You are prompted to enter a new password and then confirm it by entering the new password a second time.

```
User ID: 177000,1072
Password:
♦♦
Please Enter:

 Agreement #: FT3372

 Serial #: 11170324

Your package includes a usage credit of $15.00 (US).

As an Executive Service Option member you will be subject to a nominal $10.00
(US) per month minimum usage level starting in second month of your membership.

Your usage credit may be applied to the minimum.  However, you will not be
billed for usage until you have used up this credit.

You must register your name and billing information at this time.

If you subsequently wish to cancel your membership, simply notify us through
online Feedback or by letter.

Throughout this member registration, you will be asked to supply certain
information. Each time you see a colon (:), you are expected to make an entry.
Always follow your entry by pressing
```

Fig. 10.1. *The new user sign-on screen.*

```
!g password

CompuServe(FREE)        PASSWORD


Type your current password:

Now type a new password:

To guard against typing errors,
Please retype your new password:

Password change successful
```

Fig. 10.2. *Changing your password.*

Connecting to CompuServe

Connecting to CompuServe is only a local phone call from most cities in the United States. Your Introductory Membership packet includes a list of phone numbers for some cities, plus a toll-free number you can call to reach a full list of node numbers. To reach this list, set your communications program to use even parity, 7 databits, and 1 stop bit (E,7,1), and dial through your modem to 1-800-423-8011. When you receive a CONNECT message, press Enter. You should see a Host name prompt. Type *phones*, and a menu appears that enables you to find the access number nearest you (see fig. 10.3).

```
  02CBM

  Host Name:   PHONES


  CompuServe                 PHN-1

  Welcome to the CompuServe Phone Number Access area; a free service of the
  CompuServe Information Service. This area gives you access numbers for the
  CompuServe Information Service and allows you to report any problems you may be
  encountering with an access number.

  *** For 9600bps access, CompuServe supports CCITT standard V.32/V.42 only. ***

  Press <CR> for more !

  CompuServe                 PHN-13

    1 Find Access Numbers
    2 Report Access Number Problems

  Enter choice !
```

Fig. 10.3. The Phones access menu.

Navigating CompuServe

The CompuServe Information Service communicates with you by means of a system of menus and prompts. You respond to it with commands that you type in response to the prompts.

Throughout the system, most prompts end with an exclamation point (!). Some end with a colon (:) or with a greater-than sign (>). The context of the menu you are in should make it obvious what sort of response CompuServe expects.

Help is available from most CompuServe prompts by typing *help*; you see a series of screens that explain the available commands. As you move through the screens, the commands are explained in greater detail.

When you need help, opening a log or capture file in your communications program and then displaying the full text of the help screens generally is a good idea. Later, off-line, you can print out and browse the capture file at your leisure. In PROCOMM PLUS, for example, you press Alt-F1 to open a log file.

Table 10.1 is a summary of navigation commands that you can use at the CompuServe ! prompt. You may type these commands at any CompuServe ! prompt to move up or down menu levels, jump to CompuServe's Top menu, display help screens, exit the system, and so on. Most CompuServe commands can be abbreviated to uniqueness. One letter is sufficient in most cases. You can press T, for example, to go directly to the first page, the Top of CompuServe. For more detailed help, type *go instructions*.

Table 10.1
Summary of CompuServe Navigation Commands

Command	Function
T	Moves to Top menu page
M	Moves to previous menu
H or ?	Accesses Help
GO *word* or GO *page*	Goes directly to a service
FIND *topic*	Finds all references to that topic
OFF or BYE	Signs off
S *n*	Scrolls from *n*
R	Resends a page
F	Moves forward a page
B	Moves back a page
N	Displays the next menu item
P	Displays the previous menu item
SET option	Sets terminal option
PER	Exits to Personal File Area
Control Character Commands	
Ctrl-C	Stops the program being used
Ctrl-O	Discontinues the flow of information to your computer or terminal without stopping the current program
Ctrl-S	Suspends output from the host computer
Ctrl-Q	Resumes output at the point where it was interrupted by Ctrl-S
Ctrl-U	Deletes the line you currently are typing

Controlling Your Screen Display

When you initially join CompuServe, you answer a set of questions to let the system know how many lines of text your screen can accept before the text *scrolls* (rolls up off the top of your screen). CompuServe then displays all screens in *pages* consisting of the number of lines you have specified. Each page is followed by a prompt telling you to press a key to display the next page of information. You can change the default number of lines of text you receive before getting a prompt by going to the setup area on CompuServe (type *go terminal*) and following the menu prompts for display settings. See "Configuring Your Default Settings" later in this chapter for details and a sample screen.

You also can tell CompuServe during a given session how many lines to send you by typing the following command at any ! prompt:

SET LINES 24

You can specify any number of lines you want in place of 24.

You occasionally may want to turn off the paged display temporarily, especially if you are capturing Help screens to a log file. By prefixing the number of a menu selection (for displaying information only, not for selecting a service) with *S* and a space, CompuServe scrolls the information on-screen instead of stopping after displaying the number of lines set by your default display option. To select four articles from a menu and have each article scroll without stopping, for example, you can respond to the menu with the following commands:

S 1,3-5

This command displays articles 1, 3, 4, and 5 without pausing.

If you need to interrupt the flow of information coming to your screen, press Ctrl-S. CompuServe responds by halting the flow of characters to your screen. To continue, press Ctrl-Q.

Ctrl-O cancels the flow of information entirely. To receive that information again, you must return to the menu (or command) and request it again. Pressing Ctrl-O returns you to the CompuServe command prompt, where you can enter any valid navigation command.

Taking a Quick Tour of CompuServe

A useful first stop in your explorations of CompuServe is the Tour option. This guided tour touches on many of the most popular services on the system. You are introduced to each service with the option of getting more details in following screens. The Tour option is like a help system to explain available services, similar to the explanations of commands available from the Go Help menus. (You use the abbreviation *g* for the Go command.)

Figure 10.4 shows the beginning of a Tour session.

```
11 Computers/Technology
12 Business/Other Interests

!g tour

CompuServe(FREE)              TOUR

   The CompuServe Guided Tour takes you on a quick trip through the service
giving you an overview of what's available in each of the top menu categories.
After each section you have the option of continuing on the tour or learning
more about that section. Some of the features available to you on the tour
include:

    - Quotes and other Financial Products
    - CompuServe Mail
    - CB Simulator
    - The Electronic MALL
    - Airline Reservations
    - Grolier's Academic American Encyclopedia
    - Games & Entertainment


And many more!!!

MORE !
```

Fig. 10.4. The Tour menu.

The Find command is an easy way to find your way to any particular subject on CompuServe and is easily one of the most useful commands to know. The system gives you fast access to just about any subject you want by typing *find* followed by the subject. You can use any keyword you want with the Find command. You can type *find coins*, for example, to display all areas related to coin collecting. If your search term is not associated with any service areas on CompuServe, you see the message No matches found. Enter topic:. Figure 10.5 provides an example of a Find command.

In addition to the fast and helpful Find command, CompuServe provides several fast File Finder tools (see fig. 10.6). These tools are search utilities for finding files in a predefined group of forums. You can use any keyword or topic for the search; the search checks a keyword list associated with each file in the many file libraries. You also can search for files based on the date of submission, the ID number of the person who submitted it, the file name, the type of file (ASCII, Binary, Image, Mac, or Graph), or the file extension (ASC, TXT, ARC, DAT, or BIN). If you know in which forum the file is located, you can search on any of those search criteria in that particular forum.

```
!find help

CompuServe

    1 Billing Assistance(FREE)         [ QABILL ]
    2 CIM Support Forum(FREE)          [ CIMSUPPORT ]
    3 Command Summary(FREE)            [ COMMAND ]
    4 CompuServe Mail Help(FREE)       [ MAILHELP ]
    5 CompuServe Subject Index(FREE)   [ TOPIC ]
    6 CompuServe Tour(FREE)            [ TOUR ]
    7 Financial Documentation          [ FINHLP ]
    8 Financial Forums                 [ FINFORUM ]
    9 Forum Help Area(FREE)            [ QAFORUM ]
   10 IBM Help Files                   [ IBMHELP ]
   11 Investors Forum                  [ INVFORUM ]
   12 Logon Assistance(FREE)           [ QALOGON ]
   13 Mac CIM Support Forum(FREE)      [ MCIMSUP ]
   14 Mac Developers Forum             [ MACDEV ]
   15 Mac New Users Help Forum         [ MACNEW ]
   16 Member Assistance(FREE)          [ HELP ]
   17 Practice Forum(FREE)             [ PRACTICE ]
   18 Question & Answer(FREE)          [ QUESTIONS ]

!
```

Fig. 10.5. *A Find menu.*

```
!g ibmff

File Finder                    IBMFF

File Finder IBM

    1 About File Finder
    2 Instructions For Searching
    3 How to Locate Keywords

    4 Access File Finder

    5 Your Comments About File Finder

!
```

Fig. 10.6. *The IBMNET File Finder menu.*

You currently can use the File Finder to search for files in the IBM, Macintosh, and Atari forum networks. You enter each of these File Finder areas by typing *go ibmff*, *go macff*, or *go atariff*, respectively. The example used in figure 10.6 searches for files in the IBM network of forums.

If your search does not find any files that meet the criteria you specified, you can start a new search or modify your original search criteria. If any files in the library group match your search criteria, the files are listed with an index number and the forum in which they can be found. Simply type the index number by a file name to see a description of the file and in which library the file is located within that forum. If your search finds more than 19 files, you can display them all, do a new search, or narrow your search by adding to (and therefore qualifying) your original search parameters.

Logging Off

Exiting gracefully from any host system is always a good idea, whether the system is your local BBS or an international host such as CompuServe. The two commands for logging off are Off and Log.

Off results in a full exit from the CompuServe Information System and causes your CompuServe node to drop the connection. Most users issue this command when they are finished with their communications session.

Log logs you out of CompuServe and returns you to CompuServe's login prompt. You should see the User ID: prompt from which you can log back into CompuServe by entering your ID number.

When you use the Log command, you receive a message summarizing your connect time. Many users find this both informative and useful. The Off command disconnects faster, often before the on-line time summary can be displayed.

The distinction between these two logoff commands becomes important if you use a script to automate logging onto CompuServe. If you want your script to take some action at logoff, such as closing the log or capturing files, success depends on your script "seeing" a unique string at logoff. The on-line time summary text works well for this purpose, so you must use Log instead of Off. The section "Scripts" later in this chapter provides additional information about using scripts.

Understanding Available Service Options

The three types of service options on CompuServe consist of free, charge, and extra-charge services. You normally find each service type in any given area. When you reach the Classified Ads area, for example, you can browse the help screens with your on-line charges suspended, you can read ads at the normal connect rate (no extra charge), or you can post or renew an ad for an extra fee.

Most help screens are charge-free. You can spend as much time as you need in these areas, and the only charge is the $.35 per hour network access charge. After you have gained enough experience to try the normal services, including the forums and electronic mail, you are charged the normal on-line rate. You also can optionally use certain special services that are available for an extra charge.

The following is a summary of some of these services. This list includes only the more popular areas, and is by no means meant to be complete. Additional details about some of these services are given later in this chapter.

Free Services

You can tell when you have reached a charge-free area on CompuServe by the (FREE) message that appears after the name of the service when you first enter the area (see fig. 10.7). The following sections describe some of the areas in which the on-line charges clock is turned off.

You can get Help from almost anywhere in the system by typing *help* or *?* at any ! prompt, or */help* at most other prompts. If you type *go help*, you are taken to a set of help screens that explain the CompuServe navigation commands. See the section "Navigating CompuServe" earlier in this chapter for a list of the navigation commands and what they do.

A *forum* on CompuServe is where users with similar interests gather to share information, help one another, exchange tips, upload and download files, and chat with each other in real time. Forums provide message areas, file library areas, and conference rooms. Functionally, forums are bulletin board systems with a focus or purpose. Each forum is managed by a system operator who shares the interests of the forum users and who is not employed by CompuServe.

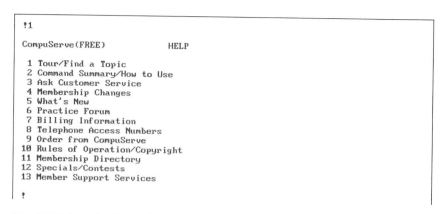

```
!1

CompuServe(FREE)                HELP

 1 Tour/Find a Topic
 2 Command Summary/How to Use
 3 Ask Customer Service
 4 Membership Changes
 5 What's New
 6 Practice Forum
 7 Billing Information
 8 Telephone Access Numbers
 9 Order from CompuServe
10 Rules of Operation/Copyright
11 Membership Directory
12 Specials/Contests
13 Member Support Services

!
```

Fig. 10.7. *Sample charge-free area.*

When you type *go practice* at any CompuServe ! prompt, you reach a practice forum that is like most forums on CompuServe except your on-line time charges are suspended. This forum is just what its name implies: a place to learn how to use forum commands, post and read messages with other users, explore the file libraries, upload and download files, and chat with other users in the conference areas. If you are a new user of CompuServe, this forum is an excellent place to start.

As mentioned earlier, a good starting place on CompuServe is the Tour option, accessed by typing *go tour*. The tour is a series of screens that summarize many of the services available on CompuServe.

Typing *go mall* at the CompuServe ! prompt takes you to the on-line shopping area. Your on-line charges also are suspended while browsing this electronic shopping center. CompuServe's Electronic MALL gives you access to a broad range of vendors and products on-line. You can browse catalogs, search for products by various categories and functions, and order while on-line. Most vendors require that you use a major credit card. Prices generally are competitive and the response time for an order usually is less than 24 hours.

Type *go new* at any ! prompt to get a menu of informational screens updating the system for you. CompuServe periodically adds and removes features and special areas. This area tells you what is current.

Basic Charge Services

You have access to most of CompuServe's resources for $12.50 per hour. The areas described in the sections that follow provide a variety of options.

Electronic mail enables you to send and receive private messages with any other subscriber to the service. All you need to know is the user's ID number. You can even respond to messages in forums and elsewhere via CompuServe Mail. This service enables you to send electronic mail across gateways to MCI Mail, AT&T Mail, Telex, and the vast Internet system. You also can fax a message, send a message as a telegram or even as surface mail to any recipient. The number of mail messages you can send and receive for the basic rate is unlimited. "Using CompuServe Mail" later in this chapter provides additional information about these mail services.

Forums are special-interest areas in which like-minded people gather to share information and ideas. CompuServe offers hundreds of forums with special interests ranging from the American Association for Medical Systems and Informatics (AAMSI) to Working at Home. Most offer message areas, file libraries, and conference areas. "Using CompuServe's Forums" later in this chapter provides additional information about CompuServe's forums.

The CB Simulator enables you to talk "live" to other subscribers on-line. It is structured to operate like a CB radio, with 72 available channels. The channels are public, and the flow of multiple conversations can be somewhat confusing at first, but it doesn't take long to get used to the ambience, and initiating private conversations with other users is easy. The CB Simulator is a good way to meet and get to know other CompuServe subscribers, and it can be great fun. "Using the CB Simulator" later in this chapter provides additional information.

CompuServe provides a broad range of services to help you keep up to date on your favorite subjects, including national and international news services, national and local weather reports, and a variety of sports-related news services, forums, and games. "Obtaining Weather Information" later in this chapter provides additional information about CompuServe's weather service.

Grolier's Academic American Encyclopedia was an extra-fee service until fairly recently. It offers fast keyword searches and full-text retrieval of 33,000 articles, updated quarterly. Considered a general interest encyclopedia, it is suitable for grade school through adult users.

EAASY SABRE provides information on the arrival and departure times for 600 airlines and enables you to book reservations on-line with 300 of them. For American Airlines, this service's sponsor, you also can obtain gate numbers and baggage claim information. You can browse among 43 million possible fares, updated at the rate of over a million a day. To complete your trip planning, you also can access information on 18,000 hotels worldwide and 45 car rental firms. You can ask the system to give you the least expensive flight connections between two points; the system will report the most favorable connection routes and fares for you.

Although this service has no charges other than your CompuServe connect rate, you are required to fill out an application including your address before you gain access. You then assign yourself a password to protect your reservations. The section "Using the Travel Services" later in this chapter provides additional information.

Extra-Charge Services

The following sections describe some of the advanced services on CompuServe that involve extra charges. As you explore the many layers and byways of CompuServe, you will likely find others as well. You can tell that a menu choice will take you to an extra-charge service by the $ symbol that appears after the menu item.

For a minimum of $10 a month, you gain access to the Executive Service option, which includes the following exclusive services: Ticker Retrieval, Disclosure II, Executive News Service, SuperSite, Institutional Broker's Estimate System, Securities Screening, Return Analysis, and Company Screening. In addition, you get volume discounts on the cost of reports from selected financial databases, six months instead of 30 days storage of personal files without extra charges, a 10 percent discount on most CompuServe products, and 50 percent more on-line storage space with options for even more at reduced rates. All fees and purchases of CompuServe products and services apply toward the $10 minimum.

"Using the Executive News Service" later in this chapter provides additional information about the Executive News Service.

IQuest is CompuServe's on-line information retrieval system. This service gives you access to more than 800 databases with sources ranging from the scholarly to the popular, from the mundane to the arcane, updated daily. CompuServe claims this system to be "the most comprehensive source of on-line information anywhere."

The fees for the use of IQuest vary with the type of searches you perform, and whether the searches are successful. Average costs may range from $5 to $25. You have additional charges if you request that printed reports of full-text searches be sent to you.

Money Matters/Markets provides stock market quotes and historical information, banking and brokerage services, information related to taxes and insurance, and a variety of other financial information. You are charged a fee, ranging from ten cents to over a dollar, for each report this service assembles and sends to you. The section "Using the Financial Services" later in this chapter provides additional information.

Obtaining a Summary of Charges

You can get a summary of your on-line charges at any time by typing *go charges* (see fig. 10.8). You can get an explanation of the billing process, your current account balance, a billing history, and reports on your current and past activities on CompuServe. You can even have the output sent to you via surface mail if you want.

```
!g charges


One moment please...



CompuServe (FREE) CHARGES
  1 Explanation
  2 Account Balance
  3 Billing History

USAGE DETAILS
  4 Current activity
  5 Previous activity
  6 Mail Hardcopy ($)
!
```

Fig. 10.8. The Charges menu.

Similarly, type *go billing* to get answers to any questions you may have about billing options or to change the method by which you are billed.

Using CompuServe's Forums

A *forum* on CompuServe is a gathering place for individuals with similar interests and needs, equivalent to a bulletin board service, complete with a message area in which visitors can read messages on various topics, leave responses, ask questions, and generally join in the flow of ideas and opinions. Most forums also include file libraries, in which files can be posted and downloaded by users, and conference areas, in which users can chat with each other live. Every forum has its unique point of view and purpose, and each has a particular culture and style, shaped by the forum administrators, known as *SYSOPs*, and by the users who frequent the forum.

Joining a Forum

When you first visit a forum, your screen fills with news announcements, a description of the forum and guidelines for its use, and other information the SYSOPs think will enhance your experience with the forum. You also receive an invitation to become a member of the forum. Joining a forum is no harder than selecting the Join option from the first menu you see and entering your name (in the case of some forums, your *handle*). Most forums on CompuServe are structured to permit full access to the forum's features only after you have joined. Joining a forum does not obligate you in any way, and it adds no additional charges to your on-line time.

Understanding the Main Menu

The forum menu shown below is for the DATASTORM forum, a typical software vendor's technical support forum in which users of ProComm, PROCOMM PLUS, Hot Wire, and other DATASTORM products meet to ask and answer technical questions. Except for the name in the top line, the main menu for all forums on CompuServe appears as follows:

```
The DATASTORM Forum Menu

 1 INSTRUCTIONS

 2 MESSAGES

 3 LIBRARIES (Files)
```

```
4 CONFERENCING (0 participating)

5 ANNOUNCEMENTS from sysop

6 MEMBER directory

7 OPTIONS for this forum

Enter choice !
```

All areas of the forum can be reached by selecting one of the options from this menu or by entering an equivalent command term at the forum prompt.

Setting Your Options in a Forum

When you join, each forum creates a profile of your defaults for the various forum functions. You can change these defaults easily at any time. Figure 10.9 shows an Options menu.

```
 7 OPTIONS for this forum

Enter choice 7

The DATASTORM Forum Options Menu

FORUM OPTIONS
  1 INITIAL menu/prompt [Forum]
  2 Forum MODE [MENU]

MESSAGES OPTIONS
  3 PAUSE after messages [Always]
  4 NAME [SYSOP Mike Robertson]
  5 Prompt CHARACTER [^H]
  6 EDITOR [EDIT*]
  7 SECTIONS [...]
  8 HIGH msg read [39154]
  9 REPLIES info [Count]
 10 TYPE waiting msgs [NO]
 11 SKIP msgs you posted [NO]

LIBRARY OPTIONS
 12 Library DISPLAY [Long]

Enter choice
```

Fig. 10.9. *A Forum Options menu.*

You can set all options directly from this screen except the Sections option (7), which brings up a second screen (see fig. 10.10).

```
Enter choice 7

The DATASTORM Forum Selection Menu


  0 [*] Sysop Section
  1 [*] Novice Nook
  2 [*] Wishing Well
  3 [*] PROCOMM PLUS 2.0
  4 [*] ProComm
  5 [*] Network Version
  6 [*] HOT WIRE
  7 [*] ASPECT scripts
  8 [*] PROCOMM PLUS 1.x
  9 [*] Hot Topic
 10 [*] Town Square
  N [*] Add new sections

Enter choice
```

Fig. 10.10. *A Forum Selection menu.*

The toggles on the Forum Selection menu are set and unset by pressing the number by each message section; they determine which section's messages will be available to you by default when you enter the message area from the main menu.

Accessing Forum Instructions and Announcements

If you select 1 from the forum main menu, you get a list of available instructions, as shown in figure 10.11.

```
The DATASTORM Forum Instructions Menu

Instructions are available for:
  1 Overview
  2 Messages
  3 Libraries
  4 Conferencing
  5 Announcements
  6 Member directory
  7 Options
  8 Miscellaneous

  9 Complete HELP facility
 10 Forum Reference Card
 11 Forum User's Guide

Enter choice
```

Fig. 10.11. *The Forum Instructions menu.*

Item 5, Announcements, gives you a list of bulletins about each area in the forum (see fig. 10.12). These announcements are individualized for the forum by the SYSOPs, whereas the other instructions are standard text files provided by CompuServe for all forums; they amount to a limited on-line manual to CompuServe.

```
Enter choice 5

The DATASTORM Forum Announcements Menu

  1 News flash
  2 General
  3 Messages
  4 Conference
  5 Library
  6 Membership
  7 Sysop roster

Enter choice
```

Fig. 10.12. *The Forum Announcements menu.*

Accessing the Message Base

From the Forum Messages menu, you can select which messages you want to read by message section (a forum may have as many as 18 message sections divided by topics), by the age of the message, or by the contents of the message's subject field (see fig. 10.13). You also can upload or compose a message from this menu.

```
  1 INSTRUCTIONS

  2 MESSAGES
  3 LIBRARIES (Files)
  4 CONFERENCING (0 participating)

  5 ANNOUNCEMENTS from sysop
  6 MEMBER directory
  7 OPTIONS for this forum

Enter choice 2

The DATASTORM Forum Messages Menu

Message age selection = [New]

  1 SELECT (Read by section and subject)
  2 READ or search messages (16 waiting)

  3 CHANGE age selection

  4 COMPOSE a message
  5 UPLOAD a message

Enter choice
```

Fig. 10.13. *The Forum Messages menu.*

Accessing File Libraries

CompuServe forums contain one or more *file libraries*, logical groupings of files the forum SYSOPs have made available for downloading. Forums may contain up to 18 libraries. Figure 10.14 shows a typical Forum Libraries menu.

```
3 LIBRARIES (Files)
4 CONFERENCING (0 participating)

5 ANNOUNCEMENTS from sysop
6 MEMBER directory
7 OPTIONS for this forum

Enter choice 3


The DATASTORM Forum Libraries Menu

0 Sysop library
1 General Help
2 Comm Utilities
3 PROCOMM PLUS 2.0
4 ProComm
5 Network Version
6 HOT WIRE
7 ASPECT scripts
8 PROCOMM PLUS 1.x
9 Not Used
10 Internal Only

Enter choice !
```

Fig. 10.14. *A Forum Libraries menu.*

When you choose a library, the library menu appears (see fig. 10.15). The Browse option displays files from the library you are in one at a time, and Directory gives you a full listing of the files, one per line.

Whether you see a brief or full description of the file depends on your setting for Library Descriptions in the Forum Options menu. If that option is set to Long, you get a full description of each file. Figure 10.16 shows a typical file description.

The top line is the unique CompuServe ID number of the person who uploaded the file. The next line contains the file name followed by the date the file was uploaded and the number of bytes in the file. The number at the end of the line indicates how many times this file has been downloaded. CompuServe software allows up to six characters plus a three-character extension for file names. The date enables you to select files for display newer than a particular date.

```
7 ASPECT scripts
8 PROCOMM PLUS 1.x
9 Not Used
10 Internal Only

Enter choice !1

Standard user assumed.


The DATASTORM Forum Library 1

General Help

  1 BROWSE Files
  2 DIRECTORY of Files

  3 UPLOAD a File (FREE)
  4 DOWNLOAD a File to your Computer

  5 LIBRARIES


Enter choice !
```

Fig. 10.15. *A Library menu.*

```
[73707,1100]
COMHEL.ARC/binary        24-Feb-89 80641            183

    Title   : Display the contents UART registers
    Keywords: COMMUNICATIONS PORT UART HELP DISPLAY REGISTERS BITS

    A good introduction to the workings of your comm port. Displays the
    contents of each of the UART's registers and allows you to reset any of
    the bits. Includes context-sensitive descriptions of each bit of each
    register, and what they indicate or do. User-provided. May not display
    correctly on systems with EGA card and monochrome displays.




Press <CR> for next or type CHOICES !
```

Fig. 10.16. *File display with full description.*

The Title line is a short description of the file. Only this line displays when you view files with the Library Description option set to Short.

Each file has a set of keywords associated with it that enables you to search the library quickly for files related to a particular topic. The section "Becoming an Expert: Speeding Things Up" later in this chapter provides hints for speeding up your searches.

A description of the contents of the file is next. This description explains the contents of a text file or tells what a program file does when it runs. It also can be a useful source of tips on how to extract additional files from an archived file or what other files you may need to download in order for this one to be useful.

If you press Enter at the final CHOICES ! prompt, you see the next file in the library list, or you are returned to the Library menu. Other possible responses to this prompt include *read* to display the file (if it is a text file); *delete* to erase it from the library or *change* to edit the text of the description (only available if you uploaded it); *display* to view the full description of the file (useful if your display is set to Short); or *download* to download the file. As elsewhere on CompuServe, typing *help* displays the options that are available to you at this point.

Downloading Files

To download a file from any library in a forum, first select the library from the Forum Libraries Menu. If you know the name of the file you want, you can immediately select number 4, Download a File to Your Computer, from that library's menu. You are asked the name of the file to download. Enter the name of the file exactly as it appears when displayed with the Browse command.

If this is the first file transfer you have done during the current session, and if you have not specified a default file transfer protocol to use under the Transfer Protocol option of the Change Permanent Settings menu (see the section "Transfer Protocol/Graphic Support" later in this chapter), you then see a menu of transfer protocol options. A *file transfer protocol* is a program that runs both on CompuServe and in your communications software. A file transfer protocol handles the low-level details of sending and receiving files, checking for errors in the transmission as it goes, and resending blocks of data until the file is transmitted successfully. Be sure to choose transfer protocols on CompuServe and in your program that are compatible (usually they have identical names) to ensure that the transfer will start and complete correctly.

CompuServe gives you the following choices when you choose to download a file:

- **1 XMODEM.** One of the earliest of the error-checking transfer protocols, this is also one of the slowest and least reliable.

- **2 CompuServe B+ and original B.** B+: The protocol of choice when available in your communications program. Fast and reliable, it also give you crash recovery, which means if you lose your connection during a file transfer, you can reconnect and begin downloading where you left off. Also sends the name and size of the file.

Original B: an earlier version, using a smaller block size. Slower and no crash recovery.

- **3 DC2/DC4 (capture).** No error-checking. DC2 (ASCII-18) and DC4 (ASCII-20) are special characters used to start and stop the flow of data between CompuServe and your program. You can use this option to dump the contents of the file to your program, capturing it into a file on your end. Provided for compatibility with dumb terminals and Teletype machines.

- **4 YMODEM.** Added recently to CompuServe, this error-checking protocol is fast and reliable. It sends the name and size of the file to the receiver, but provides no crash recovery.

- **5 CompuServe QB (B w/send ahead).** An earlier version of the B+ protocol without file name or size information or crash recovery. QB and B+ send ahead multiple blocks of data before requiring acknowledgment, which speeds up the transfer. A good choice if your program doesn't support B+.

- **6 Kermit.** This error-checking protocol, found on most computer systems, is known for its wide availability and robustness, but not for its speed. It provides a means to transfer program and other binary files to systems that require the use of parity.

Don't worry about choosing between the CompuServe B, QB, and B+ protocols. If your program supports any one of them, select either one of the CompuServe B options from the list (2 or 5). The protocols, while "handshaking" with each other, determine the level of your program's support and default to the fastest one available.

Joining Conferencess

One of the remarkable features of CompuServe is the capability, in most forums, of carrying on a conversation with a number of other members of the service in *real time*—that is, without the time lag associated with messages left in the message areas. Conference areas, generally called *rooms*, are special areas in which several users can gather to exchange live messages with each other. Forum SYSOPs occasionally may host special conferences featuring one or more special guests; these conferences usually are open to the public and are moderated by the SYSOPs.

You join a conference by selecting the Conferencing option from the forum's main menu or by typing *co* at most forum prompts. Figure 10.17 shows the Conference Rooms menu.

```
The DATASTORM Forum Menu

 1 INSTRUCTIONS

 2 MESSAGES
 3 LIBRARIES (Files)
 4 CONFERENCING (0 participating)

 5 ANNOUNCEMENTS from sysop
 6 MEMBER directory
 7 OPTIONS for this forum

Enter choice 4

The DATASTORM Forum CO Rooms Menu

Conference Rooms Available (# users):
 1 Sysop Conference  (0)
 2 Break Room  (0)
 3 Tech's Lounge  (0)
 4 Test Cube  (0)

Enter choice
```

Fig. 10.17. *The Conference Rooms menu.*

When you first enter a forum, you see a message telling you how many
people are in a particular conference room. You can interact live with
another user from any of the forum prompts using the Send command
followed by a special user number assigned when you log on to the forum.
Just type *ust* at the forum prompt, or */ust* if you already have joined a
conference, to find out who is currently logged on to the forum with you
and get the forum user numbers. Figure 10.18 shows an example of the
display you see.

```
Enter choice ust

User  User ID        Nod  Area     Name
----  -------------  ---  -------  -----------------------
   1  73217,1022     CVI  Lib      Ralph Weber
   2  71157,34       NNH  Lib      Tom Jean
   3  76506,541      GAI  Mes      terry gruver
   4  76702,1130     CBM  Mes      SYSOP Mike Robertso
   5  70233,2726     AUS  Forum    RONALD V HALL
   6  76117,3631     SLD  Mes      Ken Burton
   7  71161,2075     TPK  Lib      Patrick Nichols
   8  76556,2333     ATO  Forum    Bob Grout

Press <CR>
```

Fig. 10.18. *Issuing the Ust command.*

A forum typically has several conference areas. One room is often the main
one, and it already may have several people in it when you join. If you
discover someone you want to chat with already in the forum, you can get
that user's attention by issuing the Send command followed by the user's

forum access number (discovered with the Ust command) and a short message. To continue your chat, you may suggest going to one of the conference rooms, where you can type messages to each other in real time.

As an example, suppose that you enter a forum and type *ust* at the forum prompt. You see the following screen display:

```
User     User ID      Nod     Area      Name

1        73707,1100   CBM     Forum     John Smith

2        76702,1130   CBM     Forum     Mike Robertson
```

Because John Smith is a friend with whom you want to chat, you type the following:

send 1 Hi, John! Meet me in the Break Room if you want to chat.

You then type *co* to go to the Conference area and select Room 2, the Break Room.

You use the following commands while in a conference room. Each command begins with a slash (/).

- **/Name** (followed by any name). Identifies you in the conference. Issue this command immediately upon entering a conference to give yourself a short name or handle.

- **/Ust.** Displays the names of all participants in the conference rooms.

- **/Exit.** Takes you out of a conference, back to the forum prompt.

- **/Send.** Sends a private message to someone in a conference room with you.

Figure 10.19 shows examples of these commands. For a full list of commands that you can use in a conference, refer to "Using the CB Simulator" later in this chapter.

Using the Forum's Membership Directory

Each forum also has a membership directory that contains information about some of the members, including name, ID number, age, and general interests. You are encouraged to add to this directory because it helps participants find other members with similar interests. Figure 10.20 shows the menu for this option.

```
Conference Rooms Available (# users):
 1 Sysop Conference  (0)
 2 Break Room  (0)
 3 Tech's Lounge  (0)
 4 Test Cube  (0)

Enter choice 2

What's your name? Mike

(room) number of users


Entering Break Room room...

/name Wiz
% Your name has been changed
/ust

User  User ID      Nod  Area      Name
----  -----------  ---  --------  --------------------
   4w 76702,1130   CBM  Rm  2     Wiz

/exit
```

Fig. 10.19. Some conference commands.

```
 2 MESSAGES
 3 LIBRARIES (Files)
 4 CONFERENCING (0 participating)

 5 ANNOUNCEMENTS from sysop
 6 MEMBER directory
 7 OPTIONS for this forum

Enter choice 6

The DATASTORM Forum Directory Menu

Your current entry
 1 ADD/modify
 2 LIST
 3 DELETE

Search by
 4 USER ID
 5 NAME
 6 INTEREST

 7 AGE

Enter choice
```

Fig. 10.20. The Forum Directory menu.

Understanding the Role of the SYSOPS

SYSOP is short for system operator. In the context of a forum, the SYSOP is one or more individuals responsible for the daily maintenance and integrity of a forum; some forums have several SYSOPs. The title SYSOP usually is included with the SYSOP's name in forum message headers (From: SYSOP Mike Robertson, for example). Only persons who have been given some level of SYSOP responsibilities are able to use the SYSOP string in their forum names.

You can address a forum message privately to the SYSOP by addressing it **sysop*. Only the primary forum administrator, sometimes known as the *wizop*, can read a message addressed in this way. This person, who is responsible for managing the forum, has the power to flag other users as SYSOPs, giving them the ability to move, delete, and otherwise maintain messages and/or files. A person flagged as a SYSOP also can manage conference sessions. Among the wizop's powers is the ability to lock a user out of the forum if that user proves seriously disruptive. SYSOPs generally read all forum messages daily. If you have any questions about forum use, are having problems in the forum, or just want to drop a note to someone related to the forum's purpose, direct your message to the SYSOP.

Becoming an Expert: Speeding Things Up

Although forum menus are convenient and helpful while you are learning your way around forums, they are relatively slow. You can move from one forum function to another much faster by switching to the Command mode and typing commands directly at any of the forum prompts. In figure 10.21, the command *opt;mod com;s* changes the forum setting from Menu mode to Command mode for the current session only. This command illustrates how commands can be *stacked*, or entered on the same line with a semicolon (;) separating parts of the command.

The next commands you see in figure 10.21 access Library 1 (LIB1) and browse through the files there (bro) searching for all files with the keyword Help. You can abbreviate commands to the fewest number of characters that distinguishes them from similar commands. The CompuServe forum software is fairly forgiving in the use of spaces and punctuation in commands. The Instructions option on the forum main menu gives you a full list of supported commands that you can use to move around the forum quickly.

```
Enter choice opt;mod com;s

Forum lib1


Standard user assumed.

LIB 1 !bro/key:help


[73707,1100]
COMHEL.ARC/binary        24-Feb-89 80641              183

    Title   : Display the contents UART registers
    Keywords: COMMUNICATIONS PORT UART HELP DISPLAY REGISTERS BITS

    A good introduction to the workings of your comm port. Displays the
    contents of each of the UART's registers and allows you to reset any of
    the bits. Includes context-sensitive descriptions of each bit of each
    register, and what they indicate or do. User-provided. May not display
    correctly on systems with EGA card and monochrome displays.
```

Fig. 10.21. *Using fast forum commands.*

Examining Popular CompuServe Services

This section takes a closer look at some of the more popular services available on CompuServe, ranging from ads and games to travel and finance.

Using the CB Simulator

The CB Simulator is a very popular service that enables users to chat with each other in real-time. You have your choice of two "bands," each with 36 channels. When many users are on-line at the same time, they can spread out, reducing crowding and confusion in any given channel. Channels 1, 13, and 33 are adult-only channels. Users can chat privately with one or more CBers at any time by issuing an invitation with the command /invite, followed by the CB user's number.

Figure 10.22 shows the opening CB Simulator menu. New users should select menu options 1 and 2 first. Option 1, Welcome New CB Users, leads to a submenu with two choices: one for users of the CompuServe Information Manager program, and the other for users of any other communications program. Following that, you can choose to see either a summary of commands available to you while in CB, or a full introduction to the service and its commands.

Option 2, Guidelines for Behavior, first reminds you to read CompuServe's operating rules (Go Rules) to clarify any questions about what is generally acceptable on the system, then details the kinds of on-line behavior and courtesies expected of all users while in CB.

```
!g cb

CB SIMULATOR                    CB-10

  1 ** Welcome New CB Users **
        Read This First   (FREE)

  2 Guidelines for Behavior (FREE)

  3 Access CB Band A
  4 Access CB Band B

  5 Special Pricing - The CB Club
  6 Cupcake's CB Society Column
  7 CB Forum
  8 CB Profiles

  !
```

Fig. 10.22. The CB Simulator menu.

The command summary screens list the commands available to you while in one of the CB channels. Each command starts with a slash (/) character (for non-CIM users). Many of the commands available to you elsewhere on CompuServe are available here as well with the slash prefix.

You are encouraged to assume a handle while in CB (unlike most forums, where handles are discouraged or prohibited except in conference areas) by typing the command */handle* followed by a name up to 19 characters long.

The guidelines for behavior are brief and sensible, similar to the guidelines published by CompuServe in its service contract and manual (Go Rules).

Using the Classified Ads Service

Advertising is strictly forbidden in the forums, whether in messages to other users or in files uploaded to the forum libraries or electronic mail. The Classified Ads service, however, provides a convenient way to submit and read advertisements for equipment and services. Figure 10.23 shows the opening menus of the Classified Ads service. Guidelines for ads are similar to those imposed by most newspapers, except that personal ads, ads for firearms, or anything considered offensive to the average subscriber are not allowed (see Go Rules for CompuServe's definition of offensive). If the intention of your ad is to meet other people, the instructions for this area politely point you to the forums.

```
!g classified


One moment please...

CompuServe



Classified Ads

  1 Browse/Read Ads
  2 Age of Ads to Read (currently ALL)
  3 Reply to an Ad

  4 Submit an Ad ($)
  5 Renew Your Ad ($)
  6 Delete Your Ad

  7 Instructions/Fees
  8 Feedback to Classified Ads Manager


Enter choice!
```

Fig. 10.23. *The Classified Ads menu.*

The cost of an advertisement in the Classified Ads service depends on the length and duration of the ad. You are not charged (other than your normal on-line time charge) for browsing the ad headers or reading the ads. Charges apply only for posting or renewing an ad. Fees start at $1.00 per line for seven days.

Ads are grouped by categories. When reading ads, you are prompted to choose a category and subcategories to narrow your search (see fig. 10.24). Then ad headers with subject and location information appear. Each header has a number before it. To view an ad, just enter the number of the ad at the prompt.

A particularly handy feature is the capability of replying to an ad while reading it. Simply type *reply* at the prompt following the ad. Your response, limited to 10 lines, is sent by CompuServe Mail.

```
Enter choice!1

Classified Ads Browse/Read

CATEGORIES   (Personals NOT Accepted)
  1 Occasions/Announcements/Reunion
  2 Business Services/Investments
  3 Travel
  4 Employment/Education
  5 Real Estate
  6 MS-DOS Computers/Software
  7 Apple/Mac Computers/Software
  8 Other Computers/Software
  9 Cars/Boats/Airplanes/RV's
 10 Electronics/Hobbies/Pets
 11 Miscellaneous Merchandise


Enter choice!
```

Fig. 10.24. The Classified Ads Browse/Read menu.

Using the Executive News Service

The Executive News Service, available only to subscribers with the Executive Service option, is a unique news "clipping" service that enables you to collect articles in interest areas you have specified, gathered daily and stored in a "folder" for you to download (or send by E-mail to other users) at your convenience. News sources include the Associated Press, United Press International, the Washington Post, Reuters Financial Report, and OTC NewsAlert, an information service on over-the-counter stocks.

Playing On-Line Games

Given the size and complexity of CompuServe and the fact that the system attracts a large community of professionals, it is not surprising that its on-line games reflect the sophistication and tastes of its participants. In fact, CompuServe contains only one game designed for children, the simple but ever-popular Hangman.

Most of these games are multiple player games that enable you to compete with players from around the world. Players who rank highest in these games are accorded an especially honored *wizard* status, which in some cases comes with special privileges and powers over other players. Type *go games* at any CompuServe prompt to reach the Games area. Figure 10.25 shows the Games main menu.

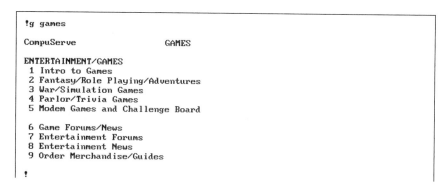

```
!g games

CompuServe                    GAMES

ENTERTAINMENT/GAMES
  1 Intro to Games
  2 Fantasy/Role Playing/Adventures
  3 War/Simulation Games
  4 Parlor/Trivia Games
  5 Modem Games and Challenge Board

  6 Game Forums/News
  7 Entertainment Forums
  8 Entertainment News
  9 Order Merchandise/Guides

!
```

Fig. 10.25. *The on-line Games menu.*

Using the Travel Services

If you travel a lot, CompuServe is an almost indispensable resource. The command Go Travel takes you to the Travel and Leisure area which includes Travelshopper, EAASY SABRE and the Official Airline Guides Electronic Edition, all of which provide airline schedules, seat availability, and reservations; car and hotel information and reservations; and an assortment of other services. You can choose your hotel through the ABC Worldwide Hotel Guide, which includes information on local attractions and other business and leisure travel services.

Using the Financial Services

The Go Money command enables you to access stock market quotes and historical information, banking and brokerage services, information related to taxes and insurance, and a variety of other financial information. You are charged a fee for each report this service assembles and sends you. Figure 10.26 shows the options menu for this area.

The heart of this service is the MicroQuote on-line database system, which provides reports and analyses based on a database of 90,000 securities as well as consensus forecasts of earnings for 3,000 companies. The service includes the following:

- Current quotes (20 minutes old) on 9,000 stocks being traded on the New York, American, Philadelphia, Boston, and National Exchanges

- Commodity pricing information, updated twice daily, along with market analysis reports

- Daily pricing history on 46,000 stocks, bonds, warrants, and mutual funds, along with dividend and interest payment histories

- Company earnings forecasts for 3 to 5 years for 1,800 companies and 80 industry aggregates

When requesting these reports, you have the option to have the output written to a file format compatible with your spreadsheet program. You then can download the file and import it directly to your program for further analysis.

You also can do your trading from your terminal by accessing on-line discount brokerage services.

```
!g money

CompuServe              MONEY

MONEY MATTERS/MARKETS
  1 Market Quotes/Highlights
  2 Company Information
  3 Brokerage Services
  4 Earnings/Economic Projections
  5 Micro Software Interfaces
  6 Personal Finance/Insurance
  7 Financial Forums
  8 MicroQuote II ($)
  9 Business News
 10 Instructions/Fees
 11 Read Before Investing

!
```

Fig. 10.26. The Money Matters/Markets menu.

Obtaining Weather Information

The Go Weather command gives you weather reports updated as many as four times daily for national, state-wide, and local areas. If your software enables you to view CompuServe's GIF (Graphics Interchange Format) files, you can view the same weather maps you see on television weather reports.

Accessing the Subscriber Directory

The command Go Directory enables you to search for the ID number of any other subscriber, based on name, state, or city. You can narrow your search by adding second and third search criteria to your original specification. CompuServe will prompt you for additional search information (first name, state or province, city) if the resulting selection is too large to be displayed. Figure 10.27 shows the Member Directory menu.

```
!g directory

One moment please...

CompuServe(FREE)        DIRECTORY

MEMBER DIRECTORY

 1 Explanation
 2 Member Directory Search
    (U.S. and Canada)
 3 Member Directory Search
    (International)
 4 Include/Exclude This User ID

!
```

Fig. 10.27. The Member Directory menu.

You can limit your search to the United States and Canada or search for a member worldwide. You also have the option of adding or excluding your own name, state, city, and ID number to the search list. After you have found the subscriber you are looking for, you can get in touch easily through CompuServe Mail.

Typing */start* at the search prompt terminates your current search and begins a new one.

Using the Mail Services

With the command Go Mail, you have access to a variety of mail services through CompuServe Mail. You can communicate with other subscribers through electronic mail, binary files, faxes, surface mail, "Congressgrams," as well as other special hard copy mail services. In addition, you can exchange electronic messages with users of several other large communications systems.

Using CompuServe Mail

The workhorse of this system is CompuServe Mail. You can send and receive ASCII messages up to 50,000 characters long, and binary files (using one of the binary file transfer protocols) of up to 512K bytes. You can read ASCII messages while on-line and then store them in your mailbox for up to 90 days. Binary files delivered to your mailbox must be downloaded to be viewed within 30 days. Figure 10.28 shows the main menu for CompuServe Mail.

```
CompuServe Mail   Main Menu

 1 READ mail, 6 messages pending

 2 COMPOSE a new message
 3 UPLOAD a message
 4 USE a file from PER area

 5 ADDRESS Book
 6 SET options

 9 Send a CONGRESSgram ($)
11 OPERATION Friendship
!
```

Fig. 10.28. The CompuServe Mail main menu.

To send a mail message to another subscriber, select the Compose option from the CompuServe Mail main menu. Type in the text of your message to the editor (or dump it to the editor if you have prepared it off-line). When finished, type /*exit* at the beginning of a line. You then are prompted for the ID number of the recipient and the subject of the message.

The method for sending binary files by way of CompuServe Mail is just as easy as sending an ASCII message. Select Upload from the menu. Choose a file transfer protocol from the menu (CompuServe B is the protocol of choice if your software supports it); then follow the protocol's prompts to start the upload.

You are not limited to exchanging messages and files with CompuServe subscribers only. The following sections explain how you can exchange ASCII messages with other systems. Size and storage time limits for CompuServe Mail apply for communicating with these systems as well. You indicate that a message is to be routed to another service by prefixing the address with the greater than (>) character.

Accessing MCI Mail

To reach a user on any electronic mail service, you need to know how to address mail correctly to them. At the CompuServe Mail `Send to (Name or User ID):` prompt, enter the MCI Mail ID in the following format:

>MCIMAIL:123-4567

where *123-4567* is an MCI Mail user address.

The surcharges for this service (at the time of writing of this book) are as follows:

Number of characters	Surcharge
1-500	$0.45
501-7500	1.00
each additional 7500 characters	1.00

Accessing the Internet System

Internet is an electronic mail system linking an international community of governmental offices, research, and academic institutions as well as various military and corporate facilities. Internet is one of the earliest and possibly the largest E-mail system in the world. You can send only ASCII text messages, and the size limits are the same as for CompuServe Mail.

An Internet address via CompuServe Mail is entered as follows at the `Send to (Name or User ID):` prompt:

>INTERNET:Jdoe@abc.michigan-state.edu

You must start the address with >INTERNET: to tell CompuServe mail to route the message to the Internet system. *Jdoe* represents the valid address for this user on the Internet system, @ indicates that the domain address follows, *abc.michigan-state* represents the organization address, and *.edu* is the domain address.

To receive mail from Internet, the sender must send it to you in the following format:

76702.1130@COMPUSERVE.COM

The sender uses your own ID number, of course. Otherwise, the address should be exactly as shown. Note the period (instead of comma) in the CompuServe ID number.

Accessing Telex

You can send and receive messages from Telex I and Telex II machines anywhere in the world through CompuServe. In the U.S., the cost is $1.15 per 300 characters. The rate varies when sending to international locations depending on the location and message length. A list of country codes and rates is available by typing *help telex international*.

The following example illustrates how to send a message to a Telex machine:

```
Send to (Name or User ID): >TLX: 1234567 ABCDEF
```

In this example, *1234567* is the machine number and *ABCDEF* is the answerback.

Accessing AT&T Mail

AT&T Mail is an X.400 E-mail system. Addresses for this system take the following format:

>x400:(c=us;a=attmail;s=SURNAME;g=GIVEN;d=id:UNIQUE ID)

The uppercase letters indicate user-specific variables that you need to supply in order to send the message. For example, if an AT&T user has the address "Surname of JONES, Given name of BOB, and AT&T ID of BJONES," you enter the following address in CompuServe Mail:

>x400:(c=us;a=attmail;s=jones;g=bob;d=id:bjones)

>x400: must always precede the address, the address must be enclosed in parentheses, and the elements must be separated by a semicolon (;).

For you to receive from AT&T Mail, the sender must address the message in one of the following ways:

> To: mhs/c=us/ad=compuserve/pd=csmail/d.id=76702.1130
>
> or
>
> To: mhs!csmail!76702.1130

As in all messages routed across E-mail gateways, the comma in your ID number must be entered as a period.

Sending Postal Letters

You can send laser-printed surface mail to anyone who can be reached by a national postal service. Just compose your message as you normally would for CompuServe Mail, making sure that your lines do not exceed 60 characters in width and that the message is no longer than 279 lines. If you type /compose/postal at the beginning of your message, the CompuServe Mail editor checks this format for you as you enter the message. When you have finished the letter, type /type/postal at the beginning of a line to see it formatted the way it is to be printed. At the Send To: prompt, enter >postal, and the system prompts you for addresses.

Your message is sent to a print site nearest its destination, where the message is laser-printed and delivered to the post office. Print sites include Los Angeles, Chicago, Atlanta, New York, Miami, San Francisco, and Washington, DC. The cost for sending domestic mail this way is $1.50 for the first page and $0.20 for each additional page.

Sending Faxes

You can send messages to any Group 3 fax machine in the world that can be reached by direct dialing. To send a fax, just compose or upload a message, making sure that lines do not exceed 80 characters and that the message length is not more than 50,000 characters (about 1000 lines). At the Send To: prompt, enter >FAX:fax address (for example, >FAX:18005551234).

The system then prompts you to enter the Attention Line information. You receive a confirmation through CompuServe Mail after the fax has been delivered; if it is undeliverable, it is returned to you. Faxes sent within the U.S. cost $.75 for the first 1,000 characters and $.25 per 1,000 characters after that.

Configuring Your Default Settings

When you join CompuServe for the first time, you are asked a series of questions on-line about your display preferences (number of lines and columns on your screen, and so on) and other questions about your terminal. Your responses to these questions determine your *profile*, or default settings for future logons. These settings are stored on CompuServe as profiles specific to each baud rate with which to connect to the system. You can change these profiles at any time by typing *go profile*, *go default*, or *go terminal* from any CompuServe ! prompt. Your on-line charges are suspended whenever you access these areas.

Type *go default* to access the Terminal/Service Options area. As figure 10.29 illustrates, you can decide whether to change your permanent settings or make changes for your current session only. If you choose to change settings for your current session, CompuServe reads and uses your permanent session defaults the next time you log on. This feature is especially useful if you use different PCs or terminals to reach the service and each requires different display options.

```
TERMINAL/SERVICE OPTIONS (FREE)

Use this area to change your terminal type/parameters and/or service options.

Note: Your permanent and session
settings match.

  1 Instructions
  2 Change permanent settings

  3 Explanation of session vs. permanent
  4 Show session vs. permanent
  5 Change current session settings




Enter choice !
```

Fig. 10.29. The global default menu.

To change your permanent display settings, select option 2, Change Permanent Settings, from the Terminal/Service Options menu. The Permanent Settings menu appears (see fig. 10.30).

```
PERMANENT SETTINGS

 1 Explanation
 2 Logon/Service options
 3 Display options
 4 Terminal type/parameters
 5 Transfer protocol/graphic support
 6 Make session settings permanent

Type EXIT when done




Enter choice !
```

Fig. 10.30. The Permanent Settings menu.

You can select the following options from the Permanent Settings menu:

- **Logon/Service Options.** Displays the Logon/Service Options menu shown in figure 10.31. The first two options on this menu determine what CompuServe displays when you log on each session. Option 3 enables you to create a personal menu. The last three options in this menu enable you to determine what happens when you type *top* to go to the Top menu, which editor to enter when you compose a message, and whether to display menus or single-word prompts when you enter a forum.

  You can select option 4, TOP Goes To, to display the normal Top menu, execute any other Go command you choose (to go immediately to an area of your choice), display a personal menu of your own creation, or go to the personal file area in Command mode.

  With the Online Editor option, you can choose between two editors on CompuServe: EDIT and LINEDIT. LINEDIT displays numbers before each line, whereas EDIT does not.

The Forum Presentation Mode option on the Service Options menu determines whether you are set to Menu mode or Command mode when you enter a forum for the first time. After you enter a forum, you can set this option and your choice of editors and save your choices as defaults for that forum.

- **Display Options.** This option on the Permanent Settings menu displays the Perm Display Options menu shown in figure 10.32. This menu enables you to tell CompuServe whether to use PAGED display (pause when the screen is filled), whether to display prompts in brief or full form, whether to clear the screen between pages, and whether to display carriage return and line feed characters.

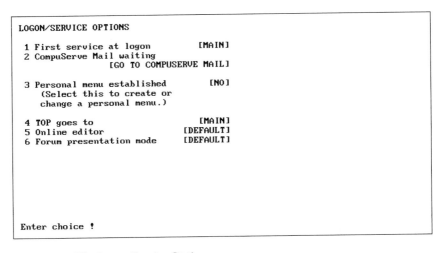

```
LOGON/SERVICE OPTIONS

1 First service at logon         [MAIN]
2 CompuServe Mail waiting
                    [GO TO COMPUSERVE MAIL]

3 Personal menu established      [NO]
  (Select this to create or
  change a personal menu.)

4 TOP goes to                    [MAIN]
5 Online editor                  [DEFAULT]
6 Forum presentation mode        [DEFAULT]




Enter choice !
```

Fig. 10.31. The Logon/Service Options menu.

- **Terminal Type/Parameters.** This option on the Permanent Settings menu displays the Terminal Type/Parameters menu (see fig. 10.33). Through the options on this menu, you tell CompuServe what kind of terminal you are using or emulating in your communications software. This information affects which, if any, escape sequences CompuServe sends your program to control the cursor position and other display characteristics. CompuServe supports VIDTEX (a graphics display compatible with Navigator and certain Microsoft Windows-based communications programs such as WinComm), ANSI, VT100 and VT52, Heath/Zenith terminals, ADM, CRT, and Other, a simple terminal option with no screen display codes.

```
PERM DISPLAY OPTIONS

  1 PAGED display                     [YES]
  2 BRIEF prompts                     [NO]
  3 CLEAR screen between pages        [YES]
  4 BLANK lines sent                  [YES]
  5 Line feeds sent                   [YES]




  Enter choice !
```

Fig. 10.32. The Display options submenu.

```
TERMINAL TYPE/PARAMETERS

   1 TERMINAL type                   [VT100]
   2 Screen WIDTH                      [80]
   3 LINES per page                    [24]
   4 Form FEEDS                  [SIMULATED]
   5 Horizontal TABS             [SIMULATED]
   6 Chars. received (CASE)           [U/L]
   7 Chars. sent in CAPS              [NO]
   8 PARITY                          [EVEN]
   9 Output DELAYS                     [0]
  10 ERASE when backspacing          [YES]
  11 Micro inquiry sequence at logon [NO]




  Enter choice !
```

Fig. 10.33. The Terminal Type/Parameters submenu.

You also can specify your default screen width and the number of lines of text your screen display supports, what to display for form feed and tab characters, and whether to send text to you in all uppercase. You also can specify the parity setting (CompuServe defaults to even parity), how much, if any, delay to add between characters, and whether to cause a character to disappear on-screen when you press the Backspace key.

- **Transfer Protocol/Graphic Support.** This option on the Permanent Settings menu displays the File Transfer/Graphics menu shown in figure 10.34. This menu enables you to tell CompuServe whether to default to a file transfer protocol when you start a download or prompt you with a menu. You also can specify which, if any, graphics standard your communications program supports. These specifications are necessary to take advantage of some graphics features on CompuServe, such as the national weather maps. "Accessing File Libraries" earlier in this chapter describes the file-transfer protocols supported by CompuServe.

```
FILE TRANSFER/GRAPHICS

FILE TRANSFER PROTOCOL
  1 PROTOCOL preference          [SHOW MENU]

GRAPHICS SUPPORT
  2 GIF SUPPORT                       [NO]
  3 NAPLPS SUPPORT                    [NO]
  4 RLE SUPPORT                       [NO]
(Note: Please consult your software and hardware documentation for graphics
support before changing these settings.)




Enter choice !
```

Fig. 10.34. The File Transfer Protocol submenu.

- **Make Session Settings Permanent.** This menu tells CompuServe that you want the current terminal/services settings to be used each time you log on to CompuServe.

Micro-Enquiry at Logon

The term *micro-enquiry* refers to a control character (hex 05) sent by CompuServe after you enter your password. Some terminals and terminal programs are coded to respond to this character with a string of characters identifying the terminal to CompuServe. You can instruct CompuServe, in the Change Permanent Settings menu, to send or not send this character. Because this character also identifies the beginning of a CompuServe B file-transfer protocol session, certain communications programs (PROCOMM PLUS, for example) can be configured to respond automatically to that character by opening a B protocol window (starting the protocol program). If you have such a program, you may want to set this option to not send the micro-enquiry character at logon.

Using Navigation Programs

A *navigation* program automates your access to a service. Executed properly, it can minimize the time you have to spend connected to the service. Because CompuServe (unlike some other public information services) charges by the amount of time you spend on-line, programmers have been stimulated to create programs of this type.

Those who use navigation programs commonly admit that they rarely reduce their on-line time charges; instead, they find themselves using the system much more than they could afford to otherwise. The following sections describe some of the more popular programs available for automating your CompuServe sessions.

TAPCIS

TAPCIS, originally written by the late Howard Benner, is a shareware program you can download from the TAPCIS Forum, where it also is supported by the current programmers. TAPCIS enables you to join forums automatically, configures your defaults for you, and captures E-mail and new messages in forums and forum sections of your choice. After you read and reply to messages off-line, TAPCIS uploads all your responses in a batch. You also can browse a forum's file libraries off-line, select files to download, and then capture them all at once. TAPCIS automates file uploads to CompuServe Mail, including the correct addressing for mail to MCI Mail, Internet, and Telex systems. TAPCIS is available for IBM-PC compatible users only.

AutoSig (ATO)

Written and supported by the SYSOPs of the IBMNET forums (a group of forums dedicated to users of IBM-PCs and compatibles), this program automates CompuServe access as TAPCIS does. AutoSig is freeware, however, with no registration fee required. This program can be found in the IBM Communications Forum, which you can access by typing *Go IBMCom*.

Navigator

Navigator is a commercial navigation program written for Macintosh users by CompuServe. It is supported by the CompuServe support staff and the Macintosh Communications Forum. You can access Navigator by entering *Go MacCom*.

Scripts

If you have a full-featured communications program (such as PROCOMM PLUS), you can create a *script*, or command file, for your program to automate part or all of your CompuServe on-line sessions. Creating a script usually requires some programming skills and some knowledge of how CompuServe works. If you can get technical support for your communications program on-line (through the DATASTORM Forum for ProComm and PROCOMM PLUS, for example) you probably can find help writing this kind of program and possibly some example scripts to help you get started. Some communications programs enable you to record an on-line session, automatically creating a session script.

The CompuServe Information Manager (CIM)

The CompuServe Information Manager is a graphics-based program designed to simplify the CompuServe interface and automate some common on-line activities. Versions are available for both IBM-PC compatibles and for Macintosh computers.

Using Gateways to Other Services

As vast and rich a resource as CompuServe Information Service is, it is only one of many on-line services you can reach by calling a local CompuServe node number with your modem. If you press Enter when your modem links with a CompuServe modem, you see the following prompt:

```
Host name:
```

You have reached an extensive data network called the CompuServe Packet Net (CPN). If you type *cis* at the prompt, you get the CompuServe Information Service's User ID prompt. The following sections describe some other services you can reach from the Host name prompt.

WELL—Whole Earth 'Lectronic Link.

This conferencing and E-mail system is quite small compared to CompuServe (about 5,000 subscribers). Nonetheless, its population of users seems disproportionately savvy and sophisticated. Described as "a lively outpost of the electronic community," this is where Mitch Kapor's Electronic Frontier Foundation—an organization dedicated to preserving the right and liberties of the community of on-line users—evolved. After you log on to WELL, you have access to mail via the worldwide *uucp* and *usenet* systems as well as Internet. (WELL is discussed more in depth in Chapter 9.)

PCMagnet

PC Magazine hosts an on-line service, called PCMagnet, with an interface similar to the CompuServe Information Service. You can communicate with many of the magazine's editors and download the various program and source files that appear in the magazine.

Chapter Summary

This chapter introduced you to the CompuServe Information Service, one of the largest and most resource-rich of the public-access information systems. Because of its size and complexity, the thought of joining and exploring this system sometimes has overwhelmed and discouraged newer users of PC-modem communications. This chapter discussed how to sign up to the service and get on-line, how to find your way around and get on-line help when you need it, and how to log off correctly. You explored some of the methods to customize your display while on CompuServe, how to get quick listings of available resources based on keywords, how to find files you need, and how to practice your on-line skills without accumulating charges. This chapter looked at some of the most popular services on CompuServe, including electronic mail, the forums, CB Simulator, news, weather, sports information, on-line databases, financial reports, and EAASY SABRE, the airline reservations system. Finally, this chapter covered some of the programs available to simplify and amplify your CompuServe access while reducing on-line costs.

11

Using Prodigy

The PRODIGY Interactive Personal Service, better known as just Prodigy, is a relative newcomer to public-access network services. Started in 1990 as a partnership venture between Sears and IBM, Prodigy has grown very fast, both in numbers of users and in services offered. This chapter is an introduction to this service and its major features. Items discussed include the Prodigy software and its installation—required in order to use the service; some of the more popular features of the service; how to get around the system; the function key commands provided by the software; how to customize and streamline your on-line sessions; how to print information from the service; and how to get customer support if you need it.

Prodigy's screens consist entirely of graphics instead of text, as found on CompuServe and most bulletin boards. To use the service, you must purchase Prodigy's communications software, designed to connect to Prodigy and download screen information, then interpret that data to produce screen displays. This process occurs continually while you are on-line, with new screen information frequently downloading to your computer's memory while you explore and use the system. The Prodigy start-up kit consists of the software and manuals needed to connect to the service and get around in it. The software is currently available for IBM-PC compatibles and for the Apple Macintosh.

Prodigy is marketed primarily for the home PC market and for business users who are less experienced or more casual in their use of information services. The graphics are bright and attractive (on a color monitor), and navigation information usually is marked clearly, regardless of what part of the service you are using.

Getting Started

Prodigy's graphic screens make directions easy to read and make good use of color to divide the screen. Figure 11.1, for example, shows Prodigy's opening screen. Approximately 40 characters are displayed per line of text, as in electronic mail. Because graphics take longer to display than ordinary text, almost everything on Prodigy is displayed more slowly than, for example, on CompuServe. This is offset partially by the fact that most graphic screens are created by executing program code already resident on your computer while updated portions of the screen are being downloaded to it. In other words, the screens you see are produced by the Prodigy software on your computer, reading and interpreting the data sent to it to update parts of your screen display without having to re-create it entirely. This results in screen changes that happen fast enough to be acceptable for most purposes.

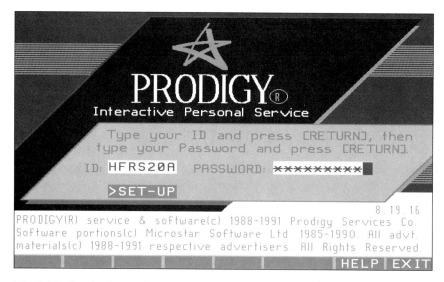

Fig. 11.1. Prodigy's opening screen.

Because Prodigy is a combination of the on-line service and your Prodigy software, getting started means installing the files from the Prodigy disks and going on-line for the first time.

Software Installation Guide

The Prodigy handbook provides you with detailed installation help. First it reviews the equipment you need: computer, disk drives, DOS operating system, monitor and graphics adapter, modem, communications port, modular phone jack and phone line, and optionally a printer and mouse. Then, as most software installation instructions do, the manual divides the installation details into those for floppy-disk only users, and those for hard drive installation. You can install for any graphics display card and monitor, from CGA (320$5$200) up to VGA (640$5$400). High-resolution graphics drivers produce sharper screens and more readable text, but take a little longer to display. To speed up performance, the installation program gives you the option to install at a lower resolution than your system supports. Because Prodigy provides a variety of colorful screens, a color monitor is recommended.

The Prodigy start-up kit comes in a compact fold-out box set with disks, installation instructions, and a phone book on one side, and the handbook and Jumpword listing on the other. (*Jumpword* is Prodigy's term for a word or phrase you type or select, which jumps you to a different part of the service. A jumpword works like CompuServe's Go command, which is covered in Chapter 10.) Between the two sides is a tear-sheet envelope with your identification number and temporary password. The package comes with two 5.25" disks and one 3.5" disk.

The Prodigy manual is small (5.5" × 5.5"), with flip-up pages and a handy function-key template attached to the front cover. It contains an installation guide, a tutorial, a set of simple example uses of the service, a troubleshooting guide, and an explanation of how Prodigy accomplishes some of what it does.

When you first run Prodigy, the installation program scans your PC for hardware information. The program finds out how many serial ports and printer ports you have and where they're located, what kind of video display adapter is in your PC, and other information (see fig. 11.2). You tell the program whether your modem uses pulse dial or tone dial, and you can adjust the level of video resolution from the default. (See the section "Change Display" near the end of this chapter for examples of video display resolution settings.)

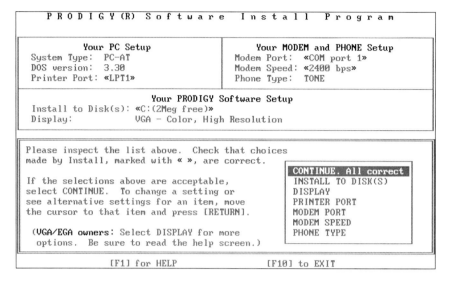

Fig. 11.2. Prodigy Install setup screen.

The Highlights screen

After the Prodigy opening screen, at which you may be prompted to enter your ID number and password (see "Autologon" later in this chapter for a way around this), the Highlights screen is the first screen you see when you go on-line (see fig. 11.3). New items of interest are mentioned on this screen. If you have electronic mail waiting, you see a `New Mail` message at the upper left of this screen when you log on.

Prodigy Software Files

This section discusses several special files that are included with the software. Two of these files—MODEMSTR.EXE and GETSCRN.EXE—are stand-alone programs you run from DOS to customize commands for your modem and to change your screen display, respectively. The other files contain screen display data (CACHE.DAT and STAGE.DAT) and provide configuration information to the program (CONFIG.SM and MODEMS.TXT). Although you normally would not change any of the data in these files outside of the Prodigy program, some understanding of these files and what they do helps dispel some of the mystery that, for some, surrounds this software and the Prodigy system.

Fig. 11.3. *The Highlights screen.*

CACHE.DAT

This file contains screen display data constantly updated while you are on-line to Prodigy. This data is interpreted by the software on your computer to produce the screens you see. As you move about the Prodigy system, new screen information is downloaded to this file and to the STAGE.DAT file.

STAGE.DAT

This file also contains Prodigy code, interpreted by your main Prodigy program, to produce the various screen displays, similar to CACHE.DAT. If Prodigy updates or changes its services, new code is downloaded automatically to this file while you're on-line. The only thing you see during this process is some delays or pauses in your screen displays.

In 1990, not long after Prodigy launched a massive subscription campaign, the updating of the STAGE.DAT file caused a brief storm of concern among regular users of modem communications software unfamiliar with Prodigy's screen update methods. On CompuServe and elsewhere, some Prodigy subscribers who had examined the contents of this file found what appeared to be personal information in the form of file names, database entries, and other information unique to them. Because of the appearance of frequent exchanges of information between the Prodigy service and their local drives, these users believed Prodigy might be uploading the contents of this file and gathering marketing information. Actually, STAGE.DAT allocates up to nearly a megabyte of disk space for screen data storage. This allocation results in a large "empty" file, which contains random data left on your hard drive when other files are deleted. Nevertheless, fears and rumors about this file persisted until Prodigy officials made public statements disclaiming any such hidden actions.

CONFIG.SM

This configuration file contains the data your Prodigy software uses to access your modem and printer and determine the correct network code needed to reach the service, for both your primary and alternate phone numbers. This file is created by the Prodigy install program, and information is added to the file when you first go on-line to Prodigy.

MODEMS.TXT

This file is Prodigy's modem data file. When you first go on-line to the service, the software queries your modem for its manufacturer ID and checksum. The program compares the answer sent back from your modem to the entries in the MODEMS.TXT file. If it matches, the entry is added to your CONFIG.SM file as the `modem-str:` command, used to initialize your modem each time you call the service.

MODEMSTR.EXE

If you want to customize the commands your Prodigy software sends to your modem before dialing, use the MODEMSTR.EXE program. This program adds modem commands to the end of the last line of your CONFIG.SM file.

For example, I needed to disable error-correction and data compression features in my modem to successfully connect to Prodigy. I also wanted to add a command to silence the modem's speaker (turned on by default) during late-night, on-line sessions. With teenagers who unexpectedly pick up the phone, I also wanted to instruct my modem to wait thirty seconds before dropping the line if a carrier was lost. The necessary commands I needed were: AT (the modem Attention command), followed by S10=30 (wait 30 seconds before disconnecting if carrier drops), M0 (disable modem's speaker), and &Q0 (disable error correction/data compression).

In the "Advanced Topics" section (under "Modifying the Modem String") of the Prodigy handbook, you are advised to append new modem commands at the end of the `modem-str:` command in the CONFIG.SM file—already quite a long string of characters—in the following form:

:ATS10=30M0&Q0

Although the handbook suggests using the MODEMSTR program to add commands, it implies that you can add those commands manually with a text editor. The text editor adds a carriage return to the command string when the `modem-str:` command exceeds 256 characters, however, so the added commands are ignored. Because the CONFIG.SM file is created during installation and the modem command line is created by going on-line for the first time, the software has to be reinstalled to correct this problem. Adding the additional modem commands with the MODEMSTR program then works correctly. If you choose Setup on the Prodigy opening screen and change one of your phone number entries, however, the software creates a new CONFIG.SM file, deleting any modifications you make with MODEMSTR. To prevent problems, back up the CONFIG.SM file after you change it.

GETSCRN.EXE

With this file on your installation disk, you can change your video display mode. On-line, you can enter the jumpword *Change Display* to get an analysis of your hardware and advice on using the GETSCRN program. To change your display driver for Prodigy, you must put the installation disk in drive A and type *A:* to log on to that drive. Then type *getscrn screen.cmp* followed by the particular screen driver you want to install, followed by the destination (for example: *getscrn screen.cmp ega640.scr c:\prodigy\driver.scr*).

Going On-Line for the First Time

Going on-line for the first time actually is a seamless part of the installation and start-up process. When you are on-line, Prodigy walks you through a series of prompts. Even though Prodigy already gathered your name, address, and phone number when you subscribed, you cannot progress past this screen without providing most of this information (see fig. 11.4). Next you are asked your sex and age, and the names of up to five other people who will use the service. These people are assigned I.D.'s similar to yours, and will be prompted to give themselves unique passwords when they sign on.

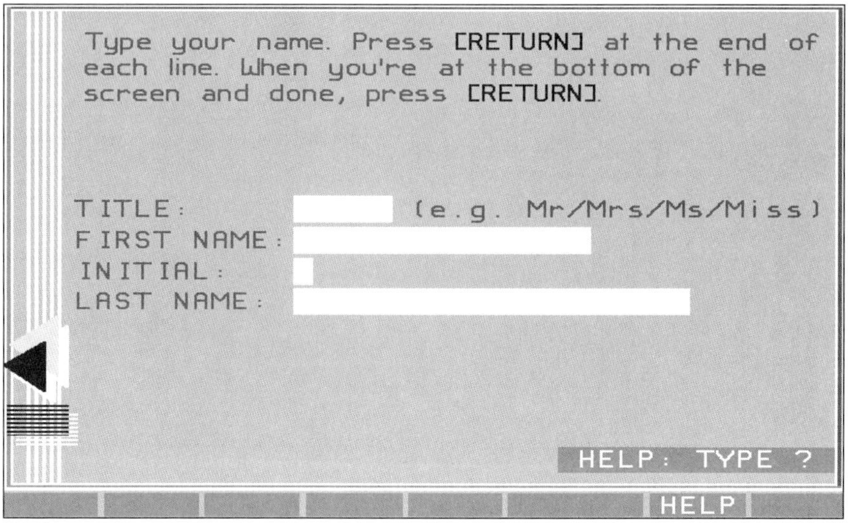

```
Type your name. Press [RETURN] at the end of
each line. When you're at the bottom of the
screen and done, press [RETURN].




TITLE:                      (e.g. Mr/Mrs/Ms/Miss)
FIRST NAME:
INITIAL:
LAST NAME:





                                        HELP:  TYPE  ?
                                         HELP
```

Fig. 11.4. Screen for supplying personal information on-line.

You also must enter primary and alternate phone numbers of the nearest Prodigy modems, provided in the Phone Book that comes with the start-up kit (see fig. 11.5.). The Phone Book also tells you how to disable call-waiting. Prodigy supports 1200 and 2400 baud modems at the time of this writing. All states except Alaska have nodes in a variety of cities. Alaskans have to call a node in the continental U.S. to reach the service. The Prodigy modems tested did not support error correction or data compression.

The last stop of your first tour of the service is the Prodigy Service Agreement. Figure 11.6 shows the opening screen of the agreement, which you can read on-line. You must choose whether or not to agree to its terms. If you choose not to agree during your initial session, you see a toll-free number to call with your questions, you are encouraged to call back and join later, and then you are disconnected. The service agreement contains nothing unusual or unexpected, but users familiar with other on-line services have been surprised by Prodigy's practice of reviewing all messages that you post before making them public.

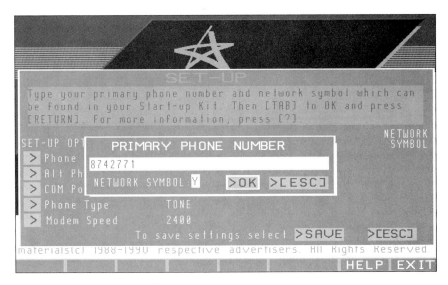

Fig. 11.5. The Phone Number entry screen.

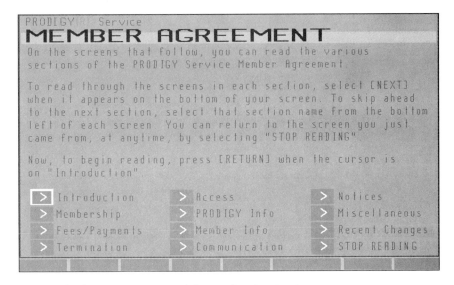

Fig. 11.6. The opening screen of the Prodigy Service Agreement.

Finding On-Line Help

Help information specific to whatever part of the service you are in is generally no farther away than the F1 function key on your IBM-PC compatible keyboard. If you use a mouse, you can double-click the Help option on the command bar—the series of command options available at the bottom of your screen. Another option is to highlight any choice field on a screen and type a question mark (?) to display help for that choice. Help screens are identified by a large question mark in the upper left corner of the screen. Because the commands on the command bar change as you move around in Prodigy, figure 11.7 shows a sample help screen and command bar. You also can press J and Enter, or double-click the JUMP option on the command bar to bring up the Jumpwindow, a special input field that enables you to enter a command word to take you to any other service on Prodigy. Next, type *help hub* to reach a menu of general Help resources for Prodigy.

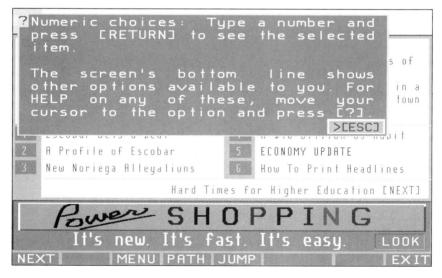

Fig. 11.7. *A command bar and Help screen.*

Exiting

You can exit from most screens by pressing E to highlight the Exit option on the command bar and pressing Enter. You then are given the option to close the Prodigy program and return to the operating system, or log back on. Any of the Prodigy users you specified during your first session then may log on. Figure 11.8 shows the Prodigy exit screen.

Fig. 11.8. Prodigy's exit screen.

Navigation Basics

Prodigy provides several ways to navigate, both within a given screen using your keyboard's cursor keys, tab, and certain letter and number keys, and between Prodigy's various services, using the options provided on the command bar. In both cases, you can use your keyboard or your mouse to get around. All references to highlighting an option (reversing its colors) with your cursor keys and selecting it by pressing the Enter key apply equally to using a mouse to highlight and clicking on the left mouse key.

The following two tables explain your keyboard options and the command bar options you can use to navigate the screens and the services. Table 11.1 shows the keys you use to highlight any of the various options on a particular screen. Table 11.2 refers to commands that most often move you from one screen to another.

Table 11.1
Navigating the Prodigy Screens

Key	Function
Tab	Moves through options on the screen from top to bottom and left to right. Each time an option is chosen, it is highlighted.
Shift-Tab	Moves backwards through the options
Arrow Keys	Moves to option corresponding to direction of chosen arrow key
Home	Highlights the first or top option on the screen
End	Highlights the last option on the screen
Letter Keys	Highlights available options represented by the first letter of the option (press N for Next, for example)
Number Keys	Selects some options represented by a number

Table 11.2
Navigating the Prodigy Services

Option	Function
Next	Displays next screen of text message when [NEXT] prompt at the end of text or the left of command bar shows message exceeds one screen
Back	Displays preceding message screen when [BACK] prompt shows preceding message exists
Look	Provides more information about the advertisement currently displayed at the bottom of your screen
Zip	From any ad screen, returns you to screen you were in when you chose the Look option

Complementing the Look option, Zip is a fast way to return from any level in an ad display. Zip is not on the command bar, but you can activate it by pressing Z whenever you are looking at an ad. The Jumpwindow appears, and the Zip option is highlighted (see fig. 11.9). Press Enter or click on the option to return to whatever screen you were in when you selected Look.

NOTE

Fig. 11.9. The Zip option in the Jumpwindow.

Changepath: Changing Your Pathlist

Prodigy provides a unique way to create a *path* through the various services each time you go on-line. You can change this path with the *Changepath* jumpword, or view the current path (and select any option on it) by entering the *Viewpath* jumpword. To use the Path option, press P at any screen with Path in the command bar. Then press Enter or click the highlighted Path option. Figure 11.10 shows the Changepath screen.

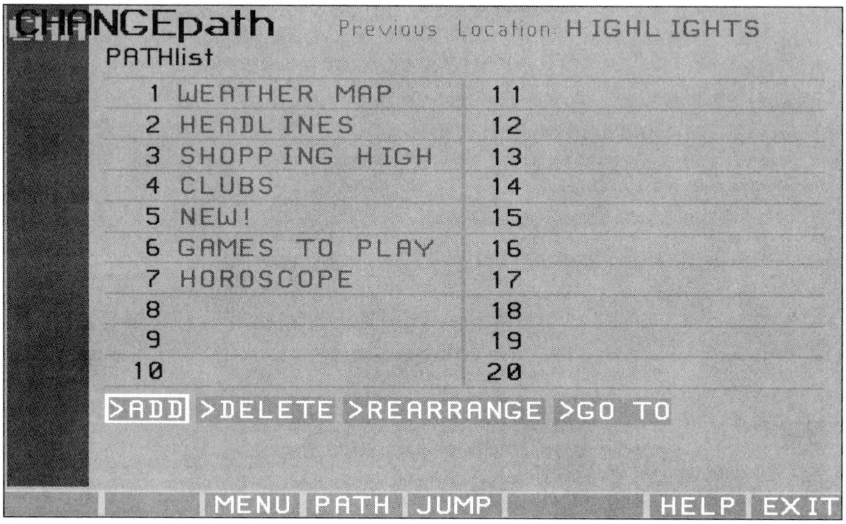

Fig. 11.10. *The Changepath screen.*

Function Key Options

If you have an IBM-PC or compatible, you can activate several function-key options—regardless of where you are in Prodigy. If you are using a Macintosh, these quick functions are available by pressing the Command key, followed by the appropriate number. The F2 function key apparently is not used by the Prodigy software and is missing from the Service Keyboard Template included with the Handbook.

F1 (Help)

At most points in Prodigy, the F1 key provides a brief help screen to explain a given option on the operative screen. You can get information about a particular option by highlighting it and pressing the ? key. Figure 11.11 shows a help screen for the Zip option. If Help appears on the command bar at a particular screen, you can highlight it by pressing H and activate it by pressing Enter or clicking it with your mouse.

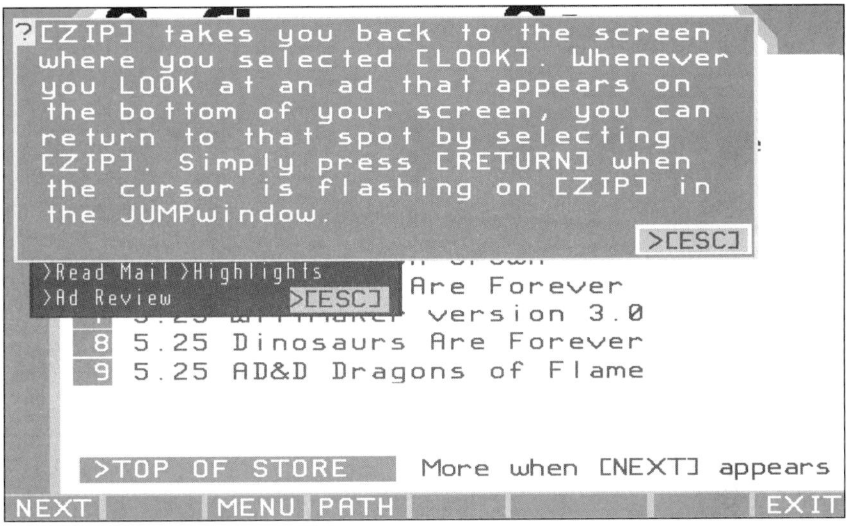

```
?[ZIP] takes you back to the screen
 where you selected [LOOK]. Whenever
 you LOOK at an ad that appears on
 the bottom of your screen, you can
 return to that spot by selecting
 [ZIP]. Simply press [RETURN] when
 the cursor is flashing on [ZIP] in
 the JUMPwindow.
                                  >[ESC]
>Read Mail >Highlights
>Ad Review          >[ESC]    Are Forever
                               version 3.0
   8  5.25  Dinosaurs Are Forever
   9  5.25  AD&D Dragons of Flame

    >TOP OF STORE        More when [NEXT] appears
 NEXT          MENU PATH                    EXIT
```

Fig. 11.11. The Help screen for the Zip option.

F3 (Viewpath) and F4 (Path)

Activate F3 (Viewpath) to view the services you have added to your custom path. The Path is a handy way to step through a series of Prodigy services that you have defined for yourself. F4 (Path) automatically takes you to the next jumpword you defined with the Changepath or Viewpath option (see fig. 11.12). This enables you to create a list of jumpwords to step you through your frequently accessed services by merely pressing F4 for each one. This appears to be as close as you can get to automating your on-line sessions with Prodigy.

F5 (Guide)

Sometimes you want to know where you are in the Prodigy menu structure, especially if you have Jumped to a particular area. Activate Guide to see a series of windows with all options listed for each menu level (see fig. 11.13). When you know where you are in these menus, you can easily move up or down the menu structure to choose different options.

Fig. 11.12. *The Viewpath screen.*

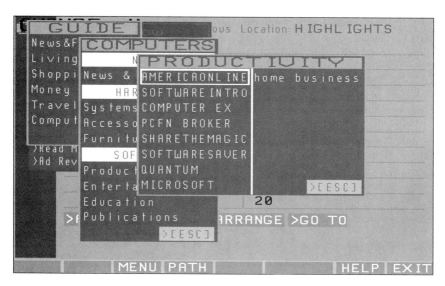

Fig. 11.13. *Sample Guide menus.*

F6 (Jump)

The Jump option is the fastest and most direct way to move from one part of the Prodigy service to another. The Prodigy software includes a *Jumpword Listings* booklet with some of the jumpwords. The most current list of jumpwords always can be found by browsing the Index (see the next section). You can enter up to 13 characters in the Jumpword search field. Although you also can select the Jump option by pressing J at most screens, the F6 option is especially helpful if you find yourself in a Forms screen, where you have to enter information before you can advance to the next screen. If you decide to exit the Forms screen, the F6 key may be the easiest (and possibly the only) way to exit without completing the form.

F7 (Index)

The Index is a full, alphabetized list of current jumpwords. You can search this list based on 1 to 4 characters, but you cannot scroll through the list to browse for available jumpwords. When you find the option you want, press Enter to activate it. See figure 11.14 for an example of the Index screen. The Jumpword Listings booklet that comes with the Prodigy software is an easier and faster method to look up jumpwords, but may not have the most current listings.

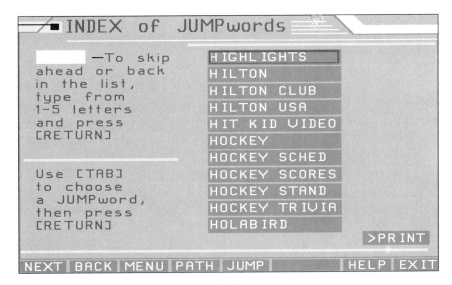

Fig. 11.14. *A section of the Jumpword Index.*

F8 (Find)

Although Jump takes you immediately to the service you've entered, Find gives you a list of jumpwords related to any keyword you enter. You then can highlight one of the options presented to go to that part of the service. According to the Prodigy manual, Find is the option to use when you know what you want, but do not know where it is. You can enter up to 20 characters in the Find search field. (If you already know where you want to go on the service, use Jump.) Figure 11.15 shows the Find screen.

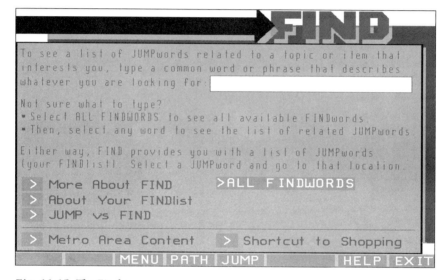

Fig. 11.15. *The Find screen.*

F9 (Menu) and F10 (Review)

The Menu function gives you at least three capabilities. It shows you where you are in Prodigy's tree of menus, enables you to move around in that tree, and enables you to back up through several levels of any service you have chosen from that menu. You can examine the context of any set of screens you may reach with a jumpword.

Review lists where you have been in Prodigy during your session. You can easily jump back to any place you visited by activating Review, scrolling up to highlight an earlier option, and pressing Enter.

Popular Prodigy Services

This section introduces some of the more popular services available on Prodigy. For a full listing, see the Jumpword Listing booklet provided with the Prodigy software, or use the Index jumpword to find more current listings. Two other useful jumpwords to help you find out about current services are *new* and *today*. *New* provides information on what's new on the service, updated weekly, and *today* gives you daily news and features.

News

Prodigy provides leading world and national news reports, complete with one or two related news items for each headline report. When a news item is longer than the screen, the `Next` prompt appears on the bottom left of the command bar. Selecting that option (press N and press Enter, or double-click the mouse) calls up the next screen. The headline of the next news item appears below the current item. You can follow several news reports as they are presented, but you cannot scan news headlines to choose only the items you want to read.

Advertisements

Advertisements usually cover the lower quarter of the screen when you navigate in Prodigy. They change each time you select a new option. Prodigy calls them *interactive* ads because you always have the option to look at an ad in greater detail, to follow a set of menus associated with an ad, and activate an on-line service so that you can browse and purchase products from advertisers. To interact with an ad, highlight and activate the Look option in the ad box. You can return to the service you were in from any level in the ad by pressing *z* to *zip* back.

Electronic Mail

The Jumpword window also contains a Read Mail option. Select this option to see if you have any mail waiting, and to reach the mailbox if you do. If mail is waiting, you receive a list showing who the mail is from and the subject of the messages (see fig. 11.16). Press the number by the message you want to read (or select it with the arrow keys or mouse) and the message text appears. If the message extends beyond the screen, the `Next` prompt appears at the end of the text and on the command bar.

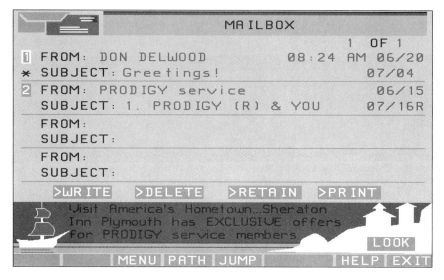

Fig. 11.16. The Read Mail screen.

To send mail to another Prodigy subscriber, select the Write option from the bottom of the mail screen. You are prompted for the address—the Prodigy I.D. of the recipient. If you do not know a user's I.D., you may be able to find it with the Member List jumpword. After you enter an address, you are prompted for a subject, followed by the text of the mail message. You can enter up to six screens of text, twelve lines of 40 characters on each screen, for a maximum of 2,880 characters, or 72 lines. The Mail service accepts anything as an address, and tries to send the mail message. If the I.D. is not a valid Prodigy I.D., however, the mail is returned to you as undeliverable.

Member List

Use the *Member List* jumpword to find information on other Prodigy subscribers. Figure 11.17 shows the Member List main screen. You can search for other users by name, by state, and by city. The entries displayed are drawn from the information you enter when you first go on-line to Prodigy. Users are not automatically added to this database, however. You must use the option to add to or delete from the list your name, city, and state.

Fig. 11.17. *Member List main screen.*

Although you can speed up a search if you know the name of the user you are looking for, a sequential search by state and city requires stepping through a series of prompts and menus. After you select the state and the city, up to eight subscriber names are displayed (see fig. 11.18). If you select one and press Enter, that user's name, city, state, and Prodigy I.D. are displayed, with the option to write a mail message to that person. Although it takes patience, this system enables you to easily find friends in your area who have subscribed to the service, and communicate with them by electronic mail.

American Academic Encyclopedia

Prodigy provides access to the Grolier American Academic Encyclopedia to all subscribers, as CompuServe does (see fig. 11.19). This nine-million word, 33,000 entry encyclopedia provides information on a wide range of popular subjects. The jumpword to reach this service is *Reference* even though it does not appear in the Jumpword Listings manual included with the software.

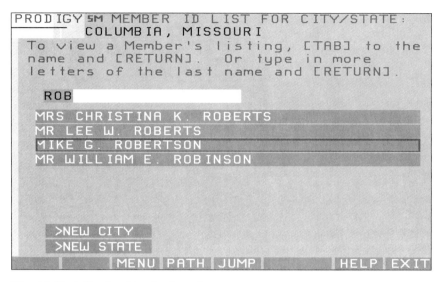

Fig. 11.18. *A Member List display of users.*

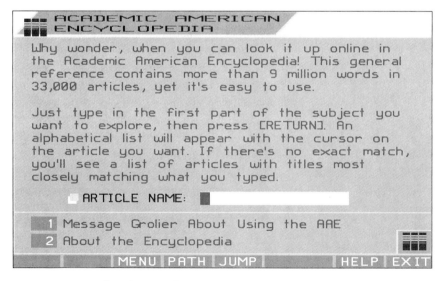

Fig. 11.19. *Encyclopedia search menu.*

Tools

The *Tools* area has options for changing or adding personal information, adding or removing members who share your account, changing your password, adding credit card information to make on-line shopping easier, adjusting your default display settings, setting the speed of your mouse while in the service, and enabling or disabling the Autologon feature. All of these options can be reached directly with jumpwords, but the Tools jumpword gets you to the Tools main menu (see fig. 11.20).

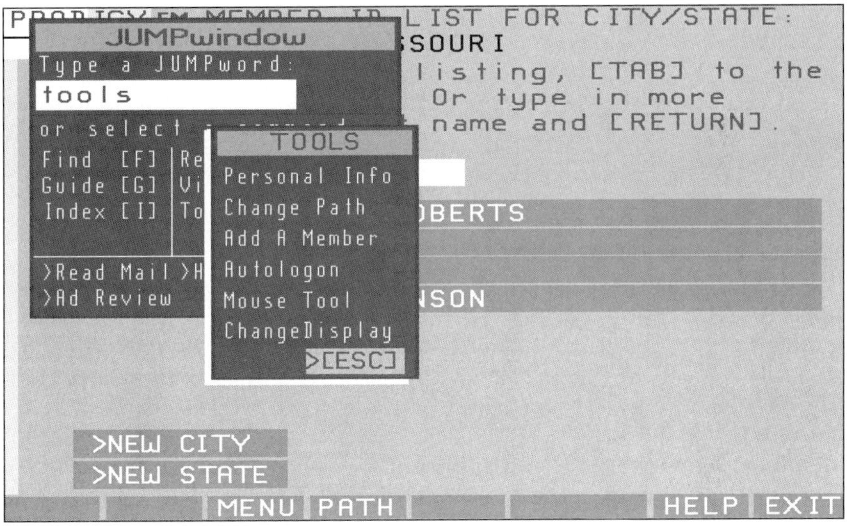

Fig. 11.20. The Tools main menu.

Personal Information

With the Tools Personal Info option you can add, delete, or change Prodigy's information about you and others in your household who use your account (see fig. 11.21). The information you can change includes your name, address, and phone number, your credit card numbers and expiration dates, and other personal information stored in your Prodigy software or on-line.

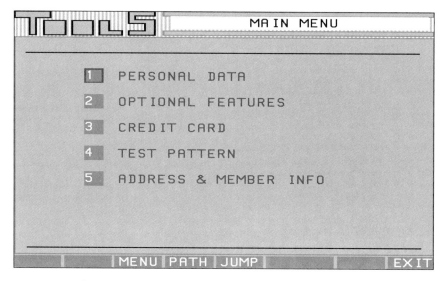

Fig. 11.21. *The Personal Info menu.*

The following options appear on the Personal Info menu:

- **Personal Data.** Whether you are the primary holder of the Prodigy account, or one of the members added to that account by the primary holder, you can edit your password, name, and other personal information from this menu option. To change any of this information, choose the appropriate prompt and enter accordingly.

- **Optional Features.** This option produces the message `Currently Being Developed`.

- **Credit Card.** You can enter a credit card number and expiration date for American Express, Discovery Card, MasterCard, and VISA in this screen. As soon as you store this information, the Prodigy software can send this information whenever you make an on-line purchase. See figure 11.22 for the Credit Card information screen.

- **Test Pattern.** Choosing this option produces a handy screen to check your screen colors. This option is simply a convenience feature to help you diagnose your screen color settings. Each bar of color is labeled so that you can see what the color should be.

- **Address & Member Info.** With this option you can check and adjust the address and phone number for your account. You also can revise information about other members of your household who use your Prodigy account. The Member Information option also offers instructions for enrolling a new member, enables you to jump to the Tools main menu, and gives you an option to print out the member's information.

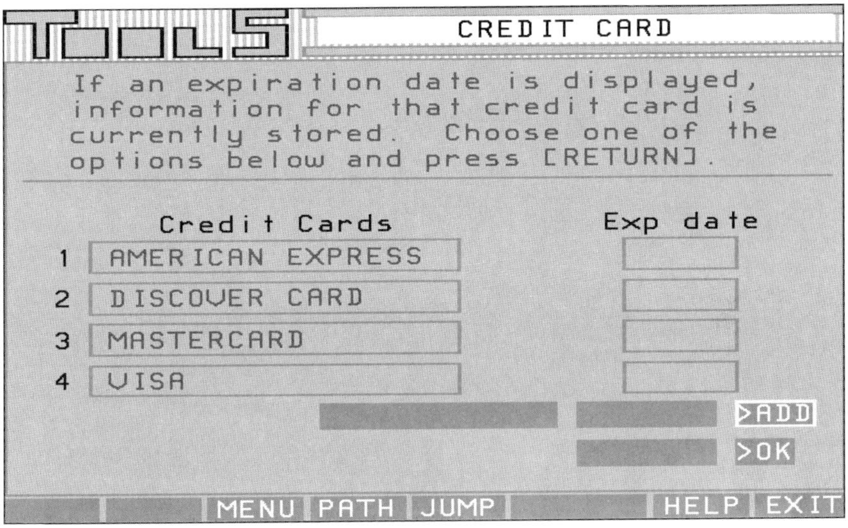

Fig. 11.22. Store credit card information.

Change Path

With the Tools Change Path option, you can edit the path—if you created one—that enables you to step through your favorite on-line options automatically by pressing P and pressing Enter.

Add a Member

This Tools option takes you directly to the Member Information menu instead of by way of the Personal Information and Address & Member Information menus.

Autologon

Prodigy provides a way to avoid the annoying I.D. and password prompt screen each time you start the program. When you are on-line, press J to bring up the Jump screen and type *autologon* in the Jumpword field (see fig. 11.23). The Autologon screen appears (see fig. 11.24). Depending on your programming preference, Prodigy logs you on automatically (sending your I.D. and password to the service), or prompts you for your password only. You are prompted to assign yourself a nickname. After enabling Autologon, start your session by typing *prodigy* followed by your nickname at the DOS prompt. To avoid logging on automatically (to be prompted for ID and password from Prodigy's opening screen), just type *prodigy* at the DOS prompt, without a nickname.

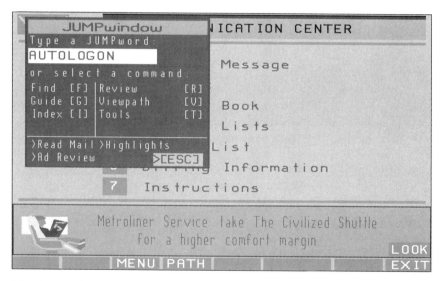

Fig. 11.23. *The Jump screen with AUTOLOGON highlighted.*

You are warned clearly, both in the manual and in the Autologon screen following the selection of item 1, that Autologon's first option reduces access security for your account. Anyone who has access to your Prodigy files can log on and use your account, possibly including your credit card accounts, your private electronic mail, and other private information. If, like most home users, you are not worried about security, Autologon enables you to easily use the Prodigy service regularly.

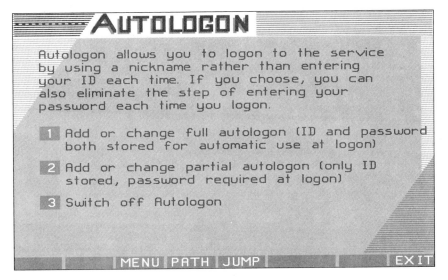

Fig. 11.24. *The Autologon screen.*

Mouse Tool

You can use the jumpword *Mousetool* to set the speed of your mouse for your Prodigy sessions. Mousetool affects the mouse cursor only while you are in the Prodigy software, and the setting only affects the computer you were on when you made the change. This is because changing the mouse speed setting changes configuration information in your computer's copy of CONFIG.SM. Like the Autologon feature, you can access and change the mouse speed at any time. See figure 11.25 for the Mouse screen.

Change Display

This option walks you through a set of instructions (which you can print out by pressing C and Enter) to change your display resolution. You normally would change the display only if you have changed your video adapter after joining Prodigy and going on-line the first time. You also may do this to speed up performance because the higher resolutions slow Prodigy down. Figure 11.26 shows the Change Display menu.

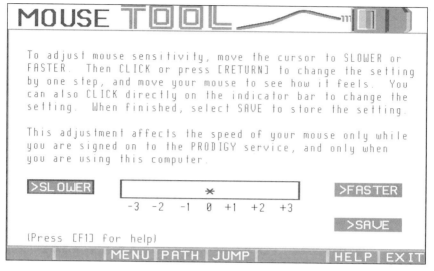

Fig. 11.25. *The Mouse tool screen.*

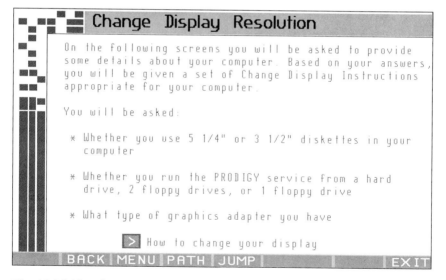

Fig. 11.26. *The Change Display menu.*

Before advising you how to change your screen resolution (using the GETSCRN program on your Prodigy installation disk), Prodigy finds out what kind of hardware you are using—including drive types and size, and type of video adapter. The GETSCRN program copies the appropriate video driver to your hard drive or Prodigy working disk. The following is a list of the drivers and the resolutions they support:

File Name	Degree of Resolution	Actual Resolution
VGA640.scr	High—color	640×480
EGA640.scr	High—color	640×350
EGA640E.scr	Medium—color	640×200
EGA320.scr	Low—color	320×200
CGA640.scr	Medium—monochrome	640×200

Printing Procedures

Many screens on Prodigy enable you to print out the information presented by pressing C (Copy), which brings up the Jumpword window with Print a Copy highlighted. Press Enter or click on the highlighted option to send a copy of the information to your printer.

Customer Support

The Prodigy Handbook is your first line of defense if you have questions or need help. Section 4, "Troubleshooting/Asking for Help/Frequently Asked Questions," provides a troubleshooting chart, suggestions and systems checklists to use before you call Prodigy, and common problems and solutions. On-line, you can get a list of phone numbers, customer service hours, and other helpful information by typing the jumpword *About Prodigy*. To send a message to Membership Services, use the Ask Prodigy jumpword. The manual also provides numbers for direct phone help: 1-800-284-5933 for help with installation, and 1-914-962-0310 for help with billing, disks and modem orders.

Chapter Summary

This chapter has taken a look at why and how to install the Prodigy software on your computer in order to connect to the service, and summarized the purpose of the program's working files. You learned how to find help in the various screens while on-line, how to activate and select various options on a given screen, how to move from one area of the service to another using Jumpwords, and how to exit normally from the service. You looked at the Path feature and how it can simplify your sessions by enabling you to move through preselected screens with a single keystroke. This chapter discussed the function key options you have in the program, and how they affect your on-line sessions. This chapter touched on some of the more popular services available to you on Prodigy, including news reports, advertisements, electronic mail, the subscriber directory, and the on-line encyclopedia. This chapter also discussed some of the services available to customize your personal data, your screen display, the speed of your mouse, and provide a way to log onto the program without having to enter your Prodigy ID and password each time. Finally, you learned how to send screen information to your printer and how to get customer support from Prodigy if you need it.

12

Using GEnie

G Enie (an abbreviation for General Electric network for information exchange) is owned and operated by GE Information Services (GEIS). The system is a subsidiary of the giant General Electric Corporation. GEnie always has been something of a maverick among commercial on-line information service providers. In spite of its giant corporate parentage, telecommunications users have been attracted to GEnie by its straight-forward, simplified menu structure and its reputation for quality informa-tion and features.

Although GEnie is structured a lot like CompuServe and many of the other nationwide, commercial on-linc information services, it has been a pioneer among text-based commercial on-line services with its rate structure and charging methods. GEnie was the first such service to set aside a group of capabilities as a basic service and price them at a flat-rate monthly fee. These services are called star services because each included menu selection is marked with an asterisk or star. This flat-rate structure has been very well received, and new members are flocking to GEnie in record numbers.

Understanding What Is Available

Among the services offered by GEnie is electronic mail—including mes-sages that contain attached files. You also can read the latest news and information from a variety of on-line press services and columnists. In addition to the AP and UPI newswires, Business Wire press releases, and NewsBytes computer news service, GEnie offers columns such as Dan

Gutman's "I Didn't Know You Could Do That With A Computer," Charles Bowen's "A Networker's Journal," and more.

GEnie also offers access to EAASY SABRE, a service that helps you plan and book all your own travel.

On-line access to the latest stock prices, from all the major exchanges, also is available on GEnie. The well-known Schwab discount stock brokerage maintains a special area on GEnie, providing important news, tips, and the ability to buy and sell stocks, bonds, and mutual funds on-line.

On-Line Chat Services

If you just like to chat with other on-liners all over the world, GEnie's LiveWire Chat and Real-Time Conferences put you in contact with over 200,000 of them. GEnie's LiveWire Chat is virtually identical to CompuServe's CB Simulator. LiveWire provides 40 different channels, or areas in which to chat. You can pick a handle when you enter the channel you want to join and jump right into the conversation. Because everyone is pretty much anonymous in LiveWire Chat, things can get pretty outrageous.

The Real-Time Conferences on GEnie are a lot like the LiveWire Chat but with a different slant. Unlike LiveWire Chat's wild and raucous party atmosphere, Real-Time Conferences are more like a lively installment of Phil Donahue or the Oprah Winfrey show. Special guests often participate with regular GEnie members who sign in for the conference. The users ask questions, typing them in on their terminals, and the guests answer questions by typing back. The guest's answers appear for all the users to see. The entire proceedings usually are directed by a moderator and often are saved to a file that is made available later for folks who missed the conference.

On-Line Shopping

You also can shop in the comfort of your computer chair. With GEnie you can browse a catalog of products from a wide variety of merchants offering everything from exotic coffee and flowers to compact discs, computer hardware and software, and almost anything else you can find in any store. Products on GEnie usually are sold at prices that are significantly lower than retail.

The GEnie equivalent of the CompuServe SIG (Special Interest Group) is the RoundTable (RT). GEnie has hundreds of RoundTables, each based upon a central theme. Most GEnie RTs offer a message bulletin board with a variety of different topics. A Software Library also is almost always available, typically with hundreds or even thousands of programs and text files that you can download. Genie RoundTables also usually feature one or more Real-Time Conference (RTC) areas, for live multiuser on-line conferences about the topics of interest to the RoundTable.

RoundTables

GEnie has over a hundred different RTs, covering literally every topic you can imagine. You can use GEnie's RTs to discuss your favorite computer, seeking help and advice or just looking for other folks with a similar taste in computers and software. Separate specialized GEnie RTs exist for literally every type and brand of computer made. Each has hundreds of shareware and public domain programs and utilities you can download, in addition to text files offering information about getting the most from your computer.

Some GEnie RTs specialize in particular brands or categories of software. For example, RTs support ChipSoft software, computer assisted learning, SoftLogik products, SuperMac product support, Enable, databases, Microsoft products, telecommunications, MUMPS, FreeSoft, ProTree, Qmodem, CP/M, and more.

In addition to technical RoundTables, more general interest RTs are available. You can find out about the latest in movies and television, including detailed reviews, in the ShowBiz RT. The Music RT is the place to discuss the latest music from your favorite artists and to read reviews of all kinds of new releases. TSR, the major publisher of numerous popular role-playing board games, also has an RT in which you can find out about the latest products and even participate in a nationwide game of Dungeons and Dragons.

If you like to laugh, you may want to try out the TeleJoke RT. It's like an on-line comedy club, in which it's open mike night every night. The Pet-Net RT can help provide valuable information for taking care of your favorite four-legged friend. The Hobby RT helps you with your favorite after-hours pastimes while the Sports RT keeps you posted on your favorite teams' activities.

Business-related RTs also are available—a Real Estate RT, an RT offering Soap Opera summaries, an RT for writers, another for fans of science fiction and fantasy, and far too many more to mention here. All of these many RTs offer the same mix of active message bases, long lists of downloadable files, and regular live conferences with notable personages in each selected field.

On-Line Games

GEnie also has games. Federation II, for example, places you in a Solar System in the future. You are the owner of a space merchant ship in search of profits. You actually build your own worlds and then trade with other GEnie users who will work with you or against you. Other GEnie games include Reversi, chess and Gomuku, the original Adventure, a multi-player version of Air Warrior, and much more.

GEnie, then, is an excellent source of news and information—not just about computers and software, but about literally any topic in which you may have an active interest. GEnie also can help you with your investments and put you in control of your travel plans. You can find help running a business, enjoying a hobby, or staying in touch with all your favorite sports. You can relax with a game, read reviews or, most importantly, you can communicate—with other people from all over the world with interests similar to yours.

Understanding Costs

One of the biggest reasons many people choose GEnie over other on-line services is its innovative flat-rate pricing structure. In October of 1990, GEnie stunned the on-line text service world by pricing many parts of its full-featured line-up of information features and activities at a flat monthly fee of $4.95 (U.S.) and $5.95 (Canadian) during non-prime-time hours.

GEnie's new flat monthly rates give you full access to what the company refers to as GEnie*Basic services. With GEnie*Basic, you can send and receive all the electronic mail you want. You also can access and participate in over 100 RT bulletin boards, read most GEnie news services and closing stock prices, use the GEnie Personal Loan Calculator (an on-line loan amortization program), play almost every game on the system, check airline schedules, and shop till you drop in the GEnie Mall.

Enhanced services that still carry hourly connect-time charges are most file downloads (uploads to GEnie RT software libraries are free), access to most computing-related RTs, participation in RTCs, and time spent on the LiveWire Chat Lines.

GEnie moved to this pricing structure as a response to the Prodigy service (see Chapter 11). If you compare the cost of the GEnie Star Services with the cost of membership on Prodigy, you find that you can have access to virtually any feature offered on Prodigy at less than half the flat-rate monthly cost on GEnie's GEnie*Basic service. And GEnie is accessible, without additional long-distance or packet-switching service charges, in many more cities than Prodigy (although that may change).

You will have to give up Prodigy's startlingly beautiful color graphic screens, but you will gain access to the many extra services for which GEnie still charges connect-time fees—such as file transfers and real-time conferencing—many of which aren't available on Prodigy at any cost. You also can use any communications software you want, unlike Prodigy.

GEnie's standard non-prime-time hourly rates, for services other than GEnie*Basic, are $6 an hour (U.S.) at 300, 1200, and 2400 baud. The prime-time rate (8 a.m. to 6 p.m. weekdays) is $18 per hour. In Canada, the rates are $5.95 per month for GEnie*Basic, $8 per hour for regular non-prime-time access, and $25 per hour for prime-time access. If you do not have a local GEnie access telephone number in your area, you may have to pay an additional $2.00 per hour whenever you use GEnie.

Understanding How To Begin

You are not charged an initial sign-up fee to get started with GEnie. To sign up for GEnie, you simply pay your first $4.95 monthly fee in advance, give them your credit card information (to cover any hourly connect-time fees you may incur outside of the GEnie*Basic service), and you're on the system. If you decide later that you don't like what you see, you can cancel your membership before the end of your first month for a full refund.

To obtain the most current pricing information to sign up for GEnie, call (800) 638-9636 (voice). GEnie will give you a local access telephone number you can call to sign up for the service with your modem and MasterCard or VISA. After you sign up, you are given a list of local telephone numbers in your area that you can call regularly to connect with the GEnie system (in most cases this will be a toll-free local telephone number).

How To Call GEnie

To sign onto GEnie, you first must make sure that you set your terminal software to operate in Half Duplex (also referred to as Local Echo) mode, then set the protocol to 7 bits, even parity, and 1 stop bit (also referred to as E-7-1). Note that the requirement for Half Duplex is different than for most other services and BBS systems. If you don't set your terminal software to Half Duplex, you will not be able to see the characters you type on your screen.

After you set up your terminal software, you next should have your modem dial the number you were given when you signed up with GEnie. You will hear the usual sounds of GEnie's modem sending out its connecting screech and yours responding, then silence as your modem's speaker shuts down. Your terminal program then will report that you are connected.

After the modem connection has been made, you should press your Enter key once or twice, if you're dialing in at 300 or 1200 baud, to establish your connection with GEnie. At connecting speeds higher than 1200 baud, you quickly should type three Hs (*HHH*) and press your Enter key to tell GEnie's computer you are establishing a high-speed connection.

Next you see a prompt asking for your user I.D. number and password. The prompt looks like U# at the top of your screen. When you signed up for GEnie access, you received your I.D. and password—two strings of numbers and letters separated by a comma. You should enter this I.D./password string following the U# prompt. Press Enter again and you see GEnie's opening welcome message.

NOTE

> Your GEnie I.D. and password is fairly long and can be cumbersome to hand-enter each time you want to log onto GEnie. This is an excellent time to use your favorite terminal program's keyboard macro capabilities, if it offers this feature. Better still, if you can, set up your entire GEnie logon as an automatic logon sequence and you will be on-line whenever you want, quickly and easily.

Following the sign-on welcoming banner, you see a line telling you how many private messages addressed to you are waiting on the system. Then a number of announcements will scroll by, announcing upcoming events on the system and any new features that have been added recently.

Understanding GEnie Menus

After the sign-on information is displayed, you enter the GEnie TOP (main) menu (see fig. 12.1). This is your gateway into the GEnie system, offering access to every part of the system via simple numeric choices. Options on this menu take you to all the other distinct parts of the system, such as the Computing section, Finance, News, Professional, Shopping, Chat, Mail, Games, Professional, and Leisure areas.

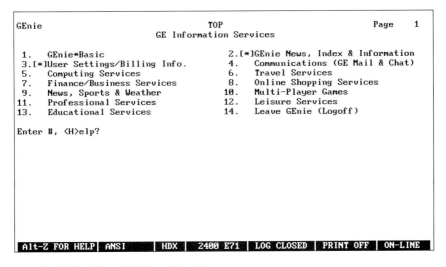

```
GEnie                              TOP                        Page     1
                          GE Information Services

   1.    GEnie*Basic                 2.[*]GEnie News, Index & Information
   3.[*]User Settings/Billing Info.  4.    Communications (GE Mail & Chat)
   5.    Computing Services          6.    Travel Services
   7.    Finance/Business Services   8.    Online Shopping Services
   9.    News, Sports & Weather      10.   Multi-Player Games
   11.   Professional Services       12.   Leisure Services
   13.   Educational Services        14.   Leave GEnie (Logoff)

Enter #, <H>elp?




  Alt-Z FOR HELP| ANSI    |  HDX  |  2400 E71  | LOG CLOSED | PRINT OFF | ON-LINE
```

Fig 12.1. *The GEnie TOP level (main) menu.*

Each of these areas is another menu of options that may lead to more menus until you finally reach each feature of the system. This menu structure enables you to step right through the system, making a series of menu choices to get anywhere you want. Each menu choice that carries no extra charge beyond the flat monthly fee is marked with an asterisk in brackets ([*]). Menu selections not marked with this symbol carry an extra connect time charge.

Understanding Rapid Navigation

In addition to being numbered choices on other menus, all GEnie menus have a keyword name and a system Page number. These are always displayed prominently across the top of each menu. After you are familiar with GEnie, you can move quickly from one part of the GEnie system to another by typing the destination area's menu name or (if you know it) the destination's specific page number. The main GEnie menu, for example, is the TOP menu, system Page 1. From anywhere else on the system, you can always move back to the GEnie main menu by typing *top* or *M 1* (Move to Page 1).

You also can move to other menus from anywhere on the system by typing the name of the desired menu. You can move to the Music RT from another location on the system, for example, by typing *music* (see fig. 12.2). Typing *showbiz* takes you to the ShowBiz RT. Typing *mail* takes you to the central menu where you can read and leave electronic mail. As you can see, GEnie's menu names are fairly easy to remember because they are logically keyed to the name of each service. If you ever lose track of where you are, just type *top* to go right back to where you started.

```
GEnie                          MUSIC                        Page   135
                        The Music RoundTable

   1.   Music Bulletin Board
   2.   Music Real-Time Conference
   3.   Music Software Libraries
   4.   About the RoundTable
   5.   RoundTable News
   6.   LOCK/UNLOCK ROUNDTABLE DOOR
   7.   MEMBER DIRECTORY
   8.   SET ONE-TIME FILES

   9.   Rocknet Entertainment News
  10.   Noteworthy Music (Mall Store)

Enter #, <P>revious, or <H>elp?




 Alt-Z FOR HELP| ANSI     |  HDX  |  2400 E71  | LOG CLOSED | PRINT OFF | ON-LINE
```

Fig. 12.2. *The Music RoundTable menu.*

GEnie's page numbering system, on the other hand, enables you to create your own shorthand scripts or macro keys for moving quickly from one area to another. After you know which page numbers get you to areas you frequently want to visit, you can enter the desired page number at the end of your I.D./password string, preceded by a comma, when logging on. This logs you onto the service and then moves you directly to the menu you specify, by page number, after a connection has been established.

The GEnie LiveWire menu, for example, is Page 400. If your I.D./password combination is *abc14633,nanoonan*, you can log directly into the GEnie LiveWire menu by typing *abc14633,nanoonan,400* at the U# prompt, after your modem has established a connection with GEnie's modem.

You also can tell GEnie to log you onto the service in a specific option on a specific menu by adding the option's number on the menu to your logon string, preceded by a semicolon. If you want to move to option 4 on the GEnie LiveWire menu, for example, you can type *abc14633,nanoonan,400;4* when asked for your ID as you log onto the system.

Understanding What Costs Extra

As indicated earlier, each menu entry that is part of GEnie*Basic's flat-rate service is preceded by an asterisk on the menus. Another way to keep GEnie*Basic services straight is to check each menu's page number. If the page number is a four-digit number beginning with 8—for example, 8010—then it's a GEnie*Basic service. Also, the keyword name of each menu (always displayed at the top of each menu page) is preceded by an asterisk if it is a GEnie*Basic service.

The first thing you should do as a new user on GEnie is to check out the various menu options that give you current information on the system itself. Notice that these are all GEnie*Basic services. They help you find the services you most want to use, offer tips on getting the most from GEnie, and keep you posted on the newest features and services.

The second menu item on the TOP GEnie menu leads you to GEnie's basic information menu (see fig. 12.3). You also can get to the information menu by typing the keyword *genie*. If you want, you also can type *M 3* to get there because its system page number is 3.

Even though the entire GEnie system is arranged in simplified menus and also can be accessed by one or more page numbers, I recommend that you get familiar with the keyword names for each menu. The keyword names help you move manually from one area to another, and the logical names usually are much easier to remember than a list of numbers.

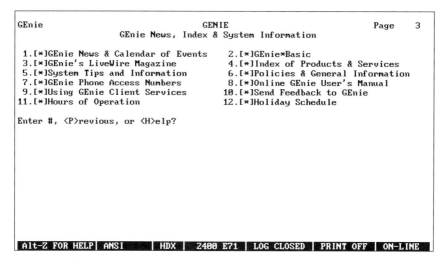

Fig. 12.3. *The GEnie Information menu.*

Obtaining Information about GEnie

Regardless of how you get there, the GEnie menu is the place to go if you have any questions about GEnie. You find everything from quick tips to a complete on-line user manual and up-to-date rate information. In fact, the GEnie on-line manual is one menu option you definitely should explore. You easily can display the manual to your computer's screen or capture it to disk using your terminal program's ASCII download or screen capture option.

You also can obtain a complete up-to-date index of all available GEnie system services from the GEnie menu. This list is always available by typing *index* at any menu prompt on the system (see fig. 12.4). You can easily search the GEnie on-line index by key word to locate specific services that match your interests. The key words *Science Fiction*, *Music*, or *IBM PC*, for example, bring up a list of all GEnie services that match the criteria you enter. Figure 12.5 shows the result of searching with the key word *Music*. The index, whether the complete list or the result of a key word search, always contains the menu name of each service and its matching menu page number or numbers.

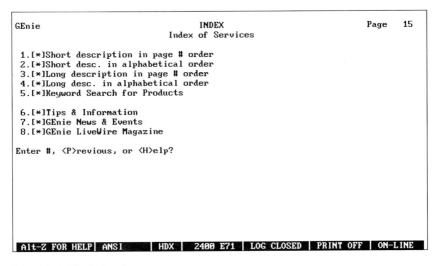

Fig. 12.4. *The Index System menu.*

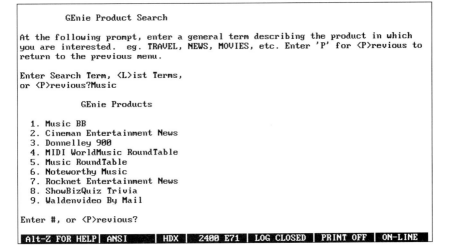

Fig. 12.5. *Result of search with the key word Music.*

NOTE

Anytime you start a display of text on GEnie, but decide you want to stop it before the end, you just type a *break*. This usually is a special function key on your terminal program. This can come in handy if you start something like the Index display command, then suddenly realize you forgot to start capturing it to disk and want to start over.

Other GEnie information options new users should check out include GEnie*Basic (to access GEnie*Basic, type *basic* or *M 8001*) for complete information about all available GEnie*Basic services (see fig. 12.6); and User Settings/Billing Info (type *set* or *bill*, or *M 8005*), where you can change your logon settings and password and other utility functions.

```
GEnie*Basic                       *BASIC                        Page 8001
                                GEnie*Basic

   1.   GEnie Users' (GENIEus) BB        2.    Aladdin Support BB's
   3.[*]GEnie News, Index & Information   4.    Send/Read GE Mail
   5.[*]User Settings/Billing Info.       6.    Entertainment Services
   7.   Travel Services                   8.    Money Matters/Personal Fin.
   9.   Hobby & Leisure Services         10.    Education Services
  11.   General Interest Services        12.    Classic Games
  13.   News, Sports and Weather         14.    Shopping Services
  15.   About GEnie Services             16.    Surveys from GEnie

  Enter #, <P>revious, or <H>elp?




 Alt-Z FOR HELP  ANSI        HDX   2400 E71   LOG CLOSED   PRINT OFF   ON-LINE
```

Fig. 12.6. *The GEnie Basic Services menu.*

GEnie News & Calendar of Events (type *new* or *M 8003*) is another important menu you should check often. This keeps you informed about any changes on the system, new features, upcoming important conferences, special events, and more.

Using a GEnie RoundTable

After you familiarize yourself with all the available features and RTs on GEnie, you naturally are going to want to start checking out some of the more interesting RTs.

One of the first RTs you should visit is GENIEus, the GEnie Users' RT. You can get to the GENIEus RT's main menu by typing *genieus* anywhere in the system or by moving to page 8001 and selecting the GEnie Users' RT option from that menu. GENIEus is a GEnie*Basic service available as part of your $4.95 monthly flat-rate fee, so no extra fee is charged for using this RT. That makes GENIEus a good place to start.

The first time you enter any GEnie RT, you are asked if you want to join. A simple *yes* answer signs you up as a member of the RT. After a brief notice telling you how many members currently are participating on-line in real-time conferences in the RT, the opening menu scrolls on-screen. Depending on the RT, you most likely will see some other brief announcements scroll by.

Each RT opening menu offers options to enter the RT's Bulletin Board, Real-Time Conferences, or Software Libraries. Figure 12.7 shows the BBS RoundTable main menu. Other menu options offer more detailed information about the RT and the latest RT news.

```
    Bulletin Board Systems (BBS) RT

Hello MUSIC-RT

  Bulletin Board Initialized.

  All messages have been updated as if you had read them,
  except for the ones in Category 1, Topic 1.

  Please take a moment to set your BBoard NAME with the NAME command.

Category   1 Welcome & Announcements

1. CATegories      10. INDex of topics
2. NEW messages    11. SEArch topics
3. SET category    12. DELete message
4. DEScribe CAT    13. IGNore category
5. TOPic list      14. PROmpt setting
6. BROwse new msgs 15. SCRoll setting
7. REAd messages   16. NAMe used in BB
8. REPly to topic  17. EXIt the BB
9. STArt a topic   18. HELp on commands
Enter #, <Command> or <HEL>p
1 ?
 Alt-Z FOR HELP| ANSI      |  HDX  |  2400 E71 |  LOG CLOSED |  PRINT OFF |  ON-LINE
```

Fig. 12.7. The BBS RoundTable main menu.

Using RT Bulletin Boards

The Bulletin Boards most likely are the part of the RT you will be visiting first. This is where some of the most interesting, genuine communication between GEnie users takes place. Most RT Bulletin Boards are available as part of the GEnie*Basic flat-rate fee service. This is an area where you can post messages electronically or read and respond to messages posted by other GEnie users. Everything you say in a GEnie RT Bulletin Board is available to be read by other users and vice versa.

Each Bulletin Board is divided into subject areas called *Categories*. The Categories within each RT depend on the interests of the group. You also can make suggestions to an RT's SYSOP, via GE Mail, for any additional Categories you feel should be added. (You learn how to send GE Mail in the section "Using GE Mail" later in this chapter.)

Each Category usually contains a series of topics that can be created by anyone. These may be in the form of a question, problem, statement, or whatever. Other Category members then can respond to the topic by entering messages. Some RTs have Categories that are private, with access restricted only to staff, owners of certain kinds of software, or others. If you ever come across a private Category to which you feel you should have access, you can always leave the RT's SYSOP GE Mail (GEnie's private electronic mail service) requesting access. Many RTs also have a Category for questions and suggestions about the RT.

To enter the RT's Bulletin Board, you select the Bulletin Board option from the RT menu. The RT's Bulletin Board has no menu. Instead, you can select from a list of possible commands or enter the commands directly. Each RT Bulletin Board displays a line telling you the Category for which you currently are set, a complete list of numbered Bulletin Board commands, and a command line at the bottom that is preceded by a number (the Category number in which you now reside) and a question mark, asking you for your next command.

At the question mark prompt, you can enter the number for a command from the list or type in the command itself. On GEnie, all commands can be shortened to the first three letters of the command and can be typed lower- or uppercase.

After you get more familiar with the system, the entire Bulletin Board prompt can be shortened by typing *PROmpt BRIef* (or its shorthand equivalent, *PRO BRI*). The command PROmpt NONe gives you simply the number of your Category and the question mark prompt. Similarly, the command PROmpt FULl returns you to the full Bulletin Board prompt, with the list of available commands displayed once again. You need only type the letters shown in uppercase to make a selection.

In any RT Bulletin Board, you most likely will first type *CATegories*, to see a list of all available categories (see fig. 12.8). The DEScribe category command gives you additional details about each available category (see fig. 12.9). The SET command asks you for a category number and sets you for access to that particular category.

```
    1 Welcome & Announcements
    2 Apple II/III BBS Systems
    3 Commodore/Amiga BBS Systems
    4 NOCHANGE BBS
    5 IBM/Compatible BBS Systems
    6 Macintosh BBS Systems
    8 BBS Ads and Notices
   12 Atari BBS Systems
   13 Tandy BBS Systems

You are currently attending Category 1

1. CATegories       10. INDex of topics
2. NEW messages     11. SEArch topics
3. SET category     12. DELete message
4. DEScribe CAT     13. IGNore category
5. TOPic list       14. PROmpt setting
6. BROwse new msgs  15. SCRoll setting
7. REAd messages    16. NAMe used in BB
8. REPly to topic   17. EXIt the BB
9. STArt a topic    18. HELp on commands
Enter #, <Command> or <HEL>p
1 ?
 Alt-Z FOR HELP  ANSI        HDX    2400 E71  LOG CLOSED   PRINT OFF   ON-LINE
```

Fig. 12.8. *Results of the CATegory command.*

```
IBM BBS SYSTEMS

This category includes discussion of BBS software, utilities, and hardware for
IBM and compatible micros. To review and discuss topics of general interest to
Sysops and BBS users of ALL types of PCs, please visit:

   1. Welcome and Introduction
   8. Public BBS Ads & Notices
   9. BBS News and Rumors
  10. BBS Law and Ethics



1. CATegories       10. INDex of topics
2. NEW messages     11. SEArch topics
3. SET category     12. DELete message
4. DEScribe CAT     13. IGNore category
5. TOPic list       14. PROmpt setting
6. BROwse new msgs  15. SCRoll setting
7. REAd messages    16. NAMe used in BB
8. REPly to topic   17. EXIt the BB
9. STArt a topic    18. HELp on commands
Enter #, <Command> or <HEL>p
1 ?
 Alt-Z FOR HELP  ANSI        HDX    2400 E71  LOG CLOSED   PRINT OFF   ON-LINE
```

Fig. 12.9. *Results of the DEScribe command.*

After you find a Category of interest, typing the command *TOPic list* displays a list of all the available topics within the Category you select (see fig. 12.10). The INDex of Topics command displays a list of all topics and in which Categories they reside, and the SEArch Topics command enables you to perform keyword searches for possible topics that match your interests.

```
  1 Reserved for future announcements!      1 Closed PINKSTONE [Mark]
  2 SCREEN SAVERS FOR PCs                    3 Open   D.BALSER1
  3 Real Time Conferences in BBS RT?         3 Open   BRAUB
  4 Find W.Seyler                            1 Open   P.APODACA
  5 Citadel Help                             2 Open   F.PELUSO
  6 National EMail & Topic Tracking System   9 Open   J.HENNING3
  7 MAC ARTIST WANTED!!!                      1 Open   B.HARNETT1
  8 IRAQ BBS ?                                2 Open   R.MUSTY [RIK]
  9 Lubbock BBS                               1 Open   J.SMOLEN
 10 dec vax vms bbs                           4 Open   NB.HKG
 11 The BBS & Civil Liberties                10 Open   M.LECCESE
 12 Unix BBS's                                5 Open   NJC
 13 WWIV Protocols                            3 Open   N.BALOG
 14 bbs's and ecology, environment            3 Open   J.RISSE
 15 Connecticut Modeming and BBS'             2 Open   W.HILL17
 16 Help Needed to locate name of Peoria Us   1 Open   CHRIS.LEMAN
 17 Other WANs                                2 Open   K.HINK
 18 TRYING TO SET UP A SMALL BBS-HELP!       12 Open   R.ROCHKOVSKY
 19  W W WHARD DRIV W W W.  QC                1 Open   E.EARVEITS
 20 HARD D WISK CRASH                         1 Open   E.EARVEITS
 21 New SYSOPS                               27 Open   G.LENZ
 22 "The Hot Shot BBS!"                       1 Open   T.FOSTER22
 23 One of the LARGEST SysOp...               1 Open   !! A.KLAUS [Pebody]
 24 Pc Pursuit boards
Alt-Z FOR HELP| ANSI  |  HDX  |  2400 E71  | LOG CLOSED | PRINT OFF | ON-LINE
```

Fig. 12.10. Results of the TOPic command.

After picking a Category and selecting a topic or topics that interest you, you can read the messages you selected by number by typing the *REAd a message* command. If you want to reply to a message you can either answer Yes to the prompt following the display of a given message, or enter the *REPly to a topic* command, then enter the number of the topic in which you want to leave a message.

The command BROwse enables you to read all new messages in all the Categories in any RT, if you don't want to limit yourself to a particular interest area (see fig. 12.11). This command, when used in conjunction with the IGNore category command, enables you to set up any RT so that you can easily read all new messages without having to wade through certain categories which don't interest you.

```
Enter #, <Command> or <HEL>p
1 ?bro

Category    1 Welcome & Announcements
 ************
Topic 1         Wed Jun 13, 1990
PINKSTONE [Mark]           at 22:36 EDT
Sub: Reserved for future announcements!

Watch this space for announcements and other news tidbits from the BBS
RoundTable SysOps...
1 new messages.
 ************
 ------------
Category 1,  Topic 1
Message 1        Wed Jun 13, 1990
PINKSTONE [Mark]           at 22:37 EDT

Watch this space for announcements and other useful information about the BBS
RoundTable!
 ------------

Topic 1 has been closed.  No replies allowed

 ↑/↓/PgUp/PgDn/Home/End:Movement keys  C:Clipboard  F:Find  W:Write  ESC:Exit
```

Fig. 12.11. Output of the BROwse command.

Additional RT Bulletin Board commands enable you to DELete a message or messages left by you, change the NAMe you use on the BBS (a sort of secondary name or handle that displays in the F r o m line in all messages you leave while in the selected RT), EXIt the Bulletin Board, or get HELp.

The HELp command can be typed anywhere on GEnie and will bring you a full descriptive text that reminds you of all the commands and capabilities that currently are available.

Using the Software Libraries

After you read messages in the RT Bulletin Board, you should examine the Software Library. Each RT has its own Software Library, and most offer a list of hundreds or even thousands of programs (shareware or public domain) and information text files that you can download (see fig. 12.12). If you are still in the Bulletin Board, you must type *EXIt* to return to the RT's menu. From the RT's menu you select the Software Library option to enter the Software Library for this RT.

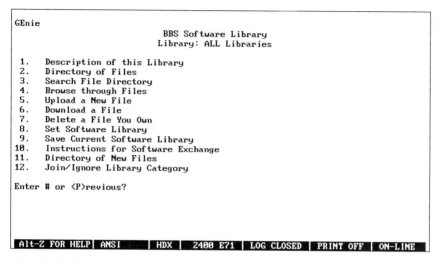

```
GEnie
                         BBS Software Library
                        Library: ALL Libraries

     1.    Description of this Library
     2.    Directory of Files
     3.    Search File Directory
     4.    Browse through Files
     5.    Upload a New File
     6.    Download a File
     7.    Delete a File You Own
     8.    Set Software Library
     9.    Save Current Software Library
    10.    Instructions for Software Exchange
    11.    Directory of New Files
    12.    Join/Ignore Library Category

Enter # or <P>revious?




  Alt-Z FOR HELP  ANSI        HDX    2400 E71   LOG CLOSED   PRINT OFF   ON-LINE
```

Fig. 12.12. *The BBS RoundTable Software Library menu.*

Each RT Software Library presents you with a menu of options. The software itself is grouped in subject libraries. The library you have selected to access always is listed at the top of the Software Library menu. When you first enter an RT's Software Library, you are set automatically for access to All Libraries.

Options on the Software Library menu enable you to obtain a description of the library for which you currently are set, see a directory of all files in the current library, search the file directory by keyword, browse through the list of available files with additional descriptions of each file as shown in figure 12.13, upload a new file, download a file, delete a file you have uploaded, set the Software Library, and examine a list of new uploads since the last time you logged on.

When you select the option Set Software Library, you see a list of all available libraries, by number. Just type the number of the library you want to access, and you then can use the other menu options to browse the list of files in that library, search the list by keywords that you specify, download a selected file, or upload a file you want to contribute to the library.

```
Start Directory backwards from what file
number or <RETURN> for ALL?

  No. File Name              Type Address        YYMMDD Bytes    Access Lib
 _____ _____  _  _____  _____ _____  _____ ___

  4232 QVIET41.ZIP                X M.CROSSON    910806  100608      9    4
       Desc: Vietnam V4.1 for Quick BBS
  4231 MEL_150.ZIP                X M.CROSSON    910806  356992      4    4
       Desc: Melee V1.50 - Gladitorial Combat
  4230 SOCAL070.LST               X C.HENDERSON8 910805   31872      2    4
       Desc: Southern California BBS List, Text
  4229 SS!25B.ARJ                 X M.CROSSON    910804   80384      8    4
       Desc: Super Slots V2.5B - Progressive
  4228 SUPRIG50.ZIP               X M.CROSSON    910804  156928      9    4
       Desc: Super Rig Door Version 5.0
  4227 SFM183.ZIP                 X S.FRERIKS1   910804  116096      4    4
       Desc: Spitfire bbs mail reader v1.83
  4226 MBU_110.ZIP                X T.CADD001    910803   50560      5    4
       Desc: NETMAIL tosser for RA
  4225 HOW2MAIL.ARJ               X M.CROSSON    910803   13184      6    4
       Desc: Set up Echo/Group Mail in Telegard
  4224 HOW2NET.ZIP                X M.CROSSON    910803   13056     10    4
       Desc: How to set up Netmail w/ Telegard
 ↑/↓/PgUp/PgDn/Home/End:Movement keys  C:Clipboard  F:Find  W:Write  ESC:Exit
```

Fig. 12.13. *The file directory listing.*

Downloading Files from a Library

Downloading a file from a GEnie Software Library is simple. Just make sure you're in the right Software Library, and then type the number of the file you want to download. You next are presented with the file's name, the name of the user who uploaded it, its size, and a more detailed description. If you still want to download the selected file, you next tell GEnie whether you want to receive the file via ASCII (straight displayed text) transfer, or using the XMODEM, YMODEM, or ZMODEM file transfer protocols. (See Chapter 3 for discussion of file transfer protocols.)

GEnie tells you when the file is ready to download. You then should set your terminal software to receive the file with the appropriate matching protocol. This usually is done by pressing the Page Down key on your terminal program and selecting the matching file transfer protocol.

Even though access to GEnie's Software Libraries costs you GEnie's regular hourly connect-time fees, you most likely will find plenty of software and text files that are well worth the nominal cost to receive them.

Uploading Files to a Library

Uploads to GEnie Software Libraries work just about the same as downloads, but in reverse. First, you have to make sure you selected a library into which you want to upload your file. If you try to upload a file with ALL selected as your library, you are prompted with a list of libraries and asked for the number of the library to which you want to upload. This is because GEnie will not know which library to place your file in if you haven't first selected one.

Next, you are prompted for a file name, a brief description of the file you want to upload, a longer description of the file, and finally the protocol you want to use for your upload.

When GEnie indicates that you should begin your upload, you then start the upload process from your terminal program. This usually is done by pressing the Page Up key, selecting the matching file transfer protocol, and entering the file name to upload.

Uploads to GEnie Software Libraries are always free during non-prime-time hours.

Using Real-Time Conferences

As you become more comfortable with GEnie and its RTs, you may find yourself wanting to participate in a Real-Time Conference. You find out about upcoming important Real-Time Conferences by watching the GEnie sign-on announcement banners from the NEWS menu, and from banners and news files in the various RTs on GEnie.

The option to enter an RT's Real-Time Conference area is located on each RT's menu. When you select the option, you first see a line that tells you how many RT users currently are in conferences, then you are given the choice of three or more conference "rooms" you want to enter (see fig. 12.14).

Real-Time Conferences usually are run by a leader who can exercise varying degrees of control over the proceedings, including throwing someone out of a room and locking the door. These on-line meetings can be scheduled or impromptu. They may include guest speakers and allow for questions and answers, and they can be either formal or casual.

```
GEnie                        Page 610;2
                 BBS RTC
                 Version 2.02

No users in the RTC.

Address of <MUSIC-RT> will be used.
What ROOM (1-3), or <Q>uit?2
Room 2,the Party room.
Type /EXIT to leave, or ? for HELP.
<MUSIC-RT> is here.
/help
/HELp        - List of commands
/BLAnk       - add blank lines
/BYE         - log-off the RTC
/CALl jj     - call on job jj
/ECHo        - echo to sender on
/EXIt        - return to menu
/JOB         - add job # to message
/KNOck rr    - knock on door of room rr
/MONitor rr  - monitor room rr
/NAMe  nn    - add name nn to address
/PRIvate jj  - go private with JOB jj
/QUIt        - return to menu
 ↑/↓/PgUp/PgDn/Home/End:Movement keys  C:Clipboard  F:Find  W:Write  ESC:Exit
```

Fig. 12.14. *Real-Time Conference entry and Help display.*

When you enter a Real-Time Conference meeting room, all the text you type is displayed for all other participants to see, preceded by a line giving your GEnie name. Similarly, all the comments and chatter typed by the other participants appear on your screen as they happen, line by line, preceded by their GE Mail addresses.

Understanding RTC Commands

A number of commands can help you when you're participating in a Real-Time Conference. Because text you type is displayed as a message after you hit the carriage return key, commands are all started with a slash.

For example, you can use the /KNOck command for a specific room number's door and request entry if the door has been locked by the meeting leader. In some meetings, the leader may require that a speaker be recognized before talking. In such cases, you can use the /RAIse-hand command to get the floor and speak. After the leader has recognized you, you then may enter your message.

You also can move from room to room by typing /ROOm and the number of the room to which you want to move. The /NAMe command enables you to add a nickname to your GEnie user name, which then is displayed with your typed messages.

Because displayed messages from other participants always print right over a message you may be entering at the time, you may want to shut out your own keyboard echo so that it doesn't get all scrambled on-screen. You can use the /ECHo and /XECho commands to turn the keyboard echo mode on or off in a Real-Time Conference.

If you want to monitor what's happening in another conference room while you participate in a different one, just type /MONitor and the number of the room you want to listen to.

Sometimes you may want to converse with another conference participant without everyone else listening in. The /PRIvate command enables you to do this with ease. Just include the Job number of the user you want to chat with and you are placed in a private one-on-one conversation with that user. Each conference participant is assigned a Job number when he or she joins the conference, and you can find out a participant's current job number with the /JOB command.

If you want, you also can simply send a private message to another conference participant with the /SEND command.

While in a Real-Time Conference, you also can send messages that are scrambled, squelch specific troublesome users so that you don't have to see what they're saying, find out the real name and location of each participant, see who's in the other rooms, and more.

Whenever you need help during a Real-Time Conference, the /HELP command displays a list of all available commands that only you can see.

GEnie's Real-Time Conferences are a great way to get to know other users, meet celebrities, obtain information from experts, or just have fun. Just remember that the hourly connect-time meter is running when you're participating in a Real-Time Conference, and make sure that you get the most from your time while you're there.

Using the LiveWire Chatlines

After you familiarize yourself with the Real-Time Conference rooms in your favorite GEnie RoundTables, you're ready to jump into the big pond—GEnie's LiveWire Chatlines (type *chat* or *M 400*). LiveWire's just like each Bulletin Board's Real-Time Conferences, except that it's much larger. LiveWire is system-wide in interest and is designed to copy the structure and feel of chatting on a CB radio.

Instead of just three rooms as in an RTC, LiveWire offers you a choice of 40 channels from which to choose. You also are asked for a handle before you log into a channel for chat, and you can pick just about anything that's not currently being used by another member.

After you choose a handle for yourself, you see a list of all 40 available channels and how many people currently are in each one. You then are asked which channel you want to enter. If you press Enter without entering a number, you automatically are assigned Channel 13. The first few times, you may want to explore all the channels with any activity.

Although no reserved channels exist, some of them may have certain characteristics. Channel 11, for example, may be the "In Search Of" channel, for people wanting dates or hoping to meet that someone special. Channel 25 could be a gathering of people interested in astronomy one night, then folks in love with stamp collecting the next.

In general, the commands you used in the RT Real-Time Conferences are the same as on the LiveWire Chatlines. The only notable exceptions are the /CHAnnel command, to move from channel to channel, and the /HANdle command, to change your handle.

The GEnie LiveWire Chatlines menu is also a RoundTable, offering a menu of options that take you to a Bulletin Board and numerous information files, including a schedule of upcoming Chatline events.

A system-wide Real-Time Conference area with 40 rooms (RTC or Real-Time Conference on the Chat menu) also is available on GEnie. This is very similar in nature and operation to each individual Bulletin Board's RTCs and the LiveWire Chatlines, but you usually find more structured, serious conversations rolling along there.

Keep in mind that the Chatlines and all Real-Time Conferences generate hourly connect-time fees. Real-Time Conferences and the Chatlines, however, are the best way possible to get to know other users and make friends on the system.

Using GEnie Mail

After you get to know a few folks, or if you already have friends and acquaintances who are GEnie members, you soon will start sending and receiving GE Mail, the private electronic message service. One day, when you log onto the system, you will get a message `You have 1 (or more!)` `message waiting` and GE Mail is how you must access it.

To get to the GE Mail menu, select the GE Mail item from the Top Menu (or type *mail* from anywhere on GEnie or *M 200* to move to page 200). The GE Mail menu appears (see fig. 12.15). GE Mail is sent and received using each member's GE Mail address. You were assigned a GE Mail address when you first signed up for the service, and so was every other user on the system.

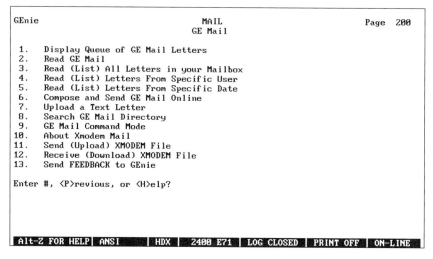

```
GEnie                          MAIL                      Page   200
                              GE Mail

   1.    Display Queue of GE Mail Letters
   2.    Read GE Mail
   3.    Read (List) All Letters in your Mailbox
   4.    Read (List) Letters From Specific User
   5.    Read (List) Letters From Specific Date
   6.    Compose and Send GE Mail Online
   7.    Upload a Text Letter
   8.    Search GE Mail Directory
   9.    GE Mail Command Mode
  10.    About Xmodem Mail
  11.    Send (Upload) XMODEM File
  12.    Receive (Download) XMODEM File
  13.    Send FEEDBACK to GEnie

Enter #, <P>revious, or <H>elp?




 Alt-Z FOR HELP  ANSI       HDX    2400 E71   LOG CLOSED   PRINT OFF   ON-LINE
```

Fig. 12.15. The GE Mail menu.

The GE Mail menu includes options to check your mail (telling you how many letters you have on hold, read, or unread), read your mail, send a text letter, upload a text letter, search for an address, enter the GE Mail command mode, send a paper letter, send an XMODEM file, or receive an XMODEM file.

If you are notified that you have GE Mail waiting for you, simply move to the GE Mail menu and select option 2 to read your mail. Each letter you have not yet read will be displayed to you, one at a time, with options to delete the message or reply to it, at the bottom of each message. Ctrl-S stops a letter as it scrolls by and Ctrl-Q starts it scrolling again (this will stop and start any scrolling text display on GEnie).

After a letter has been read, it is removed from the queue list. The letter is placed in a holding file for five days, after which time it is discarded.

You also can check waiting mail manually by selecting the Display Queue of GE Mail Letters option from the GE Mail menu. Any letters you have waiting for you then are listed, one by one, numerically ordered as they were received. This list includes each letter's queue number, a system item number, the GE Mail Address of the person who sent the letter to you, the name of the receiver, the date the letter was sent, and a brief description of the letter's subject.

To read any letter waiting in your Queue, just select the Queue number and it is displayed.

You can send a GE Mail Letter in several ways. You can select the Compose and Send GE Mail Online option from the GE Mail menu, or you can use the GE Mail Command Mode. Either way, you first are asked for the name of the GEnie user you are addressing and given the opportunity to send a carbon copy of your letter to another GEnie user or users. You then enter the subject (up to 30 characters) and type your letter. When you finish entering the text of your letter, just type *S on a line by itself to send it.

GEnie Mail commands abound to help you edit your text on-line, but you also can choose to upload a letter you write with your favorite word processor and save it to disk as a straight ASCII text file. Just select the Upload A Letter option from the GE Mail menu and then, after you're prompted for the To, Carbon Copy and Subject lines for the letter, you see a prompt that reads READY FOR INPUT. At this point, you should start transferring the letter from your computer to GEnie. The file you upload must be sent in a line-by-line ASCII format and must have a carriage return at the end of each line.

After your download is complete, press the Break function key and you see a line number prompt that reflects the last line in the letter you just uploaded. Enter *S to then send the letter or *L if you want to List the letter first, to make sure that it was received properly by GEnie.

Sometimes you will want to send a letter to another GEnie user but you will not know his or her GE Mail address. The Search GE Mail Directory option on the GE Mail menu enables you to search the entire list of GE Mail addresses for a specific user's name.

Binary program files and even large text files can be attached to a GE Mail letter by selecting the Send (Upload) XMODEM File option on the GE Mail menu. With this option you first fill out an entire GE Mail letter, then upload any file you want using the XMODEM file transfer protocol. The file then is attached to the letter. When the recipient of your electronic mail receives your letter, he or she is prompted that a file is attached and is given the opportunity to download the file after reading the attached note.

After you gain some experience with GEnie and with GE Mail in particular, you also can enter the GE Mail Command Mode, which offers a lot of additional flexibility and power. The GE Mail Command Mode enables you to cancel a mailed letter, defer unread mail you don't want to read, and check the status of a letter (for example, to find out if a letter you sent has been received). The GE Mail Command Mode also offers a variety of List commands that enable you to custom-tailor your mail. The Command Mode's Tbatch command even enables you to enter your letters off-line, including all the usual To and Subject lines within the file itself.

Another important GE Mail menu option is Send Feedback to GEnie. This option enables you to send your comments and suggestions directly to the GEnie staff.

You can even choose to send GEmail—printed on paper and mailed with a stamp in an envelope via regular first class mail—to someone who isn't a member of GEnie. This option, however, requires that you pay an additional surcharge.

A new GE Mail option enables you to upload files that then can be sent to a fax machine. Again, this is an added value option that carries a surcharge.

Regular GE Mail, however, can be sent and received as part of the GEnie*Basic flat-rate service. The only exception is when a file is attached to GE Mail, which is billed at the regular hourly connect-time charges for the upload and download.

Using GEnie Information Services

While GEnie's RoundTables, Real-Time Conferences, and Mail covers all the bases regarding nationwide on-line communications, its information offerings also are extensive and easy to use. Possibly one of the most interesting information resources available on GEnie is the complete Grolier's Electronic On-Line Encyclopedia (type *groliers* or *M 8365*).

You browse Grolier's articles using search terms. The Grolier's program searches its 10-million word database of more than 31,000 articles for words that match the same set of letters you enter as your search term, then presents you with a list of articles. To read a selected article, you simply enter the number of the article from the selection list and it is displayed on-screen.

Longer articles first are displayed with a submenu of choices that enable you to display the entire article or only portions of it.

Other reference works you find on GEnie include Cineman Entertainment (type *cineman*), Hollywood Hotline (type *hotline*), Rainbo Electronic Reviews (type *rainbo*) and Photosource International (type *psi*).

Getting the News on GEnie

If it's news you want, GEnie's got a number of resources at your disposal (type *news* or *M 300*). NewsGrid Headline News is a collection of stories compiled daily from seven different international agencies plus other specialized sources. Stories cover general U.S. and world news, business news, sports news, and features. Sources include The Associated Press, United Press International, Agence France Presse, Deutsche Press-Agentur, Kyodo, PR Newswire, Xinhua, INTEX, Garin Guybutler, Security Trader's Handbook, and Panhandle Eastern Pipeline Corporation.

Option 3 on the Newsgrid menu gives you a menu of current news stories from which you can select by number. You also can search all available news by keyword (countries, place names, newsworthy people, and topics such as movies, TV, congress, airports, and so on). The service presents you with a list of all stories that match only the whole keyword or words you enter.

You also can create a custom news-clipping service tailor-made just for you. It's an option on the Newsgrid Headline News menu. Just select the Custom Clipping Service option and enter up to 50 key words, then GEnie will clip all news stories that match them and present them in a list when you choose the menu option to check news stories in your Profile. GEnie now can even send all clipped stories to you as GE Mail that will be waiting when you log on.

Other news services available on GEnie include USA Today DecisionLines, The Home Office Newsletter, a vast assortment of late-breaking press releases from PR Newswire, personal computer news, Fight Back with David Horowitz, and FCC Proposal News.

Using the NewsRoom

Virtually all of GEnie's news and reference resources are available as part of the GEnie*Basic flat-rate monthly fee service. The one exception to this rule is the LiveWire Newsroom, where you can go if you're interested in discussing current news events or if you want to download the latest news and later read it off-line.

The LiveWire Newsroom is designed like a Real-Time Conference. All the latest news of the day is displayed in the first four rooms, updated continuously as it comes over the wires. Rooms 5-8 are discussion areas, each divided by interest type (U.S. and World News, Business, Sports, Features). You can enter these to discuss late-breaking news with other GEnie members. You can even monitor rooms 1-4 and discuss recent developments as they happen.

Finding Financial Services

If money and high finance are of interest, you will want to take advantage of GEnie's variety of financial services. These are found on the Finance menu (type *finance* or *M 600*) and include Dow Jones News Retrieval, GEnie Quotes Securities Database, VESTOR 24-hour Investment Advisor, and GEnie Loan Calculator.

Several of these features are value-added services, requiring that you pay a surcharge above and beyond the usual hourly connect-time fees for access.

As a new user, you first should select the About Financial Products option on the Finances menu and get more facts about each available service before proceeding here.

Shopping on GEnie

When you're ready to spend your money, the large assortment of merchants in the GEnie Mall give you good value for your money. The Shopping menu (type *shopping* or *M 700*) is a doorway into a fascinating array of goods and services, from magazine subscriptions and flowers to complete computer systems and even antique cars. Figure 12.16 shows the GEnie Mall opening screen.

Many GEnie merchants offer extensive, fact-filled, on-line catalogs that you can browse or download. Most also enable you to order products directly on-line, automatically charged to your credit card, and shipped quickly. You also usually can send feedback directly to participating merchants and get more details on any product.

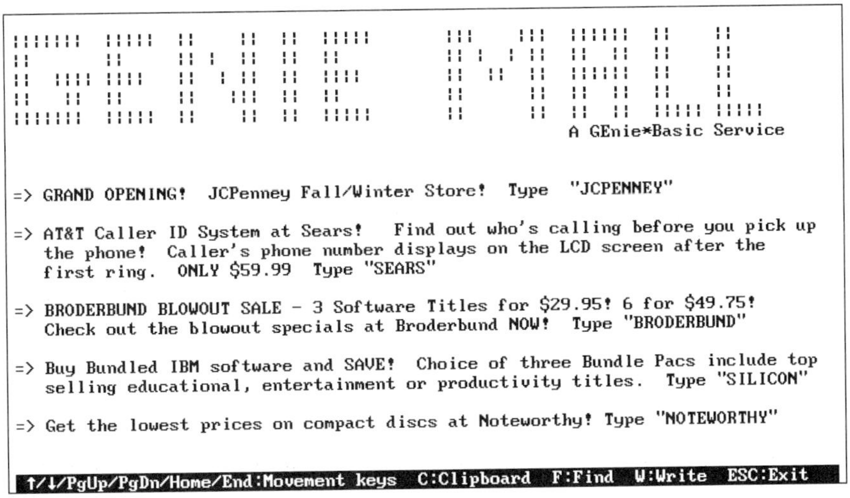

```
!!!!!!!  !!!!!  !!   !! !!  !!!!!       !!!    !!!  !!!!!!  !!   !!
!!       !!    !! !! !! !!  !!         !! !!  !! !! !!  !!  !!   !!
!! !!!! !!!!   !! ! !! !! !!  !!!!      !! !!  !! !! !!!!!!  !!   !!
!! !! !! !!    !! !!! !! !!  !!         !!    !! !! !!  !!  !!   !!
!!!!!!!  !!!!!  !!   !! !!  !!!!!       !!       !! !! !! !!!!! !!!!!
                                             A GEnie*Basic Service


=> GRAND OPENING!  JCPenney Fall/Winter Store!  Type  "JCPENNEY"

=> AT&T Caller ID System at Sears!   Find out who's calling before you pick up
   the phone!  Caller's phone number displays on the LCD screen after the
   first ring.  ONLY $59.99  Type "SEARS"

=> BRODERBUND BLOWOUT SALE - 3 Software Titles for $29.95! 6 for $49.75!
   Check out the blowout specials at Broderbund NOW!  Type "BRODERBUND"

=> Buy Bundled IBM software and SAVE!  Choice of three Bundle Pacs include top
   selling educational, entertainment or productivity titles.  Type "SILICON"

=> Get the lowest prices on compact discs at Noteworthy! Type "NOTEWORTHY"


 ↑/↓/PgUp/PgDn/Home/End:Movement keys  C:Clipboard  F:Find  W:Write  ESC:Exit
```

Fig. 12.16. *The GEnie Mall entrance.*

Other shopping services on GEnie include Computer Express Products (type *express* or *M 720*), which offers a list of numerous merchants, each specializing in products for a specific kind of computer; Compu-U-Store On-Line, the well-known on-line service that offers prices that are 10-50 percent off manufacturer's suggested list prices on over 250,000 name-brand products; and Gift of Time, a unique service that enables you to buy additional GEnie time for friends.

You also can shop with AT&T, Broderbund Software, Contact Lens Supply, Inc., Gimmee Jimmy's Cookies, Engraving Connection, Ford, Hawaii General Store, J.C. Penney, Sears, Walden computer books, Walden Video, and too many more to mention here.

Another menu also enables you to order GEnie products. The Order menu (type *order* or *M 730*) presents you with an extensive list of products, including user's guides and the full-text GEnie User's Manual.

Of course, access to all of GEnie's mall merchants is available as part of the GEnie*Basic package.

Using the Classified Ads

If you have something to sell, you can log into GEnie's Classified Ads menu (type *ads* or *M 740*). You can browse ads left by other GEnie users at no additional charge, as part of the GEnie*Basic service. If you want to leave an ad, however, you must pay a small additional surcharge. You can choose to have your ad run for 7, 14, or 30 days.

Each ad prompts you for a reply after you read it. If you answer Y, your reply automatically is sent to the user who posted the ad. Ads are posted automatically with your GE Mail address unless you choose to leave your ad anonymously. You pay an extra charge for anonymous ads.

You also can search all the ads on GEnie by keyword, so you only have to read ads for items of interest to you.

Understanding GEnie's Travel Aids

The next time you plan a trip, you may want to try EAASY SABRE. This is a gateway product, which means you connect, via GEnie, to American Airline's computer. Regular GEnie commands no longer work and you are charged an extra fee in addition to GEnie's regular hourly connect-time rate. EAASY SABRE is item 1 on the Travel menu (Page 750). It enables you to get quick information and make immediate reservations for flights, hotels, and rental cars. EAASY SABRE is not cheap and takes a little getting used to, but it is a fast, complete service for the traveler.

GEnie's Travelers Information Service RoundTable (TIS) enables you to discuss travel with other GEnie members, participate in Real-Time Conferences, download articles about travel, read an on-line travel newsletter, and look up information about travel services worldwide.

You also can shop for special travel adventures on GEnie's Adventure Atlas (type *atlas* or *M 775*). With this service you can choose an adventure by category (Backpacking, Bicycling, Canoeing, and so on), by area, or by date. You then are presented with a list of adventures that matches the criteria you specified. Each item on the list includes details about the trip, the mode of transportation involved, number of days, type of lodging provided, prices, travel agency contact information, and other necessary information.

Understanding Advanced Options

In addition to all the regular menu options and commands you can use for getting around on GEnie, a number of other commands give you added power and flexibility anywhere on the system. These commands always are available at the Enter # prompt, which appears on most GEnie menus.

Whenever you want to find another GEnie user, just type *locate* and the GE Mail address of the user in question. GEnie responds with the person's present location on the system if he or she currently is using the system, and the message Not Currently On GEnie appears if he or she currently is off-line.

After you use the LOCate command to find another user, you can send a message directly to the user with the NOTify command, followed by the GE Mail address and the message you want to send. The message will be displayed to your friend the next time they move to a menu prompt. If the user being notified is in the Chatlines, one of the Real-Time Conferences, in a game, or on one of the GEnie gateway products, then a note will be sent to you. Your friend will receive your message upon leaving the game or gateway or whatever.

When a notice you send has been received, you get a message announcing that your message has been received. If the user you're trying to contact isn't on-line at the time, GEnie will tell you that as well.

In case you're wondering if you've recently been notified, just type *Notify* * and any incoming messages addressed to you will be displayed. If you have important work to do on GEnie and you don't want to be disturbed while you're doing it, just type *nonotify* and any messages sent to you will not be displayed. Instead, the person who sent the message will receive a message saying you currently are not available.

Any time you want a list of all the users currently logged on to GEnie, along with their GE Mail addresses, just type *users all* at any menu prompt. If you simply type *users*, you get a list of all current on-line users in the same product category you currently are occupying. If you type the command while in a RoundTable, for example, you get a list of all GEnie members currently logged on and in the RoundTable you're in. You also can type the *users* command followed by the name of any GEnie product and receive a list of all GEnie members currently logged onto the product area you specify. Typing the command *users mail*, for example, lists all GEnie members currently logged onto the system and located in the GE Mail menu.

You also can read and send mail anytime and anywhere on the system with the REAd and SENd commands.

The PREvious command provides another way to move around the GEnie system. Whenever you type the command, you are moved to the last menu you accessed.

If you ever need to know how long you have been connected to GEnie, along with the current date and time, just type *time* at any menu prompt.

The PORT command displays the number of the port through which you currently are accessing GEnie. This can be important because you will be asked for the access port number you were on whenever you seek customer assistance for communications problems.

Finally, the BYE command logs you out of any GEnie menu or product and signs you off the system.

Understanding Aladdin

All this information, communication, friendship, and fun, combined with flexibility, ease of use, and low cost, is nice, but it can be time-consuming. As you get accustomed to using GEnie and get more involved in happenings there, you're sure to discover countless services and features that you will be using frequently. You also will find, however, that keeping up with all your mail and areas of interest on the system takes time.

Even though you can access most of GEnie's features for a flat monthly fee, you may be reluctant to tie up your phone lines for the time it takes to find and capture 500 new letters in 10 RTs, download 15 new interesting files, and upload a few files of your own.

An easier way to handle GEnie's features is available, and its name is Aladdin. Aladdin is a special software program, written for IBM PCs and compatibles, that enables you to read and respond to all your GE Mail and messages in GEnie RoundTables and message areas, and even download and upload files. It's just like being on GEnie, except it's much easier to move around the various message bases and less complicated to receive and send files. In addition, the Aladdin message editor is full-screen and a lot simpler to use.

Even better, with Aladdin you can do almost everything you do on GEnie off-line, without tying up your phone lines or without the pauses and delays associated with moving around on a huge mainframe on-line service loaded with dozens or even hundreds of other users. The actual transactions with GEnie happen automatically, and you don't have to wait for them.

The complete Aladdin software package is available right on GEnie, in the Aladdin RoundTable's Software Library (type *PCAladdin*, *Aladdin*, or *M 8016*). The RoundTable is very similar to other GEnie RoundTables, offering a support Bulletin Board, Real-Time Conference area, and Software Library. The Aladdin RoundTable's Bulletin Board always is available as part of GEnie's GEnie*Basic flat-rate service.

You can always download the latest edition of Aladdin from GEnie, as well as the complete Aladdin user's manual. You are charged no additional fee to use the program other than the connect-time fee to download the necessary files. The complete Aladdin user's manual also is available on-line.

Aladdin brings you the power of menu-driven, full-screen software while you access GEnie. After you download the program, it will automatically decompress into all the separate files you need to run the program.

Configuring Aladdin

The Aladdin configuration menu enables you to enter your logon I.D. and password, local access telephone number, baud rate, and other particulars of your computer hardware and GEnie account. You then can view a list of all available GEnie RoundTables and select the ones in which you want to participate.

Aladdin also can be configured to automatically monitor selected RT Software Libraries, presenting you with updated lists of new programs and text files.

After Aladdin is all set up and ready to go, Aladdin will dial your phone, connect to GEnie, pick up any waiting GE Mail, check all your selected RTs for new messages, create a list of all new files in the Libraries you're monitoring, then log off.

Using Aladdin

You then can run Aladdin and enter the RTs you selected to monitor, exactly as if you were on-line. The software enables you to scroll back and forth through all the messages in each RT you selected. Replying is as easy as pressing the Esc key and typing in your response. Unlike GEnie, however, Aladdin's message editor is a full-screen text editor. You can move your cursor all over the screen, insert text, correct your message, go back, and change something later—like any good word processor.

You also can browse through any GE Mail messages that were waiting for you when Aladdin called, then reply to them in the same way.

Finally, you can scan the list of new files Aladdin found in the Software Libraries you told it to scan, then tell the program to download any files you found interesting next time it calls GEnie. With Aladdin you can specify the file names you want to use for downloads when they're saved on your computer. You also can tell Aladdin to upload a file to GEnie, start new Topics on RT Bulletin Boards, and almost anything else you can do when normally connected to a GEnie RT real-time.

When you tell Aladdin to make its next automatic pass, the program again will call GEnie and pick up any new GE Mail that's waiting and grab all new messages in the bulletin boards that interest you. This time, and on each future call, the program also will send all messages and GE Mail you entered off-line, download files you selected off-line, and upload any files you chose to upload.

Aladdin also offers options to simply log onto GEnie, pick up any waiting GE Mail and drop off any GE Mail you've written, just check into a given RT or set of RTs, only download files, only upload files, and so on. In fact, Aladdin also will log on to GEnie and then turn all actions over to you, acting as your GEnie-specific terminal program.

In short, the program is fast and flexible, and definitely will keep you from tying up the phone for hours on end while you stay connected to GEnie. Just remember, Aladdin is a large program and runs very slowly on floppy disk drives. Aladdin will run on a two-floppy system (720k drives), but it wasn't really designed to do so. Searches of large message bases literally will seem to take forever on a floppy-based computer.

You also are severely limited in the number of RTs you can monitor with Aladdin if you don't set it up to run on a hard disk—especially if you don't plan to call GEnie every day. (Remember, every day you don't call GEnie is another day's worth of messages and mail traffic that is building up on the system for Aladdin to grab all at once later.)

With Aladdin installed on a hard disk drive-equipped PC, however, you truly will harness all the power of GEnie, easily and without confusion. Aladdin will not do a number of things automatically, such as checking stock quotes and monitoring the news wires, but it is all the terminal program you need to do everything on GEnie.

Chapter Summary

GEnie is one of the most rapidly growing on-line information services available. It has a wide range of on-line services and information, making it a full-featured system. It offers electronic mail, special interest bulletin boards, software file libraries, and on-line chat as well as information services and on-line games.

Even without the help of Aladdin, GEnie will give you your money's worth because of its innovative flat-rate pricing structure. This low cost for the basic service even makes GEnie a serious contender for a second system if you already are a member of one of the other available national on-line services. It's a cost-effective way to stretch your modem's reach.

13

Using Electronic Bulletin Board Systems

Computer bulletin board systems (BBSs) represent the "grass roots" of computer communication. A BBS enables you to exchange messages and files through a central PC over dial-up telephone lines. The public access computer BBS is one of the most exciting areas of computer communications. In 1978 only one BBS existed in the world. Today more than 25,000 public access bulletin board systems exist in the United States alone, and that number is growing at a rapid rate.

Because a BBS consists only of software installed on a PC, anyone can set one up. (You learn how to set up a BBS for yourself in Chapter 14.) Some public bulletin board systems are set up as a hobby while others are set up specifically to sell information or entertainment to the public. Many bulletin board systems are hybrids—systems that began as a hobby but now request donations to support the equipment and monthly telephone bills—much as a club or users' group asks for dues.

BBS software has matured to the point that nearly any electronic information application can be addressed with a BBS. Many companies are setting up BBSs to provide better service, on-line shopping, and other benefits to their customers. Much of the information in this chapter also applies to this type of business BBS, but the focus is on the public access BBS. Public access bulletin board systems often are referred to as *hobby BBSs*, but they can be much more than that term implies.

Computer BBSs are a tremendous resource. You can find information on subjects ranging from birds to boats on special-interest BBSs. Technical information on computer hardware, software, shareware, and public domain software also is available. (Chapter 4 discusses shareware and public domain software.) If you become active in the BBS world, you quickly learn more about computers than you can learn in nearly any other way. You also gain access to a network of experts that you can call on when you have problems.

The BBS world is not all wonderful, however—precisely because the BBS provides a grass roots, free market of information. The free flow of information on a BBS includes antisocial and destructive expressions. Although public BBSs are a robust source of information, they are also a reflection of society in general.

You will find that each BBS has its own personality, much like any gathering of people. Some personalities you will like, and other personalities you will not like. You can be sure, however, that a BBS with the information you need exists somewhere and includes the kind of people with whom you want to communicate. In the process of looking for that BBS, you will find many BBSs that interest you. You also are likely to be hooked by the excitement and energy you will find in the BBS world.

In this chapter you learn how to locate the telephone numbers of public access BBSs and how to configure your computer properly to call a BBS. You learn how to log on and find your way around on a BBS as well as what proper etiquette is in the BBS world.

Locating BBSs

If you are new to computers, the biggest hurdle you face is locating the telephone numbers of BBS systems. Many BBSs post the telephone number of other BBSs; after you call a few systems you can learn the numbers of other systems and build your own list of favorite BBSs. First, however, you need a place to begin. (This situation is like being unable to get a job without experience and being unable to get experience without a job!)

Two good sources of BBS numbers in your area are local computer stores and computer users' groups. Many computer users' groups even sponsor their own BBS to provide information to their members. At any computer users' group meeting, you can find several people who are active on BBSs in your local community. These people are your best source of local BBS information.

If you don't have access to a local users' group and the local computer store doesn't have any information, you are not out of luck. You can obtain printed sources of BBS information. Most newsstands that carry computer publications have *BoardWatch Magazine*, which lists BBS numbers. This magazine provides monthly information on the BBS community along with reviews of BBS systems around the world. If you are serious about keeping up with the BBS scene, *BoardWatch Magazine* is a valuable resource. *Computer Shopper* magazine also publishes a national BBS phone number list each month, and many other magazines also list BBS numbers from time to time.

When you use a BBS number list, you should keep in mind that the BBS community is extremely dynamic. In any given week, several hundred BBS systems change telephone numbers. Additionally, any list of numbers is subject to misprint. This means that phone numbers from even the most carefully maintained BBS list may be inaccurate.

You should make your first call to a BBS at a reasonable hour and leave your modem's speaker on. If a person answers the phone, you can assume that the number you have is incorrect. The person that answers the phone is not likely to have any information about the BBS you are calling, so don't bother them. Think for a moment about how you would feel if your telephone number was published accidentally in such a list. You should inform the person or publication from which you got the incorrect number so that the error can be corrected.

Calling a BBS

Placing your first call to a BBS can be a scary experience. You know that you have no idea what you are doing, and to make matters worse you are dealing with someone else's computer! You should remember, however, that these systems exist because their owners want you to call. You cannot do anything that will hurt the computer you are calling, so don't worry about pressing the wrong keys. Everyone has to start at the beginning, and you will be welcome if you are a good participant.

To connect successfully to a BBS, set your serial port parameters to 8 data bits, no parity, and 1 stop bit. This setting is used by most BBSs (see Chapter 1 to learn more about these parameters and Chapter 3 to learn how to set them in your terminal program). Next, tell your terminal program to dial the number of the BBS.

Many popular BBSs are extremely busy during evening hours; therefore, you may need to dial several times to get through. If you connect to the BBS and nothing is displayed on your terminal, press the Enter key a few times. Some BBSs require this step to determine your computer's speed before sending information.

Examining the Logon Screen

After you connect, the BBS sends you one or more opening screens. This opening information usually identifies the system you are calling and often contains information about what you must do to obtain full system access.

After the opening screen is displayed, you are prompted for your logon identification. Some systems ask for your first name and last name separately. Other systems ask for your full name on a single line. Look closely at the prompt to see what information is being requested. If you don't answer this question properly, you will not be granted access to the system.

Figure 13.1 shows a logon screen for a TBBS system that asks for your first and last names separately. Press Enter after you type each portion of your name. Figure 13.2 shows the same TBBS system configured to ask for your full name as a single entry. Figure 13.3 shows a logon screen for a PCBoard system that asks for your first and last names separately. Note the opening question about graphics in figure 13.3. Answering yes implies that your terminal program is set to support ANSI terminal emulation and the IBM PC graphics character set (see Chapter 16 to learn more about terminal emulation).

Notice, too, that your password echoes as dots, asterisks, or some other character so that observers cannot see your password easily. In some systems, the password is *case sensitive*. In those situations, you must enter not only the proper letters for your password, but you also must use the proper capitalization on the letters to match your initial password entry. If you are certain you are entering the proper password, but the system is not accepting it, you should verify that you have made each letter upper- or lowercase the same way you did when you first entered (or were assigned) your password.

```
Welcome to the Jungle.

This BBS is open to all who are willing to fill out the registration.

We welcome you to our discussion groups. The current "hot topic"
is the future of the IBM PC.




First Name? Phil
Last Name? Becker
Searching User File ...
Calling From AURORA, CO
Enter Your Password: *****
```

Fig. 13.1. A TBBS system logon that requests your first and last names separately.

```
Welcome to the Jungle.

This BBS is open to all who are willing to fill out the registration.

We welcome you to our discussion groups. The current "hot topic"
is the future of the IBM PC.




Enter your FULL Name? Phil Becker
Searching User File ...
Calling From AURORA, CO
Enter Your Password: *****
```

Fig. 13.2. A logon showing the same TBBS system configured to request your full name as a single entry.

```
CONNECT 2400 / 08-12-91 (12:55)

«« PCBoard Premium Bulletin Board »»
PCBoard (R) - Version 14.5/E3

Do you want graphics (Enter)=no? y

Welcome to the Jungle.

This BBS is open to all who are willing to fill out the regstration.

We welcome you to our discussion groups.  The current "hot topic"
is the future of the IBM PC.

What is your first name? Phil
 What is your last name? Becker
Password (Dots will echo)? ( .....        )




 Alt-Z FOR HELP| ANSI    |  FDX  | 38400 N81 | LOG CLOSED |  PRINT OFF | ON-LINE
```

Fig. 13.3. A logon showing a PCBoard system that asks for your first and last names separately.

Configuring the BBS to Your Display

After you enter your name, the system may ask you a few more configuration questions. These questions enable the system to adjust to your terminal program and give you the optimum display and the maximum features. This initial configuration step is often a source of confusion for a user who is new to PC communications. You may be asked questions that have no meaning to you, leaving you frustrated at the very beginning of your BBS experience. The reason the configuration is required is an extension of what you learned about serial communications in Chapter 1. Different computers have different characteristics, and the BBS needs to know enough about your computer and terminal program to present a proper display.

You learned in Chapter 3 that terminal emulation is required to enable formatted displays and color. Even the simple act of clearing the screen requires many different character sequences for different terminal emulations and computers. Although most computers have screens that display 80 characters per line and 24 lines of display space, some computers do not. Older computers, for example, may have only 40- column screens or fewer than 24 lines. Newer display devices can display more than 40 lines of up to 132 characters each.

During the configuration step you probably will be asked questions about the following:

- Line feeds

- Nulls

- ANSI emulation (Do you want color?, for example)

- Graphics display (whether your terminal can display IBM graphics characters)

You should be prepared to answer these questions before you call. After you have connected to a BBS, quickly finding the answers to these questions is difficult and guessing rarely works out well.

Line Feeds

Most computer printers and displays separate the act of moving to a new line on the printer or display into two parts. One part, called the *carriage return* (from the days of typewriters), moves the cursor to the beginning of the current line. The other part, called a *line feed*, advances the cursor to the next line. To advance the display to a new line, then, two characters are required—carriage return and line feed.

Because the need to perform one of these functions without the other is rare, some terminal emulations assume that a line feed is implied by the presence of a carriage return. This assumption saves one character per line sent, which saves time on any serial connection. The BBS line feeds configuration question is asking whether your system requires the sending of both characters to advance your display to a new line or whether only the carriage return is required. Answer yes if you don't know the answer.

If the line feed setting is incorrect, everything is printed on a single line (if you indicated that you didn't need line feeds when you actually did) or everything is double-spaced (if you indicated that you needed line feeds when you actually didn't). In such cases, you need to reconfigure the setting to correct the condition. Usually a BBS will provide a way to reset the terminal configuration after you log on. Look for a menu entry named Terminal Configuration, User Profile, or something similar to make this change.

Null

When serial communications were used primarily for printing terminals, some method was required to delay sending characters after each line. The delay was needed because the mechanical parts needed time to travel from the end of one line to the beginning of the next line. The *null* character was chosen to do nothing at the receiving end so that it could be used for such delay timing. Because a null is a full character like any other, it takes time to send. When a null is received, however, it is discarded, so the only effect of sending a null is to use up time.

Today's computers rarely require such delays—even at very high speeds. You therefore should begin by indicating that you require no nulls. If you notice that the first one or two characters of several lines are missing, however, reconfigure to indicate that you need a number of nulls equal to the number of characters you are losing from the beginning of the line. If you configure for a large number of nulls when you don't need them, the computer seems slow for no obvious reason.

ANSI Emulation

Most BBSs today use color and screen cursor positioning to provide a display that is more visually interesting and informative than is possible by simply typing plain black and white text on your screen. To use color and screen formatting, however, your terminal program must be able to interpret command sequences that go beyond letters and numbers. Chapter 16 explains terminal emulation in detail, but for this discussion you only need to know that in order for you to properly see these fancier displays, your terminal program must have ANSI terminal emulation set on.

If the BBS asks about color or ANSI, it is asking you whether your terminal program is currently configured for ANSI terminal emulation. The BBS needs to know whether or not it can send you the color and formatted screens ANSI emulation allows, because the ANSI display screen will look like garbage characters with no meaning if you don't have ANSI emulation configured.

Most BBS programs have an alternate format for each display, which enables you to access all of the information without requiring ANSI emulation. That is the format you will see if you answer no to this configuration question. Because the dominant feature you will notice about ANSI screens is their color, some software asks `Do you want color?` for this question. The BBS is still asking you whether your terminal program supports ANSI emulation.

One note about ANSI emulation—it requires far more characters to be sent to your terminal program to produce the display than are required by the plain text screens. If you are using a slow modem (300 bps or 1200 bps), therefore, you may want to answer the ANSI emulation question with no to speed up your transmission.

Graphics Display

In another effort to produce displays that are more informative visually, many BBS programs use the IBM character graphics set to draw boxes and other line displays. If the BBS asks you about Graphics or IBM Graphics it is asking whether your terminal program supports the display of the IBM extended graphics character set. If you answer no to this question, then any such lines will be drawn using text dashes and vertical bars that are not as visually attractive.

Many computers that are not IBM PCs now are supporting this extended character set because of its popularity. Many terminal programs on computers such as the Apple Macintosh can be configured to display IBM Graphics characters. If your computer can display this extended character set, you should answer yes to this question as there is no extra transmission time penalty to obtain the better looking display (unlike ANSI mentioned above).

You may have to configure some terminal programs specially to allow these characters to print properly on your screen. If your terminal program is not set for 8-bit display, these characters will show as capital letters instead of lines.

Obtaining Full Access

After you answer all of the configuration questions, the system informs you of the requirements for obtaining full access. Because each BBS is an independent operation, each system operator (known as a *SYSOP*) has different procedures for granting full access. Some systems still enable a new caller to gain full access immediately, but because of abuses, this practice has become quite rare.

SYSOPs who run systems open to the public usually require that you register before you can gain full access. The fact that you are being asked to register does not mean that you will be denied access or charged to have access. Almost all systems now require registration because of the incidence of caller abuse, which has become common through the years. Such registration usually requires you to give your full name, address, and telephone number. You should be prepared to supply this information on your first call to any BBS.

Such registration may seem unnecessary, but its primary purpose is to screen out the few users who enjoy causing trouble on public systems. Some systems give you the benefit of the doubt and provide you with immediate full access when you input the requested information. Other systems impose a 24-hour (or longer) waiting period, which enables the SYSOP to verify that the information you gave is correct.

Subscribing to a BBS

A growing number of bulletin board systems operate on a subscription basis. When you first call this type of BBS, you are given information explaining the subscription rates. Subscription systems usually give you access at a demonstration level so that you can evaluate the information offered.

Subscribing to a BBS is much like any other buying decision. You decide whether what is being offered is worth the price being asked. If you do not want to subscribe, you don't have to do anything to avoid being billed—simply don't call back.

Subscription systems often offer extremely up-to-date resources and information. If a BBS specializes in shareware or public domain software, that BBS usually has an up-to-date set of files. Many subscription systems also have a group of users who are extremely knowledgeable about a particular subject. The subscription fee helps pay for the costs of running the BBS and offsets the cost of maintaining the information and file databases.

Learning BBS Etiquette

As with all social interactions, both proper and improper behavior is possible when you're using a public BBS. Most BBS etiquette is just common sense and follows the same rules as any social interaction.

If a BBS has an active message system, you should try to contribute to the ongoing discussions. As a new caller, the best approach is to leave an introductory message in a public area asking what the rules are for the system. On your first call, do not leave messages containing dry humor that may appear antagonistic (dry humor is easily misconstrued in a printed medium and people need to get to know you so that they have more context for what you write before you can be sure they won't misunderstand). You should spend a bit of time learning about each system to see what is considered appropriate in each setting.

One area of BBS etiquette that isn't obvious to many new BBS users is file transfers. Many systems charge for file downloads. On these systems, the fee enables you to download as many files as you want. On many systems, however, the SYSOP does not explicitly charge for file downloads but operates on the basis of trading files. You are expected to contribute some public domain or shareware files to the system to retain download rights. You should learn what the rules are on each system before you begin to download files. Taking a large number of files without making a contribution is considered rude on some systems, but on other systems that practice is acceptable. Some SYSOPs are very sensitive about this subject.

If you go slowly at first and ask questions when you have them, you will quickly become a seasoned user of BBSs. You will find that you are welcomed in most settings, and you will build access to a valuable information resource.

Finding Your Way Around a BBS

Each BBS is set up differently. Almost all systems, however, are *menu driven*, which means that you are shown a menu of choices that you can make at each location within the BBS. To avoid getting lost, think of these menus as locations or *rooms* within the BBS.

Two types of menu selections are available. The first type deals with moving from menu to menu within the BBS. These options usually are labeled clearly so that you can tell you will be moving to a new menu or functional area of the system. The second type of menu option deals with actually performing a function, such as listing files, displaying information, and so on.

If you read each menu carefully, you should be able to avoid getting lost. The best approach is to go slowly with each new BBS that you call. If you understand each area before moving on, you can avoid being overwhelmed and getting lost.

Locating Files

One of the largest attractions of bulletin board systems is the library of files offered for downloading. Often, so many files are available on a single BBS that even a list of their names can be overwhelming. A strategy is necessary to locate files of interest to you.

Although the procedure varies for each type of BBS software, a way to search for specific files almost always exists. On some systems, for example, the option is named Zippy DIR Search, and on others an option is given when you list a file directory. When you select the file search option, you are asked for text strings to match in file names and descriptions. For still other systems, you may have to specify such search strings as an option on the list command. Spend a few minutes becoming acquainted with file searching methods on the systems you call because file searching methods can save you a great deal of time.

Figure 13.4 shows the screen output of a PCBoard Zippy DIR Search command. You select the command from the main menu and then PCBoard asks you to input a text string to search for. Next PCBoard asks whether you want it to search all file areas it has for the text string you specify, or whether it should search only one specified directory. You can enter *a* to ask PCBoard to search all directories. In this example, I asked PCBoard to find files that had the text LAN in their descriptions. Only one file on the system matched my search criteria, so it was the only file listed. After you find the files you want to download, you can use the download command from the main menu to transfer them to your system.

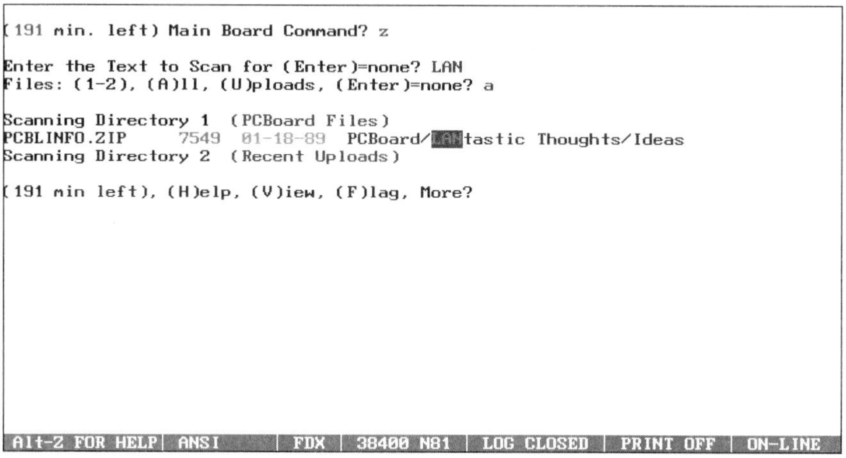

```
(191 min. left) Main Board Command? z

Enter the Text to Scan for (Enter)=none? LAN
Files: (1-2), (A)ll, (U)ploads, (Enter)=none? a

Scanning Directory 1  (PCBoard Files)
PCBLINFO.ZIP    7549  01-18-89  PCBoard/LANtastic Thoughts/Ideas
Scanning Directory 2  (Recent Uploads)

(191 min left), (H)elp, (V)iew, (F)lag, More?
```

```
Alt-Z FOR HELP| ANSI      | FDX | 38400 N81 | LOG CLOSED | PRINT OFF | ON-LINE
```

Fig 13.4. *Using PCBoard Zippy DIR search to locate files.*

To illustrate the different methods used by different BBS software to search for files, figure 13.5 shows a TBBS selective list command. In this example, I asked for all files having the keyword *kermit* in their name or description by entering the command *l kermit*. TBBS responded by listing all files in this file area meeting that search criteria. Three files met my search criteria and were listed. In TBBS I immediately could enter the name of the file I wanted and the download sequence began. In this example, my download protocol was set to XMODEM.

Downloading files from any BBS requires you to know how your terminal program does file transfers. Most terminal programs require you to press the PgDn key and select the protocol. Some protocols require you to repeat the file name on your computer before the transfer will begin, and others do not. One protocol, ZMODEM, requires no action on your part at all because it starts automatically.

```
>>> Files Posted on or after 08-12-91

* none *

<D>ownload, <P>rotocol, <E>xamine, <N>ew, <H>elp, or <L>ist
Selection or <CR> to exit: l kermit
>>> Selection string = kermit

MLMOD30         2432    6-14-90  Fix KERMIT "expanding file" bug
MLMOD20         2304    7-13-89  Autobaud and KERMIT packet size prob
MLMOD11         2304    2-26-89  Auto-baud, KERMIT, & "Stranded File" fix

<D>ownload, <P>rotocol, <E>xamine, <N>ew, <H>elp, or <L>ist
Selection or <CR> to exit: d mlmod30

File Name: MLMOD30
File Size: 19 Records
 Protocol: XMODEM
Est. Time: 0 mins, 03 secs at 9600 bps

Awaiting Start Signal
(Ctrl-X to abort)

 Alt-Z FOR HELP| ANSI      |  FDX | 38400 N81 | LOG CLOSED | PRINT OFF | ON-LINE
```

Fig 13.5. *TBBS selective list to locate files.*

In the example shown in figure 13.5, the BBS is waiting for me to tell my terminal program to start the download. If I am using PROCOMM PLUS as my terminal software, I then press the PgDn key followed by X to select XMODEM protocol. XMODEM requires me to enter the file name again and press enter after which the file transfer begins.

When you are trying to find files on a BBS, you also should look for a list of files made available in downloadable form. Such a file is often provided and if it is, you should download it first. You then can examine this directory of the files available on that BBS at your leisure on your computer—not while you are connected to the BBS. After you choose the files you want to download, you can call back and efficiently select the desired files.

Using Off-Line Reader Programs

The message areas of a BBS are where social interaction occurs. The BBS message area is a unique communication structure, and after you participate in BBS message areas for awhile, you will see that they offer unique capabilities. You can ask a question of a very large audience, and each reader can decide whether to answer. This process enables you to find expertise when you don't know exactly where to look. The social aspects of the BBS message base also are quite interesting.

Many BBSs, however, have many message areas and only one or two telephone lines. If each caller spends the time to read and respond to messages on-line, the system can service very few callers. In addition, if the message bases become quite large, you may not have time to read and digest all the messages on a single call.

To address these problems, a category of software known as the *off-line reader* has been created. Off-line reader software runs on your computer and re-creates the on-line environment of a BBS without actually being connected to another computer. You can call a BBS that supports such a reader and indicate through special commands that you would like the system to extract and package certain messages for you.

If a BBS supports off-line readers, it has a menu option clearly indicating that it is the entrance to the off-line reader interface. When you select that option, you are asked a series of questions to determine which messages you want transferred to your off-line reader. The messages you select then are packed in a specially formatted file and downloaded to your computer. You then can disconnect from the BBS and run your off-line reader program.

While you are in the off-line reader program, you can read and reply to messages just as if you were connected to the BBS. You have the option of reading some of the messages, stopping, and then continuing later. When you are finished, call the BBS again, upload any packaged responses that you have made, and then repeat the process.

The off-line reader gives you total control over your reading and replying time. At the same time, the off-line reader frees up a tremendous amount of time on the BBS so that more callers can use the system.

Many different brands of off-line readers are in use, and you should learn about the off-line readers that are used by your favorite bulletin board systems. The most popular off-line reader interface format today is the QWK format and is supported by most off-line reader programs. The off-line reader programs usually are shareware and can be downloaded from the BBSs that use them. Registration fees vary but are usually in the $25 range. After you have used an off-line reader and experienced its convenience, you will want to use it at every opportunity.

Chapter Summary

BBSs represent a valuable but inexpensive resource. On BBSs you can find programs and information that are not easily available anywhere else. You also meet people who can help you with your computer problems and teach you about your computer as well. In addition to providing information, BBSs provide rich entertainment. The time you spend learning to call and use BBSs will be the best investment you can make in your computer.

14

Becoming a
BBS SYSOP

O nly one computer BBS existed in 1978. Today more than 100,000 bulletin board systems (30,000 public and 70,000 private) are in operation. BBS growth has been extremely rapid, and growth still is accelerating. More than 1 million BBSs are expected to be in use by the year 2000. Assuming that the average BBS has at least 50 users, more than 50 million people will be using a BBS by that time.

SYSOP is the abbreviation for *SYS*tem *OP*erator. In this chapter, you learn what is involved in becoming the operator of a BBS system. In Chapter 13 you learned how to use a BBS from the calling side. In this chapter, you learn what is involved in setting up and operating a BBS. You learn what you can expect BBS software to do for you and what you have to do to make BBS software solve your problems. You learn how to evaluate BBS software, how much time you must spend designing and setting up a BBS system, and what the hardware requirements are.

Who Becomes a SYSOP?

BBSs initially were run by people who loved BBS technology. During the early period, only highly technical people could create and operate a BBS. Several philosophies of how BBS software should operate and what services the software should provide were developed and explored. BBS software represented a technological frontier, and these pioneers devoted themselves for the pure love of the enterprise.

As BBS technology developed, people with less expertise were able to participate. Many people were drawn by dreams of what BBS technology could accomplish. Some users were drawn by the almost magical feelings associated with knowing that a computer in your home can communicate instantly with people anywhere in the world. Other people were motivated by the system's capability to access large amounts of knowledge at their request. Still other people started using a BBS to solve data delivery problems in their businesses.

The ratio of private to public BBS systems currently is greater than two to one. As this ratio indicates, the business use of BBS software is becoming commonplace. BBS software installed and controlled by a single individual, however, still can provide public access and entertainment.

People choose to invest time and money in setting up and operating a BBS for many reasons. Many users still find starting a BBS to be one of the most informative and pleasurable experiences possible in computer communications.

After you have used BBSs and learned what BBSs can do, you may find that you want to set up a BBS. Whether you are looking to use BBS software in a business setting or planning to set up a personal system for hobby use, you must consider several issues. The first task is to understand what is involved in setting up and operating a BBS.

Where Do I Start?

A caller does not realize what is going on behind the scenes at a BBS—being a BBS system operator (SYSOP) is an experience totally different from being a BBS caller. As the SYSOP, you select and set up the BBS software to create the system. You also design the BBS layout, obtain and catalog any files you make available for downloading, and maintain the message bases.

With most types of software, you first look for magazine reviews and articles to give you an idea of what is available. From these articles you get an idea of available programs and their features. During this process you also develop a feeling for the types of tasks you face in using a particular kind of software.

Because BBS software is a niche category and quite difficult to review, however, not many of the traditional resources are available to help you understand the tasks you are facing. Simply finding out what software is available can be difficult. As a result, many people begin setting up a BBS with no idea of what is involved; these users obtain their education at the school of hard knocks.

This chapter helps you avoid that problem as you learn what types of software are available and what being a SYSOP is all about. You learn how to ask the right questions when designing your BBS, how to locate BBS software, and how to get information. In addition, you learn several rules of thumb to help you evaluate BBS software.

What Am I Getting Into?

People who start a BBS usually expect to accomplish certain goals. These expectations often are based on what these users have learned from using one or more BBSs. When you call a BBS, however, you are connected to a computer you never see. You are using software that you never understand from the operator's viewpoint.

You also are using a finished product, and you cannot see how much time and effort the SYSOP put into creating that BBS. From this perspective, concluding that the BBS software has almost magical powers is easy. Although BBS software indeed has become very powerful, BBS software cannot do every remote access job you can imagine. In addition, learning how to use BBS software effectively takes time.

Setting realistic expectations of what BBS software can do is difficult. Most users, therefore, go to an extreme. They expect to find software that does everything they can imagine (with very little work on their part), or they expect almost nothing and are content to struggle with software that is hard to use and unreliable.

Your first step in setting up a BBS should be to develop a firm vision of what you want your BBS to do. To develop this vision, you need to know what BBS software is capable of doing. A wide range of BBS software is available, and tradeoffs can become complicated if you start comparing features directly. Before you begin comparing BBS software, decide which tasks are most important to you. Some tasks are much easier to implement than others with today's BBS software. If you know what is important to your application before you begin, you can better evaluate how much a particular feature is worth to you.

What Are Realistic Expectations?

With today's BBS software, you realistically can expect to collect and organize information for remote access. You also should be able to achieve multiuser capability for up to 50 or more telephone lines with good reliability. To put these expectations in perspective, a 50-line system easily can provide access to more than 300 users an hour (3000 users per ten-hour day)—with no caller ever getting a busy signal!

With today's state-of-the-art software and well-maintained hardware, you can expect a system to run 24 hours a day, seven days a week, with greater than 95 percent availability. If you use good software, you should not have to put up with unexplained crashes or system malfunctions.

What is not realistic is to expect a BBS to spring forth ready to use, embodying your ideas, with only a few hours work on your part. You should expect a learning curve for the software you choose. The time frame can vary from a few weeks to a couple of months (if you have had no experience with BBS software). The learning curve often is proportional to the power and capability of the software you choose; that is, you will experience a longer learning curve for a more powerful package.

After you have learned how to use a particular BBS package, you should be able to design a system quite quickly. You also should be able to learn another BBS package fairly rapidly.

Developing a BBS usually is an evolutionary process. First, you set up a system according to your initial vision. As the system is used, you see that certain changes may enable the system to do a better job. As you go through this cycle, your BBS installation becomes more "tuned" to the exact tasks you had in mind.

When you see an easy-to-use BBS that performs its function well, you usually can conclude that the BBS has gone through several refinement iterations. Keep in mind that the "tuning" begins when you create your BBS; be sure that the software you choose accommodates this refinement cycle.

If you want to establish a BBS "on the cheap," you can combine the BBS computer with your regular computing work. This situation requires the use of a DOS multitasker and greatly reduces both the performance and reliability of your BBS. Avoid this approach if at all possible. If you do go this route, be aware that you are making a very large compromise that will be reflected in your BBS.

Information Presentation versus Remote Hosting

An area that can cause confusion when you plan a BBS is the difference between information presentation and remote hosting of software. These two functions are quite different and require a different set of software capabilities.

Information presentation is any task whose end result is passing a data file between the host computer (BBS) and the caller. These tasks can be quite complex (for example, searching a file library by keywords) but do not cause the file to be changed before being sent. Information presentation tasks involve only selection and transfer of data files and messages. Applications in this area are data collection, file upload and download, and electronic mail.

Remote hosting of software is the use of any program that uses input from the caller to produce a specific result in real time. In the general case, remote hosting enables you to run any program you choose from a remote location. Remote hosting has several subcategories, the largest of which is known as transaction processing.

Transaction processing refers to any application in which the caller enters selection parameters and the BBS calculates and produces a response specific to those parameters. Tasks of this type range from something as simple as an on-line calculator to something as complicated as an airline reservation system.

Other remote hosting applications include running a word processor or spreadsheet from another location. Each category of remote hosting places drastically different demands on the BBS software; therefore, you should determine early what portion (if any) of your desired task requires remote hosting. The need for unique remote hosting tasks complicates BBS installations faster than any other parameter.

BBS software started out simply presenting information. Messages and file transfer were the software's only jobs. Later, remote hosting was desired, and various forms were implemented. The most common extensions of remote hosting have found their way into BBS software remote database operation and on-line games. Today, this set of remote hosting applications can be handled smoothly with several existing BBS software packages. Most other remote hosting applications still are searching for package solutions and presently tend to be unique programming exercises.

With a DOS-based BBS, general purpose remote hosting is simple only when you are using a single telephone line. After you move to a multiline system, you must build a multiuser remote host capability—a complicated process. If your job requires extensive, general purpose remote hosting, expect to spend much more time and money to get the desired results. You also can expect a more technically challenging project.

How Long Will Setting Up a BBS Take?

The amount of time you spend setting up your BBS depends a great deal on how well you define your task before you start. If you plan well, you can choose software and hardware that enable you to build a BBS quite quickly and reliably. If you are working with BBS software for the first time, you should add time to account for the learning curve of the software you choose.

A BBS project (especially your first project) rarely is completed in one shot. You usually develop a system that reflects your best guess of what you want. As you use the system, you learn more about the technology, and you can fine-tune the design. With most of today's software, you can have a BBS up and operational in a couple of hours. Fully creating the system you had in mind, however, may take two or three months. You probably will return to the design process and update your system several times during the following few months.

How Much Money and Time Are Required To Set Up a BBS?

One of the most attractive features of a BBS is that a BBS can be started with very little money and then expanded incrementally. This situation requires a very low capital investment to start and planned incremental expenditures to add capacity as usage grows.

A BBS also enables you to trade time for money over quite a wide range. In general, if you spend more money, you spend less time setting things up and ensuring that everything is reliable. The method of hardware and software implementation you choose have an effect as well. The two primary methods available to implement a multiuser BBS are discussed later in this chapter under "Understanding Multiline Systems."

The cost to implement a BBS usually is very low. You quite possibly could buy everything you need (computer, modems, and software) for a 16 user BBS for less than $8,000. This total cost for all hardware and software represents approximately $500 per line. A few years ago, you would have paid that much for the modems alone!

When you compare the price of different implementations, you have to compare the total cost of hardware plus software. Comparing total cost is necessary because different BBS implementation methods trade off hardware versus software quite differently. Some methods are heavy on hardware costs but light on software costs; other methods are the opposite. If you compare only software to software costs or hardware to hardware costs, you are not making a full comparison.

One BBS approach, for example, requires a $300 software investment and a $5,000 hardware investment. A less expensive approach costs $1,000 for software and $3,500 for hardware. If you look only at the software prices, however, the $300 (versus $1000) price may give you the opposite impression. You also should factor in the complexity of each approach, because complexity affects the time and expertise you need to set up the system.

How Do I Choose the Right BBS Software?

Choosing software to build your BBS is not as simple as it may seem. A wide variety of BBS software is available, and each package provides a different set of capabilities. Before you choose a BBS software package, you need to make some decisions about what you want your BBS to do.

As you narrow down your choices, try to find and talk with SYSOPs who use each kind of software you are considering. You will find that each SYSOP is convinced that he or she has made the best choice, but if you listen closely you can learn about factors that are important to you. You also can get a good idea of the capabilities and difficulties associated with each approach.

Choosing Single-Line versus Multiline Software

One of the most important decisions you have to make is the number of telephone lines or terminals you want your BBS to support. You should think not only about how many lines you want initially but also about how many lines you expect to need in the future.

You first need to examine the difference between one line and more than one line. Several BBS packages cannot run more than one line; therefore, if you know you will need more than one line, you can rule out such packages early.

You also need to examine the use of four lines. Many BBS software packages provide excellent service and capability for two or three lines but become prohibitively complicated or expensive if you need more than a few lines. Knowing what you need up front can help you avoid a conversion effort later, which can save you a tremendous amount of time and money.

If you need more than one telephone line (and most active BBS applications do), you need a separate modem for each telephone line. Hooking up multiple lines to a BBS is similar to hooking up multiple lines for a voice system. You can give out one number to callers, however, if you ask the telephone company to connect several phone numbers in a *hunt group*. When users call that number, the telephone company automatically rings the next line in the sequence that is not busy. At your end, you have individual phone jacks to hook up to individual modems—one for each BBS line.

Estimating Line Loading

Most people do not have experience in estimating line usage; however, estimating line usage isn't difficult. You begin by guessing how much time an average user of your system will require on an average call. If you don't know what is average for your system, you can use the following rules of thumb:

Type of system use	Average call length
General Usage	15 minutes
File Downloads 2400 bps	60 minutes
File Downloads 9600 bps	15 minutes
Multiuser Chat Usage	45 minutes

Suppose that an average call lasts 15 minutes. Estimate how many users will call your system per day. For this example, assume that 100 users will be calling each day.

Armed with these two numbers, you can see how much available telephone time your callers will need each day. In this example, 100 users need 15 minutes each, which totals 25 hours.

The next item you need to pin down is how many clock hours represent a "day" or calling period. In some cases, you may have all 24 hours in a day available. In other applications, you may have to provide service to all callers during a smaller time period (say a working business day) or an evening time slot, which is less than 24 hours.

For this example, suppose that you need to provide access to all 100 callers within an 8-hour business day. That parameter means you must provide 25 hours of available calling time in an 8-hour period. To achieve this goal, you would require 25/8 or 3.125 telephone lines. You therefore would have to install a four-line system.

The following formula summarizes the estimating process:

$$\text{\# of lines} = \frac{\text{hours/call} * \text{callers/period}}{\text{hours/period}}$$

In this example, the following numbers are used:

hours/call = 0.25 (15 minutes)

callers/period = 100

hours/period = 8

Use the numbers in the formula to produce the following result:

$$3.125 = \frac{0.25 * 100}{8}$$

You should choose the number of lines for your system based on peak requirements. Because these numbers involve some guessing, you should start smaller and see what the actual requirements become. After you have some experience with how your system is being used, you can plug the real numbers you have found you need into the formula to arrive at the size you need for your system.

> Most people who never have run an on-line system tend to overestimate drastically the number of telephone lines they need. A system with 50 lines can handle a very active community of 10,000 to 30,000 users or a less active community of up to 100,000 users. If you use the formula provided in this section, you can avoid making this common mistake.

Understanding Multiline Systems

Although the details of multiuser software can be quite complicated, BBS software uses only two methods to provide multiple access to a single BBS.

The first method is the *one-copy-per-node* method. In this method, the BBS software requires that a multitasking environment be established. The software then executes multiple copies of the program at the same time—one copy on each node of the multitasking environment. By using the file-sharing services of the multitasking environment, programs of this type provide multiple telephone line access to a common BBS. These programs therefore require a multitasking operating system (for example, DESQview or UNIX) or a LAN, which connects multiple computers to build a multiuser platform.

The advantage of the one-copy-per-node method is that programs running outside the BBS itself (commonly called *doors*) can be run on a multiuser basis. If your application requires such general purpose remote hosting, using this method is necessary.

The disadvantage of the one-copy-per-node method is that a great deal of software must be coordinated, and a great deal of hardware must be provided. The expense of the hardware becomes much greater than the expense of the software after a small number of lines. The complexity level of this system also rises rapidly because of the underlying multitasking platform.

The other method that BBS software programs use to provide multiline access is the *integrated multitasker* method. Software that uses this method currently can handle up to 64 users on a single computer using only DOS. The BBS software shares itself internally so that nothing but DOS and a single computer is required. A method of providing multiple serial ports is the only special hardware required. This method usually involves the use of a multiport serial card.

The advantages of the integrated multitasker method are lower total system costs—especially at higher line counts—and fewer complications during setup and use. Because much less hardware is involved and only one piece of software needs to be installed and kept operating, integrated multitasker systems tend to be much more reliable than one copy per node systems. (Less hardware means less equipment that can break or malfunction.) A single CPU is inherently more reliable than 20 computers connected by a LAN—simply because a single computer has fewer components.

One disadvantage is that these programs are integrated and cannot run outside software or doors—although most integrated multitasker packages provide expansion modules that can overcome much of this disadvantage.

To decide between the one-copy-per-node and integrated multitasker methods of achieving multiline operation, you should understand the impact of each approach on hardware.

Hardware Considerations for the Integrated Multitasker

The integrated multitasker approach requires only one computer. For 8 or more lines, you should use an 80386 computer. The major impact of more lines in this system is that you need more memory and more serial ports. For a system handling up to 64 users, you may need as much as 8M of RAM. The only special hardware required is multiport serial cards to connect the large number of modems. These cards plug into a standard IBM PC expansion slot and come in sizes up to 16 lines on a single card. Prices range from approximately $150 for a simple, nonexpandable 4-line card to about $1000 for a 16-user card.

The integrated multitasker approach requires no extra software beyond DOS and the BBS software. These two elements, therefore, represent the cost of an entire system.

Hardware Considerations for One-Copy-Per-Node

The one-copy-per-node approach requires a multitasker underneath it. For up to 4 lines, using a very powerful computer (such as an 80486 or 33 MHz 80386) and multitasking software (such as DESQview or Multi-Link) is possible. This capability holds the hardware cost to a single computer and a four-port serial card for four lines. You have to add the cost of the DOS multitasker to the DOS and BBS software; in addition, the setup is much more complicated because you must install and debug the multitasker as well as the BBS itself. If you use a system such as UNIX, you get into a much more complicated setup and much greater hardware requirements.

For more than four users, you must use multiple computers and a LAN to provide the multitasking base for the one-copy-per-node software. For line counts up to 10 or so, low-cost LANs such as LANtastic are very effective. For higher line counts, you need to use large system LAN software such as Novell NetWare.

When using the LAN approach, you have to buy a computer for each telephone line in addition to the LAN hardware and software to connect them. This makes the cost of the LAN approach rise very rapidly when you use many lines. By the time you have eight or more lines, the LAN approach costs more than five times as much in hardware as the integrated multitasker. From that point on, the hardware costs escalate rapidly.

The benefit you gain with the one-copy-per-node approach is that you can run nearly any DOS program remotely as part of your BBS; however, the program may need to be modified to handle multiuser access. If this capability is essential to your task, then you must use the one-copy-per-node approach.

Understanding BBS Software Styles

Different software packages offer you different methods to set up a BBS. Some software packages have a standard interface approach; other packages seek to give you maximum flexibility in designing the presentation of your BBS.

Software that has a *standard interface* has a fixed structure for presenting on-line information. The philosophy behind standard interface software is that a common menu format enables users to learn one way to use a BBS—

no matter how many systems they call. In return for standardization, your ability to change the appearance and structure of the BBS is limited. Standard interface software exchanges ease of installation and configuration for limited flexibility in BBS layout and operation.

Software that attempts to give you maximum flexibility in presentation design falls into two categories. The first category provides the source code for the BBS so that you can rewrite the interface any way you want. The second category uses a menu template approach to provide maximum flexibility in design without changing any program code.

If you are a programmer, having the source code may be desirable. For most people, however, having the source code does not help because today's BBS software design is extremely complicated to program. Template systems offer tremendous flexibility to nonprogrammers, and large and complex systems routinely are built by people who have never written a program.

The major tradeoff between these approaches is that you spend more time learning how to build templates with a template system, but you are able to build the exact system you want. Standard interface systems are appropriate if their design meets your needs. If you need something the standard interface systems do not provide, you have to add external programs or rewrite the source code—an extremely difficult and error-prone process.

Understanding Commercial BBS Software versus Shareware

BBS software is available in many forms. An active shareware market exists for BBS software—especially single-line BBS software. (Refer to Chapter 4 for in-depth discussion of shareware.) All kinds of commercial BBS software also is available. True public domain BBS software is free. Shareware BBS software usually is priced from $50 to $150 per copy. Commercial BBS software is priced from $100 for basic systems to several thousand dollars for high-end systems. How should you decide where to obtain your software?

If you are experimenting with BBS software and want to learn without much expense, the shareware route clearly is the way to go. If you have a commercial need for a BBS or are planning a large installation for a specific purpose, however, the choice is not as easy.

Shareware BBS software is easy to obtain. You usually can find shareware available for downloading from public BBSs. Shareware also is available from many user groups. All shareware BBS software that can handle multiple lines uses the one-copy-per-node method described earlier in this chapter under "Understanding Multiline Systems." Support for shareware BBS software, however, may be available only on an erratic basis. You therefore should be sure that the software available today meets your needs, because you cannot count on updates in the future.

The following tables list commercial and shareware BBS software, the methods used to do multitasking, and the philosophy of presentation design. For commercial software, the companies' addresses and telephone numbers are provided as well.

Table 5.1
Commercial BBS Software

Name	Multiuser method	Presentation method	Customization method(s)
TBBS	Integrated	Template	Templates & expansion modules
The Major BBS	Integrated	Standard	Source code available & expansion modules
WildCat!	1 Copy/Node	Standard	Menu appearance & external "doors"
PC Board	1 Copy/Node	Standard	Menu appearance & external "doors"

You can get the commercial software listed in table 5.1 at the following addresses:

TBBS—eSoft, Inc.
15200 E. Girard Avenue
Suite 2550
Aurora, CO 80014
(303) 699-6565

Wildcat!—Mustang Software
915 17th Street
Bakersfield, CA 93301
(805) 395-0713

Major BBS—Galacticomm, Inc.
4101 S.W. 47 Avenue
Suite 101
Ft. Lauderdale, FL 33314
(305) 583-5990

PC Board—Clark Development
6000 Fashion Ave.
Suite 101
Murray, UT 84107
(801) 261-1686

Table 5.2
Public Domain and Shareware BBS Software

Name	Presentation method	Customization methods
RBBS-PC	Standard	Menu appearance, source code available, external "doors"
Opus	Standard	Menu appearance, menu macros, external "doors"
Remote Access	Template	Templates, external "doors"
QuickBBS	Template	Templates, external "doors"
SuperBBS	Template	Templates, external "doors"
WWIV	Standard	External "doors"

Keep the following points in mind when considering public domain and
shareware BBS software:

- All shareware and public domain programs use the one-copy-per-
  node multitasking method.

- You can obtain these programs by downloading them from a
  public access BBS or a commercial service such as CompuServe.

- External doors are programs that can be remote hosted; doors are
  not part of the BBS software.

- Public domain software is free, but shareware is not. Shareware is a
  "try before you buy" sales process, not a source of free software.

Table 5.3
Comparison of BBS Features

Name	Maximum message areas	Maximum FidoNet capability	Number of users	Text search	True database	Doors
TBBS	50,000	Yes	64	Yes	Yes	No
WildCat!	26	Yes	250	No	No	Yes
Major BBS	200	No	64	Yes	Yes	No
PC Board	65,000	3rd party	99	Yes	No	Yes
RBBS	unlim.	3rd party	99	No	No	Yes
Opus	255	Yes	1	No	No	Yes
Remote/ Access	200	Yes	1	No	No	Yes
QuickBBS	200	Yes	1	No	No	Yes
SuperBBS	200	Yes	1	No	No	Yes
WWIV	32	3rd party	1	No	No	Yes

When comparing BBS features, keep in mind the following items:

- Features may be added in future releases.

- TBBS and Major BBS have expansion modules that serve many functions for which other systems require doors.

- TBBS and Major BBS use the integrated multitasker method to run all users on a single computer. All other programs use the one-copy-per-node multiuser method.

Selecting System Hardware

The computer hardware you need is determined largely by the software you choose. For a single-line BBS or a multiuser BBS that uses the integrated multitasker approach, you need only one computer. For small line counts, a PC/XT-style machine with an 8088 CPU works well. In fact, this type of computer can run up to 8 lines at 2400 bps with some software.

Unless you are just dabbling, you should plan to dedicate a computer to running the BBS. If you are using high-speed modems (9600 bps or faster), you need at least a PC/AT-level 80286 computer. One 9600 bps modem represents the same load as four or five 2400 bps modems; therefore, you need more CPU power to handle this load.

For larger systems that use the integrated multitasker programs, you need the power of an 80386 computer. A 20 Mhz computer can handle up to 16 lines—even with a few 9600 bps modems. For larger systems, you should use a computer that runs at 25 Mhz (or faster).

The most important part of a BBS computer is the disk system. The efficiency and speed of your disk affects BBS performance more than any other single item. You therefore want to get the fastest disk that you can. The disk size is determined primarily by the files you want to make available on-line. Approximately 3M for DOS and BBS software is about right in most cases.

If you choose software that uses the one-copy-per-node multiuser method, you need to add LAN software and multiple computers.

When you purchase hardware for a BBS, remember that the equipment will be running 24 hours a day, seven days a week. Buy hardware that is conservatively rated and runs cool—heat is the largest source of hardware failure over time. Choose modems that are rated as powered on 100 percent of the time, and be sure the modems are well ventilated.

How Do I Make My BBS Pay for Itself?

Although you may start a public BBS as a hobby, you eventually may want to expand the BBS beyond the size you want to fund yourself. You also may want to run the BBS as a business—an enterprise that has made several people good incomes. In either case, the secrets described in the following paragraphs can help your BBS bring in money.

Secret 1: Accept Credit Cards

This reality cannot be overstated. When people call your BBS, they may see a great deal they like. They may want to pay you, and they may intend to send you a check first thing in the morning. Most people will not send you that check. Morning is another day, and the time just isn't available to stop and send you money. If that same person can type in a credit card number, phone number, and address while on-line with your system, he or she will. If you accept credit cards, you can expect to recognize a minimum of 10 times as much income as if you don't. No BBS will make money without accepting credit cards.

Getting yourself in a position to take credit cards can be difficult. Many banks no longer issue merchant numbers to small businesses that do not have a store front. Because your main business will be telephone orders, getting a bank to issue you a merchant number can be tough.

You can take several steps to combat this problem. First, be sure that you have talked to every bank in your town. Just because three banks have turned you down does not mean the next bank will. If you cannot convince anyone that you are a good credit risk, you may have to put down a deposit of as much as $5,000 to get a bank to grant you a merchant number.

If every bank in town refuses to issue you a merchant number, you need to find a partner. Getting a partner means that you have to convince a business in your area that you are a good credit risk and pay them a percentage to handle your credit card requests. Regardless of the method you choose, you must be able to accept credit cards if you want your BBS to make money.

Secret 2: Provide On-Line Ordering

If you are selling anything through your BBS, permit callers to order on-line. This option enables you to cater to impulse buying habits. As with your basic subscription fees, you must take credit cards. The on-line environment is quite conducive to impulse buying. If you have information, books, audio tapes, data files, or hardware to sell, you should present these items on-line in a catalog with an option to order. Make the caller's job of finding out (and buying) what you have to offer easy!

Secret 3: Conduct Written Follow-Up

Any business thrives on repeat customers and word-of-mouth advertising. To increase both benefits, you should provide some sort of written follow-up when a customer subscribes or purchases from your BBS. In an electronic world, you easily can forget that written documents are important. Your customers, however, will appreciate your "thank you for subscribing" notice if they receive it in written form. If this follow-up letter contains information on how to use your system or what is available, callers will keep your note around and reference it often. Written follow-up is one more way to remind your subscribers what they paid for.

Secret 4: Have a Theme

You don't need a large number of ideas to make a BBS pay for itself—you just need one good idea. You can offer entertainment with chat systems and computer graphics pictures. You can provide file downloads with massive amounts of public domain and shareware software. You can offer on-line games. You also can provide stimulating message conferences that have interesting people and topics.

A theme for your BBS helps you stand out in the crowd. Even if you have many things to offer, a theme is a good idea. If you had to select from "The World's Largest BBS with Lots of Great Stuff," "The Tall Ships BBS," or "The Sports Corner," which BBS would you call last? The first theme definitely is too vague. In general, a theme attracts more initial callers and gets your system talked about. Even calling your system something like "Harold and Maude's Corner" is better than a nebulous title—the name provides your BBS with an interesting personality.

Secret 5: Promote Your BBS

The electronic nature of the BBS community makes promotion relatively easy. You should draft a short advertisement about your BBS in electronic form. Include your system's two or three most attractive or unusual features along with the telephone number.

You then should call as many BBS systems as you can and post this message. Avoid being openly annoying, but post your advertisement as widely as you can. Thousands of people will read such an advertisement in just a few days.

Finally, spend a bit of money and place a classified advertisement in your local newspaper. Again, focus on the top two or three interesting items that your BBS has to offer.

With this sort of low-cost advertising, you should have many callers investigating what you have to offer. After that, your system can sell itself to those who call.

Secret 6: Run Your BBS Like a Business

Even if you are hoping just to recoup your costs when running a very informal BBS, you always should conduct transactions in a business-like manner. Because setting up a BBS as a business is so inexpensive and casual, you easily can lose track of the fact that any time you are asking for money, you are in business. If you conduct yourself professionally at all times and if you constantly think of ways to make a better and more professional impression on your users, your BBS is likely to be successful.

The last secret is that no secrets really exist to making money with a BBS—all tips that apply to any small business apply to a BBS. The key to making a BBS pay for itself is enabling people to give you money at the moment they are acutely aware of the value provided by your BBS. In addition, if you give callers reasons to remember your system (and you) in a positive light, they will recommend your system to other BBS users.

Chapter Summary

Operating a BBS is very different from using a BBS. The software you choose makes a big difference in the cost and performance of your system and in how much time installing and maintaining your BBS takes.

The best source of information on being a SYSOP comes from the people already operating their own BBSs. You should ask questions about BBS setup and operation in public message areas and commercial information systems. SYSOPs usually are glad to talk to you and tell you what you are getting into.

Today, BBS software has become so powerful that most of what commercial on-line services did a few years ago now can be done with BBS software on a PC. If you have information-handling problems in your business, are looking to start a BBS for your personal education or enjoyment, or are looking to run a BBS as a business, you can find the software you need. You will not need to be a programmer or highly technical person to do the job.

15

Examining Store and Forward Networks

O ne of the most rapidly growing segments of telecommunications is *store and forward networking*. Store and forward networks provide a method of linking many separate computers spread over a wide area so that electronic mail and files may be exchanged. In addition, most store and forward networks enable groups of users to connect through distributed message conferences. Examples of large store and forward networks are FidoNet, UUCP, USENET, and Internet.

Today these networks connect millions of computer users into shared conferences and electronic mail. On-line communities form within networks based on common interests. The information sharing present in the commercial on-line services and community bulletin boards described in previous chapters reaches new heights in large store and forward networks.

Although these networks may appear complicated and mysterious, they really are quite easy to understand. After you understand how these networks work, you will see that they are another valuable resource PC communications makes available to you directly through your personal computer.

In this chapter, you learn what store and forward networks are as well as how to use the more common ones. You learn how to locate these networks and obtain access to them. You also learn how these networks connect to the commercial services and bulletin board services discussed in previous chapters.

Understanding Store and Forward Networks

Store and forward networks usually are just called networks because the longer name is unwieldy. They should not be confused, however, with local area networks (LANs) or wide area networks (WANs), which are quite different.

A store and forward network enables you to enter on your local computer (or computer service) a message that is addressed to someone at a distant location. This message is *stored* on your local computer for some period of time and then *forwarded* to the computer at the address to which you directed the message. From this process comes the name *store and forward* for this type of network. This forwarding process may occur in a single step, or your message may be sent via a series of intermediate stops. If your message is sent through intermediate stops, the message is stored at each stop for some period and then forwarded to the next stop and ultimately to the final destination.

This electronic store and forward process is analogous to sending a letter through the post office. You first write the letter at your desk and put it in an envelope addressed to the person to receive the letter. Next, you put the letter in your outgoing mailbox. After it is "stored" in the mailbox for awhile, the mail carrier picks up the letter and "forwards" it to the addressee's location.

When you mail your letter, it does not go directly to the address on the envelope. The letter goes first to your local post office where it is sorted into bundles bound for large clearinghouses. From these clearinghouses, the letter is sent to the main post office in the city where the recipient lives. From there, the letter is sent to the local post office closest to the recipient and then to the recipient's office or home. Similarly, electronic store and forward networks often transmit electronic mail to many intermediate locations before it finally is delivered.

When you use a store and forward network, however, you normally don't need to be concerned with how a message gets from where you mailed it to its destination (a process called *routing*). You simply need to know the correct electronic address of the recipient—just as with paper mail, for which you need to know only the street address. The network takes care of routing automatically.

Network Components

A store and forward network requires three components to operate correctly:

- An electronic addressing method

- An established file transfer session protocol

- A directory of the systems in the network and the way mail is to be routed

The most obvious of these three components is the electronic *addressing* method. As a user, you first become aware of this part of the network because you need to know the address of anyone you send mail to. Any successful addressing method must uniquely identify every user of the network so that the network can determine from the address the one user to receive the message.

Networks use many different addressing methods. Most attempt to make the address structure hierarchical so that the address is a combination of simple pieces based on locations. This approach enables users with the same name in different locations to have unique addresses, much the same way that making your street name and city part of your postal address enables several John Smiths to have unique mail addresses.

Addresses formed in this way also facilitate routing. With a paper mail address, the first routing task is to get the mail to your city, a defined portion of the address. This part of the trip takes place without concern for the street address. After the mail reaches the city, the street address enables the postal service to route the mail to your home. Finally, the name routes the letter to you, not others in the home. Electronic addresses often are created to be read easily by computers; therefore, they frequently contain many numbers.

The second component required in a network is a file session *transfer protocol*. This protocol is required because a network must make all its deliveries automatically. Any computer in the network must be able to connect to another computer in the network and transfer files. Such a connection may involve dialing a telephone, connecting via a public or private packet data network (Telenet or Tymnet, for example), and so on. This type of connection most likely involves a security check, such as a logon, to verify that the connection is allowed. The computers then transfer the files and check the results for errors.

The file session transfer protocol may be as simple as a terminal program script that dials another computer, logs on, and uploads or downloads files with particular names. A more complex protocol normally is used that includes a logon verification as well as the capability of handling errors during all parts of the transaction and verifying that the files have been delivered. In any case, this portion of a network comprises the "nuts and bolts" of transferring information automatically from one computer to another without human intervention.

The third component of a network is the *directory*. This directory lists the systems that comprise the network and the addresses of all network users. The directory also indicates when and how mail is transferred from one computer to another. If a network is very large, the directory function may be large and complex, but its job is quite simple. It must find out how to send mail from the current computer to the given address.

For a simple network, the directory may exist as a single entity. Most large networks, however, distribute the directory because it is too large to be assembled in one place—much as it would be impossible to have a single telephone book for everyone in the world. The major task of administering a store and forward network is maintaining an accurate network directory.

When these three components are in place, a network exists. Because each user has a unique address, you can enter a message and address it to anyone on the network. Your computer (or the computer on which you enter the message) consults the directory to determine where and when to send the message you entered. When that time comes, the computer follows the established transfer protocol to send the message to the next computer. If this computer is not the final destination, it repeats the process to forward the message until it is delivered to its ultimate destination.

Although a store and forward network can appear large and complicated, it is simply many independent systems carrying out a few simple tasks. As you learn later in this chapter, hooking up to a store and forward network is quite easy.

A Brief History of Networks

Computers originally accepted their input on paper tape or punched cards and then printed output on paper, handling a single job at a time (called a job or *batch*). Because computers were so big and expensive, owners wanted them to perform more than one task at a time. In 1965, IBM released O/S, the first widely used multiuser batch operating system. This development increased the need for convenient ways to feed these large systems in order to keep them busy, creating the need for remote access to computers.

To meet this need, modem technology, which enabled computers to be connected via leased lines to remote batch terminals, was developed. These terminals contained a punched card reader and printer and could be placed at a remote location. This setup, known as *remote batch computing*, enabled users to submit programs and print results without being at the computer.

In 1966, low-speed modems and teletype terminals were introduced, which provided interactive time-sharing access to large central computers. Time-sharing services such as GE Time-Sharing (General Electric) and Tymshare (Tymshare, Inc.) sprang up, marking the beginning of the first computer communications industry, called the time-sharing industry. These developments spurred the first period of explosive growth in the computer industry. Companies appeared, became huge success stories, and then disappeared overnight. The computer industry would never be the same again.

These events initiated the development of better computer communications hardware and techniques. After this technology was in place, the next step was to hook computers together. Although many small projects were carried out during this time, the first packet-switched data network of any size was implemented in Britain by the National Physical Laboratories (NPL) in 1968. About the same time, a larger network was implemented in the United States by the Defense Advanced Research Projects Association (DARPA). This network (ARPANET) used high-speed leased lines to connect several large computers together in a store and forward network.

Time sharing exploited the idea that computer resources could be shared between several users (if the software knew how), allowing multiple use of the then very expensive computer hardware. The idea of sharing also made sense with expensive high-speed data communications links. After a high-speed data link existed between two locations, users had an economic incentive to try to send every bit of computer communications over that one link. Even if only a single batch terminal was at the other end of the data link, the need may have existed for both a card reader and a control terminal keyboard to send data to the computer and for the computer to send data to a printer, card punch, and data terminal screen at the same time. If a remote location had several remote batch stations, trying to connect them all to the central computer over a single high-speed data link made sense.

This idea of time sharing an expensive data link is called multiplexing at its simplest levels, but when provided in a general purpose way, this time sharing is known as a *packet switched network*. In a packet switched network, several low-speed local data streams are connected to a single high-speed data stream, but the purpose is to retain the individual circuit connections as shown in figure 15.1. The input from each local data link (*circuit*) is placed in data packets (from which the name derives). Each data packet also has a destination address (which identifies its circuit) added to it. All of these data packets are placed on the single high-speed data link and sent to the other end. When they are received, each data packet is sent to the local data link (circuit) indicated by the address in its packet header.

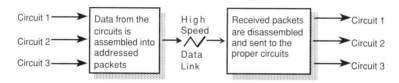

Fig. 15.1. The multiplex structure of a packet switched network link.

When several such circuits are connected to each other, a packet switched network is formed. In a true packet switched network, any of the local connecting circuits can request a connection through one or more of the high-speed lines to any other local connecting circuit on the network. This operation is analogous to dialing a telephone number to connect your telephone to any other telephone in the world via the voice switched telephone network, except that the resulting connection is a data connection instead of a voice connection.

Each type of packet switched network has a protocol that defines its internal operation. The pioneering packet switched networks used a variety of custom designed protocols, and a few of these still survive in isolated systems. Today, however, packet switched networks predominantly use the CCITT standard X.25 protocol or the ARPANET-developed TCP/IP protocol. The ISDN digital telephone service that is becoming available has a data packet switching capability as well.

Although these packet switched data networks were formed primarily to facilitate remote batch computing, moving messages and files over these networks soon became a focus all its own, and Electronic mail was born.

ARPANET eventually grew into a national U.S. "backbone" network on which a large and active user base developed many of the techniques used today in store and forward networks. ARPANET is largely discontinued today, but its functions have been absorbed by ARPA-Internet, which administers several interconnected networks.

Meanwhile, telephone modems were becoming more common. In 1976, Mike Lesk of Bell Laboratories created a program known as UUCP (UNIX to UNIX Copy) for UNIX computer systems. UUCP enabled a UNIX computer to dial a second UNIX computer and transfer files. UUCP quickly grew into a system that provided multiple hop (called multihop) transfers and remote execution of programs. Small networks of several groups of UNIX computers formed, using this software.

The capability of sending mail to users on other UNIX systems was a dramatic development that tied the UNIX community together and led to a rapid growth in UNIX systems. Over time, several of these UUCP networks combined and, through a more carefully managed administrative group, assigned addresses over a wider area.

A 1978 memo published by Lesk notes that one user was using UUCP to call up ARPANET and copy his mail to UNIX. This is probably the first documented case of connecting systems to ARPANET by dial-up modem. Today such connections have created a much larger "metanetwork" that is expanding rapidly.

Following these dial-up modem connections, software soon was added to provide news groups and distributed message areas. In 1979 in North Carolina, two universities began a bulletin board publishing scheme called the User's Network or USENET. USENET grew rapidly and today is probably the largest electronic distribution medium in the world. The capability of USENET to distribute so much information in a timely fashion was exciting, and its presence quickly became a major driving force in the development of store and forward networks.

In 1981, the City University of New York (CUNY) connected its IBM computers in a small network called the "Because It's Time Network," or BITNET. The capability of sending electronic mail between IBM computers caused BITNET to expand; today BITNET is a network of more than 2,500 computers.

While these and many other networks were developing on large computer systems, the personal computer revolution was taking place. In 1984, computer enthusiast Tom Jennings developed FidoNet technology, enabling IBM PCs to accomplish the same kind of automatic transfer functions that UUCP provided for UNIX systems. He combined this technology with his Fido bulletin board software (named after his computer), and FidoNet was born. Jennings initially administered FidoNet by himself, but in 1985,

Ben Baker developed a series of programs that automated and distributed the administration of the FidoNet directory (known as the *nodelist*). FidoNet then could grow without relying on any single individual, and today FidoNet has more than 10,000 systems and 500,000 users worldwide, making it by far the largest amateur store and forward network in the world.

With electronic mail systems increasing and multiple separate networks forming, the need developed for all these systems to be interconnected. As in the case of UNIX systems, many users were accomplishing such interconnections individually. In order to establish such interconnections easily and make the entire scheme workable, a common addressing method had to be established.

One of the most important early unifying developments in networking electronic mail was the development of the RFC822 message format. This standard was developed to enable external networks to connect to ARPANET. Over time, this message format has become the "native" message format of many store and forward mail systems, facilitating the process of moving a message from one network to another.

Most early internetworking of electronic mail was centered around ARPA-Internet because it was the largest and most active network. In the 1980s, discussions on a way to develop a unified addressing method began within the ARPANET community. These discussions proceeded through the mid-1980s until 1987, when the Domain Name System (DNS) was devised and adopted by DARPA. This system provides modular addressing and accepts most existing addressing methods as subsets. With the adoption of DNS, the integration of networks began in earnest.

Gateways are used to connect separate networks together. A gateway is software that understands the formats of messages on each network it connects and converts the messages as it moves the messages from one network to another. Gateways between networks can be entire separate computers that just perform the gateway function. Frequently, however, gateways are just programs that run as a separate task on a computer which is part of one of the networks. The DNS makes addressing across such gateways easy.

In the 1980s, it also became clear that interconnected networks would be severely limited in size and usability if their administration and use could not be decentralized. The Internet Domain Name System provides such administrative decentralization.

The DNS establishes a top set of domains administered by Internet. Each domain may be a separate administrative unit and subdivide itself. DNS provides a protocol for establishing *name servers*, software which handles the tasks associated with address lookup and routing in a distributed directory.

In 1987, the DNS became developed fully enough in software that it could achieve widespread implementation. The DNS has done its job, and the resulting distributed metanetwork is growing at a blinding rate. From 1980 to 1987, the size of this metanetwork doubled. With the acceptance of the DNS in 1987, the next doubling occurred by 1988. In 1989, the size more than doubled again, and by 1990, the metanetwork was doubling every few months. We are rapidly approaching the day when every modem user can send mail to any other modem user in the world using interconnected store and forward networks.

Currently these networks are primarily government subsidized. Portions are private networks, but the backbone systems that carry the bulk of the mail traffic are primarily educational or governmental research facilities. A funding crisis may arise as these metanetworks grow, but for now the cost to use them is very low—often free.

An ethic has developed within the community of people who use these networks regularly that they should remain free. The networks cost a tremendous amount of money to administer and operate, however, and current conditions may not be able to last long.

The good news is that technology is developing so rapidly that a single individual can afford to install and operate a large mail gateway system. The resulting metanetwork therefore is taking on much more of a "grass roots" structure, in which new links spring up when old links disappear because the users have the resources and desire to keep the mail flowing.

The history of store and forward networks shows that electronic mail system users have an inherent drive to connect with one another. This drive, coupled with the DNS addressing structure, enables individual and small groups to continue to expand the metanetwork that is forming. This grass roots structure has likely already reached the critical mass needed to keep the mail moving through the metanetwork even if the current large backbones are discontinued.

Understanding FidoNet

FidoNet began in 1984 when Tom Jennings wanted a way to send messages and files to his friends automatically. He created the FidoNet software as an add-on to his Fido BBS software.

Initially each system had an assigned number, referred to as a *node number*, that served as its address. By 1985, FidoNet had grown enough that some sort of routing system was required. The address was broken into two portions at this time so that a *net number* was added to the node. The net was envisioned as a local administrative group. Net numbers were assigned nationally, and the local net administrator assigned node numbers within the net.

Even though FidoNet address assignment was decentralized, the structure of the network required a single static directory, called a *nodelist*, that contained every address in the network. This directory had to be distributed to each FidoNet user. As FidoNet grew, this task became quite difficult.

In 1985, Ben Baker created a set of software that enabled the local net coordinators to send the changes for their local areas automatically to the central network administrator. The software then collected all these changes and updated the nodelist. Finally, it produced an updated file containing all changes to the nodclist and distributed it automatically to all net coordinators. The net coordinators then distributed the updated list to all the systems in their nets.

This system has continued; therefore, each system in FidoNet can maintain a nodelist that is updated weekly. In addition, the administrative load of updating this list is distributed to the local net coordinators so that no one person has a very large task to keep the nodelist current. Keeping the process of updating the nodelist simple is important because FidoNet is an all-volunteer network.

FidoNet initially provided only point-to-point electronic mail and file transfer. If you wanted to send a message to another network user, you sent the message to the FidoNet address that corresponded to that user's system. In 1986, Dallas computer enthusiast Jeff Rush devised a distributed conference capability known as *echomail*. This system enabled the participating BBSs in FidoNet to designate certain message areas as being shared across the network. Any message entered in such a message area is repeated to all systems that subscribe to that echomail message area, creating a national or international message conference.

Echomail fueled explosive growth within FidoNet. Over the next few years, the software to handle it improved and an administrative system to coordinate its distribution was established. Today, an active FidoNet echomail backbone of systems carries more than 3 megabytes of compressed echomail messages per day. This system has become quite efficient; a message entered in any of the hundreds of echomail conference areas carried on the backbone propagates around the world in less than 24 hours.

Using FidoNet

Accessing a FidoNet echomail conference is quite easy. You simply call a BBS that carries one or more echomail conferences. These conferences look the same as a local BBS message area. Any message you enter becomes part of the echo. At the end of each message, you see a system identification indicating where that message was entered.

Using FidoNet direct electronic mail is reasonably simple as well. You enter the message on your local FidoNet-participating BBS and add the destination user's address. Because mail calls are made directly from the BBS where you leave the message, the SYSOP may levy a fee for long-distance net mail use to cover the telephone call.

FidoNet addresses now have the following four numeric components:

zone:net/node.point

The *zone* is the largest administrative area. Zone 1 is North America, Zone 2 is Europe, and so on. The *net* and *node* numbers are as described in the preceding section. The *point* is a subdivision that enables further distribution of FidoNet connections to small self-contained units not listed in the FidoNet nodelist. The point is omitted from the address if it is zero.

For example, the address for sending mail to Phil Becker at node 23 in net 104 in a FidoNet system in North America is as follows:

Phil Becker 1:104/23

In this FidoNet address, the numbers identify the computer to which the mail is to be sent. The name also is part of the address because it identifies who the message is for, and many computers have more than one person using them. Therefore, the user's name and the computer's address are required in an electronic mail address. The point is omitted because it is zero.

Several networks other than FidoNet also use FidoNet technology. These networks maintain separate nodelists for themselves, and most are internetworked to FidoNet through a gateway. In order to administer this connection, most FidoNet technology systems have an assigned zone number within FidoNet. Their addresses thus appear in FidoNet as coming from that zone, and their addresses therefore do not conflict with the FidoNet addresses.

Joining FidoNet

As with most store and forward networks, you must find someone who is currently part of FidoNet in order to join it. That person can refer you to the net coordinator for your geographical area, who provides your point of entry to FidoNet. If you have a BBS in your area that is part of FidoNet, you can leave a message to the SYSOP asking how to contact the net coordinator.

You must obtain and install software that supports the FidoNet protocols on your computer. You can download suitable software at no charge from a BBS or from the communications libraries of commercial services such as CompuServe or GEnie. Commercial software packages that support FidoNet interconnections also arc available.

After you install your FidoNet software, you must create a nodelist that contains a single entry—the name, address, and telephone number of your net coordinator. You assign your system a dummy node number, usually 9999, within the net to which you are applying for a number. If you want to join net 104, you can give yourself the following temporary node number:

> 1:104/9999

Next, you send a message to your net coordinator asking for a FidoNet address. This message must include your name, address, telephone number, and the telephone number of your FidoNet system. Sending this message proves to the net coordinator that you have installed your FidoNet mailer software correctly and are capable of handling mail.

The net coordinator assigns you a FidoNet address and sends in the update to be added to the nodelist. Within a week (two weeks at the most), your node address appears in the national FidoNet nodelist, and you are part of FidoNet.

If you want to subscribe to an echomail conference, your net coordinator can tell you the procedure for your net. You may be charged a fee to connect to the echomail system in your network. Such fees are levied based on the cost of making the long-distance telephone calls to send the echomail to and receive it from the backbone systems each day.

Understanding USENET

USENET is the original UNIX-based distribution network. It is based on the UNIX UUCP program and generally uses domain name addressing. It is split into *news groups* that are similar to FidoNet echomail areas but are one-way and moderated. USENET news groups include news, recreation, computer discussion, and science fiction groups. New news groups are added from time to time.

USENET was begun in 1980 by students at Duke University and the University of North Carolina. These students used the UUCP program (part of their UNIX operating systems) not as it was normally intended (to send messages to private mailboxes) but rather to establish a broadcast to a specific file directory established on each machine. This special directory could be viewed by any user of either system and thus a distributed message system was created.

More sophisticated interface software was developed that permitted the users of these systems to not only review, but comment on items in the conference as well as add original material of their own. This software was distributed freely to other universities through the already existing UUCP network. Other installations were added to the system and USENET (an abbreviation for USEr's NETwork) was born. Today USENET is probably the largest distributed conference network (as measured by volume of message text) in the world.

USENET constantly is growing in the topics it covers. Over 250 individual conferences (called News Groups) now are being broadcast, and new ones are added regularly. USENET has become a valuable source of technology and information exchange between universities and research facilities. The amount of information circulated in USENET is prodigious—often over 15 megabytes of uncompressed text per day for all news groups combined.

The USENET structure can appear chaotic, because it is a totally grass roots structure. The inherent redundancy in this setup means that it has become a system which literally has attained a life of its own. New systems add themselves to USENET and other systems disconnect, all without any overall planning or supervision. It appears as though it could never be reliable, but in fact it shows the total strength of "guerilla" networks. Because there is no central structure to fail, the network becomes extremely reliable.

This grass roots nature of USENET occurs in many store and forward networks (for example, FidoNet) but nowhere is it as pure as in USENET. This self-defining structure can make such networks appear difficult to understand because you keep looking for a formal structure which isn't there. But the structure also makes such networks almost a life form of their own with an energy and excitement little else in the world of computing can match.

USENET has no administration; no one directly measures it or keeps track of its architecture. If you have access to the UNIX UUCP program (or a compatible program), you can receive USENET. Volunteers keep the network running, but no one controls the growth of the system.

Software has been developed that enables an IBM PC to connect to USENET. Such software is available for downloading from many BBSs and communications software libraries on commercial systems such as CompuServe or GEnie.

To connect to USENET and receive news groups, you have to locate someone that is currently part of USENET and agrees to connect you. No other administrative steps are required; that user connects you to the system and you become part of USENET.

Understanding the UUCP Network

The UUCP network is the oldest UNIX electronic mail network. More electronic mail is sent via this network than any other, although this fact may be obscured because of the internetworking of systems. The only requirement for joining this network is to have the UUCP programs (included as part of the UNIX operating system), a modem, and a telephone line.

UUCP addressing describes the full path a message travels to reach its destination. Each computer system has a name, and the address is a list of these names separated by exclamation points. Thus, the sender indicates the routing of each message. The following is a sample UUCP address:

hosta!hostb!hostc!host!user

This address represents a list of all the computers a message travels through to get to its destination. Because the sender (the source of the message) specifies the complete route the message travels to reach its destination, this type of addressing is called *source routing*. The exclamation point often is called a *bang*, and this UUCP-style addressing is sometimes referred to as a *bang path* or *bang path addressing*.

Newer software that adds intelligent addressing to UUCP now is available at some UNIX sites. If this software is available, full source routing is not required because this software can fill in some of the intermediate computer names automatically. If you use a UNIX site, be sure to ask your system administrator how you address outgoing UUCP mail. With the passage of time, all UUCP sites probably will migrate to a complete Domain Name System (DNS) addressing capability (see next section) as new software is added, but at this time each site is at its own point in the evolution of the UUCP addressing software.

As with USENET, the UUCP network has no central administration. A central directory (called a *map*) is compiled by a volunteer group, but you do not have to be listed in it. You can obtain this map from a USENET news group that publishes it every month.

UUCP is probably the most distributed network in existence, surpassing even FidoNet in this regard. Because of this highly distributed nature, no one knows how many people or computers participate in the UUCP network, but it is very large.

Understanding Internet DNS Addressing

As mentioned at the beginning of this chapter, the Internet Domain Name System (DNS) has become the de facto standard addressing method for connecting networks. Although the ISO/CCITT X.400 standard is proposed for this purpose and the X.500 standard moves towards a directory for addressing, the level of DNS implementation makes it by far the most commonly encountered electronic mail addressing scheme. Because DNS addresses were first used by ARPA-Internet, DNS addresses often are called Internet addresses.

The metanetwork that has formed from interconnecting most major store and forward mail networks can be reached even from commercial services such as CompuServe and MCI Mail. More networks are installing gateways to this metanetwork every day, and many systems that do not use the actual DNS software for mail routing have adopted the DNS syntax for their addressing.

The *Domain Name System (DNS)* decentralizes the administration of the mapping of computers to addresses. First you examine the structure of a DNS address. Then you examine how the DNS actually works to resolve an address so that it can determine to which computer it should send a message next. The DNS is so successful because of the capability for each computer in the DNS to know only the full addresses of the few computers it directly connects to, but still know how to route mail to arrive properly at the computer for which it is intended.

The form of a DNS domain name address is as follows:

local@domain

The *domain* is the fully qualified name of the last computer in the DNS directory to which the message is to be delivered. (A fully qualified name uniquely identifies a single computer site.) The local address provides directions for reaching the local user's mailbox from the gateway specified by the domain. If the local portion of this name contains multihop routing information, this information is separated by percent signs. If the domain name contains several levels, they are separated by periods. The top level domain is the rightmost portion of the domain name.

The top level domain is a two-letter country code or site-type designation in the U.S. or Canada such as *edu* (educational), *com* (commercial), or *org* (nonprofit organization). These top level domains are further divided into subdomains as required to specify uniquely a single site.

For a hypothetical computer in an engineering lab at Vanderbilt University, for example, the top level domain is *edu* because the site is part of a university. The university name can identify it within that group. The engineering school then may use the abbreviation *eng* to further indicate where the computer is. Finally, the computer may be named *frank* to differentiate it from all other computers in the engineering department. The resulting domain name has the following components:

Top level domain: edu

Subdomain 1: vanderbilt

Subdomain 2: eng

Subdomain 3: frank

To address a message to user *jsmith* on this computer, the DNS address is as follows:

jsmith@frank.eng.vanderbilt.edu

If the university elects to distribute mail internally from a single DNS entry point, the name of the computer may be indicated as an extended part of the local address as follows:

jsmith%frank%eng@vanderbilt.edu

Remember that DNS domain names are specified from right to left, with the top level domain at the right. The @ character separates the domain name from the local user information. If this local user information normally contains a further bang path address, the exclamation points are converted to percent signs.

Understanding Gateway Addressing

The Internet DNS addressing allows gateways between systems as well. One such gateway connects the FidoNet system—with its addressing scheme of zones, nets, and node numbers—to the DNS metanetwork. This gateway between FidoNet and the DNS metanetwork is given here as an example of how the DNS address accommodates addressing of incompatible networks.

For this example, consider the following FidoNet address:

Phil Becker 1:104/23

The first problem encountered here is the local name, which consists of two words separated by a space. The DNS maps this name by replacing the space with a period. Thus, the first portion of this address becomes PHIL.BECKER@ in DNS. The syntax 1:104/23 is incompatible with the DNS syntax; therefore, each portion of the FidoNet address is treated as a subdomain in order to map it to DNS. In this structure, the domain name takes the following form:

Top level domain: org

Subdomain 1: fidonet

Subdomain 2: zone

Subdomain 3: net

Subdomain 4: node

The top level domain is *org* because FidoNet is a non-profit organization. Subdomain 1 is *fidonet*, the name of the organization. The *zone*, *net*, and *node* become subdomains with the same hierarchy as in FidoNet; however, because the DNS name hierarchy is listed right to left, they occur in reverse order from the FidoNet address.

One final complication is that the DNS system requires the first character of a domain name element to be a letter. Thus, a letter is prefixed to each number. The letter *z* is used for zone and *n* for net. For node, the letter *f* is chosen because *n* already is selected. The full DNS version of the FidoNet address, therefore, is as follows:

> phil.becker@f23.n104.z1.fidonet.org

Using Gateways

A multitude of DNS gateways have combined to form a very large metanetwork. When you use this metanetwork, you rarely have to consider what really is happening to your message as it travels. FidoNet and the commercial services CompuServe and MCI Mail, for example, have gateways to the DNS metanetwork. The following sections explain how you use these gateways to send messages via the Internet DNS system.

CompuServe to FidoNet via Internet

You can send a message from the CompuServe mail system to any DNS address. To show how gateways work in both directions, the following example illustrates how to send a message from CompuServe to my FidoNet address. When you enter a message on CompuServe mail, you see the prompt TO: asking for the destination of the message. At the prompt, you enter the following:

> TO: >INTERNET: phil.becker@f23.n104.z1.fidonet.org

For the rest of the message entry, editing, and sending procedure, you follow the normal CompuServe mail system procedure just as if you are sending mail to another CompuServe user. When you send the message, it leaves CompuServe, enters Internet, finds the FidoNet gateway, enters FidoNet, and finds its way to the computer in my den automatically. This process always occurs within 24 hours, but normally it is much faster.

You can use this method to send a message from CompuServe to any person in the world with an Internet DNS address.

MCI Mail to FidoNet via Internet

You also can send mail to Internet DNS addresses from MCI Mail. This procedure is somewhat different, but equally straightforward. Again, the following FidoNet address serves as an example:

Phil Becker 1:104/23

As shown in the preceding section, this address translates to the following Internet DNS address:

phil.becker@f23.n104.z1.fidonet.org

In MCI Mail, you enter the CREATE option to send a message. At the TO: prompt, you enter the name of the person to receive the message followed by (*EMS*), including the parentheses. When you see the EMS: prompt asking you to specify the mail system name, type *internet*. Next, you receive the prompt MBX:, to which you respond with the Internet DNS address. This prompt repeats to ask for multiple addresses, so press Enter when you are done. From this point on, you enter and send the message as with any other MCI Mail message. The specific process for sending a message from MCI Mail to my FidoNet address is as follows:

TO: Phil Becker (EMS)

EMS: Internet

MBX: phil.becker@f23.n104.z1.fidonet.org

MBX: (Press Enter at this prompt)

Again, when this message is sent, it leaves MCI Mail, enters the Internet, finds the FidoNet gateway, enters FidoNet, and is sent to my FidoNet mailbox. The trip takes less than 24 hours.

CompuServe to MCI Mail

You also can use the Internet gateway to send mail between commercial services. For example, my MCI Mail number is 460-2871. The MCI Mail DNS address is MCIMAIL.COM because MCI Mail is a commercial service. Thus, the Internet DNS version of my MCI Mail address becomes the following:

0004602871@mcimail.com

The local mailbox convention adopted by MCI Mail leaves out the hyphen in the MCI Mail number and adds three leading zeroes. You also can try using names for the local mailbox, but names work on MCI Mail only if the user name is unique. Thus, if my name stays unique on MCI Mail, you also can use the following address:

pbecker@mcimail.com

If the name isn't unique, MCI Mail returns a message indicating all the MCI users with that name and their MCI Mail numbers. You then have to send the message again using the proper MCI mailbox number.

To send a message from CompuServe to me at MCI Mail, you enter the following in the CompuServe mail system:

TO: >INTERNET: 0004602871@mcimail.com

The message leaves CompuServe, enters Internet, finds the MCI Mail gateway, and is placed in my MCI mailbox within 24 hours.

MCI Mail to CompuServe

You also can send a message in the other direction, from MCI Mail to CompuServe. When you are forming a DNS address for a CompuServe user, you need to know the user's CompuServe ID number. For this example, my CompuServe ID is 73165,1062. Because the DNS system doesn't allow commas in its addresses, the comma in the CompuServe ID is changed to a period to create the local portion of the DNS address. The DNS address of CompuServe itself is COMPUSERVE.COM; the full Internet DNS address is the following:

73165.1062@compuserve.com

To send a message to this address from MCI Mail, you create a message and enter the following as you are prompted:

TO: Phil Becker (EMS)

EMS: Internet

MBX: 73165.1062@compuserve.com

MBX: (Press Enter)

When you finish entering the message, the message leaves MCI Mail, enters Internet, finds the CompuServe gateway, and is placed in my CompuServe mailbox. Again, this process takes 24 hours or less.

FidoNet to CompuServe

When you leave a message on FidoNet destined for an Internet DNS address, the procedure is somewhat different. You may have a UUCP gateway in your local net, in which case the message is a local call. Ask your local FidoNet net coordinator if you have a local UUCP gateway. If not, you can use the national UUCP gateway—1:1/31—but it may be a long distance call.

Again, my CompuServe address in DNS format is as follows:

73165.1062@compuserve.com

To send a message to this address from FidoNet (using the national gateway), you address the message to UUCP at the gateway you are using. Then, in the text of the message itself, you type the Internet DNS address on the first line preceded by the keyword TO:, as follows:

TO: UUCP

NET/NODE: 1:1/31

SUBJECT: Test Message

TO: 73165.1062@compuserve.com

The last line is an example of the first line of the text in your message. Follow the last line with the message itself, and then send it in the usual way. The message leaves FidoNet and enters Internet, finds the CompuServe gateway, and ends up on my CompuServe mailbox as in the other examples.

Understanding RFC822 Message Format

You have learned the power of Internet DNS addressing in the metanetwork growing up around its use. When using this network, you also may need to know the RFC822 message format, which is commonly encountered. In this format, you must enter all portions of the message directly as text. Although you are in a free-form text entry setting, the format is not free form. You must enter it correctly, or your message will not be sent properly. You normally encounter the RFC822 message format directly only in UNIX systems. Other systems usually prompt for the important parts and error-check them, but have their own message formats.

An RFC822 message is divided into the header and the body. The body contains the text of the message and is free form. The format of the header, however, is important. The header ends and the body begins with a blank line. You therefore cannot insert any blank lines in your header.

Each header line begins with a keyword followed by a colon (:) and then the appropriate information. The permissible keywords and their meanings are as follows:

Keyword	Meaning
From:	Sender's address. This line, in DNS format, indicates where the message originated.
To:	Primary recipient(s) of the message. This line, also in DNS format, can have multiple addresses separated by commas.
Cc:	Copy recipients. This line can have multiple addresses separated by commas. Each address listed receives a copy of the message.
Subject:	The subject of the message, in free-form text.
Date:	The date and time the message was sent.
Message-Id:	A unique identification for the message. This ID is used by the software to relate groups of messages.
Received:	A trace line inserted by the software to show how a message traveled.
Resent-From:	The address of the person or machine that forwarded this message to you. The message originally was sent to this individual or machine.
Reply-To:	The address of the person you are to send any replies to, normally the same as the sender.

You probably never will input most of these entries, but you may see them and wonder what they mean. The keywords you likely will have to enter directly are To, Subject, and Cc. The following sample message illustrates how these elements may appear.

```
From: Phil Becker <pbecker@mcimail.com>

To: John Smith <jsmith@frank.eng.vanderbilt.edu>

Cc: jjones@fred..eng.vanderbilt.edu,
   bbrown@mcimail.com
```

```
Subject: Linear Accelerator Problems

Date: Fri, 12 Apr 91 14:23:30 +0100

Dear John,

I think those last results you sent were not really an
indicator of a new physical discovery. I suspect you
forgot to clean the walls of the accelerator tunnel.
Check the second column in the input data, and I think
you will see what I mean. It would have been nice to
get the Nobel prize for this discovery, but I think it
will have to wait until next time.

Regards,

Phil
```

This example illustrates that lines that need to be continued (such as the Cc line) can be put on the next line with leading white space.

You probably don't need to know the RFC822 message format in detail, but you should know it well enough to understand the components. You may need to find the DNS address of the sender of a message, for example, so that you can address return mail.

Understanding Security in Store and Forward Networks

When using store and forward networks, you should be aware that your messages are not secure. Messages in transit are protected from casual exposure to public view, but system operators of the systems through which your mail passes can read your messages if they want. So much mail traffic occurs that such reading rarely happens, but you should not use a store and forward network to transmit information that you consider harmful if it were compromised.

Some networks enable you to encrypt the body text of your messages while others do not. Even in these situations, however, you should consider such measures as only a prevention against casual compromise of your data. Computers are capable of making copies of anything they handle, and you should assume that someone is listening at all times and can access your messages if they want to badly enough. A store and forward network is not an appropriate medium for truly confidential information.

In this respect, these networks are similar to cellular telephones that can be listened to easily with a scanner purchased at an electronics store. Cellular telephone conversations are protected by the communications act, however, which makes disclosure of anything monitored (without permission) a crime. The legal ramifications of electronic mail still are not set in either law or court precedent. At this time, therefore, disclosing anything that has come through your computer as part of a store and forward network is not clearly illegal. It is clearly unethical, but that doesn't prevent it from happening. The legal implications of this will be a topic for discussion and court cases for quite awhile before it is all resolved.

Chapter Summary

Store and forward networks are one of the most exciting and rapidly growing parts of telecommunications. They literally connect you to the world from your personal computer.

If you operate or use a BBS, you can join networks such as FidoNet and participate in international message conferences. A more recent development in this technology is the capability for individuals to participate in wide-area conferencing from their own homes.

The Internet DNS addressing method, which makes these connections possible, appears at first to be complex and mysterious, but it is fairly simple to understand. If you know the DNS address of another person, you have several options for sending electronic mail to that person. You can send it through a commercial electronic mail service or from your personal computer at home.

Store and forward networks will be one of the most rapidly growing areas of computer communications in the years ahead. If you remember that these networks are not appropriate for proprietary or confidential messages, you will find them to be a tremendous resource for information exchange. These networks also are connecting individual commercial information services and both private and public BBS systems into a truly global information community. After you experience a world-wide distributed message conference, you never again will want to be without the capability to participate in them.

Part III

Examining PC Communications

Includes

Understanding Terminal Emulators

Understanding PC to Mainframe Connections

Understanding Network Communications

Programming PC Communications

16

Understanding
Terminal Emulators

The ASCII code used in PC telecommunications does not contain any codes for screen position, attributes such as brightness or color, or extended function keys such as the cursor-movement arrows. Therefore, something beyond the ASCII code is required if you want a program to display formatted output on a remote terminal screen or to accept function keys from that terminal to control program operation.

Computer data terminals make such formatted output and extended input possible and provide a model of how to accomplish these functions on a PC. By encoding these extended function requests as a sequence of characters, data terminals extend the capability of the communications link to format output on the CRT screen. They also allow the transmission of many special functions keys that have no ASCII character set equivalent codes. As long as the sending and receiving programs agree on what functions each extended code represents, a wide variety of extended keyboard keys and CRT screen functions may be supported. By the emulation of one or more of these data terminals, the PC software can control a remote display and produce formatted screens.

Data terminals transmit requests for these extended functions as a special sequence of codes. Although these codes are made up of characters that individually may be from the ASCII character set, these codes are not ASCII themselves because several characters as a group represent a single function—a function beyond the encoding of the ASCII character set. Usually a single ASCII control character (for example, the Esc character) is used to indicate the start of such a sequence, but not always. What always occurs is that a specific multiple-character sequence represents a function instead of the normal one-to-one character to function mapping of the ASCII control character set.

The problem, however, is that several models of data terminals have been available over the years, and terminals often implement the same general functions through procedures that are not compatible with each other. This incompatibility makes the process of terminal emulation a bit more difficult because you must match the proper emulation mode with the software to which you are connected on the remote computer. If you do not take care of this matching, the screen output does not look correct (often it appears nearly meaningless), and any extended function keys do not operate correctly.

Not many official standards exist in the data display terminal arena. Instead, you find de facto standards that came about because a particular brand of data terminal was used frequently and had enough functionality to be adopted widely as a standard by software programmers. Official standards have been proposed for data terminals, but even the ANSI X3.64 standard, which is the most widely adopted formal data terminal standard, has several variations in practice.

In this chapter, you learn what a data terminal is and how your PC terminal software emulates one. You also learn about the specific codes for extended functions on several of the more popular, commonly emulated terminals. You then can use these codes to generate formatted output screens. You don't need to be a programmer to want to generate formatted output screens. If you are posting display files for a computer bulletin board or other communications service, for example, you may need to understand terminal emulation codes in order to know how to format your screens. And for this task, you need to understand the principles of terminal emulation.

Defining a Data Terminal

A *data terminal* is a device that has a keyboard and a CRT display screen. A data terminal also has enough memory to hold a screen's worth of display characters and attributes, the control logic to translate the characters in the memory into video for the CRT, and an RS-232 interface to connect the data terminal to a computer or modem. Figure 16.1 shows this structure.

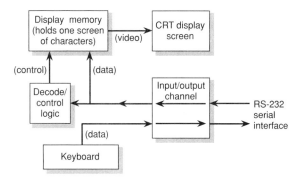

Fig. 16.1. *The internal structure of a data terminal.*

All data terminal operations are controlled by the data that is sent to the terminal from the computer. When a character is sent to the terminal, the character is displayed at the current cursor position with the current display attributes. These display attributes are variations in the video such as high intensity, reverse video, blink, or color. Special character sequences are used to indicate to the control logic that it should change the settings of the cursor position or the display attributes.

The data terminal keyboard may have more keys than just the normal typewriter keys. In that case, the data terminal transmits special character sequences when those extended function keys are pressed. These extended character sequences for display control and extra keyboard keys are the focus of terminal emulation. To emulate a specific brand and model of data terminal, software must decode these special sequences when they are received and then alter the display accordingly. The software also must generate the extended sequences on output when any extended function keys on the keyboard are pressed.

Exploring the History of Data Terminals

In the early days of computers, all display devices were printing terminals, such as Teletypes. These terminals had no capability to format data except for spacing the paper a line at a time and returning the print head to the first column. In some cases, a terminal also could backspace or feed the paper to the top of a page. For this reason, the ASCII computer code contains no provisions for anything beyond the printing of alphanumeric text and the most rudimentary output control functions.

In the late 1960s and early 1970s, CRT data display terminals were developed. Because CRT terminals had no moving parts, they had several advantages. They were quiet (much appreciated in an office setting) and with no moving parts they required much less periodic maintenance and broke down less often. The biggest advantage was that with no moving parts, a CRT had no physical speed limit. Mechanical printers are limited by the force and momentum required to move the print head. This limit typically results in quite slow display speeds (in the order of a few hundred characters per second for even very fast mechanical printers). A CRT has no such mechanical limit and is limited in speed only by the speed of the electronics themselves. This meant that even very early CRTs had display speeds which were many times faster than printing terminals. As a result of these advantages, CRT terminals soon became the most commonly used operator data entry terminal device.

Early data display terminals were quite expensive. In 1970, prices ranged from $5,000 to $10,000. Adjusted for inflation, those prices would be $20,000 to $40,000 in today's dollars—and for a CRT and a keyboard that only accepted operator input and displayed computer output!

The first display terminals emulated printing devices. These terminals often were referred to as "glass Teletypes" to indicate that they operated the same as printing Teletypes but with their output on the glass face of the CRT. The primary advantage offered by a glass Teletype terminal was speed. Because a CRT beam has no mass, these terminals could display characters as fast as the electronics could process them. Also, because the keyboards were electronic rather than mechanical, they could be made much easier for a data-entry operator to use. The fact that data display terminals were popular even at those high prices gives you some indication of how much more useful they were than the slower printing terminals.

Because the display device included no moving parts, the extra motion required to format a display screen became only cursor movement. CRT users quickly wanted to take advantage of this and do advanced display formatting. This function required at least the capability to clear the screen when needed. The capability to position the cursor to a specific spot under program control enabled the display terminals to produce output patterns that no printing terminal could duplicate. Having several different display attributes, such as blink, high intensity, and reverse video, also was possible. Each of these capabilities made the CRT data display terminal capable of more meaningful display formats. These advancements in turn eased the data-entry task and made errors less likely.

In this age of microcomputers and large-scale integrated circuits, you may have trouble realizing the difficulty involved in producing a data display terminal 20 years ago. Early data display terminals had several large circuit cards with far more electronic parts than most personal computers have today. An integrated circuit didn't hold many transistors in those days, and many transistors were required to make even relatively simple circuits. Because each added feature took a significant amount of electronic circuitry, many engineering techniques were devised by terminal manufacturers looking to pack more functions into a terminal as well as to reduce the cost. As is usual during the early part of any computer technology, each terminal maker did things its own way, and each terminal was different from the others.

As terminal designs proliferated, programmers had to adapt their software to each model and manufacturer. In 1974, the American National Standards Institute (ANSI) released the first standard, which attempted to bring some order to the chaos of data terminal designs. This standard (ANS X3.41-1974) showed how to encode extended functions by using the standard ASCII character set. Terminals still were too expensive to design or change, however, and this standard had little immediate effect.

When microcomputers came on the scene, data display terminals quickly began to use them as part of their control logic to implement more functionality at lower cost. The term "smart terminal" was coined to describe these full-featured data display terminals. These terminals were not computers as you know the term today, but they used the power of the microcomputer to reduce drastically the parts count in the terminal electronics while at the same time increasing the number of features available.

Normally, because IBM is the largest computer company, it sets many de facto industry standards. In the field of data terminals, however, IBM used primarily synchronous communications with the EBCDIC character set, which is completely different from the ASCII character set. As a result, IBM's data display terminals could not be used easily in other computer applications and therefore did not set many de facto standards. With the rise of minicomputers, which used asynchronous communications, Digital Equipment Corporation became a major force in the data terminal industry as well as the minicomputer field. DEC's most popular terminal at the time was the DEC Model VT-52.

When other manufacturers' terminals began using microcomputers, adding an option to make those terminals perform like a VT-52 rather than in their normal mode of operation became possible. This feature was the beginning of terminal emulation—one maker's terminal emulating another make or model of terminal. Soon most terminals could emulate the functions of the largest-selling terminals of other manufacturers.

In 1977, ANSI tried again to create a standard set of terminal functions. The standard X3.64-1977 was created after much industry input, argument, and compromise. In 1978, DEC produced the first terminal that implemented most aspects of this standard: the DEC Model VT-100 terminal.

As early true microcomputers emerged, the demand also arose for a low-cost data display terminal. *Popular Electronics* magazine is widely known for publishing the first article on how to build a personal computer and thus starting the personal computer revolution. *Popular Electronics* also published a series of articles by Don Lancaster in which he showed how to build a "TV Typewriter" data display terminal, and the era of ultra-low-cost data display terminals also began.

These low-cost displays used a wide variety of hardware interface techniques, but in most cases programmers wrote software that made the display terminals appear to operate like the popular data terminals of the day. This function also was terminal emulation. But this time it was being done in software on a computer rather than in the terminal itself. Thus began the era when data terminal functions were removed partially from the terminal itself and put in the host computer.

Most data terminal development has been evolutionary rather than revolutionary. A series of refinements over time has led to much more powerful data display terminals at much lower cost. One event in display terminal history, however, deserves mention because of its tremendous impact. That event is the development in 1977 of the ADM-3 data display terminal by Lear Siegler, Inc.

The ADM-3 terminal went against all the trends toward high functionality. In an apparent retrograde move, this terminal provided only a display, a keyboard, and a serial interface with the bare minimum of screen-formatting functions. Lear Siegler coined the name "dumb terminal" for the ADM-3 and advertised it directly to the emerging home computer market. The vendor even produced a kit version of this terminal. With a price of only $600 in kit form, the ADM-3 dumb terminal was a significant price breakthrough in the data display terminal market. With the catchy name, the low price, and a strong marketing program, the ADM-3 became the largest-selling data terminal in history. Lear Siegler had surmised correctly that computer technology had changed enough to make moving the intelligence from the terminal to the host computer a feasible approach. The company also proved that a large segment of the data terminal buying public had become price-sensitive.

Today, true data terminals rarely are encountered outside the mainframe and minicomputer arenas. Some multiuser UNIX systems on personal computers do use data terminals, but you rarely see them in a PC environment. Instead, terminal emulation is used to obtain the functionality of these data terminals directly from your personal computer.

Knowing When Terminal Emulation Matters

You must turn to terminal emulation whenever you want to use any extended display or keyboard features beyond those of the printing Teletype. For the screen display, terminal emulation enables you to control cursor positioning and display attributes. If you can fulfill your display application by simply displaying text in a forward scrolling manner on-screen, you don't need terminal emulation. On the other hand, if your display application requires you to position the cursor at a specific spot on-screen or change display attributes such as brightness, reverse video, color, or underlining, then terminal emulation is required.

On the input side, terminal emulation gives you the capability to send extended function keys and cursor-control keys. If your application doesn't require any keyboard input beyond the normal text and control keys, you don't need terminal emulation of the input. If you want to have your application use the cursor keys, the Ins or Del key, or other extended function keys from the remote terminal, however, terminal emulation of the keyboard is required.

Which data terminal model you choose to emulate to accomplish your task often is reasonably arbitrary. If you are using software that was written by others, or operating systems that provide terminal emulation for their programs, you must choose to emulate a terminal supported by this software. Otherwise, you can choose to emulate any terminal that provides all the specific functions your program requires.

Emulating Display Screen Handling

The degree to which PC terminal software may emulate correctly the display screen operation of a particular brand of terminal is limited only by the capabilities of the PC display itself. If a particular data terminal has a screen that is 132 characters wide, for example, a PC with a display incapable of more than 80 characters per line cannot emulate that terminal fully. The PC can come close, and in many applications it may even come close enough, but it never can emulate the terminal 100 percent.

The other display emulation area that depends on the capabilities of the PC video display is the group of display attributes being emulated. The PC display can provide reverse-video, blink, high-intensity, and low-intensity display attributes no matter what display type is in use. But only monochrome and EGA/VGA PC displays can provide underlining, and only color displays can provide colors. Only some advanced PC video displays can provide more than 80 characters on a single line or more than 25 lines on a screen. When you choose a terminal to emulate, you must keep in mind the attributes of the PC display you are using.

The PC always is capable of properly emulating any cursor-positioning and character-manipulation capabilities of a given terminal. Such commands as Insert Line, Delete Line, Clear to End-of-Line, Clear to End-of-Screen, and so on always can be handled correctly by a PC terminal emulator. Later in this chapter, in the section "Describing Specific Data Terminals," you learn about specific terminal emulations and their capabilities.

Emulating Extended Keyboard Functions

The IBM PC keyboard has keys for the most common extended functions. But it may not have all the keys that a particular data terminal possesses. In those cases, a terminal emulation program has to assign alternate key configurations for the extended keys that are not present directly on the IBM PC keyboard.

To handle this operation smoothly, most terminal emulators enable you to "map" the PC keyboard to the emulated keyboard. Usually the terminal emulator software provides a method for assigning each extended key function to a specific key on the IBM PC keyboard, as you prefer. Also provided in most cases is a default keyboard mapping. In some applications, the default mapping is suitable for your use, but in other situations, you may have to change this mapping before you can perform your task. In still other cases, you may want to change this mapping just to make the program easier to operate.

In the specific terminal emulations that are described in the following sections, the keyboard mapping often used by terminal emulators is given in detail. You should check closely the manual for your particular terminal emulation software, however, because this area is one in which a great deal of discretion is used, and each program is somewhat different from the others. In the end, choosing which IBM PC keyboard key is mapped to which key of the terminal being emulated can be a fairly arbitrary decision.

The important thing to remember about terminal keyboard emulation is that you have no direct way to send any extended key (such as cursor-movement keys, function keys, and so on). The terminal emulator and the host software must make an agreement about which character sequences represent each extended key, and only if they match on both ends can you use these keys correctly.

Describing Specific Data Terminals

The remainder of this chapter is devoted to describing the specific operations of several data terminals you may need to emulate. To give you a basis for comparing the functionality of these terminals and also to enable you to program output to them, a standard capability description is developed first.

In table 16.1, the basic cursor-control and clear functions for the terminals discussed in this chapter are defined. Some terminals do not have all these functions, but this general description can help you use and understand the information given in the following sections.

Table 16.1
General Screen-Control Functions for Data Terminals

Command	Description
Cursor Up	Moves the cursor to the same column of the previous line on the display. In general, if the cursor is already on the top line of the display, this command is ignored.
Cursor Down	Moves the cursor to the same column of the next line on the display. In general, if the cursor is already on the bottom line, the command is ignored. Some terminals, however, scroll the display.
Cursor Left	Moves the cursor to the next display position to the left of its current position. If the cursor is already in the first column of the display, some terminals ignore this command, and other terminals place the cursor at the last position of the previous line.
Cursor Right	Moves the cursor to the next display position to the right of its current position. If the cursor is already in the last column of a display line, some terminals ignore this command, and other terminals place the cursor at the first position of the next display line.
Cursor Home	Moves the cursor to the first display position on-screen. This position, called the *home position*, is in the upper left corner of the display.
Absolute Cursor Position	Places the cursor at an exactly specified coordinate position on-screen. If the specified coordinates lie outside the display screen, most terminals ignore the command. Some terminals place the cursor in an invisible position in this case.

Command	Description
Clear Screen	Clears the entire screen to blanks. With most terminals, this command also places the cursor at the home position. With some terminals, however, the cursor is not moved when the Clear Screen command is sent.
Clear Line	Clears to blanks the entire line containing the cursor. Normally the cursor does not move, but some terminals place the cursor in the first column of the newly cleared line.
Clear to End-of-Line	Clears to blanks all characters from the current cursor position to the end of the line containing the cursor. The cursor is not moved.
Clear to End-of-Screen	Clears to blanks all characters from the current cursor position to the end of the display screen. The cursor is not moved.
Clear from Beginning-of-Line	Clears to blanks all characters from the beginning of the line to the current cursor position. The cursor is not moved.
Clear from Beginning-of-Screen	Clears to blanks all characters from the home position to the current cursor position. The cursor is not moved.
Delete Line	Removes from the display all characters on the line currently containing the cursor. All lines below the cursor's line move up one. The cursor position usually is not changed.
Insert Line	Moves down one line all characters on the line containing the cursor. All lines below also are moved down one line. Usually this action begins at the cursor's position, and if the cursor is not in the first column of a line, the line is "split." Some terminals, however, move the entire line down regardless of the cursor position.

On all terminals, cursor position is given by the row and column numbers, with the upper left corner as the origin or home position. The row is the vertical position on-screen (line number), and the column is the horizontal position on-screen relative to the left edge of the screen. Some terminals number rows and columns beginning with 0, and others number them beginning with 1. (The numbering method used by each terminal described in the following sections is indicated in the discussion of that terminal.)

Some terminals have other important cursor-control functions, and these are explained in the individual terminal definitions that follow. The general functions described in this section, however, are not explained again each time they are listed. Attribute controls (such as reverse video) are generally self-explanatory enough that they are explained only when they aren't obvious.

The keyboard keys provided by each terminal are listed along with the code sequences they generate. Also listed, where appropriate, is the most common mapping to PC keyboard keys used by terminal emulators.

All character sequence listings are given in hexadecimal notation.

Emulating the VT-52 Terminal

The DEC VT-52 was the first widely emulated data terminal. It has only the most basic functionality, as listed in table 16.2.

Table 16.2
The VT-52 Terminal Screen-Control Functions

| | Code sequence | |
Command	*Hexadecimal*	*Character*
Cursor Up	1B 41	Esc A
Cursor Down	1B 42	Esc B
Cursor Right	1B 43	Esc C
Cursor Left	1B 44	Esc D
Cursor Home	1B 48	Esc H

| | Code sequence | |
Command	Hexadecimal	Character
Clear to End-of-Screen	1B 4A	Esc J
Clear Screen (combination)	1B 48 1B 4A	Esc H Esc J
Clear to End-of-Line	1B 4B	Esc K
Absolute Cursor Position	1B 58 r c	Esc X r c

r and *c* are row and column as hexadecimal numbers with 1F added. Row and column numbers begin with 1, and thus the home absolute position command is 1B 58 20 20.

NOTE

For cursor control, the VT-52 keyboard has arrow keys that generate the same codes as the Cursor Up, Cursor Down, Cursor Left, and Cursor Right commands given in table 16.2. In addition, the VT-52 keyboard has four program function keys positioned above the numeric keypad and labeled PF1 through PF4. They generate the following codes:

Function key	Hexadecimal	Character
PF1	1B 50	Esc P
PF2	1B 51	Esc Q
PF3	1B 52	Esc R
PF4	1B 53	Esc S

You also can put the numeric keypad in a mode known as the *keypad application mode*. In this mode, the numeric keypad keys generate the codes listed in table 16.3.

Table 16.3
Using the VT-52 Keyboard in Keypad Application Mode

| | Code sequence generated | |
Numeric Key	Hexadecimal	Character
Dash	1B 3F 6D	Esc ? m
Comma	1B 3F 6C	Esc ? l
Period	1B 3F 6E	Esc ? n

continues

Table 16.3 *(continued)*

| Numeric Key | Code sequence generated | |
	Hexadecimal	Character
Enter	1B 3F 4D	Esc ? M
0	1B 3F 70	Esc ? p
1	1B 3F 71	Esc ? q
2	1B 3F 72	Esc ? r
3	1B 3F 73	Esc ? s
4	1B 3F 74	Esc ? t
5	1B 3F 75	Esc ? u
6	1B 3F 76	Esc ? v
7	1B 3F 77	Esc ? w
8	1B 3F 78	Esc ? x
9	1B 3F 79	Esc ? y

Most terminal emulations map the PF1 through PF4 keys to the PC keyboard F1 through F4 keys. How the keyboard application mode keys are mapped varies widely. You should check your terminal emulator's documentation. You usually can remap these keys to meet your preferences.

Emulating the VT-100 ANSI Terminal

The DEC VT-100 family of data terminals is one of the most widely used and emulated terminals in the computer world. In addition, it is similar to implementing the ANSI X3.64 terminal standard.

The DEC VT-100 terminal was the first member of a family of data terminals from Digital Equipment Corporation. The original VT-100 terminal was created in 1978, and improved versions have been released at regular intervals since then. All these terminals support the VT-100 ANSI-compatible command set, which is described here.

All VT-100 ANSI-compatible commands begin with the character sequence Escape followed by a left bracket ([). This sequence indicates to the terminal that the characters which follow are display-control commands rather than data. Most terminal emulators discard any sequence that begins with these characters, but some drop only the leading two characters and display the remainder of any invalid sequence.

The VT-100 implements the basic functions described earlier under "Describing Specific Data Terminals" as well as many extended functions. The basic screen-control functions are listed in table 16.4.

Table 16.4
The Basic VT-100 Screen-Control Functions

	Character sequence	
Command	*Hexadecimal*	*CSI*
Cursor Up	1B 5B 41	CSI A
Cursor Down	1B 5B 42	CSI B
Cursor Right	1B 5B 43	CSI C
Cursor Left	1B 5B 44	CSI D
Cursor Home	1B 5B 48	CSI H
Clear to End-of-Screen	1B 5B 4A	CSI J
Clear Screen (combination)	1B 5B 48 1B 5B 4A	CSI H CSI J
Clear to End-of-Line	1B 5B 4B	CSI K
Absolute Cursor Position	1B 5B r 3B c 48	CSI r ; c H

r and *c* are row and column as decimal numbers in ASCII. Row and column numbers begin with 1, and therefore the Absolute Cursor Position command to move the cursor to the home position is 1B 5B 31 3B 31 48.

If you compare these basic commands to those of the VT-52, you see that they are similar except for the addition of the 5B character ([) to each VT-100 command. This addition of the left bracket was one of the moves made by the ANSI committee to prevent the VT-52 from becoming an instant ANSI-compatible terminal and gaining an unfair marketing advantage over other terminal makers.

The VT-100 ANSI command set also added the concept of *parameters* separated by the semicolon character (3B). These parameters are always ASCII numbers in decimal and may be more than one character in length. All these parameters also have a default value of one or zero (depending on which is the smallest parameter value for the specified command).

A series of parameters separated by semicolons is called a *parameter string*. The basic commands in table 16.4 therefore may include a parameter that enables you to move the cursor more than a single position. To move the cursor right 15 positions, for example, use the command 1B 5B 31 35 43. Because these ANSI/VT-100 commands are so complex, this text uses an alternate method to specify them. Each command begins with a *control sequence introducer (CSI)*, the specific sequence of the two characters Esc and [. Each control sequence of parameters ends in a *final character*, which defines the command and the action that should be taken. Final characters always are printable letters.

As an example of this notation, the following paragraphs examine an Absolute Cursor Position command that sends the cursor to row 10 and column 25. The hexadecimal byte sequence for this command is as follows:

> 1B 5B 31 30 3B 32 35 48

Using the alternative method, the command is represented as the following:

> Esc[10;25H

Esc[is the control sequence introducer, which is 1B 5B in hexadecimal. 10;25 is the parameter string. This particular string lists the cursor position by row and column. The hexadecimal representation is 31 30 3B 32 35. H is the final character.

In the VT-100 ANSI notation, this command is indicated as Esc[P1;P2H, where P1 equals the row number, and P2 the column number. Because any parameter may be omitted, the shortest form of this command is Esc[H, which is the equivalent of the Cursor Home command. Table 16.5 describes how the basic commands are depicted with this new notation.

Table 16.5
The Basic VT-100 Commands with Parameters Added

Command	Character sequence
Cursor Up	Esc[P1A
Cursor Down	Esc[P1B
Cursor Right	Esc[P1C
Cursor Left	Esc[P1D
Cursor Home	Esc[H
Clear to End-of-Screen	Esc[J
Clear Screen (combination)	Esc[HEsc[J
Clear to End-of-Line	Esc[K
Absolute Cursor Position	Esc[P1;P2H

P1 is the row number, and P2 is the column number. Position numbers begin with 1.

In addition to the basic functions, the VT-100 ANSI terminal adds many extended functions, which are described in table 16.6.

Table 16.6
The Extended VT-100 Commands

Command	Character sequence
Erase in Display	Esc[P1J

If P1 = 0, erases from the cursor position to the end of the screen. If P1 = 1, erases from the start of the screen to the cursor position. If P1 = 2, erases the entire display but does not reposition the cursor.

Erase in Line	Esc[P1K

If P1 = 0, erases from the cursor position to the end of the line. If P1 = 1, erases from the start of the line to the cursor. If P1 = 2, erases the entire line containing the cursor but does not move the cursor.

continues

Table 16.6 *(continued)*

Command	*Character sequence*
Save Cursor	Esc[s

Memorizes the current cursor position.

Restore Cursor	Esc[u

Places the cursor back at the last position memorized with a Save Cursor command.

Select Graphics Rendition	Esc[P1;....Pnm

Enables you to set several display attributes. You may string several attributes in a single command and separate them with semicolons. Allowable parameter values and their meanings are as follows:

Parameter value	*Description*
0	Normal display mode; all attributes off
1	Bold or high-intensity attribute on
4	Underline attribute on
5	Blink attribute on
7	Reverse-video attribute on

Any other parameter value is ignored. The command to set high-intensity normal video regardless of the current attribute settings, for example, is Esc[0;1m.

Set Mode	Esc[?P1h

If P1 = 7, sets wrap at EOL On. Any display past the end of the line gives an automatic new line. The meaning of other mode parameters depends on the exact model of the terminal (refer to your terminal manual).

Reset Mode	Esc[?P1l

If P1 = 7, sets wrap at EOL Off. Any display past the end of the line is ignored until a carriage return or cursor position command places the cursor back in the active area of the screen.

The VT-100 keyboard has the same keys as the VT-52. The arrow keys for cursor control generate the same codes as the Cursor Up, Cursor Down, Cursor Left, and Cursor Right control keys listed earlier in this chapter under "Describing Specific Data Terminals." In addition, the VT-100 keyboard has four program function keys, labeled PF1 through PF4, which are positioned above the numeric keypad. These keys generate the following codes:

Function key	Hexadecimal	Character
PF1	1B 4F 50	Esc O P
PF2	1B 4F 51	Esc O Q
PF3	1B 4F 52	Esc O R
PF4	1B 4F 53	Esc O S

You also can put the numeric keypad in a mode known as the *keypad application mode*. In this mode, the numeric keypad keys generate the codes listed in table 16.7.

Table 16.7
Using the VT-100 Keyboard in Keypad Application Mode

| Numeric Key | Code generated | |
	Hexadecimal	Character
Dash	1B 4F 6D	Esc O m
Comma	1B 4F 6C	Esc O l
Period	1B 4F 6E	Esc O n
Enter	1B 4F 4D	Esc O M
0	1B 4F 70	Esc O p
1	1B 4F 71	Esc O q
2	1B 4F 72	Esc O r
3	1B 4F 73	Esc O s
4	1B 4F 74	Esc O t
5	1B 4F 75	Esc O u
6	1B 4F 76	Esc O v

continues

Table 16.7 *(continued)*

	Code generated	
Numeric Key	Hexadecimal	Character
7	1B 4F 77	Esc O w
8	1B 4F 78	Esc O x
9	1B 4F 79	Esc O y

Most terminal emulations map the PF1 through PF4 keys to the PC keyboard F1 through F4 keys. How the keyboard application mode keys are mapped varies widely.

Emulating the IBM PC ANSI Terminal

In most cases, terminal emulation involves imitating the operation of a specific brand of terminal. Knowing exactly what the emulation has to do in those cases is easy: the computer must behave as identically as possible to the way the particular brand of terminal operates. ANSI emulation is a bit different, however, in that the ANSI X3.64 terminal standard leaves room for interpretation. This variability is reflected most noticeably in the level of implementation (some ANSI emulations are subsets of others). But some outright differences exist in the way ANSI terminal emulations interpret certain operations, most notably the Clear Screen and Set Color operations.

The version of ANSI X3.64 terminal emulation you will encounter most often on a PC is the one implemented by IBM in the ANSI.SYS driver under DOS. Here again, however, some ambiguities exist. In versions of DOS prior to Version 3.0, the ANSI.SYS emulation didn't clear the screen to current set colors but instead cleared it always to black. In DOS 3.0 and all subsequent versions, this function was changed to clear the screen to the current set colors.

The resulting terminal emulation is similar to the VT-100 discussed previously in this chapter, except that IBM chose to omit some of the Clear Screen and Clear Line functions. The Clear Screen function also has been redefined to include a Cursor Home function. Finally, IBM added options to the Select Graphics Rendition (SGR) command to provide the capability to set screen colors. With these options, IBM also introduced another incompatibility—this time among its own emulations. In early versions of DOS, the Clear Screen command always cleared to black no matter what colors were set. IBM realized that this feature made doing full-color screen backgrounds difficult (because more than 2,000 characters had to be written to fill the screen). In DOS 3.0, therefore, IBM changed the Clear Screen command to clear to the color that was set most recently.

The IBM emulation is identical to the VT-100, with the following exceptions:

- The Erase in Display command, with the character sequence Esc[2J, erases the entire display and moves the cursor to the home position. (Homing the cursor is a difference between this emulation and the VT-100 terminal operation.)

- The Erase in Line command, with the character sequence Esc[K, erases from the cursor position to the end of the line.

- The Select Graphics Rendition command, with the character sequence Esc[P1;....Pnm, enables you to set several display attributes. You can string several attributes in a single command and separate them with semicolons. Allowable parameter values and their meanings are as follows:

Parameter value	Description
0	Normal display mode; all attributes off
1	Bold or high-intensity attribute on
4	Underline attribute on
5	Blink attribute on
7	Reverse-video attribute on
30	Black character color
31	Red character color
32	Green character color
33	Yellow character color
34	Blue character color
35	Magenta character color
36	Cyan character color
37	White character color

Parameter value	Description
40	Black background color
41	Red background color
42	Green background color
43	Yellow background color
44	Blue background color
45	Magenta background color
46	Cyan background color
47	White background color

Any other parameter value is ignored. The command to set high-intensity normal video regardless of the current attribute settings, for example, is Esc[0;1m.

Note that most of the ANSI IBM emulations include the full VT-100 set of parameters for the Esc[J and Esc[K commands. IBM, however, documented only Esc[2J and Esc[k. Some IBM ANSI emulations do not honor the other variations of these two command sequences.

The ANSI emulation has no standard keyboard. Usually some variation of the VT-100 keyboard is used, but substantial variations occur from emulator to emulator.

Emulating the Televideo 920 Family

This family of data terminals includes the Televideo 910, 920, 925, and 950 terminals. Variations are small among these models when it comes to emulation. The basic functions are listed in table 16.8.

Table 16.8
The Televideo Family Screen-Control Functions

Command	Code sequence	
	Hexadecimal	Character
Cursor Up	0B	Ctrl-K
Cursor Down (910,920)	0A	Ctrl-J

| | Code sequence | |
Command	Hexadecimal	Character
Cursor Down (925,950)	16	Ctrl-V
Cursor Right	0C	Ctrl-L
Cursor Left	08	Ctrl-H
Cursor Home	1E	Ctrl-^
Clear to End-of-Screen	1B 59	Esc Y
Clear Screen	1A	Ctrl-Z
Clear to End-of-Line	1B 54	Esc T
Absolute Cursor Position	1B 3D r c	Esc = r c
Start Reverse Video	1B 6A	Esc j
End Reverse Video	1B 6B	Esc k
Start Underline	1B 6C	Esc l
End Underline	1B 6D	Esc m
Start Blink	1B 5E	Esc ^
End Blink	1B 71	Esc q

r and *c* are row and column as hexadecimal numbers with 1F added. Row and column numbers begin with 1; thus the Absolute Cursor Position command is 1B 3D 20 20.

The Televideo 925 keyboard has 11 function keys. These keys generate the codes listed in table 16.9.

Table 16.9
The Televideo 925 Function Keys

| | Code generated | |
Function key	Hexadecimal	Character
F1	01 40 0D	Ctrl-A @ Ctrl-M
F2	01 41 0D	Ctrl-A A Ctrl-M
F3	01 42 0D	Ctrl-A B Ctrl-M

continues

Table 16.9 *(continued)*

| | Code generated | |
Function key	Hexadecimal	Character
F4	01 43 0D	Ctrl-A C Ctrl-M
F5	01 44 0D	Ctrl-A D Ctrl-M
F6	01 45 0D	Ctrl-A E Ctrl-M
F7	01 46 0D	Ctrl-A F Ctrl-M
F8	01 47 0D	Ctrl-A G Ctrl-M
F9	01 48 0D	Ctrl-A H Ctrl-M
F10	01 49 0D	Ctrl-A I Ctrl-M
F11	01 4A 0D	Ctrl-A J Ctrl-M

Most emulators directly map these function keys to the IBM PC keyboard function keys. The following special keys also are included:

Function	Hexadecimal	Character
Line Erase	1B 54	Esc T
Page Erase	1B 59	Esc Y
Char Delete	1B 57	Esc W
Line Delete	1B 52	Esc R
Char Insert	1B 51	Esc Q
Line Insert	1B 45	Esc E

These keys are mapped in various ways to the IBM PC keyboard by different terminal emulation software.

Emulating the ADM-3 Terminal

The ADM-3 is the original dumb terminal, and as such it has few functions. Its limited capabilities for remote control are listed in table 16.10.

Table 16.10
The ADM-3 Screen-Control Functions

| | Code sequence | |
Command	Hexadecimal	Character
Cursor Up	0B	Ctrl-K
Cursor Down	0A	Ctrl-J
Cursor Right	0C	Ctrl-L
Cursor Left	08	Ctrl-H
Cursor Home	1E	Ctrl-^
Clear Screen	1A	Ctrl-Z
Absolute Cursor Position	1B r c	Esc r c

r and *c* are row and column as hexadecimal numbers with 1F added. Row and column numbers begin with 1; thus the Absolute Cursor Position command is 1B 20 20.

The ADM-3 or ADM-3A keyboard has no special function keys. The keyboard was marked for use of the control key plus a letter to generate the control functions in table 16.10. Most PC emulators of the ADM-3 map the arrow keys to generate the Cursor Up, Cursor Down, Cursor Left, and Cursor Right codes given in table 16.10 as well as the Home key. The mapping of the Clear Screen code and the Home Cursor code to the keyboard is arbitrary and varies.

Chapter Summary

Terminal emulation is the act of imitating the operation of a specific brand of data terminal. You must be able to emulate some terminal in order to have control over the cursor and display attributes of the remote screen or to use the extended function keys from a remote terminal as input. In this chapter, you learned the history of terminal development, how to know when terminal emulation is necessary, and how display screen functions and extended function keys are emulated. You also learned the specific details of emulating several popular data terminals, including the VT-52, the VT-100, the IBM ANSI, the Televideo 920 family, and the ADM-3.

When you emulate a terminal on the IBM PC, you have to make some choices about extended function keys because the PC keyboard is likely to be different from the keyboard of the data terminal you are emulating.

ANSI terminal emulation is the most commonly used terminal emulation in the PC communications world, followed closely by the VT-100 terminal. Properly used, these emulations can enhance on-line functionality tremendously.

17

Understanding PC to Mainframe Connections

A t some point, you may want to use your PC to access information generated or processed by a mainframe or minicomputer. The application requirements for sharing information with a mainframe vary greatly. In some cases, you may need to transfer small files between the mainframe and your PC; in other cases, you may need real-time access to database output files. This range of requirements has spawned a variety of responses to the problem of PC to mainframe connections.

In this chapter, you learn several methods for transferring files between a PC and a mainframe. You learn how to access mainframes over telephone lines with a modem and how to connect your PC to various mainframe and minicomputers to share files directly. You learn about 3270 emulation, SNA networks, and TCP/IP networks. You also learn how to make a portion of the mainframe's disk system appear to be an extension of your PC's disk system.

Mainframes and Minicomputers

Originally, all computers were mainframes—large, expensive, and centrally structured. Many companies made mainframe computers in the 1960s and

1970s. Among the most well-known companies were RCA (Spectra), General Electric, Sperry Rand (UNIVAC), NCR, Control Data Corporation (CDC), Xerox, Burroughs, Honeywell, and IBM. Over the years, some of these companies merged, other firms left the computer business, and IBM became the dominant player in mainframe computers.

Meanwhile, Digital Equipment Corporation (DEC) appeared on the scene with a smaller *minicomputer*. Unlike other large companies, DEC didn't attack IBM directly; rather, DEC built a new market with smaller, less expensive computers. Other companies, such as Perkin Elmer, Data General, ModComp, and Gould rose up to compete with DEC in the newly created minicomputer arena. After withstanding a challenge from IBM and Honeywell, however, DEC emerged as the leader of the minicomputer market.

As computer hardware technology advanced, distinguishing between mainframes and minicomputers became more difficult. At first, the newer minicomputers were called *super-minis*, indicative of their increased power. Super-minis eventually became more powerful than low-level mainframe computers.

Microcomputer technology soon appeared, however, and the cycle began again. Today's microcomputers can be more powerful than some minicomputers and mainframes. Gigabytes of disk storage can be connected to the now inappropriately named *personal* computers.

Raw hardware capability no longer defines a mainframe system. What remains intact, however, are the organization and use of the hardware—the most important elements in defining a mainframe computer today.

Although some older mainframe and minicomputer systems remain in service, when you speak of connecting a PC to a minicomputer or mainframe, you probably are talking about one of the following types of systems:

- An IBM mainframe computer

- A DEC VAX-technology minicomputer

- A minicomputer or supermicro running the UNIX operating system (or a variation of UNIX)

This chapter discusses methods you can use to connect a PC to each type of central computer system.

Asynchronous Connections

All mainframe and minicomputer systems provide a way to connect ASCII terminals via RS-232 connections. RS-232 connections are this book's primary focus and are detailed in Chapters 1, 2, and 3. If you connect a PC to a mainframe using this method, your PC must use a terminal emulator to simulate a data terminal that the mainframe can be configured to support. (For more information about terminal emulators, see Chapter 16.) Asynchronous connections can be used easily to provide dial-up modem access to mainframes, and asynchronous connections don't require special PC hardware—only a standard serial port and possibly a modem.

The disadvantage of these simple connections is that mainframe environments typically are not set up to handle binary file transfers through this type of link. The usual method is to install on the mainframe a version of KERMIT, which enables file transfers over the serial connection. ZMODEM also is becoming available for micro- and mainframe computers and sometimes can be used instead of KERMIT. (For more information on KERMIT and ZMODEM, see Chapter 3.) If you want to transfer small amounts of data infrequently, using KERMIT or ZMODEM is satisfactory. This method's throughput speed is often poor, however, because the mainframe environment is not designed to move large quantities of information through asynchronous serial data connections.

If you need greater performance or flexibility than the standard asynchronous serial port connection provides, you have to go deeper into the mainframe communications structure. The next available layer usually is a complete data communications network. This layer is far more complex than the simple asynchronous serial connection. These communications interfaces, however, enable you to connect your PC to a mainframe and provide a high degree of integration between the two computers.

The IBM Mainframe: 3270 and SNA

The IBM mainframe is the oldest and most widely used mainframe technology. In many ways, the newest IBM mainframes (and compatibles such as Amdahl) operate much like the original Model 360 and 370 computers, which began the line more than 25 years ago. Because the IBM mainframe has existed for so long and has such a large user base, a great deal of technology and terminology that is unique to the system has developed. Any explanation of IBM mainframe communications inevitably leads to a discussion of IBM equipment model numbers, because IBM uses numbers to name much of its system architecture. IBM uses the number *3270* to refer to its mainframe communications equipment.

All IBM mainframe communications devices (terminals, controllers, and so on) are part of the 3270 family of equipment. Originally, each piece of equipment was given a model number that began with 327. Over the years, however, that numbering system was not sufficient. Nevertheless, mainframe devices that do not have model numbers beginning with 327 still are known as 3270-family products. Terminals that are part of this family are said to use the *3270 protocol* to communicate with the IBM mainframe computer.

Unlike the asynchronous ASCII communications that you learned about in Chapters 1 and 2, the 3270 terminals use synchronous serial communications with a special protocol to integrate them into the mainframe environment. (As you may recall from Chapter 1, synchronous serial communications always require a protocol.)

IBM even uses its own character coding system rather than the ASCII character code used by all other modern computer systems. IBM's code system is called *EBCDIC*, which stands for Extended Binary Coded Decimal Interchange Code. IBM developed this character code many years ago and never converted its mainframe systems to ASCII. The EBCDIC code serves the same purpose as ASCII—EBCDIC enables text to be represented in computer memory, as described in Chapter 1.

IBM also devised its own communications interconnection protocol—*Systems Network Architecture (SNA)*. SNA first appeared in 1974 with the goal of enabling IBM to interconnect all of its communications products with its mainframes. Over the years, SNA's capabilities have grown and today it is a very robust interconnection protocol with much more than just 3270 terminal connections. You also can connect an SNA network directly to an IBM Token-Ring local area network. This type of connection often is called a *TIC* connection (pronounced tick)—referring to the Token-Ring Interface Coupler hardware that enables a PC to make this type of connection.

The following sections explore in detail each segment of the IBM communications system. The goal is to help you understand how to use these segments to connect PCs to an IBM mainframe computer. In the process, you have to work with many IBM equipment model numbers. Paying careful attention to the numbers and acronyms, however, makes understanding the system much easier.

Understanding IBM 3270

The IBM 3270 terminal system is organized in clusters. The 3270 terminals actually are the Model 3278 (monochrome) and the Model 3279 (color). These terminals connect to a Model 3174 or Model 3274 cluster controller via a coaxial cable. These *cluster controllers* (CCs) act as concentrators and also manage much of the local terminal handling.

One or more cluster controllers are connected to a *communications controller*, which also is called a *front end processor* (FEP). The connection between the cluster controller (CC) and the communications controller (FEP) typically is made through leased telephone lines and modems; however, a local hard-wired connection is possible if the two units are located near each other. Commonly used IBM front end processors are the Model 3705 and the Model 3725. IBM encounters serious competition from *plug compatibles* in this area, however, and several companies market compatible front end processors. Figure 17.1 shows a typical 3270 terminal hookup.

Understanding SNA

As you can see from the preceding section, a typical 3270 communications configuration with a computer in each controller offers a great deal of computing power. IBM uses this power to implement most of the Systems Network Architecture (SNA) protocol in the controllers. The host mainframe computer only has to implement the SNA interface functions that perform high-level control tasks.

IBM refers to the application program interface for any complex hardware—such as a disk system or communications network—as an *access method*. Access methods provide an application program on the host computer with an *Application Program Interface (API)* through which the program communicates with the hardware—in this case with the 3270 terminals. The API also provides required security and network data flow control.

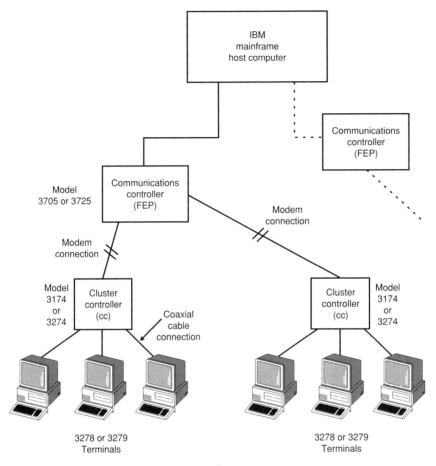

Fig 17.1. The 3270 terminal and controller structure.

The most frequently encountered host computer access method for SNA is the Advanced Communications Function/Virtual Telecommunications Access Method (ACF/VTAM), often called *VTAM*. In addition to supporting 3270 terminals, VTAM also provides the capability to interconnect two or more SNA networks.

A communications controller (Model 3705 or 3725) executes software called Network Control Programs, which manages the routing and flow of data between the controller and connected devices. The most common communications controller program is Advanced Communications Function/Network Control Program (*ACF/NCP*), which can perform error recovery and control the configuration of lines to the communications controller.

Understanding SNA Nodes

The entire SNA system is based on the concept of node connectivity. A *node* is the smallest identifiable functional block of an SNA network, and each node is located uniquely by a network address. All hardware and software in an SNA network are grouped into nodes based on the function that each piece supports. The three SNA node types are host nodes, controller nodes, and peripheral nodes.

A *host node* consists of a host computer (the IBM mainframe) and its attached communications hardware. A host node contains the application interface (the *access method* in IBM terminology) and the network supervisory control interface. Host nodes usually provide the ACF/VTAM access method (API) and are attached to other SNA equipment through the mainframe's peripheral I/O channel. An IBM system 370 I/O channel is a high-speed connection that enables data to flow at speeds in the megabyte-per-second range.

A *controller node* consists of a communications controller (Model 3705 or 3725) and a Network Control Program (NCP). Controller nodes provide connectivity between host nodes and peripheral nodes. A controller node is an intermediate node in any SNA connection.

SNA defines each end of an SNA connection as an *end user*. An SNA end user, therefore, can be the person who interacts with network resources through a 3270 terminal as well as the host application program with which that person is interacting. SNA end users include the source and the destination of information that passes through the SNA network. An end user in an SNA network doesn't have to include a person. In some cases, two computers in the network can establish a connection and make an intermediate communications link.

A *peripheral node* consists of the hardware and software that directly support an end user. The typical peripheral node contains a cluster controller and its attached 3270 terminals. For special connectivity applications, however, peripheral nodes can consist of intelligent devices capable of connecting directly to a communications controller.

SNA acts as a computer controlled switch panel that connects peripheral nodes and provides a complete data circuit. Figure 17.2 shows the 3270 Information Display System shown in figure 17.1 in terms of the three types of software nodes.

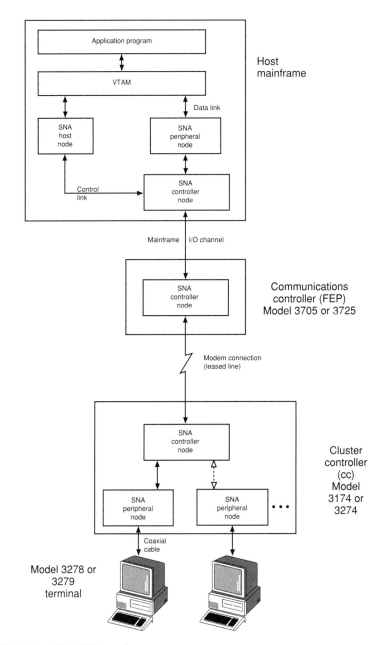

Fig 17.2. Simplified SNA node structure.

In figure 17.2, you can see each segment of the connection between the application program and the 3278 (or 3279) data terminal. Although figure 17.2 takes a few liberties with SNA node terminology, careful examination of the diagram can acquaint you with the system's structure and operation. You will learn the SNA terminology of each segment after you understand the system shown in figure 17.2. The system is explained in the following paragraphs.

Establishment of a data connection between the program and the 3278 (or 3279) data terminal begins when the application program requests that the VTAM access method software open a data link between the program and the specified data terminal's peripheral node. The peripheral node in the cluster controller is the last item in the chain between the host computer and the terminal, which is known to the SNA network. Because the peripheral node is connected to one terminal only, the address (or name) of the node is sufficient to locate the terminal. The SNA network is not concerned with how the peripheral node connects to the terminal—that connection is an issue only within the cluster controller. The next section explains how that portion of the interface operates.

To open the data link to the terminal's peripheral node, the VTAM software makes a control request of the SNA host node to establish the circuit. The host node software module then negotiates with the SNA network (via its control link with its associated controller node) to establish the connections through the FEP and CC to the data terminal's peripheral node. During this process, the two peripheral nodes and all intermediate controller nodes negotiate and agree upon the data path.

After a data path is established, data can flow from the host's peripheral node to the terminal's peripheral node through the intermediate portions of the SNA network.

Understanding SNA Software Modules

Each node shown in figure 17.2 represents a software module within the SNA network. In SNA terminology, a software module is a *unit*. Because each module (unit) has an address (or name) that enables its unique location on the SNA network, these software modules are called *Network Addressable Units (NAUs)*.

Figure 17.2 shows these software modules (NAUs) as nodes. In SNA terminology, the nodes have special names based on their functions and their original software and hardware implementation.

The host node of figure 17.2 controls the network and provides system services to the VTAM access method. In SNA jargon, this type of software module is called a *System Services Control Point (SSCP)*.

The controller node of figure 17.2 was implemented originally as the controlling portion of a remote device controller. As a result, this type of software module is known as a *Physical Unit (PU)*.

The peripheral node represents the end of a logical data circuit. Several peripheral nodes usually can be found in a single controller (such as a cluster controller), and each peripheral node represents a possible end point of a logical connection. In the SNA lexicon, therefore, these peripheral nodes are called *Logical Units (LUs)*.

The SNA feature called *path control* establishes the connections between software modules and enables the SNA network to provide end-to-end data flow control and point-to-point data routing and pacing. Figure 17.3 provides the official names for each of the SNA Network Addressable Units (NAUs) shown in figure 17.2.

By examining figure 17.3, you can see that the Logical Unit (LU) software modules provide end user access to the SNA network (in that they perform the peripheral node function). Because the SNA network begins and ends at an LU module, the actual SNA network user is not visible to the network—only the Logical Units, which provide end user services, have SNA addresses.

This structure enables the LU software to interface any of several types of hardware or software to an SNA network. As long as the LU software presents the correct LU protocol to the SNA network, it can convert that protocol to and from anything that it wants on the other side of the connection. In figure 17.3, the LU in the cluster controller converts the SNA protocol to the signals needed to drive a 3278 (or 3279) data terminal.

To summarize, an SNA network makes and controls connections between pairs of LU programs to provide an end-to-end connection. This connection is called a LU-LU session.

Logical Unit software modules (LUs) are given a type number to indicate their capability and functionality. Each type of LU has its own protocol based on the types of connections the LU is designed to handle. As a result, an LU can establish sessions and communicate only with another LU of the same type.

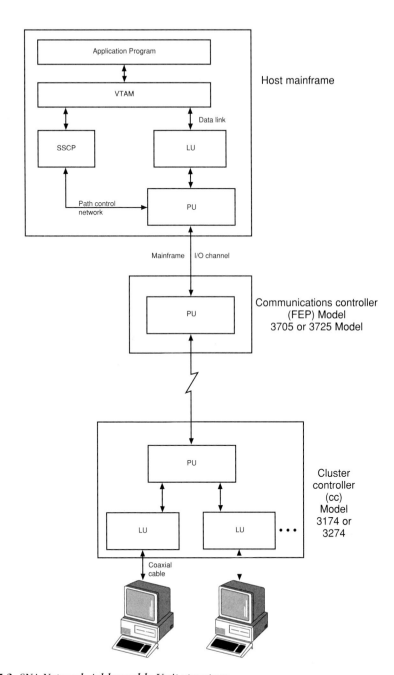

Fig 17.3. *SNA Network Addressable Unit structure.*

The most interesting LU from the PC standpoint is LU 6.2, which implements Advanced Program-to-Program Communications (APPC). APPC enables peer-to-peer communication between any two computers connected to the SNA network—without the need for a mainframe host to control the LU-LU sessions.

The *Physical Unit* (PU) software modules manage the resources of physical devices, including processors, controllers, printers, and other end user devices (in that PUs perform the controller node function). The PUs control the routing by establishing or modifying the links within the network. A Physical Unit program resides in each of the network's discrete processing points.

A PU module's purpose is to handle issues associated with an intermediate portion of the network data circuit. The negotiation between Physical Units is known as the Path Control Network. Although the primary purpose of the PU software module is to negotiate and route connections, a PU module also can provide trace information and performance data.

The *System Services Control Point (SSCP)* software modules reside in the host computer. The SSCP activates, controls, and deactivates network resources (host node function). The SSCP establishes a session with a PU (an SSCP-PU session) to activate, monitor, or deactivate system resources. The SSCP initiates sessions with LUs (SSCP-LU sessions) to initiate and control a LU-LU session—the ultimate purpose of the network.

In large SNA networks, SSCP-SSCP sessions are possible—enabling one SSCP to act as a backup. If one host fails in such systems, the other host's SSCP can take control of both networks.

Now that you have learned the correct terminology, the operation of an SNA connection can be stated in its proper form. An SNA session begins when an LU software module (representing an end user) requests a session with another LU on the network. A System Services Control Point software module (SSCP) handles this request and activates the LU-LU session. The Path Control Network (a function of the PU software modules) routes control information that initiates and terminates the session and manages the routing of messages between the two LUs. After the LU-LU session is established, data flows between the two LUs.

Understanding SNA Capability

After learning the overall structure of an SNA network, you may wonder what is gained by such complexity. The primary benefit of a complex data network structure (such as SNA) is the capability to connect a variety of computers and devices through physical links—without affecting the applications software. The connection between the FEP and the cluster controller in an SNA network, for example, may be Binary Synchronous (Bi-Sync or BSC) protocol, X.25 protocol, or Synchronous Data Link Control (SDLC) protocol. These connections are drastically different in format, hardware, and speed requirements, but because the SNA software modules adapt to them, the application and the 3278/3279 terminals never have to know which protocol is in use.

The capability to conceal how data is transported enables the addition of new data communications methods to an SNA network—without changing existing applications. As you learn in the next section, SNA even enables the addition of full LAN interconnection capability. Applications written before this capability was developed also can use it without change.

The second major benefit provided by complex data network schemes (such as SNA) is redundant data paths. In SNA, this feature takes the form of *transmission groups*, which can assign several communications links in parallel and make them appear to be one link. If one mode fails or one telephone line is damaged, the network uses the remainder of the assigned links to pass the data. The application and the terminal user never know that the failure occurred. If the data path was extremely busy, the process may slow slightly; however, the slowdown represents the entire effect of a physical transmission link failure.

Since its creation, SNA architecture has evolved from a rigid host-to-terminal structure to a versatile computer-to-computer structure. Because of IBM's prominence in the mainframe environment, SNA has grown and developed as technology has advanced. Although SNA is an expensive architecture to implement, SNA provides reliability and tremendous flexibility in data connections.

Connecting a PC to an SNA Network

You may need to connect a PC to an SNA network for several reasons. You may need to run programs that reside on the mainframe, using your PC as a data terminal. You may want to extract data from the mainframe databases and transfer it to your PC. You may want to transfer files you created on your PC to the mainframe for processing. To accomplish any of these tasks, you need to establish a connection between your PC and the mainframe computer, which requires connecting to the SNA network.

An early method of connecting a PC to an SNA network used a special computer known as a *protocol converter*. This special computer was connected to the SNA network and simulated a cluster controller. The protocol converter also had several asynchronous ASCII RS-232 ports, which could be connected directly to a PC or modem.

The PC was connected to the asynchronous ports of the protocol converter with a modem or hard-wired null modem cable, as discussed in Chapter 2. The PC used terminal program software that emulated one of several popular asynchronous data terminal models. (See Chapter 16 for details about terminal emulation.) The same PC setup was used to connect to commercial data services and BBSs, as described in Part II.

Using this method, the protocol converter changed ASCII character codes to and from EBCDIC data codes to solve the data compatibility problem. The protocol converter also emulated the 3270 terminal keyboard from the ASCII terminal input and converted 3270 display functions to the asynchronous terminal control functions, which produced the same display as a 3270 terminal would have produced. This method of interface is expensive and provides only a minimum level of functionality. This method, therefore, is not used widely in new installations.

An extremely popular method of connecting a PC to an SNA network involves the PC's emulation of a 3278 or 3279 terminal. An adapter card is used in the PC—enabling the PC to connect to the 3174 or 3274 cluster controller via the same coaxial cable link that the 3270-family data terminals normally use. Software on the PC emulates the 3270 terminal functions and communicates through the PC adapter card to the cluster controller. Products such as DCA's IRMA and Attachmate's 3270 emulators operate on this principle.

The advantage of using this method to connect a PC to an SNA network is that the process does not require special action on the mainframe. The PC simulates a true 3270 terminal; therefore, the mainframe doesn't have to alter its normal processes.

A PC emulating a 3270 terminal can operate in several modes. A Control Unit Terminal (CUT) can have only one session with the mainframe. A more advanced mode is the Distributed Function Terminal (DFT). A DFT can have up to five concurrent sessions with the mainframe. If the PC is plugged into a 3174, yet another mode is available: Multiple Logical Terminal (MLT), which enables multiple sessions with a CUT-mode terminal. PC terminal emulation software for 3270 coaxial adapter cards will supply one or more of these modes.

With the advent of PC emulation of 3270 protocol, IBM expanded the 3270 application interface through LU 6.2 (APPC). The 3270-PC interface and the High Level Language Application Program Interface (HLLAPI) enable a high degree of integration between emulation software on the PC and the mainframe, all from a simple coaxial cable 3270 connection. Several PC 3270 emulators use this capability to establish sophisticated connectivity between the mainframe and the PC. Some newer versions of 3270 emulators use Microsoft Windows to enable simultaneous mainframe sessions to be opened in overlapped windows.

The coaxial cable connection method is easy and requires no special action on the mainframe. The emulation software on the PC enables you to interact with the mainframe (as you would on a 3270 terminal) and to transfer files between the PC and mainframe. In addition, this method's performance—throughput and response time—is good. This method can be expensive, however, if you want to attach many PCs to the mainframe. (This method requires that a coaxial cable be connected to each PC that is attached to the mainframe—regardless of whether the PC is active.) Along with the 3270 emulation software, a 3270 coaxial adapter card must be installed in each PC.

Connecting SNA to a PC LAN

Another IBM connection solution called TIC directly connects a PC to an IBM mainframe through a Token-Ring local area network. The name *TIC* comes from the *Token-Ring Interface Card* that is plugged into the PC to make this connection. After being connected, the mainframe and the PCs interact as peers on the same network—as long as the appropriate software is installed on each end.

To make this connection, Token-Ring Interface Cards are plugged into the PC and the 3174 cluster controller. The cluster controller then is wired into the token ring LAN—just like any other PC. Special software must be installed on the mainframe, and each time a PC is added to or deleted from the network, the change must be coordinated closely with the mainframe system administrator. This coordination is required because the mainframe must define each connected PC's Physical Unit (PU) software address to the SNA network before the PC can participate in the network.

TIC is an expensive connection path because the token ring network requires several expensive cables; however, TIC's performance is nothing short of astounding. Files transfer at speeds up to 50K per second—many times faster than with any other connection method described in this chapter. With good PC software, the integration of the PC and the mainframe can be very tight indeed.

The VAX: DECnet & PCSA

The personality of a company—even a very large company—usually reflects its founders. The IBM mainframe environment discussed in the preceding section certainly is reflective of Thomas Watson—always a salesman first as he built the world's largest computer company. Digital Equipment Corporation (DEC) has a different personality, which also is an outgrowth of its founder's vision.

DEC began in the 1950s, when Kenneth Olsen felt that computers had a mission beyond accounting and payroll—the predominant jobs of computers at that time. Olsen had a vision of computers built for scientists and engineers. In his vision, computers would not require an expensive, airconditioned environment and several people to operate. Instead Olsen felt that a computer should fit the way that engineers and scientists work. This vision demanded a computer that could be placed anywhere and be operated by a scientist or engineer without outside help.

In 1959, DEC released its first minicomputer—the PDP-1. The PDP-1 was not very small by today's standards but was revolutionary at the time. The PDP-1 had a CRT integrated into the operator's console and came in a cabinet the size of a refrigerator. In addition, the PDP-1 required only normal electrical power and air conditioning. Olsen nurtured his dream over the following years, and in the late 1960s—with the release of the PDP-8 series—DEC finally saw the minicomputer become a successful product.

The rapid rise of DEC began, however, with the introduction of the PDP-11 series. Electronic technology had advanced to the point where the PDP-11 series provided substantial price for performance benefits, and many companies rapidly adopted the PDP-11. By the mid 1970s, the PDP-11 was the dominant minicomputer in the world, and DEC had become a large company.

In 1977, DEC announced the long-awaited follow-up to the PDP-11 series—the VAX 11/780. The VAX line provided an entirely new and much more powerful series of computers. In addition, the VAX series contained two other firsts in the computer world: a compatibility mode that ran PDP-11 software without modification, and a compatible bus structure that enabled custom-built PDP-11 interfaces and existing peripherals to be used without modification.

This upward compatibility enabled PDP-11 users to move smoothly into the new technology. The arrival of the VAX also enabled DEC to make two important moves. First, the VAX series was a full 32-bit architecture (whereas the PDP-11 was a 16-bit computer). Second, the VAX series had full hardware support for virtual memory. (If the system evolution sounds familiar, you may recognize the pattern from current events in the Intel-based microcomputer world). To support these new features, DEC released the Virtual Memory System (VMS) operating system—DEC's first true multitasking, multiuser operating system supported fully by hardware and software.

One other announcement was made when VAX 11/780 was released. At the time, the world didn't consider the announcement very important; however, the announcement heralded what may be the most important move DEC has ever made. DEC announced a computer networking product called DECnet to connect its computers in a multicomputer network.

Understanding DECnet

DECnet originally was designed for parallel hardware interfaces. (For more information on parallel hardware interfaces, see Chapter 1.) DECnet was intended to connect computers located not more than 25 feet apart. DECnet was designed to enable the sharing of peripherals and files and was supposed to introduce an era of distributed processing. In *distributed processing*, a computer task is broken into several parts that are executed simultaneously on several connected computers. The dream of distributed processing was not realized by DECnet, but a robust way of connecting computers to share resources did emerge.

As a software-based product, DECnet has been capable of adapting to changing technology. In May 1980, DEC—along with XEROX and Intel— announced its commitment to EtherNet as DECnet's connection method. Today, DECnet running over EtherNet cabling forms the basis of DEC's networking solutions.

DEC has become the second largest computer company in the world, and DECnet itself has become quite important. When the subject of directly connecting IBM PCs to VAX computers arises, DEC's size and influence make DECnet a natural choice. As you will see in the following sections, DECnet has been up to the task.

Understanding PCSA

Personal Computing Systems Architecture (PCSA) is an integrated hardware and software approach that connects an IBM PC (or a compatible) to a VAX computer via DECnet. In PCSA, the VAX becomes a file server for one or more PCs, enabling a very high level of integration between the two systems.

Installing PCSA on a PC involves adding an EtherNet adapter card and installing the PCSA client software. The EtherNet adapter may be DEC's own DEPCA (Digital EtherNet Personal Computer Adapter) or an EtherNet adapter marketed by Western Digital, 3COM, Interlan, or other vendors. After the EtherNet adapter is connected to the EtherNet cable of a DECnet network, the PC becomes part of that DECnet network.

Special workstation software can be installed in any VAX to handle the PCSA file server functions, but DEC makes a special hardware platform for this task. The small platform (approximately the size of an IBM PC/AT) is a specially configured micro-VAX called the PCLAN/Server 3100. The computer in the PCLAN/Server is a normal micro-VAX that uses the VMS operating system. (As a footnote on the advancement of computer technology, this micro-VAX computer has almost four times the computing power of the original VAX 11/780.)

The PCLAN/Server 3100 computer has an EtherNet port built in so that connecting the PCLAN/Server to the network is straightforward. The PCLAN/Server also comes with all PCSA server software—designed to provide shared printer access and to supervise connections to other DEC and non-DEC machines. Additional hardware and software can be added to provide SNA links or 3270 terminal emulation for connection to an IBM mainframe. (For more information, see "Understanding IBM 3270" and "Understanding SNA" earlier in this chapter.)

The PCSA client software is based on Microsoft's DOS Redirector, which also is used by IBM and other DOS local area network companies. The PCSA network provides NETBIOS compatible services so that the PCSA network can support most PC network products. The client PC software is easy to configure and leaves a PC more than 500K of RAM free for application program use (with DOS Version 3.3).

The PCSA enables the client PC to contact multiple servers through the EtherNet network while the PCLAN/Server 3100 acts as a gateway to other DEC machines on the DECnet network. This integration of files on a VAX-VMS system with DOS-based PCs provides a smooth solution to this problem of integrating a DOS-based PC with a VMS-based VAX computer.

The UNIX Operating System and TCP/IP

In this chapter, you have examined the IBM mainframe and the DEC minicomputer. You have seen that the interconnection architecture in these two cases reflects the corporate vision of the two largest computer companies in the world. Although many other types of large computer installations exist today, all are rapidly converging on some variant of the UNIX operating system. Unlike the DEC and IBM operating systems, UNIX is not the outgrowth of a single individual's vision; rather, UNIX is the product of many contributors in the areas of research and education. The TCP/IP protocol's origins are similar, and the histories of UNIX and TCP/IP are intertwined.

The UNIX operating system is more than 20 years old. UNIX was developed in 1969 as a platform for research work by Dennis Ritchie and Ken Thompson at AT&T's Bell Labs. Originally written in FORTRAN on a PDP-7 computer, UNIX was rewritten completely in 1973 using C, a new language that Ritchie and Thompson developed specifically for this task. Because AT&T was restricted by government regulations from entering the computer business, the company could not sell copies of UNIX. Instead, AT&T made the UNIX operating system and its source code available to colleges and universities at no charge. This action made UNIX a popular computer science curriculum focus and spawned significant independent development of the operating system and its utility programs.

UNIX's development has received little formal guidance and has been subject to the chaotic influences of a democratic process. UNIX's development, however, has benefited from the tremendous energy that participation gave to its early developers. In recent years, attempts have been made to produce a standard UNIX, but UNIX's development history makes this outcome unlikely. Too many people feel that standardization will limit development, and UNIX always has been primarily a development environment. As a result, proprietary variants of the UNIX operating system have become dominant commercially, but UNIX itself continues as a development platform for research and educational use.

Like UNIX, TCP/IP (Transmission Control Protocol/Internet Protocol) is a collection of protocols, or a "protocol suite," that was developed in a research environment more than 20 years ago. In the case of TCP/IP, the research environment was the U.S. Department of Defense's ARPANET project. The ARPANET project of the Defense Advanced Research Projects

Agency (DARPA) is the world's largest continuous network development project. The project began in the late 1960s and has served as a pioneer in many networking techniques. ARPANET is slated to be terminated in the early 1990s, but computer networking will be indebted to the project forever.

Initially, only the Internet Protocol (IP) portion of TCP/IP existed. IP was the outgrowth of early ARPANET networking experiments and reached a form recognizable as IP around 1973. The goal of IP was to provide an architecture that was robust in the face of partial network failure and faulty (or error prone) data transmission facilities.

In 1974, Robert Kahn and Vint Cerf proposed the idea of adding a Transmission Control Protocol (TCP) so that more networking functionality could be achieved. The combined TCP/IP protocols were refined and improved in the ARPANET environment over the next several years. To spur conversion to the new protocols (which in turn would encourage the protocols' continued development), DARPA paid Bolt Beranek and Newman, Inc. (BBN) to develop a TCP/IP for the UNIX operating system. DARPA also funded the University of California at Berkeley to integrate this protocol into its popular BSD UNIX distribution.

Because TCP/IP and UNIX were free to universities and were the networking method for the U.S. Department of Defense, TCP/IP developed rapidly. The Department of Defense developed an official testing program that certified various TCP/IP implementations as compatible. Although the program was disbanded in 1987, all TCP/IP implementations now are standardized and can communicate freely. By the mid-1980s, TCP/IP was being integrated into other commercial operating systems. Today TCP/IP is the most widely implemented vendor independent protocol in the world.

Since 1988, TCP/IP has become available for use with EtherNet and Token-Ring local area network hardware—in addition to the dial-up or leased-line modem connections for which the system originally was designed. TCP/IP represents the most universal computer interconnection method available today.

Understanding TCP/IP

TCP/IP establishes an addressable data path between any computers that are connected to the network. Unlike the SNA and DECnet protocols discussed earlier in this chapter, TCP/IP alone does not provide any file transfer or application capability. In this regard, TCP/IP is much like the modems used in asynchronous communications—TCP/IP provides the data connection, but more software is required to use the connection to transfer data.

The TCP/IP protocol can be thought of as a network building block. Below TCP/IP is the hardware transmission medium—EtherNet, high speed modems, satellite or radio communications links, wire, or any other hardware data transmission link. TCP/IP in turn provides a standard interface to any overlying software. The interface looks the same regardless of the actual physical connection; therefore, the details of the communications link are hidden from the application software. In addition, TCP/IP provides the capability to link several networks and integrate an entire set of diverse computers into one *internetwork*. In such an internetwork, any computer can connect to any other computer and send and receive data.

To further standardize TCP/IP networks, the Department of Defense developed three high-level services that were codified as Military Standards (MIL-STDs). The first service is MIL-STD 1780, a File Transfer Protocol known as FTP. The second service is MIL-STD 1781, a Simple Mail Transfer Protocol known as SMTP. The third service, MIL-STD 1782, is a terminal emulation program known as TELNET. TELNET enables remote logon through a TCP/IP network. These three protocols are standardized and are provided by nearly every TCP/IP network. The following paragraphs discuss these protocols in greater detail.

TELNET provides simple remote terminal capability. With TELNET, a user at one computer site can establish a TCP/IP connection to a login server at another computer site, and keystrokes are passed from one computer to the other. TELNET also supplies a network virtual terminal interface that provides a way to negotiate such options as seven- or eight-bit binary data. TELNET is a simple but important program that enables a TCP/IP network to be used as a generalized packet switched data network under operator control.

The File Transfer Protocol (FTP) enables an authorized user to select any other computer connected to the TCP/IP network, log into it, list file directories, and send or receive files between the computers. FTP can be used interactively—with a person typing the commands on-line—or under program control for automatic file transfers. FTP also enables third-party transfers in which one computer initiates a file transfer between two other computers connected to the TCP/IP network.

The Simple Mail Transfer Protocol (SMTP) transports electronic mail between two computers on the TCP/IP network. In the SMTP protocol, the sending computer identifies itself, identifies the intended mail recipient computers, and then sends the mail messages across the TCP/IP network. SMTP generates acknowledgment responses for each transaction so that the sender knows whether the transmission was successful.

TELNET, FTP, and SMTP form the basis of the internet described in Chapter 15. These standardized services reflect the initial purpose of TCP/IP networks: to provide wide-area connections between systems composed of different computers. Today, however, TCP/IP is combined with local area network software to provide the type of PC-UNIX interconnections on which this chapter focuses.

Understanding TCP/IP Addressing

Each computer in a TCP/IP network is given an address. The address is a 32-bit number written as four decimal numbers separated by periods—for example, 128.10.0.5. Each number is the decimal value of one 8-bit byte of the 32-bit address. Each computer connected to a TCP/IP network has to have a unique address.

The 32-bit address is broken into two components—the *network ID* and the *host ID*. The network portion specifies the network of which the host computer is a part. The host portion uniquely identifies that computer within its network. This structure accommodates the setup in which computers are grouped into networks and networks are combined into internets.

Because so many network and internetwork combinations exist, the TCP/IP address has three forms. The three forms leave space for more addresses in the network or host portion—as required by a network's structure and size. Figure 17.4 shows the binary structure of the three TCP/IP address forms, or *classes*.

The Class A address form is used when many host computers exist on a few networks. The 7-bit network portion of the address can support up to 127 networks; the 24-bit host address has space for more than 16 million computers per network. The Class B address form is used when both the number of networks and computers per network are large. The 14-bit network portion can accommodate more than 16,000 networks; the 16-bit host address can handle more than 65,000 computers per network. The Class C address form has space for more than 2 million networks but can accommodate only 255 computers per network.

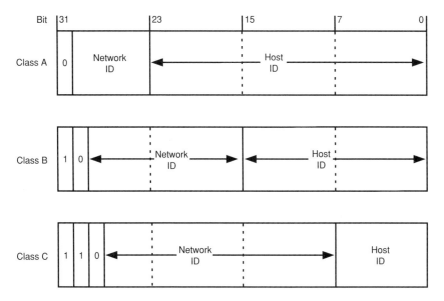

Fig 17.4. *The three forms of internet addresses.*

Regardless of the address form used, the notation of four decimal numbers separated by periods remains constant. Each decimal number represents one byte of the internet address. The internet address 128.10.0.5, for example, encodes into a 32-bit hexadecimal address of 0x800A0005 and is a Class B address. A central authority must assign each address in a TCP/IP network to ensure that no two computers have the same address.

Connecting a PC to UNIX

AT&T and DEC market an implementation of Microsoft's LAN Manager that runs on the UNIX operating system. A PC can be connected to such a system using any TCP/IP connection, but EtherNet seems to be the preferred connection method. The EtherNet adapter is plugged into the PC, and then the Microsoft LAN Manager client software is installed on the PC. (The same software is used in LAN Manager networks in which the server is running the OS/2 operating system.)

The client PC can map the UNIX system's disk areas to look like DOS drives. When the PC accesses the networked disk, the UNIX-based server software translates the data between the UNIX and DOS file formats. This process results in slightly slower operation than an all-DOS or all-OS/2 network in which the file systems are similar; nevertheless, the performance is sufficient for most applications. In addition, the high level of file directory integration between UNIX and the DOS PC enables easy file sharing between the two systems.

Another widely used method for connecting a PC to UNIX via a TCP/IP link is a Sun Microsystems program called Network File System (NFS). With this connection method, the PC runs a program called PC NFS that connects to the NFS server software on the UNIX system. NFS enables the PC to have simultaneous access to multiple files stored in the UNIX computer's file system. Many companies marketing UNIX license NFS from Sun Microsystems and include NFS as an option with their UNIX software.

Currently, a movement is underway to standardize the NETBIOS interface used by many PC-based local area networks. The goal is to enable NETBIOS to be implemented on UNIX with TCP/IP networks. If this effort succeeds, you will be able to use most PC network software through a TCP/IP link to a UNIX system—when the proper server software is in place on the UNIX computer. At this time, the PC-to-UNIX NETBIOS interface is not a commercial reality; however, the interface may materialize soon.

Chapter Summary

In this chapter, you have learned that you can connect a PC to a large central computer system in several ways. You can use a modem and terminal program to create a loose link. Although this process is easy, you can achieve only limited functionality, which may not be sufficient for your application. You have learned about the most commonly encountered mainframe and minicomputer data networks (3270 SNA and DECnet) and how you can connect a PC directly to them.

You also have seen that the most universal interconnection protocol is TCP/IP, which is available for almost every brand of computer and every operating system made today. You also have seen, however, that few standards for high-level file server software on TCP/IP exist—although the Microsoft LAN manager and Sun NFS are emerging as de-facto standards. TCP/IP provides a high-speed way to exchange data with computers of different architectures, but in its present state of development, using TCP/IP requires more technical knowledge than other connection methods.

18

Understanding
Network
Communications

I n Chapter 7, you learned about using PCs on a network and were
introduced to concepts related to using asynchronous serial communications in a network environment. Local area networks (LANs) provide a
tremendous amount of flexibility in connecting serial communications
ports to workstations. This flexibility does come with a price, however, in
the form of extra hardware and software requirements as well as complexity
of installation and use. In addition, the performance of serial communications routed through a network can degrade rapidly if they are not installed
and used properly.

In this chapter, you learn about asynchronous servers—how they operate,
and what their strengths and weaknesses are. This knowledge will enable
you to approach your serial communications problems with a full understanding of the benefits and trade-offs of using this technology. Every area
of computing also has its own jargon or buzzwords, and in this chapter you
learn the language of LAN communications.

Understanding How LAN Servers Work

In Chapter 17, you learned how data communications can take place over a wide area. You saw that protocols are used to package the serial characters you want to send into packets and then transport them from one computer to another. Your LAN works the same way. When you have a modem pool or other serial data communications server on your network, the programs you run on your workstation must establish a logical data channel between your workstation and the serial port on the server's computer, as shown in figure 18.1.

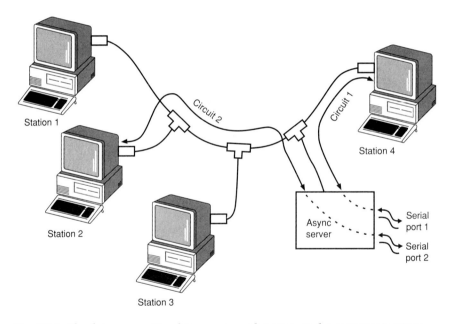

Fig. 18.1. The data connection between a workstation and a communications server.

Figure 18.1 shows a LAN with an asynchronous server that has two serial ports. These ports can be connected to modems or any other serial communications device. In figure 18.1, the users at stations 2 and 4 of the LAN are connected to serial ports 1 and 2 of the asynchronous server. Each connection is made by a software data circuit through the LAN. This connection enables the program on each workstation to function as though the serial port on the server is actually part of the workstation itself.

In order to make the server's ports appear to be connected to the workstations, special network interface software is required. This interface software establishes the data circuits shown in figure 18.1 so that each character the program on the workstation sends to the "serial port" is actually received by the network interface software. The interface software then sends the character across the network to the server, which in turn sends the character to the serial port.

The process happens in reverse when a character is received by the serial port. The server sends received characters across the network to the workstation LAN interface software, which then imitates a local serial port to the program on the workstation. Figure 18.2 shows this data flow.

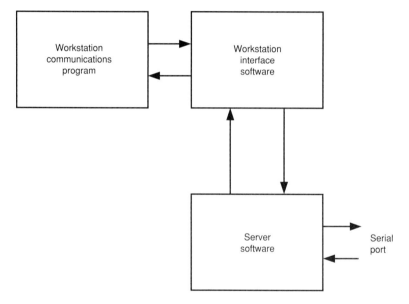

Fig. 18.2. Data flow between workstation program and server serial port.

This data flow structure is the same as that on a LAN when you access files on another computer in the network. In the case of LAN file access, the LAN interface software (called a *redirector*) substitutes for the software at the DOS function call level. This redirector software intercepts disk calls, sends requests for file data to the other computer, receives the file data over the LAN, and returns it through the same DOS interface that a local disk access would use.

Such an interface point between an application program and the system services it requires is called an *application program interface*, or API. Because the DOS file system API is a complete interface, your programs require no changes to operate correctly when the LAN redirector interface software is inserted.

With serial communications, however, the DOS interface is very limited. As a result, most communications programs manipulate the serial port hardware directly in order to provide the features and performance you need; a DOS communications program has no natural boundary where the LAN interface may be inserted. Communications programs that provide for the use of a LAN communications server, therefore, must be different from those that use local serial ports; these programs must use a software API instead of going directly to the serial port hardware registers.

This chapter covers the most commonly encountered communications server interfaces—APIs. You learn what the differences are in each of them and what issues you face in choosing an API. You also learn what the performance considerations are when you use each type of interface. But before you learn how an asynchronous server is implemented and what specific APIs are available, you need to understand how adding the LAN interface software affects the performance of a serial data link.

Understanding Performance Issues

When a serial port is on the same computer as the communications program, a character sent by the program arrives at the serial port in a single step. Likewise, a character received by the serial port arrives at the communications program in a single step. Figure 18.2 illustrates that when the serial port is placed on a different computer from the communications program, two extra data handling steps are introduced to move the characters across the LAN.

In a LAN server environment, a character leaving the communications program goes first to the workstation interface software. That software takes the character and sends it across the LAN to the server software. When the server software receives the character, it sends it to the serial port. These intermediate programs cannot send the character on to the next location until they receive the character from the previous location and therefore add delay to the transmission.

These transmission delays (interface to server and server to serial port) add to the time it takes to get the character into the serial port. The same delays occur in the other direction when a character is received by the serial port. These delays can be increased or decreased, but they can never be eliminated. Therefore, the presence and length of these delays are a primary factor in LAN asynchronous server performance.

One factor that affects the impact of these delays is the number of characters sent over the network in a single batch. Each network transaction has a minimum delay time known as *overhead*, which is a fixed time added to the time required to transmit the data itself. The difference in the time required to send one character versus several characters over the LAN is quite small if the characters are sent in a single LAN transaction.

An example can help demonstrate this effect. Assume that it takes 50 milliseconds of start-up time (overhead) to send any packet on the LAN. Also assume that you are using a LAN that transmits characters at 10 megabits/second (a common speed for LAN connections). Transmitting an actual character on the LAN therefore takes about 1 microsecond. To send a single character over the LAN takes 50 milliseconds (overhead) plus 1 microsecond. To send 100 characters across the LAN in a single packet takes the same 50 milliseconds plus 100 microseconds, or 50.1 milliseconds. In the first case, therefore, a data rate of only 20 cps can be sustained (20 transactions × 50 ms = 1 second), whereas in the second case, a data rate of 2000 cps can be sustained on the same network because each transaction transmits 100 characters.

This overhead number is exaggerated to make the example clear, but this effect is very real in a LAN environment. If several characters are sent between the workstation interface software and the server software in a single transaction, the overhead time represents a smaller delay for each individual character, making the effective transmission rates much higher.

Another problem occurs, however, when too many characters are sent in a single batch. If the workstation interface software waits for several characters to be typed before it sends them to the server, you feel this delay in response time in the form of a delayed *echo* from the remote system you are calling.

The terms response time and throughput often are confused with each other. *Response time* is the end-to-end time it takes for you to receive data as a result of when you enter characters. An echo of a character you type is a response that requires the character you type to travel to the other end of the circuit and back. The response time of this transaction is the time from when you press the key until the echoed character is displayed on your terminal. *Throughput*, on the other hand, is the rate at which data transfers can be sustained over a long period of time. If you can obtain a transfer speed of 9600 bps, then your throughput is 960 characters per second.

Different applications make optimizing either response time or throughput more important. If a person is doing a lot of typing into a remote program, then the response time should be optimized to prevent the program from "feeling" very sluggish. If file transfers are the primary goal, however, then you will want to optimize throughput so that you can send the data over the link in the shortest possible time. The transaction overhead in any data network makes you sacrifice throughput to obtain faster response time and sacrifice response time to obtain higher throughput.

This inherent, unresolvable conflict between response time and data throughput rates leads to differences in how LAN communications servers appear to perform. Some servers enable you to "tune" settings to favor throughput or response time. Other designs try to manage the settings automatically. Still others optimize for what is expected to be an average system load. In all cases, a communications program rarely feels the same when connected to a LAN asynchronous server as it does when it is connected directly to a local serial port.

Understanding Communications Server APIs

As explained in "Understanding How LAN Servers Work" earlier in this chapter, the standard DOS communications interface (API) is not adequate for most communications programs. In fact, no real DOS communications API exists; only the interface offered by the PC BIOS is available.

This standard DOS/BIOS communications API is accessed by the computer instruction *INT 14* and therefore has acquired the name INT 14 interface. Most asynchronous server software and most network versions of communications programs support the INT 14 API because it is the only standard available. But it usually does not provide high speeds or good performance.

Because the INT 14 API is limited, LAN manufacturers have developed their own APIs to support high-performance asynchronous servers. This chapter describes the Novell NASI/NCSI communications API, the Ungermann-Bass UBCI/NETCI communications API, and the IBM ACSI communications API. In addition, the standard INT 14 API is discussed in some detail and common extensions made to it arc described.

If you are buying communications software to run with a particular network asynchronous server, make sure it supports the interface your server provides. Otherwise, you cannot use that interface with the program.

You can learn enough about each communications server API from the following sections, which discuss the capabilities and limitations of each API, to decide whether you need one or another to accomplish your tasks. After you understand these APIs, you also learn how they are used to implement software or hardware asynchronous servers.

Understanding the INT 14 API

The INT 14 API is present on every IBM PC ever made because it is in the IBM PC BIOS ROM. This API is extremely limited, however. In its original form, it supports only four functions: initialize the serial port, read port status, send a character to the serial port, and read a character from the serial port. Because multiple characters cannot be sent together, this API cannot easily avoid the high overhead of sending characters across the LAN one at a time. Some implementations use sophisticated programming tricks to improve the situation somewhat, but this interface usually is not satisfactory for high speed links.

The PS/2 added two extra functions to the INT 14 API, extending the initialization control and adding to its capability to control the port. Some INT 14 LAN APIs also add other functions beyond those defined by the BIOS, but these functions transform the INT 14 into a proprietary interface. Usually these extensions are called an EBIOS API (extended BIOS). Some agreement exists on these extensions, but in general, one EBIOS API is not compatible with another EBIOS API beyond the basic IBM-defined functions.

One extended INT 14 API is worthy of special note because of the number of installations—3COM's Bridge Applications Protocol Interface (BAPI). BAPI extends the INT 14 API by adding a read and write block command as well as a way to enter the Command mode to set the LAN link. BAPI is not discussed here in detail because this protocol will be encountered much less often in the future; it is, however, an extended INT 14 API.

The basic INT 14 API uses the computer registers to pass input parameters and return results. The entire functionality of the standard IBM INT 14 API is given in the following paragraphs.

To initialize the communications port, use the following function call:

    AH = 0 - Initialize the Communications Port
    DX = Serial Port (0=COM1, 1=COM2, 2=COM3, 3=COM4)
    AL = Parameters for initialization
         Bits 7, 6, 5 - Baud Rate
                0 0 0 - 110 bps
                0 0 1 - 150 bps
                0 1 0 - 300 bps
                0 1 1 - 600 bps
                1 0 0 - 1200 bps
                1 0 1 - 2400 bps
                1 1 0 - 4800 bps
                1 1 1 - 9600 bps
         Bits 4, 3 - Parity
                0 0 - None
                0 1 - Odd
                1 0 - None
                1 1 - Even
         Bit 2 - 0=1 Stop Bit, 1=2 Stop Bits
         Bit 1, 0 - Word Length
                1 0 - 7 bits
                1 1 - 8 bits

    On Return:

    AL = Modem Status
         Bit 7 - Carrier Detect
         Bit 6 - Ring Indicator
         Bit 5 - Data Set Ready
         Bit 4 - Clear to Send
         Bit 3 - Delta Carrier Detect
         Bit 2 - Trailing Edge Ring Detector
         Bit 1 - Delta Data Set Ready
         Bit 0 - Delta Clear To Send
    AH = Line Status
         Bit 7 - Time-out
         Bit 6 - Transmitter Shift Register Empty
         Bit 5 - Transmitter Holding Register Empty
         Bit 4 - Break Detect
         Bit 3 - Framing Error
         Bit 2 - Parity Error
         Bit 1 - Overrun Error
         Bit 0 - Receive Character Ready

To send a character, use the following function call:

    AH = 1 - Send Character
    DX = Serial Port (0=COM1, 1=COM2, 2=COM3, 3=COM4)
    AL = Character to Send

    On Return:

    AL = Preserved
    AH = Line Status (same as AH=0)

To receive a character, use the following function call:

    AH = 2 - Receive Character
    DX = Serial Port (0=COM1, 1=COM2, 2=COM3, 3=COM4)

    On Return:

    AL = Character Received (unless Timeout Shown in AH)
    AH = Line Status (same as AH=0)

Note: This routine waits a brief, unpredictable time for an input character and then times out if one is not present.

To read port status, use the following function call:

    AH = 3 - Read Port Status
    DX = Serial Port (0=COM1, 1=COM2, 2=COM3, 3=COM4)

    On Return:

    AL = Modem Status (same as AH=0)
    AH = Line Status (same as AH=0)

These four functions comprise the entire standard IBM INT 14 API. The IBM PS/2 added two extended functions that often are implemented in LAN INT 14 APIs as well.

The first added function allows more detailed initialization of the port, as follows:

    AH = 4 - Extended Initialize
    DX = Serial Port (0=COM1, 1=COM2, 2=COM3, 3=COM4)
    AL = Break: 0-No Break, 1-Break
    BH = Parity: 0-None, 1-Odd, 2-Even
    BL = Stop Bits: 0-One Stop Bit, 1-Two Stop Bits
    CH = Word Length: 0-5 bits, 1-6 bits, 2-7 bits, 3-8 bits

CL = Baud Rate:
    0-110 bps
    1-150 bps
    2-300 bps
    3-600 bps
    4-1200 bps
    5-2400 bps
    6-4800 bps
    7-9600 bps
    8-19200 bps

On Return:

AL = Modem Status (same as AH=0)
AH = Line Status (same as AH=0)

The second added function allows control of the modem port handshake signals, as follows:

AH = 5 - Extended Port Control
DX = Serial Port (0=COM1, 1=COM2, 2=COM3, 3=COM4)
AL = 0: Read UART Modem Control Register
AL = 1: Write UART Modem Control Register
BL = Value to write to modem control register if AL=1
   (Format same as return BL)

On Return:

BL = Modem Control Register
    Bits 7, 6, 5 - Reserved
    Bit 4 - Loop
    Bit 3 - Out 2
    Bit 2 - Out 1
    Bit 1 - Request To Send
    Bit 0 - Data Terminal Ready
AL = Modem Status (same as AH=0)
AH = Line Status (same as AH=0)

Most network communications programs operate using this limited API. If you have a choice, however, using one of the proprietary extended APIs generally gives you more control and better performance.

Using the INT 14 API

To use the INT 14 API on a LAN workstation, you first have to install a terminate-and-stay-resident (TSR) interface program, which establishes the API. This interface usually is installed from the DOS command line (or in a batch file), but in some cases, a device driver may be included in the DOS CONFIG.SYS file in order to load the driver automatically.

Because the INT 14 API does not include the capability to connect to a specific server port, you usually have to run a special command interpreter program, which is part of the LAN software and which enables you to connect your workstation to the server. This connection must be made before you enter your communications program. Each INT 14 implementation handles this connection process differently, so you need to read the instructions for your particular network asynchronous server software to learn the exact procedure for connecting your workstation to the desired serial port on the server.

After this logical connection has been made, you simply run your communications program. Most operations work as they do with a local serial port.

Understanding NCSI/NASI

Novell NetWare is one of the most widely used PC LAN systems. To support asynchronous serial communications, Novell offers the NetWare Asynchronous Communications Server (NACS), which consists of a standard IBM-compatible PC with a LAN interface card, special serial interface hardware, and special software.

The NACS provides dial-out and remote dial for modem operations. Each Novell NACS can handle up to 16 serial ports. A variation on the NACS is called the NetWare Access Server. This server has the same hardware as a NACS but runs software that provides dial-in access to the network.

Serial ports are attached to the NACS in groups of four using an interface card called a wide-area network interface module, or WINM. Each WINM provides four RS-232 serial ports. One Novell network can have several NACSs installed to provide more than 16 serial ports.

To support the best throughput when using the NACS as a modem pool or other outbound serial connection, Novell provides the NetWare Asynchronous Services Interface (NASI). This API provides workstation software with a high-performance, full-featured interface to the NACS.

Network Products Corporation also makes an asynchronous server for Novell LANs (and many NETBIOS LANs as well) called ACS^2. To support this server, they provide a workstation API called NCSI that is compatible with the NASI interface but provides higher performance. For all practical purposes, however, the NCSI and NASI API protocols are identical and therefore are discussed as one.

A Novell NACS using the NASI workstation API can support serial communications speeds up to 19,200 bps. A Network Products Corporation ACS^2 using the NCSI workstation API can support serial communications speeds as high as 38,400 bps.

The NASI/NCSI protocol supports a *name service* that enables each serial port on each communications server to be assigned a name. In fact, each port has at least two names—a general name and a specific name. The general name groups serial ports by type as a single resource, whereas the specific name uniquely identifies a specific serial port. All 2400-bps modem ports can be given the same general name, for example, enabling a user at a workstation to ask for the general name only and be given the next available 2400-bps port. This type of naming frequently is used in modem pools in which the type of port, not the exact port assignment, is important.

The NASI/NCSI protocol is activated at a workstation by loading a TSR program. This program installs the NASI or NCSI API so that application programs run on that workstation can use the protocol. The NASI/NCSI API provides a command interpreter that can be used to set communications parameters and make or break LAN connections to the NACS serial ports. Some applications may use this interpreter, but the NASI/NCSI protocol also provides commands enabling a program to have its own internal service routine perform these same functions.

Some of the network communications programs that support the NASI/NCSI protocol are PROCOMM PLUS, Relay Gold, ASCOM IV, and PC Anywhere.

The Basic NASI/NCSI Protocol

The basic NASI/NCSI protocol is a superset of the Ungermann-Bass NETCI (UBCI) API protocol. This API communicates requests through the computer interrupt INT 6B using the registers to pass parameters and return completion codes. This basic NASI/NCSI API interface may be used in conjunction with the NCSI command interpreter, which provides session management functions (connecting to specified ports, for example). This basic protocol easily enables you to adapt a communications program to the network environment.

Each NASI/NCSI API INT 6B function call follows the same format. The AH register contains the function number, and the AL register contains the logical port or virtual circuit number for the workstation. This port number is not related to the port number or connection reference on the server; it refers to one of the three logical ports NASI supports on each workstation.

All the basic functions affect the logical circuit that is currently active. The application program (or the command interpreter) has the responsibility of ensuring that the correct logical circuit is active before the function calls are made.

If a function call requires a *pointer* (such as to a memory buffer), this pointer address is passed in ES:BX.

If a function call requires a byte count, the count is passed in CX. For some functions not requiring a byte count, CX may contain other input data.

Each function call indicates an error by setting carry and placing the error code in the AL register. Successful calls return the carry clear. The possible error codes are as follows:

Error Code	Meaning
AL = 0FFh	NASI busy
0FEh	Server not responding
0FDh	Unable to open session (usually indicates a port is already in use)
0FCh	Transmit buffer full
0FBh	Transmitter in X-OFF state
0FAh	Circuit not active
0F9h	Request not supported by the port. Returned when the function cannot be provided, as when attempting to set a 2400-bps port to 9600 bps.
0F8h	Request refused. This error generally is returned when the target port is not logged to the requesting circuit.
0F7h	Unable to allocate a circuit
0F6h	Illegal buffer size

The basic NASI/NCSI API consists of the following five functions:

Function	Meaning
AH = 00h	Writes data to a logical port
01h	Reads data from a logical port
06h	Issues a control request
07h	Gets local status
08h	Checks for break

The remainder of this section examines these five basic NASI/NCSI API function calls in detail.

To write data to a logical port, use the following function call:

    AH = 00h - Write Data
    AL = Logical Port
    CX = Number of bytes to write
    ES:BX = Address of buffer to write

    On return: If Carry Clear CX = number of bytes written.

Up to 498 bytes may be written on a single call. Upon returning from this function, the buffer can be modified immediately by the application if desired.

To read data from a logical port, use the following function call:

    AH = 01h - Read Data
    AL = Logical Port
    CX = Maximum number of bytes to read
    ES:BX = Address of read buffer

    On return: If Carry clear CX = number of bytes read (may be zero)

The maximum number of bytes returned by this function call is 498. This function does not wait for data; it returns CX=0 if no data is available. If this call returns a full buffer of data, making an immediate call for more data is advisable in order to avoid overrunning the network input buffers.

To issue a control request, use the following function call:

    AH = 06h - Issue Control Request
    AL = Logical Port
    CL = 2 - Break

4 - Disconnect
6 - Hold
8 - Enable XON/XOFF flow control
10 - Disable XON/XOFF flow control

On Return: Carry clear, AL=0 function performed.

You can use this function call to transmit a break through the port on the server, disconnect from a port and server, or to enable or disable the XON/XOFF flow control at the server.

An application that does its own flow control from the workstation program can disable XON/XOFF flow control at the server. The client, however, then must clear the input buffers out of the server fast enough to prevent data overrun.

To obtain the local status for a logical port, use the following function call:

AH = 07h - Obtain Local Status
AL = Logical Port

On return: Carry Clear CL = number of open logical circuits.

CH: 0FFh = Port Idle
00h = In Service by command interface
01h = In Service through session
06h = Port on hold

This function determines whether or not a particular circuit is available. If the port reports that the circuit is *idle*, it is available for use.

To check if a break signal has been received on a logical port, use the following function call:

AH = 08h - Check For Break
AL = Logical Port

On return: Carry Clear no break received

Carry Set: AL=0 if break, not 0 if error code.

The NASI/NCSI interface increments an internal counter each time a break is received by a serial port. When the Check for Break request is sent, this counter is checked to determine whether a break has occurred since the last Check For Break was issued. If one or more breaks have been received, the carry flag is set and the counter is cleared.

The Extended NASI/NCSI Protocol

You can use the basic NASI/NCSI protocol detailed in the last section (which is equivalent to the Ungermann-Bass NETCI protocol) to implement a simple communications program to operate with a Novell or NPC communications server. The NASI/NCSI protocol, however, also has several extended function calls that provide more control over the target serial port or its attached device.

This extended protocol enables the application to establish several virtual circuits between the workstation program and one or more communications servers. This protocol also enables the program to have much more control over each circuit than the basic protocol allows. For most applications this extended protocol has far more capability than you require, but it does enable an application to have total control of the serial port on the server as well as total control of the data flow on the network itself.

This extended protocol also enables an application to integrate itself smoothly with the network name service so that the application can provide a total network communications control interface to the workstation user. This means that the user of such an application will not have to learn how to use the specialized network connection service software because the application may integrate this function into its own interface.

The following function call returns the current state of the NASI/NCSI API:

    AH = 10h - NASI/NCSI status
    ES:BX = Return Buffer Location

On return: If carry clear the buffer has the following data:

    Byte 0 - Number of virtual circuits in use
    Byte 1 - Reserved
    Byte 2 - State of virtual circuit 0
    Byte 3 - State of virtual circuit 1
    Byte 4 - State of virtual circuit 2
    Byte 5 - State of virtual circuit 3
    Byte 6 - State of virtual circuit 4
    Byte 7 - State of virtual circuit 5
    Byte 8 - State of virtual circuit 6
    Byte 9 - State of virtual circuit 7
    Byte 10 - State of virtual circuit 8

Each state byte has the following meanings:

Byte	Meaning
00h	Circuit idle: no process or connection
01h	Circuit active: in Command state
02h	Connected: circuit established with server
03h	Hold: but active
04h	Busy: arbitrating a process
05h	Super: in supervisory process
06h	Down: shut down, unavailable

The following function enables an application to allocate a virtual circuit:

AH = 11h - Allocate a Virtual Circuit

On return:

Carry Clear, AL = Virtual Circuit Number (0-8)

If a virtual circuit is available, the circuit number is returned in AL. This virtual circuit number then is used for all subsequent communication for a connection established using the assigned virtual circuit.

To get the status of a virtual circuit, use the following function call:

AH = 12h - Get Virtual Circuit Status
AL = Virtual Circuit Number (0-8)

On return: Carry Clear, AL = virtual circuit number

CL = State of virtual circuit as follows:
    00h = Idle
    01h = Active
    02h = Connect
    03h = Hold
    04h = Busy
    05h = Super
    06h = Down

To locate a named server port of a desired type or to produce a list of such names, use the following function call:

AH = 13h - Set/Retrieve Request/Reply Service Name
AL = Virtual Circuit Number (0-8)
CL = Function Requested
    0 - sets service request name
    1 - returns requested service name
    2 - returns current server name
    3 - returns current general name
    4 - returns current specific name
ES:BX = Address of buffer with request or to receive reply

On return: Carry Clear, AL = 0

CL = 0 if name is valid
    > 0 then name is returned in ES:BX

The format of the request/reply buffer is as follows:

Bytes 0-1: Length of name
Bytes 2-16: Service Name Text null terminated

This call enables the application to resolve server names. The application provides NASI with a 1- to 14-byte service name. This name may contain wild-card characters (?) to match if desired. NASI establishes a connection with the first known NACS server and first asks the server if it has a specific server name that matches the request. If no such name is found, then NASI queries each NACS in turn until a service is located or until all NACS have been checked.

If no such specific service is located and the name supplied has more than 8 characters, an error is returned. If the requested name has fewer than 8 characters, however, NASI asks each NACS in turn if it has a general service name that matches the request. If none is located, an error is returned.

When CL=0, this call is analogous to the CONNECT command in the NASI command interpreter. When the other subfunctions are used, the buffer is filled in with the requested information.

The following function is used by an application to display the IPX address of a NACS communications server. Note that the address length requires that the full address be used, including leading zeros.

AH = 14h - Set/Receive Service Address
AL = Virtual circuit number (0-8)
CH = Address Type: 00 (IPX)
CL = Function: 0-set address, 1-retrieve current address
ES:BX = Address of request or reply buffer

On return: Carry clear, AL=0 function call successful.

Request/Reply Buffer:
    Byte 0: Address Type (00h)
    Byte 1: Address Length
    Byte 2-n: Address (Address Length Bytes)

To set/retrieve virtual circuit configuration, use the following function call:

AH = 15h - Set/Retrieve Virtual Circuit Configuration
AL = Virtual circuit number (0-8)
CL = 0-Set configuration, 1-Retrieve configuration
ES:BX = Address of request or reply buffer
Set Buffer:
    Byte 0-1: Length of buffer (0Ch maximum)
    Byte 2: Logical server port ID
    Byte 3: Receive baud rate
    Byte 4: Receive word length
    Byte 5: Receive stop bit
    Byte 6: Receive parity
    Byte 7: Transmit baud rate
    Byte 8: Transmit word length
    Byte 9: Transmit stop bit
    Byte 10: Transmit parity
    Byte 11: DTR control (0=off, 1=on)
    Byte 12: RTS control (0=off, 1=on)
    Byte 13: Reserved, must be 0
Retrieve Buffer:
    Byte 0-1: Length of buffer (0Ch maximum)
    Byte 2: Logical server port ID
    Byte 3: Receive baud rate
    Byte 4: Receive word length
    Byte 5: Receive stop bit
    Byte 6: Receive parity
    Byte 7: Transmit baud rate
    Byte 8: Transmit word length
    Byte 9: Transmit stop bit
    Byte 10: Transmit parity
    Byte 11: DTR control (0=off, 1=on)
    Byte 12: RTS control (0=off, 1=on)
    Byte 13: XON/XOFF flag (0=disabled, 1=enabled)

This function enables the application to change the NASI configuration table without relogging and reintializing the port. When a configuration is retrieved, it is taken from the server port that is mapped currently to the indicated virtual circuit.

Note that if BX=0, then the default configuration for the circuit is set. The values for the transmit and receive parameters are the same as follows:

Baud rate:

0 - 50 bps	6 - 600 bps	12 - 4800 bps
1 - 75 bps	7 - 1200 bps	13 - 7200 bps
2 - 110 bps	8 - 1800 bps	14 - 9600 bps
3 - 134.5 bps	9 - 2000 bps	15 - 19,200 bps
4 - 150 bps	10 - 2400 bps	
5 - 300 bps	11 - 3600 bps	

Word length	*Parity*	*Stop bit*
0 - 5 bits	0 - None	0 - 1 stop bit
1 - 6 bits	1 - Odd	1 - 1.5 stop bits
2 - 7 bits	2 - Even	2 - 2 stop bits
3 - 8 bits	3 - Mark	
	4 - Space	

To initialize and log a virtual circuit, use the following function call:

AH = 16h - Initialize and log a virtual circuit
AL = Virtual circuit number (0-8)
ES:BX = Address of request/reply buffer

On return: Carry Set: AL - error code. Standard codes plus
15h = Port not available.

Carry clear: AL = port assignment.

Request buffer:
    Byte 0-1: Length of buffer
    Byte 2: Logical server port ID
    Byte 3: Receive baud rate
    Byte 4: Receive word length
    Byte 5: Receive stop bit
    Byte 6: Receive parity
    Byte 7: Transmit baud rate
    Byte 8: Transmit word length
    Byte 9: Transmit stop bit
    Byte 10: Transmit parity
    Byte 11: DTR control (0=off, 1=on)
    Byte 12: RTS control (0=off, 1=on)
    Byte 13: Flow Control (0=none, 1=XON/XOFF, 2=RTS/CTS)
    Byte 14: Reserved, must be 0.

The server port ID indicates the target port on the target server and is not related to the circuit number. This call provides an application with total control over the serial port in use. Ports may be configured dynamically. If BX=0, then the default NASI configuration for the port is used. The transmit and receive parameter values are the same as listed for function AH=15h.

The following function call disconnects the indicated virtual circuit from the communications server. All buffers are cleared, and the circuit allocation is released.

> AH = 17h - Disconnect
> AL = Virtual circuit number (0-8)
>
> On return: Carry Clear, AL=0, circuit disconnected.

To write data to a virtual circuit, use the following function call:

> AH = 18h - Write Data on a Virtual Circuit
> AL = Virtual Circuit Number (0-8)
> ES:BX = Address of data buffer
> CX = Number of bytes to write
>
> On return: Carry Clear, CX = number of bytes written.

Up to 512 characters can be written in a single function call. When the call returns, the buffer is immediately free for reuse by the application.

To read data from a virtual circuit, use the following function call:

> AH = 19h - Read Data on a Virtual Circuit
> AL = Virtual Circuit Number (0-8)
> ES:BX = Address of data buffer
> CX = Maximum number of bytes to read
>
> On return: Carry Clear, CX = number of bytes read (may be 0)

This function enables an application to read up to 512 bytes of input. It does not wait for data; a read of zero bytes indicates that no data is waiting. If you receive a full buffer of data, issue another read call immediately to avoid overrunning the network data buffers.

The following function call enables the application to check whether any input data characters are waiting to be read:

> AH = 1Ah - Receiver Status
> AL = Virtual Circuit Number (0-8)
>
> On return: Carry Clear

AL = 0 - No input data waiting
AL = 1 - One or more characters waiting

The following function call enables an application to determine whether the serial port transmitter is busy sending data:

AH = 1Bh - Transmitter Status
AL = Virtual Circuit Number (0-8)

On return: Carry Clear

AL = 0 - Transmitter available
AL = 1 - Transmitter is busy

The following function call deletes all data currently in the receiver buffer:.

AH = 1Ch - Receiver Buffer Control
AL = Virtual Circuit Number (0-8)

On Return: Carry Clear, AL = 0, receive buffers cleared

The following function deletes any data in the server's transmit buffers that has not yet been sent to the serial port:

AH = 1Dh - Transmit Buffer Control
AL = Virtual Circuit Number (0-8)

On Return: Carry Clear, AL = 0, transmit buffers cleared

To issue a port control request, use the following function call:

AH = 1Eh - Issue Control Request
AL = Virtual Circuit Number (0-8)
CL = Control Function:
    2 = Break
    4 = Disconnect
    6 = Hold
    8 = Enable XON/XOFF
    10 = Disable XON/XOFF

On Return: Carry Clear, AL=0, Function has been performed.

If CX=2, the serial port generates a break signal. If CX=4, then this function is similar to function 17h, except the virtual circuit remains allocated instead of being released. For CX=8 or CX=10, the XON/XOFF flow control is enabled or disabled at the server.

To get the external status of a serial port on the server, use the following function call:

AH = 1Fh - Get External Status
AL = Virtual Circuit Number (0-8)

On return: Carry Clear

AL = Status:
    Bit 7, 6, 5; Reserved
    Bit 4: Carrier Detect
    Bit 3: Data Set Ready
    Bit 2: Clear To Send
    Bit 1, 0: Reserved

This function returns a bit map in AL that indicates the status of the connection to the server serial port. This call should be used to query the state of the port connection, not of a specific event.

To set the buffers used internally by NASI, use the following function call:

AH = 20h - Set Receiver Buffer Pointers
AL = Virtual circuit number (0-8)
ES:BX = Receive Buffer Address
CX = Size of receive buffer (0=use internal buffer)

On return: Carry Clear, AL=0 function performed

This call sets or resets the buffers used by NASI for receiving data. It may be used by an application that can perform its own buffer management. This call enables data to be placed directly in an application's receive data buffer by NASI itself.

ES:BX points to the front of the buffer, and data is filled from that point on. The size of the buffer allocated in memory is limited only by the memory available to the application.

To query the name service for a specific name, use the following function call:

AH = 21h - Query Name Service
AL = Virtual circuit number (0-8)
CL = 0-Search First, 1-Search Next
ES:BX = Request Match Pattern

On Return: CL = 0, ES:BX = next matching name

CL = 1, No match found
Request Buffer:
    Byte 0-1: Length of buffer (30)
    Byte 2-9: Server Name
    Byte 10-17: General Service Name
    Byte 18-31: Specific Service Name

The request buffer pattern must be 30 bytes long. Any unused bytes in each name field should be padded with underscore characters—if the name is not ambiguous—or with wild-card characters (?) if the name is ambiguous.

Unlike the Set Service Name call (AH = 13h), which uses a specific hierarchical polling sequence to locate a name service, this function issues pattern queries to the name service. Using this function, an application can provide a user with a list of full names of all serial ports and servers throughout the network.

NASI/NCSI Summary

As you can see, the NASI/NCSI interface gives an application far more complete control of the serial ports in the server than the INT 14 API does. In addition, the NASI/NCSI interface enables the application program on the workstation to adjust the number of characters sent in a single network transaction to provide the best trade-off of performance and response time. For example, the program can change the size of the data reads and writes for typing versus doing a protocol file transfer over the serial port. The terminal program therefore can preserve response time for typing and maximize throughput for a file transfer.

The NASI/NCSI communications server protocol provides the program with the capability to integrate the server port name service of the LAN smoothly into the program's user interface. In this regard, NASI/NCSI clearly is superior to the other API protocols in use today. With a well-written communications program (such as PROCOMM PLUS Network Version), the workstation user does not have to learn how to use the LAN command interpreter interface in order to use serial ports on a server.

Understanding the IBM LANACS/ACSI

The IBM Local Area Network Asynchronous Connection Server (LANACS) is available for either the PC Network LAN or for the IBM token-ring LAN. LANACS can use the server's standard serial ports. To allow more than two ports on a server, however, LANACS uses the special IBM intelligent serial port cluster called the Real-Time Interface Co-Processor (RTIC) card, which provides four or eight serial connections to the PC bus. With the proper number and type of RTIC adapter cards, a LANACS server can support up to 32 asynchronous ports.

The API provided by the LANACS server software is the Asynchronous Communications Server Interface (ACSI) protocol. The ACSI protocol uses the NETBIOS interface directly. The ACSI protocol consists of special message formats sent using the NETBIOS protocol itself, making the ACSI protocol very involved to program for and susceptible to the variations in NETBIOS implementations.

The ACSI API does provide much better performance than the INT 14 interface, but it does not provide the total level of serial port control that the NASI/NCSI protocol does. ACSI does allow the application direct access to the network connection services so that the application can provide a complete user interface.

The IBM LANACS server also supports the INT 14 API sometimes known as EBIOS. For use with this API, the LANACS software provides a workstation program for use at the DOS command line, making the server and port connections before you enter the communications program.

Understanding Dedicated versus Shared Servers

Asynchronous communications servers are simply special software installed on a computer that has serial ports and is connected to the LAN. If a computer is running only the communications software and cannot be used as a normal workstation, then it is called a *dedicated* server. If the server also is used as a workstation (in addition to its job as an asynchronous server), then it is called a *shared* server.

The most obvious distinction between dedicated and shared servers is that a dedicated communications server ties up an entire PC just to supply the serial interface to the network. The shared server, on the other hand, does not require an extra computer because it shares a workstation. A shared serial communications server, however, suffers a tremendous drop in performance.

Because of their nature, serial communications require much higher computer response times than most computing tasks. As a result, a shared communications server usually is not adequate for use above 2400 bps or for more than two serial ports. Depending on what the shared workstation is doing, shared asynchronous servers often cannot support data rates above 1200 bps.

In contrast, dedicated asynchronous servers can easily support many lines and high data rates with good response times. A dedicated server can perform much better because it can "dedicate" the entire attention of its CPU to the task of managing the serial port data flow.

Chapter Summary

As you learned in Chapter 7, local area networks connect together many users in order to share resources. An asynchronous communications server adds the capability of sharing modems or other serial PC communications connections among the workstations on a LAN. LAN communications servers also have much in common with the PC-to-mainframe connections described in Chapter 17. In particular, they can provide a variety of connections between different computers.

You have learned that different interfaces are available for the communications programs you use with an asynchronous server; these programs now must communicate with the LAN software instead of directly with the serial port hardware. You have seen the details of these application program interfaces (APIs) with a special emphasis on the standard INT 14 API and the widely used NASI/NCSI API. You also have learned how the communications server function is implemented on a network and what the performance considerations are for different configurations.

LAN asynchronous servers provide a tremendously flexible option for integrating serial communications and a local area network. You need to understand the trade-offs that these servers must make, however, in order to use them properly with your communications applications.

19

Programming PC Communications

E ven the best programmers find that writing code to control the serial ports on an IBM PC takes careful planning and thought. The DOS and BIOS support for serial communications is totally inadequate for any serious application, and therefore you must program the serial port hardware and PC interrupt structure directly in order to have successful communications software. You must learn how this hardware works at a level of detail that you do not need to deal with in most programming tasks.

In this chapter, you learn how to program both the serial port itself and also the IBM PC interrupt structure. You learn the techniques required to build software that drives the serial port hardware effectively. You also learn why only two normal serial ports can be used on an ISA bus PC at the same time.

First, the chapter examines the serial port hardware and PC interrupt structure in detail. This material provides the background you need in order to understand how to program the serial port. The text then shows you how to develop a complete serial port driver, which serves as a replacement for the standard BIOS communications support routines. By examining this specific application in detail, you learn the nuances of programming the PC serial port.

Programming IBM Serial Ports

When the IBM PC was designed in 1980, IBM chose to use a large-scale integrated circuit called a *UART* (Universal Asynchronous Receiver-Transmitter), which contains all the software programming control of the port. The UART is the heart of the IBM PC serial port, and a complete understanding of how the UART is programmed is the first step toward being able to write serial port driver software.

The UART chosen was an Intel 8250 chip that contained not only the serial port logic but also the baud rate clock circuitry. More than a decade has passed since IBM designed this serial port hardware, and technology has moved forward a great deal in that time. Today you can find vastly improved versions of the 8250 UART in the form of the 16450 and 16550 UART chips. From a programming standpoint, however, they are largely the same as the original 8250 UART. The differences are noted in this chapter as the programming of the UART is described, but you can write a single piece of driver software that can work with all three versions.

The standard IBM serial port makes the internal registers of the UART available as a series of eight I/O addresses. More than eight registers are contained inside the UART, and the UART makes multiple use of these I/O addresses to enable you to access all the registers. Some of these registers are read-only or write-only. Two examples are the *input data* and *output data registers*. The UART accesses such a pair of registers with a single I/O address. If you write to that address, you access the write-only register; and if you read from the same I/O address, you access the read-only register. The results are different from what you normally expect, because reading an I/O address does not give you back the data you last wrote to that same address.

Figure 19.1 shows the layout of all of the registers in all variations of the standard UART used in an IBM PC port. You can use this figure as a reference as you learn the meaning and use of each of the registers in the following sections. You can easily become lost when you are using a device as complex as a UART. Figure 19.1 will keep you oriented as the discussion proceeds. The figure also will later serve as a valuable reference chart when you are programming the UART itself.

The UART makes multiple use of two addresses to access the 16-bit *Baud Rate Divisor Register*. Because you don't need to change the baud rate setting that often, a bit in one of the registers (named DLAB) acts as a switch. If this switch is set on, two of the I/O addresses of the port access the Baud Rate Divisor Register. If this switch is set off, the same two addresses access other registers in the UART.

Knowing about this type of multiple register access eliminates some of the confusion that may arise as you see how each register in the UART operates.

UART Register Name	I/O Address and Access Conditions	Bit 7	Bit 6	Bit 5	Bit 4	Bit 3	Bit 2	Bit 1	Bit 0
Receive Data Buffer Register (RBR)	Base+0 Read Only DLAB=0	Last Character Fully Received by the input Shift Register (input data)							
Transmit Hold Register (THR)	Base+0 Write Only DLAB=0	Next Character to place in output Shift Register (output data)							
Interrupt Enable Register (IER)	Base+1 Read/Write DLAB=0	Reserved / 0	Reserved / 0	Reserved / 0	Reserved / 0	Enable Modem Status Interrupt (EMSI)	Enable Line Status Interrupt (ELSI)	Enable Transmit Buffer Empty Int (ETBEI)	Enable Receive Data Available Int (ERBFI)
Interrupt Identification Register (IIR)	Base+2 Read Only	FIFO Enabled (16550 only)	FIFO Enabled (16550 only)	0	0	Interrupt ID Bit 2 (16550 Only)	Interrupt ID Bit 1 (IIDB1)	Interrupt ID Bit 0 (IIDB0)	0 = Interrupt Pending (IP)
FIFO Control Register (FCR)	Base+2 Write Only (16550 Only)	Rcvr FIFO Trigger MSB (RTMSB)	Rcvr FIFO Trigger LSB (RTLSB)	Reserved	Reserved	DMA Mode Select (DMS)	Transmit FIFO Reset (TFR)	Rcvr FIFO Reset (RFR)	FIFO Enable (FEWO)
Line Control Register (LCR)	Base+3 Read/Write	Divisor Latch Access Bit (DLAB)	Send Break (SBR)	Stick Parity (STP)	Even Parity Select (EIPS)	Parity Enable (PE)	Number of Stop Bits (STB)	Word Length Select Bit 1 (WLS1)	Word Length Select Bit 0 (WLS0)
Modem Control Register (MCR)	Base+4 Read/Write	Reserved / 0	Reserved / 0	Reserved / 0	Loop Back Test	Out 2	Out 1	Request To Send (RTS)	Data Terminal Ready (DTR)
Line Status Register (LSR)	Base+5 Read/Write	Error in Rcvr FIFO (ERF) (16550 Only)	Transmit Shift Register Empty (TSRE)	Transmit Hold Register Empty (THRE)	Break Interrupt (BI)	Framing Error (FE)	Parity Error (PE)	Overrun Error (OE)	Receive Data Available (DR)
Modem Status Register (MSR)	Base+6 Read/Write	Rcvr Line Signal Detect (RLSD)	Ring Indicator (RI)	Data Set Ready (DSR)	Clear To Send (CTS)	Delta Rcvr Line Sig Detect (DRLSD)	Trailing Edge Ring Indicator (TERI)	Delta Data Set Ready (DDSR)	Delta Clear To Send (DCTS)
Scratch Register (SCR)	Base+7 Read/Write (Not on 8250)	Temporary Storage of any data byte you choose - Rarely used							
Divisor Latch LS Byte (DLL)	Base+0 Read/Write DLAB=1	Least Significant Byte (Low Order) of Baud Rate Divisor							
Divisor Latch MS Byte (DLM)	Base+1 Read/Write DLAB=1	Most Significant Byte (High Order) of Baud Rate Divisor							

Fig. 19.1. *The internal registers of the UART in an IBM PC serial port.*

NOTE Although the IBM serial port allocates eight I/O addresses for control of the serial port, only seven of them are used. The eighth accesses a *scratch register*, which has no real use.

The IBM serial port allocates I/O addresses in order from a *base address*. Because of the way serial port hardware is built, base addresses always end in either 0 or 8. For the COM1 port, the base address is always 0x3f8. The eight I/O addresses allocated begin with this base address and go through 0x3ff (0x3f8, 0x3f9, 0x3fa, 0x3fb, 0x3fc, 0x3fd, 0x3fe, and 0x3ff). In the following sections, each register's I/O address is indicated as base+0, base+1, and so on. In the case of COM1 (in which the base address is 0x3f8), an address of base+3 would be 0x3fb. For COM2 (in which the base address is 0x2f8), base+3 translates to 0x2fb. For a serial port with a base address of 0x100, the address base+3 would be 0x103. The following chart can help you translate the relative I/O address notations used in this chapter to describe the UART registers into the actual I/O addresses you would use as a programmer to access each register:

address	base=0xnnn0	base=0xnnn8	base=COM1(0x3f8)
base+0	0xnnn0	0xnnn8	0x3f8
base+1	0xnnn1	0xnnn9	0x3f9
base+2	0xnnn2	0xnnna	0x3fa
base+3	0xnnn3	0xnnnb	0x3fb
base+4	0xnnn4	0xnnnc	0x3fc
base+5	0xnnn5	0xnnnd	0x3fd
base+6	0xnnn6	0xnnne	0x3fe
base+7	0xnnn7	0xnnnf	0x3ff

In general, any software that directly accesses an I/O port should be written to use these relative offsets from a base I/O port address so that the software can work with any serial port the user chooses.

To program the serial port, you must know what each register of the UART is for and how to program or read each register. Before delving into actual programming examples, therefore, you need to examine the standard IBM serial port programming structure in detail.

Examining the Receive Data Buffer Register (base+0)

The *Receive Data Buffer Register* is a read-only register. If you read address base+0, you access this register, which contains the last character received by the UART. Although you can read this register as many times as you want, the first time you read each character, the UART clears an internal latch (receive data available) that was set when the character was placed into the receiver buffer. You can read this latch directly in the Line Status Register (described later in this chapter under "Examining the Line Status Register"), and the latch also is used to set the Receive Data Available Interrupt from the UART. This interaction is detailed in this chapter in the descriptions of each of these items, but you need to be aware that the UART "knows" the first time you read a character from the UART and clears certain status based on that action.

The Receive Data Buffer Register is at the computer end of the UART's serial input logic. The input section of the UART consists of a Serial Shift Register, which assembles the character B.1 by bit from the serial data stream, and the Receive Data Buffer Register. Figure 19.2 shows this structure.

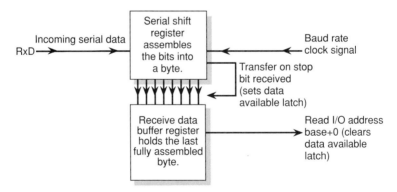

Fig. 19.2. The input section of the IBM PC serial port UART.

The UART input section shown in figure 19.2 is conditioned by the setting of the Line Control Register (LCR) to know how many bits of data are present in an incoming serial character. When the Serial Shift Register detects a start bit, it begins the clock (whose speed is set by the value in the Baud Rate Divisor Latch) which divides the incoming serial data stream into successive bits based on time. Each bit is shifted into the Serial Shift Register to assemble the bits in the proper order into a data byte. When the Serial

Shift Register has shifted in all of the data bits, the Serial Shift Register verifies that a stop bit exists which "frames" the end of the character. If a stop bit isn't there at this point, the Framing Error status bit is set in the Line Status Register (LSR). If the LCR indicates that the parity of the incoming data byte should be checked, then the input Serial Shift Register logic verifies that it received the proper parity and sets the Parity Error bit in the Line Status Register if the parity is not correct.

After these data integrity checks have been performed, the Serial Shift Register transfers the data byte to the Receive Data Buffer Register and sets the Receive Data Available Latch. If a character already was in the Receive Data Buffer Register which had not been read out by the computer, then the Overrun Error bit is set in the Line Status Register (LSR) to indicate that one or more data bytes were lost.

You also need to understand this structure later when you examine the 16550 UART, which handles high-speed serial ports.

Examining the Transmitter Holding Register (base+0)

When you write data to I/O address base+0, the data is placed in the *Transmitter Holding Register*. This register holds the next character to be transmitted. When no character is being sent (shift register is empty), the character from the Transmitter Holding Register is transferred to the Transmitter Shift Register. From there, the character is sent out one bit at a time to the serial line with a start bit and stop bit plus any indicated parity added.

The Line Status Register (described later in this chapter under "Examining the Line Status Register") has two status bits that show whether the Transmitter Holding Register is empty and also whether the Transmitter Shift Register is empty. If you write a character to the Transmitter Holding Register when a character is already there, you overwrite the character in the register (the character has no place to go). Because the character had not been placed in the Transmitter Shift Register yet, the overwritten character is lost forever and is not sent. For this reason, you must use the status bit to see whether the Transmitter Holding Register is empty before you store another character in it. Figure 19.3 shows this structure.

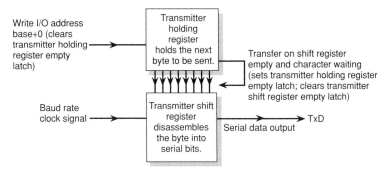

Write I/O address
base+0 (clears
transmitter holding
register empty
latch)

Transmitter holding register holds the next byte to be sent.

Transfer on shift register empty and character waiting (sets transmitter holding register empty latch; clears transmitter shift register empty latch)

Baud rate clock signal

Transmitter shift register disassembles the byte into serial bits.

Serial data output

TxD

Fig. 19.3. The output section of the IBM PC serial port UART.

You can see in figure 19.3 that when the Transmitter Shift Register is empty, any character in the Transmitter Holding Register is transferred into the Transmitter Shift Register. This shift register then generates a start bit, sends the character data bits, appends any parity bit requested, and then adds a stop bit that forms the character on the serial line itself. The clock rate at which bits are generated is given by the setting of the Baud Rate Divisor latch, and any parity to be generated is indicated by the Line Control Register. The Line Status Register (LSR) has status that reflects the empty or full status of the Tranmsitter Holding Register and the Transmitter Shift Register.

You also can arrange to have an interrupt trigger when the Transmitter Holding Register becomes empty (this process is described later under "Handling Interrupts"). Your program then can do other things while it is waiting for the buffer to become empty.

Examining the Line Control Register (base+3)

The *Line Control Register* is used to set the format of the serial port. This register sets the number of bits, the parity, and the number of stop bits the serial port will use. Because the Line Control Register can be read and written to, you can always read the Line Control Register to see how you (or another program) last set it.

In addition to setting the format of the serial port data, the Line Control Register also contains the switch that provides access to the 16-bit Baud Rate Divisor Register. The format of the Line Control Register is as follows:

Bit	Description
7	Divisor latch access bit (DLAB)
6	Set break
5	Stick parity
4	Even parity select (EPS)
3	Parity enable (PEN)
2	Number of stop bits (STB)
1	Word length select bit 1 (WLS1)
0	Word length select bit 0 (WLS0)

Bit 7 of the Line Control Register accesses the Baud Rate Divisor Register. When this bit is set to 1, the addresses base+0 and base+1 access the 16-bit Baud Rate Divisor Register rather than their normal UART registers—Receive Data Buffer Register, Transmitter Holding Register, and Interrupt Enable Register.

To set the baud rate of the serial port, you first program the Line Control Register bit 7 to 1. You then store the baud rate divisor value, which sets the speed you want, in addresses base+0 and base+1. Finally, you reset bit 7 of the Line Control Register to a 0 to restore normal UART register access. The following code illustrates this process and sets the baud rate to 1200 baud:

```
MOV     DX,[BASE]     ;obtain port base address in DX
ADD     DX,3          ;point to line control register
IN      AL,DX         ;read LCR
OR      AL,80h        ;set DLAB to access divisor
OUT     DX,AL
MOV     AX,0060h      ;get 1200 baud divisor in AX
SUB     DX,3          ;point back to divisor address(base+0)
OUT     DX,AL         ;set divisor register (low byte)(base+1)
INC     DX            ;point to high-order divisor reg
MOV     AL,AH         ;put high order in AL
OUT     DX,AL         ;output divisor (high byte)
ADD     DX,2          ;point back to LCR
IN      AL,DX         ;read LCR
AND     AL,7Fh        ;set off DLAB access switch
OUT     DX,AL         ;put back in LCR
```

The values you should store as a baud rate divisor to obtain the standard serial port speeds are as follows:

Baud rate	Baud rate divisor value
110	0x0417
134.5	0x0359
150	0x0300
300	0x0180
600	0x00C0
1200	0x0060
2400	0x0030
4800	0x0018
9600	0x000C
19200	0x0006
38400	0x0003
57600	0x0002
115200	0x0001 (maximum rate possible)

If you want to calculate nonstandard baud rates, divide the number 115,200 by the bit rate to find the divisor value. Thus for 9600 bps the formula is 115,200/9600. The result is 12 (decimal), which translates to 0x000C in hexadecimal.

Bit 6, the set break bit, is used to generate a break on the outgoing serial data line. When this bit is set to 1, the serial output of the UART is forced to the spacing condition (logical 0), which begins a break sequence. The software must time the length of the break and set this bit back to 0 again to end the break condition. This bit overrides other transmitter activity, so to prevent output data loss you should set this bit only when no data is in the Transmitter Shift Register or the Transmitter Holding Register.

The parity of the serial port is configured by the setting of bits 3, 4, and 5 as described in the following chart:

Bit 5	Bit 4	Bit 3	Resulting parity
0 or 1	0 or 1	0	No parity
0	1	1	Even parity
0	0	1	Odd parity
1	0	1	Mark parity
1	1	1	Space parity

The resulting parity is the parity that is generated on transmitted serial data and checked on incoming received serial data.

If bit 2, which controls the number of stop bits (STB), is a 0, only one stop bit is transmitted or checked on received data. If bit 2 is a 1, two stop bits are transmitted.

The Word Length Select (WLS) bits 0 and 1 specify the number of bits in each transmitted or received serial character as described in the following chart:

Bit 1	Bit 0	Word length
0	0	five bits
0	1	six bits
1	0	seven bits
1	1	eight bits

Examining the Line Status Register (base+5)

The *Line Status Register* is used to keep track of the progress of the serial data transfer. When you read this register, you obtain the current status of both input and output serial transfers. For output, the register tells you whether the Transmitter Holding Register is empty and also whether the UART currently is shifting a character out to the serial port. For input, the register tells you whether a character is waiting in the Receive Data Buffer Register. The Line Status Register also tells you whether any errors were found on the last character received and whether a break signal was received. The format of the Line Status Register is as follows:

Bit	Description
7	Unused; permanently set to 0
6	Transmitter Shift Register Empty (TSRE)
5	Transmitter Holding Register Empty (THRE)
4	Break Interrupt (BI)
3	Framing Error (FE)
2	Parity Error (PE)
1	Overrun Error (OE)
0	Data Ready (DR)

Bit 6, the Transmitter Shift Register Empty (TSRE) bit, is set to a 1 when the UART has emptied the Transmitter Shift Register and no character is waiting in the Transmitter Holding Register. This setting indicates that the UART has sent all transmit data it was given and is idle with respect to output.

Bit 5, the Transmitter Holding Register Empty (THRE) bit, is set to a 1 when the UART has placed the character that was in its holding register into the Transmitter Shift Register for sending. This setting means that the UART is ready to accept another character for output. When bit 5 is set, the UART also issues a Transmitter Register Empty Interrupt if requested by the programmer (see "Examining the Interrupt Registers" later in this chapter). This bit is set to a 0 when a character is written to the Transmitter Holding Register.

The Break Interrupt (BI) bit, bit 4, is set to a 1 whenever the UART receive data line is held in a space (logical 0) condition for longer than a full character time. This setting indicates that a break condition was received.

The Framing Error (FE) bit, bit 3, is set to a 1 whenever the received character did not have a valid stop bit. This setting indicates that the data was framed improperly by start and stop bits.

Bit 2, the parity error (PE) bit, is set to a 1 if the UART is programmed to check parity and the received character did not have the proper parity.

Bit 1, the overrun error (OE) bit, is set to a 1 if the data in the Receive Data Buffer Register was not read by the computer before another character was received and placed in the Receive Data Buffer Register. This setting indicates that the UART's input was overrun by the received data, and one or more characters were lost.

> The four flags just discussed (BI, FE, PE, and OE) are the *receiver status flags*. These flags are all reset when the Line Status Register is read. They also generate a *receiver line status interrupt* if the UART is programmed properly (see Interrupt Enable under "Examining the Interrupt Registers").

The Data Ready (DR) bit, bit 0, is set to a 1 when the UART places in the Receive Data Buffer Register a character just assembled in the Serial Shift Register. This bit is reset when the character is read by the computer from the Receive Data Buffer Register (base+0).

Examining the Modem Control Register (base+4)

Writing to the *Modem Control Register* controls the RS-232 modem signals DTR and RTS. The register also controls a diagnostic feature of the UART as well as two output pins from the UART chip labeled OUT1 and OUT2. You can read and write to the Modem Control Register, so you can read back at any time the current settings of these output signals. The format of the Modem Control Register is as follows:

Bit	Description
7	Unused; permanently
6	Unused; permanently
5	Unused; permanently
4	Loop-back diagnostic
3	OUT2
2	OUT1
1	Request To Send (RTS)
0	Data Terminal Ready (DTR)

Bit 4, the loop-back diagnostic bit, normally should be set to a 0. A program that wants to test the UART, however, may set this bit to a 1, which connects the serial output to the serial input internally. The four modem control inputs (CTS, DSR, CD, and RI) are disconnected from their pins and internally connected to the four modem control outputs (DTR, RTS, OUT1, and OUT2), enabling a full loop-back test of the UART.

Because of the design of the IBM PC interrupt bus (see "Handling Interrupts" later in this chapter), a PC serial port does not connect its interrupt control line immediately to the bus. Setting bit 3, the OUT2 bit, to a 1 enables the UART interrupt line for this serial port to connect to the PC bus. Setting this bit to a 0 prevents any interrupt from this serial port from being seen by the PC bus.

The output of the UART bit 2, the OUT1 bit, generally is not connected to anything on a standard IBM serial port. Some internal modems use it as a special reset line.

The Request To Send (RTS) bit, bit 1, directly sets or resets the RTS signal that appears on pin 4 of the port's serial connector. Setting this bit to 1 asserts the RTS signal; setting the bit to 0 deactivates the RTS signal.

Bit 1, the Data Terminal Ready (DTR) bit, directly sets or resets the DTR signal that appears on pin 20 of the port's serial connector. Setting this bit to 1 asserts the DTR signal; setting the bit to 0 deactivates the DTR signal.

Examining the Modem Status Register (base+6)

By reading the *Modem Status Register*, your program can monitor the current status of the four RS-232 input signals CTS, DSR, CD, and RI. The program also can tell whether these signals have changed since the last time they were examined. The format of the Modem Status Register is as follows:

Bit	Description
7	Carrier Detect (CD); pin 8 of DB-25 port connector
6	Ring Indicator (RI); pin 22 of DB-25 port connector
5	Data Set Ready (DSR); pin 6 of DB-25 port connector
4	Clear To Send (CTS); pin 5 of DB-25 port connector
3	CD changed since last read
2	RI changed since last read
1	DSR changed since last read
0	CD changed since last read

For each of the status signals (bits 7, 6, 5, and 4), the port shows a 1 if that signal is currently asserted and a 0 if it is not currently asserted. For each of the status change reporting bits (bits 3, 2, 1, and 0), a 1 is shown if the status of the bit has changed since the Modem Status Register was last read.

The signal may have changed many times since the register was last read and may now be in the same condition it was then. The change bits still indicate, however, that the signal did not stay constant for the entire interval. The status change bits are all reset to 0 when the Modem Status Register is read by the computer.

Examining the Interrupt Registers

Two registers in the UART control and identify interrupt operations. These are the *Interrupt Enable Register* (base+1) and the *Interrupt Identification Register* (base+2).

With the Interrupt Enable Register, you can request the UART to generate an interrupt when specific conditions occur. You can request four different interrupt conditions, and you are free to ask for all, none, or any combination of them. The Interrupt Enable Register can be read and written to, so you can check at any time to see what interrupt conditions currently are requested of the UART.

If multiple interrupt conditions occur at the same time, the UART has a priority for each interrupt and generates them in that order until they all have been generated. As each interrupt is cleared, the next is generated immediately.

Table 19.1 summarizes the possible interrupts, their priority within the UART, the conditions that cause them, and the way your program must reset them.

Table 19.1
UART Interrupts

Name	Priority	Source	Reset Action
Receive Line Status	Highest	OE, PE, FE, or BI	Reading the Line Status Register
Receive Data Available	Second	Data received	Reading the Receiver Buffer Register
Transmitter Interrupt Holding Register Empty	Third	Moving data out of the Transmitter Holding Register	Reading the Identification Register or writing to the Transmitter Holding Register
Modem Status Modem	Fourth	Change in one or more of CD, DSR, RI, or CTS	Reading the Status Register

The structure of the Interrupt Enable Register is as follows:

Bit	Description
7	Always 0
6	Always 0
5	Always 0
4	Always 0
3	Enable Modem Status Interrupt
2	Enable Receive Line Status Interrupt
1	Enable Transmitter Holding Register Empty Interrupt
0	Enable Receive Data Available Interrupt

When the UART generates an interrupt, the Interrupt Identification Register indicates which interrupt is being generated. The structure of the Interrupt Identification Register is as follows:

Bit	Description
7	Always 0
6	Always 0
5	Always 0
4	Always 0
3	Always 0
2	Interrupt ID bit 1
1	Interrupt ID bit 0
0	0 if interrupt is pending; 1 if no interrupt

Table 19.2 lists the settings of the Interrupt Identification Register and describes the interrupts identified.

Table 19.2
Using the Interrupt Identification Register

Bit 2	Bit 1	Bit 0	Interrupt identified
0	0	1	No interrupt is pending
1	1	0	Receive Line Status
1	0	0	Receive Data Available
0	1	0	Transmitter Holding Register empty
0	0	0	Modem Status

Handling Interrupts

Interrupts are the means by which an external device (such as the serial port hardware) obtains the attention of the computer when the device needs servicing. Interrupts tell the software when critical events happen so that the program can respond to them as needed but does not have to check the devices constantly. Interrupts are particularly important in handling serial communications because as each character is received or transmitted, the program has to service the UART again. This servicing occurs frequently, but you don't want to stop the computer from doing all other tasks while you are sending and receiving characters. Besides, you cannot tell when characters will be received by the serial port, so interrupts are often the only choice you have to avoid losing incoming serial data.

Interrupts on the IBM PC are controlled by an Intel 8259 interrupt controller chip. On some newer computers, this function is built into larger chips to reduce parts count, but from a software perspective interrupts still function the same way in those cases. The IBM PC originally had eight external interrupts that the cards plugged into the I/O slots could use. The IBM AT computer added a second eight interrupts, and newer computers have added even more. On every PC-compatible machine, however, you can count on having only the original eight interrupts.

From these eight interrupts, IBM assigned interrupts 3 and 4 (sometimes called IRQ3 and IRQ4 for Interrupt ReQuests 3 and 4) to serial communications. As part of the standard COM1 and COM2 port layouts, COM1 was assigned IRQ4, and COM2 was assigned IRQ3. All standard COM1 and COM2 ports use these interrupts.

The original IBM PC was designed in 1980, when the cost of a single integrated circuit was still high. Apparently, a designer managed to save a chip by not placing inverting receivers on each of the interrupt request lines from the I/O card bus. This design probably reduced the list price of the computer by $50 or so at that time, but today you have to live with the result of this move.

The result is that the interrupt bus triggers an interrupt by bringing the bus signal high rather than low as is normally done. Because a signal handled this way is more noise-susceptible, the IBM PC interrupt bus is triggered by detecting only the rising "edge" of the interrupt request signal from the bus. Thus an interrupt is signaled at only one specific point in time on an ISA bus, and if it is missed, the interrupt bus is lost until the hardware is reset.

Because ISA bus interrupts are triggered by a rising edge, you cannot have more than one I/O slot connected to the same interrupt. (*Note:* The Microchannel bus in the PS/2 computer series corrected both of these problems, but if your software is to run properly on an ISA bus machine, you have to program all PC serial I/O handlers as though the interrupts are active high and edge triggered anyway.)

In summary, because only two interrupts are allocated to serial I/O and because only one I/O port can use an interrupt, an ISA bus IBM-compatible computer is limited to two normal serial ports. Special multiport serial cards get around this limitation by sharing a single interrupt among all ports on the card and by presenting the interrupt to the I/O bus as a single interrupt.

The 8259 interrupt controller is accessed at I/O addresses 0x20 and 0x21. These registers, called *operational control words*, enable you to set the 8259 to a variety of modes. This chapter covers only the mode to which the IBM PC is programmed by its BIOS when it is initialized. Refer to the Intel 8259 data sheet for complete information on the interrupt controller chip.

How To Enable Serial Port Interrupts

The 8259 register accessed at 0x21 contains the eight-bit interrupt enable mask. For an interrupt to occur, its corresponding bit in this register must be set to 0. This register is both read- and write-accessible, so you can read it in, change only the bits you want, and write it back out. Because you are concerned only with IRQ3 and IRQ4, you change only bits 3 and 4 in this register.

In addition, the IBM serial port does not connect its interrupt circuit to the I/O bus until you program the OUT2 signal in the UART. This step is necessary so that the interrupt can be shared by other I/O cards as long as they all don't want to use it at the same time.

Finally, you must program the UART Interrupt Enable Register to request the UART to generate the interrupts you want.

In summary, to enable a serial port to generate an interrupt, you must follow three steps:

1. Set the UART Interrupt Enable Register (base+1) to request the UART to generate the desired interrupts.

2. Set the OUT2 signal in the Modem Control Register (base+4) to a 1 to connect the serial port's interrupt line to the I/O bus.

3. Set bit 3 (for IRQ3) or bit 4 (for IRQ4) in the 8259 Interrupt Controller Enable Register (0x21) to a 0 to enable the appropriate interrupt line.

How an Interrupt Occurs

When an interrupt occurs in an IBM PC or compatible, the CPU stops between instructions and executes a specific interrupt entry sequence. This sequence keys off of an *interrupt vector* that is unique for each possible interrupt. In the CPU mode that DOS uses (called real mode), the interrupt vectors for IRQ3 and IRQ4 are connected to interrupt locations 11 and 12 (the first eight interrupt locations in an 8088 computer are used for internal interrupts such as divide overflow). Because an interrupt vector in this mode is a four-byte full segment:offset address, the vector for IRQ3 is located at 0000:002C, and the vector for IRQ4 is located at 0000:0030 in the computer's memory.

The 8088 family of computers keeps an interrupt vector table which begins at the very first location of its memory. This has a segment address of 0000:0000. Because each interrupt vector is four bytes long, the interrupt vector offset location for a particular interrupt is found by multiplying the interrupt number by 4. You should understand, however, that the IBM PC bus interrupt numbers are not the same as the internal 8088 interrupt numbers. The 8088 has several internal interrupts that occur first in the interrupt table, so the first external interrupt (IRQ0) is connected to the 8088 interrupt number 8. IRQ1 is connected to the 8088 interrupt 9, IRQ2 is connected to interrupt 10, and so forth in order through IRQ7, which is connected to 8088 interrupt 15.

The interrupt vector for IRQ3 therefore is calculated by taking the IRQ number (3) and adding 8 for the internal interrupts that precede the IRQ number, and then multiplying by 4 (because there are four bytes per interrupt vector). Therefore, the vector for IRQ3 begins 33 bytes into the computer's memory, which in hexadecimal is 2C, resulting in a vector address of 0000:002C for IRQ3.

When an interrupt occurs, the computer places on the stack the address of the next instruction that normally would be executed, along with the condition code settings (such as zero, carry, and so on). The computer then sets the interrupt inhibit status flag and begins execution at the address to which the interrupt vector points. This vector address is the beginning of the code you need to write to handle an interrupt.

When an interrupt routine is entered, the 8259 interrupt controller shows that interrupt as active. This setting tells the interrupt controller to ignore any new requests on the interrupt line until it is reset with a special End Of Interrupt command (abbreviated EOI). You execute this special command by writing the value 0x20 to the 8259 at port address 0x20. (In this case, note that the data you write and the address to which you write it are the same, but the match is just a coincidence.)

The interrupt routine must do each step in the proper sequence because of the edge-triggered nature of the ISA bus PC interrupt system. You must be sure that the 8259 will be enabled whenever an edge or trigger condition occurs, or you can lose interrupts. This procedure is discussed in detail as the interrupt handler code is developed later in the chapter in the section "Understanding Interrupt Operation."

When an interrupt routine is complete, it executes the IRET instruction that removes the address and condition codes from the stack. The CPU then is restored to the point at which it was interrupted so that it can continue whatever it was doing before the interrupt occurred. Notice that because the 8088 CPU doesn't automatically save any registers when an interrupt occurs, the interrupt routine must save and restore any registers it uses to prevent interfering with the program it interrupted.

Building a Serial Driver

The PC BIOS does have rudimentary serial port support. This support comes in the form of a set of INT 14 service calls that enable you to set up a serial port's format and baud rate, send or receive a character, and read the UART status. These routines are not very useful, however, because they do not use interrupts and they do not prevent input overflow by using RTS/CTS flow control. The INT 14 provides four service calls.

To initialize the serial port, use the following service call:

AH = 0 (initialize the serial port)
AL = Parameters for initialization as listed in table 19.3
DX = 0 for COM1 or 1 for COM2

On Return (from the BIOS service call): AH and AL are set as in the read status call that follows. If the time-out bit is set, the requested port is not installed in hardware.

Table 19.3
Standard IBM PC BIOS Initialization Parameters

Bits	Parameter setting
7,6,5	*baud rate*
000	110
001	150
010	300
011	600
100	1200
101	2400
110	4800
111	9600
4,3	*Parity*
00	None
01	Odd
10	None
11	Even
2	*Stop bits*
0	1 stop bit
1	2 stop bits
0,1	*Character length*
10	7 bits
11	8 bits

To initialize the send character, use the following service call:

AH = 1 (send character)
AL = character to send
DX = 0 for COM1 or 1 for COM2

On Return:

AL is preserved
AH = line status
　　　Bit 7 = time-out (other bits meaningless if set)

Bit 6 = Transmitter Shift Register empty
Bit 5 = Transmitter Holding Register empty
Bit 4 = break detected
Bit 3 = framing error
Bit 2 = parity error
Bit 1 = overrun error
Bit 0 = data ready

To initialize the receive character, use the following service call:

AH = 2 (receive character)
DX = 0 for COM1 or 1 for COM2

On Return:

AL = character received

AH = line status as listed for send character routine

Note: This routine waits for a character to appear if none is ready when the routine is called.

To initialize the read status, use the following service call:

AH = 3 (read status)
DX = 0 for COM1 or 1 for COM2

On Return:

AL = modem status
Bit 7 = carrier detect (CD)
Bit 6 = ring indicator (RI)
Bit 5 = data set ready (DSR)
Bit 4 = clear to send (CTS)
Bit 3 = CD changed since last status
Bit 2 = RI changed since last status
Bit 1 = DSR changed since last status
Bit 0 = CTS changed since last status
AH = line status as listed for send character routine

This set of calling and return sequences details the operation of the standard BIOS INT 14 service calls. As a serial port programming example, you develop in the following sections a replacement handler that provides these same services, using the same calling sequences, but that uses interrupts to avoid losing input data and to enable constant full-speed output data rates. In addition, this routine fully respects RTS/CTS hardware flow control so that you can use it in buffered applications.

This example shows how to manage circular buffers for input and output interrupts. The example also shows how flow control responsibility is divided between the interrupt routine and the main program. Finally, the example shows all considerations you face when handling interrupts because it handles many types of interrupts for two serial ports simultaneously.

To begin with, the driver has to load as a TSR (terminate and stay resident) program so that it can replace the BIOS INT 14 handler. The driver does not enable the interrupts for a serial port until the first time the initialize port call is made for that port to configure it for operation.

A program listing the code for the driver in Microsoft MASM assembler format (Program Listing 19.1) is placed at the end of this chapter. The first column of the listing contains line numbers, which are referenced in the following discussions. This program is simplified in many areas so as not to complicate the code. You then can see the serial I/O handling considerations more clearly. In the following sections, each portion of the code is examined in detail to see how it works.

Loading as a TSR

Lines 562 through 585 of this program (refer to the program listing at the end of this chapter) are executed when you invoke it from the DOS command line. These lines provide the minimum TSR capability possible. Lines 568 to 570 replace the INT 14 service vector with a pointer to the BIOS replacement handler at label INT_14. Lines 572 to 574 simply print the message that this program has been installed. Finally, lines 576 to 583 calculate the size of the program in paragraphs (16 bytes is 1 paragraph) and tell DOS to leave that code resident and active. This step enables the replacement routine to stay in memory while you return to the DOS prompt. The routine's memory is protected from other programs by DOS.

You can add many enhancements to this portion of the code. You can have the program detect whether it is already active, for example, to prevent the program from loading over itself. You also can enhance it to detect and unload itself to release the memory later. Many fine reference books on TSR programming are available if you want to explore this portion of the program further.

Understanding the Data Structures

Lines 17 through 104 contain the data structures for this driver program. You may notice right away that each structure has two entries. This design enables the program to control two serial ports at the same time, using separate data structures for each port. The two EQU commands are present in lines 17 and 18 so that you can easily change the size of the input and output buffers. The given sizes work fine for speeds up to 9600 bps in most applications, but you can change the sizes to anything you want if necessary. *Note:* Because of the limits imposed for flow control (described later in this chapter in the section "Understanding Flow Control"), the input buffer may not be smaller than 50 bytes.

For each COM port, an input and an output buffer exist. The input buffer receives characters from the serial port and holds them until the INT 14 read character routine takes them. The output buffer holds characters from the INT 14 send character routine until the output interrupt routine places them in the UART.

The input buffer prevents the loss of input characters when the program is running and not asking for input. These characters are stored in the input buffer until the program requests them. The output buffer enables the program to continue to run instead of waiting for each character to be transmitted to the UART. The program thus can overlap other processing with serial data output flow. Because the program can be preparing characters for output before they are sent, output buffering enables full output speed to be maintained without choppiness, a problem that is especially likely at higher speeds.

Both buffers are managed as circular buffers. Because memory cannot be organized in a circle, the organization has to be simulated with pointers. Each buffer has four pointers: the start, end, insertion, and extraction pointers. For the input buffer, these pointers are called IB_START, IB_END, IB_IPTR, and IB_EPTR, respectively. The start and end pointers are static and simply define the limits of the buffer. The end pointer is the address of the end of the buffer plus one to make its usage easier. The insertion and extraction pointers indicate how much data currently is in the buffer and where the data is located. The diagram in figure 19.4 shows how these pointers are used to implement a circular buffer.

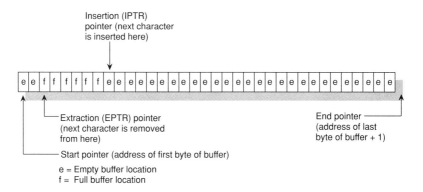

Fig. 19.4. *Operation of a 35-byte circular buffer with 6 bytes currently used.*

From this diagram, you can see that the pointers manage the buffer. If the extraction pointer is greater than the insertion pointer, the buffer currently is being used "around the corner." Thus the buffer effectively becomes a circle.

When a character is inserted into the buffer, the character is placed at the current value of IPTR, and IPTR is incremented to the next location. When a character is removed from the buffer, the character is taken from the current value of EPTR, and EPTR then is incremented to the next location. When the pointers are the same, the buffer is empty. If a pointer is incremented, its value is checked with the END pointer. If the pointer's value has reached the END value, the pointer is reset to the START value pointer so that it travels constantly "around" the "circular" buffer.

You see this circular buffer management in several places in this routine, and you should understand the concept well to prevent confusion. Because two sets of pointers exist, based on the COM port being used, the Index Register SI is used to select the proper pointer. The following code fragment demonstrates how a character from the AL register would be inserted into the input buffer:

```
        MOV       BX,[IB_IPTR][SI]    ;fetch insertion pointer
        MOV       [BX],AL             ;put AL in next location
        INC       BX                  ;bump insertion pointer
        CMP       [IB_END][SI],BX     ;at the corner?
        JNZ       NOTURN              ;no, normal case
        MOV       BX,[IB_START][SI]   ;yes, reset to start
NOTURN: MOV       [IB_IPTR][SI],BX    ;update insertion pointer
```

In the program, this process often is accompanied by checks for buffer full (detected if the incremented insertion pointer equals the extraction pointer) or buffer empty (detected if the insertion pointer equals the extraction pointer at the beginning). The circular buffers are managed this way in the example code.

Understanding Output Operation

Doing serial output using interrupts has one consideration that makes the process appear more confusing than you might expect. The *transmit interrupt* occurs when the Transmitter Holding Register becomes empty. For this reason, the interrupt has no way to start itself. When you want to send a buffer full of characters to the UART, you must send the first one explicitly from the main program level to "prime the pump" and get things started. After that, the Transmitter Holding Register Empty Interrupt can send the remaining characters until the buffer is empty.

To control this process, this program uses a flag named XMIT_ACT, which has one entry for each port. When a character is placed in the output buffer by the INT 14 send character routine, this flag is checked. If it is set, then placing the character in the buffer is all that must be done. The interrupt routine finds and sends the character in its proper sequence. If this flag is not set, however, the INT 14 routine must set the flag and send the character to the UART. When finding the buffer empty, the interrupt routine resets the flag and exits, so the process is repeated when the next character is fed to the INT 14 send character routine.

In this program, the routine START_XMIT (lines 524 to 559) handles the "pump priming" task and the interrupt routine TX_HOLD_EMPTY (lines 434 to 465) empties the buffer, character by character.

Understanding Input Operation

Serial input is handled by the *Receive Data Available Interrupt* from the UART. This interrupt occurs every time the UART receives a character. The interrupt routine reads the character from the UART and places the character in the input buffer. The INT 14 read character routine examines the input buffer. If any characters are in this buffer, the routine returns the next character.

In this program, the interrupt routine RECV_DATA (lines 475 to 514) receives characters from the UART and places them into the input buffer. The routine DO_READ (lines 180 to 216) handles the INT 14 request for a character from the buffer.

Understanding Flow Control

The output must be stopped if the modem (or connected device) lowers the Clear To Send (CTS) signal. The lowering of the CTS signal is the hardware flow control request for the computer to stop sending. Sending should resume when the CTS signal returns. Stopping is the easy part. The routine that is putting characters in the buffer doesn't send one if CTS is low, and the Transmitter Holding Register Empty Interrupt halts output by not sending any more if CTS is low. The problem is knowing when to start output again. To handle this problem, you can use the *Modem Status Interrupt* feature of the UART. This feature causes an interrupt any time one or more of the four signals—CD, CTS, RI, and DSR—change. When this interrupt is received, therefore, the program looks to see whether CTS is high. If it is, and the output isn't running, the interrupt routine sends the next character waiting in the output buffer, which restarts the output interrupt process.

Input flow control is used when the input buffer gets full. The input interrupt routine lowers RTS to indicate to the sending device that data will be lost if the input doesn't stop. As characters are removed from the buffer by the INT 14 read character routine, space becomes available again. If RTS was lowered, then the INT 14 read character routine must raise RTS again to resume input when the buffer has space.

In this program, output flow control is handled by the routine MODEM_STATUS (lines 418 to 426), which handles the CTS signal from the device. Lines 434 to 437 in the TX_HOLD_EMPTY routine stops output when CTS is removed by the modem if the output interrupt routine is active at that time. Input flow control is handled by lines 485 to 506 in the RECV_DATA interrupt routine, which lowers RTS when fewer than 10 characters remain in the input buffer. Lines 197 to 216 of the DO_READ routine raise RTS to restart input when only 40 characters remain stored in the input buffer.

Understanding Program Operation

In this program, lines 109 through 132 form the INT 14 *service call handler*. First, the STI instruction enables interrupts to continue, because all INT instructions turn them off. Next, the registers used by this routine are saved on the stack, and the data segment is set to point to the program's data area.

The calling DX register (which indicates the COM port being used) then is turned into a table index in the SI register. Use of this index register enables the same code to operate on both serial ports at the same time. If the DX register asks for an invalid serial port, an error return is given.

Lines 137 to 150 form the *function decoder*. The DX register is loaded with the COM port base address. (Because all routines want this value, it is loaded one time here.) Then the AH register is tested to see whether it is the initialize port function. If AH is 0, the routine DOINIT is called. If AH is not 0, the decoder checks the flag PORT_INIT to see whether an init call has been done on this port (because all other functions require an initialized port). If the port has not been initialized, an error is returned. If the port has been initialized, the decoder sends the program to DOSEND if AH is 1 and DOREAD if AH is 2, and falls through to lines 155 to 157 to return status if AH is 3 (the read status function). If AH is greater than 3, it is an invalid function code, and an error is returned.

Lines 160 to 175 handle the INT 14 *send character function*. The output buffer is checked to see whether it has room to hold another character. If no room is left, the program loops to label DOSND1 to wait for the interrupt routine to send a character and make space available. When space is available, the character from the AL register is placed in the output buffer, and the routine START_XMIT is called to send a character if needed to "prime the pump" and start output (as described in the section on "Understanding Output Operation"). After the character is sent, the line status is loaded in the AH register, and the program exits the INT 14 call.

Lines 180 to 216 handle the INT 14 *read character function*. The input buffer is checked to see whether a character is available. If one isn't available, the program loops to label DORD1, waiting a short time (counting CX 65,535 times) for a character. If a character doesn't appear in that delay interval, the program returns the time-out status and exits.

If a character is available in the input buffer, the program removes the character from the buffer and places the character in the AL register. The flag INPUT_FLOW then is checked to see whether the input interrupt routine lowered RTS because the input buffer got full. If it did, the input buffer is checked to see whether fewer than 40 bytes are now full. If so, RTS is raised, and the flag INPUT_FLOW is cleared. If the buffer still has more than 40 characters in it, RTS is left low. Finally, the AH register is loaded with the line status and then exits the INT 14 call.

> The input interrupt routine lowers RTS when 10 characters are left free in the input buffer (a few more may appear while the external device responds to the RTS signal). The INT 14 read character routine doesn't raise RTS again, however, until only 40 characters are left in the buffer. The reason for this feature is to prevent RTS from "thrashing" on a single character when the buffer gets full. This process is known as *hystereis* and is used to damp out the rate at which RTS is raised and lowered. By making RTS lower when the buffer is full but not enabling RTS to raise again until it is mostly empty, you guarantee that when RTS is raised you can accept several characters quickly before you have to lower RTS again. This design gives the remote device plenty of time to respond to RTS without running the risk of data loss.

Lines 224 to 297 handle the INT 14 *initialize port function*. Lines 224 to 231 convert the baud rate bit mask in AL to a 16-bit baud rate divisor value (from the table BRG_DIVISOR). Lines 232 to 241 switch the UART to make addresses base+0 and base+1 access the baud rate divisor latch. The selected divisor value then is loaded into the UART divisor latch. Finally, lines 242 to 245 set the word length, parity, and stop bits as well as switch back the first two port addresses to normal. At this point, the UART is programmed for the proper baud rate and serial data format.

Lines 246 to 276 connect and enable the UART interrupt routine. The flag PORT_INIT is checked to see whether this port has already had its interrupts connected and enabled. If not, then lines 252 to 261 change the interrupt vector address to point to this program's IRQ handler (GOTIRQ3 or GOTIRQ4 depending on the interrupt).

Lines 263 to 266 enable the UART interrupts for Receive Data Available, Transmitter Holding Register Empty, and Modem Status Register. The Receiver Line Status Interrupt is not enabled because you don't need it for this application. Lines 267 and 271 turn on the OUT2 line to connect the UART's interrupt line to the bus. They also raise RTS and DTR to enable input to begin. Finally, lines 272 to 276 enable the proper IRQ line in the 8259 interrupt controller chip so that it will pass to the CPU. All these steps are required so that interrupts from the UART can pass all the way into the computer.

Lines 278 to 289 ensure that the input and output buffer pointers are reset. These lines are executed with the interrupts inhibited in case characters already are being sent to the UART. Inhibits generally are not needed because the interrupt routine and the INT 14 main code alter only the pointers for their respective sides of the buffers. But here the INT 14 routine is resetting the pointers normally managed by the interrupts, so an inhibit is needed to prevent an attempt by both to change the same pointers at the same time.

Finally lines 291 to 296 load the modem and line status and exit the INT 14 service call. Note that line 295 saves the modem status in the cell MODEM_SR. This step is necessary because if the INT 14 routines read the Modem Status Register directly when output is active, a Modem Status Interrupt may be lost. This cell therefore is maintained by the Modem Status Interrupt routine with the correct value of the UART register so that the INT 14 routines can use this cell rather than read the register. By having the interrupt routine be the only code that reads the true register, you ensure that no interrupts are lost.

Lines 304 to 329 form the subroutine that loads the line status into the AH register. Lines 305 to 306 save registers this routine will use. The true UART Line Status Register first is read into the AH register by lines 307 to 310. The Transmitter Holding Register Empty and Receive Data Available bits, however, are cleared at line 311 because they must reflect the status of the buffers that are now the logical equivalents—not the status of the UART itself. Lines 312 to 316 check to see whether the input buffer has any characters. If the input buffer is not empty, the Receive Data Available status bit is set. Lines 317 to 325 check to see whether the output buffer is full. If it is not full, the Transmitter Holding Register Empty bit is set to indicate that another character can be sent immediately. Lines 326-329 restore the saved registers and return the character read in AL.

Lines 336 to 339 load the Modem Status Register into AL. As noted previously, to avoid interrupt loss, this routine returns the latest value of the register that was saved by the Modem Status Interrupt.

Understanding Interrupt Operation

Lines 346 to 408 are the interrupt entry and decoder logic. Lines 346 to 352 form the entry point for an IRQ4 interrupt. Lines 357 to 362 form the entry point for an IRQ3 interrupt. These entry points save the registers used by the remainder of the interrupt routines, set the SI Index Register to show which interrupt is being serviced, and then enter the common interrupt routine code.

Lines 369 to 377 set the data segment to point to the program's data area. Then the DX register is loaded with the port base address. Next, an End Of Interrupt (EOI) command is sent to the 8259 to enable it to recognize another interrupt immediately. This part of the program is a compromise to handle the fact that ISA bus interrupts are edge triggered. For this reason, the 8259 must be ready to recognize them when the edge occurs, or they are lost forever. The remainder of the interrupt routine loops to clear out all interrupts pending in a UART (status, input, and output interrupts may all be waiting at the same time). This process can lead to a case in which the 8259 generates an "empty" interrupt because it "saw" a second or third interrupt during this loop process. But this approach is the only alternative to leaving a window of time in which an interrupt can occur and be lost forever.

If you are running only on a Microchannel computer, you can move lines 376 and 377 so that they follow the label IRQXT1 (line 402) to prevent empty interrupts, but you do not have that alternative if you're using an ISA bus machine. The good news is that if interrupts are occurring rapidly, empty interrupts will occur only rarely. These empty interrupts tend to happen when things are slow and when plenty of time is available to handle the situation without a slowdown. In the end, then, the problem of empty interrupts is more a nuisance than the performance problem it at first may seem to be.

Lines 381 to 392 read the UART's interrupt identification register and jump to the routine that handles this type of interrupt. Lines 387 and 388 detect when no more interrupts are pending and then cause the program to exit the interrupt handler. This code handles a loop back to check for more interrupts and to detect empty interrupts. Lines 397 to 400 provide a safety check. Because the initialization routine didn't enable the Line Status Interrupt, this interrupt should never occur. If this interrupt does occur, however, the program clears it by reading the Line Status Register and then ignores the empty interrupt.

Lines 402 to 408 restore the interrupted registers and then return to the interrupted program. This step ends the interrupt process.

Lines 418 to 426 handle the Modem Status Interrupt. The Modem Status Register is read and stored in the MODEM_SR cell. Because the Modem Status Interrupt occurs every time the Modem Status Register changes, the memory cell is maintained as an accurate reflection of the UART register.

Also, the routine START_XMIT is called in case this interrupt was caused by the CTS flow control signal being raised to ask output to restart.

Lines 434 to 456 handle the Transmitter Holding Register Empty Interrupt. Lines 435 to 444 exit the transmitter interrupt routine immediately if CTS is low (flow control asking output to halt) or if no more characters are in the output buffer to send. In either case, the flag XMIT_ACT is reset to indicate that the interrupt routine halted. Lines 448 to 456 fetch the next output character and send it to the UART Transmitter Holding Register to keep output running.

Lines 460 to 465 exit the Transmitter Interrupt Routine when another character was not sent to the UART. The flag XMIT_ACT is cleared here to indicate that the XMIT interrupt has quit operation and the next output operation again will need to "prime the pump" (see "Understanding Output Operation" earlier in this chapter).

Lines 475 to 514 handle the Receive Data Available Interrupt. Line 476 reads the character from the UART. Lines 477 to 484 check to see whether the input buffer has room for the character. If the device connected to the serial port honors RTS/CTS flow control, the buffer always should have room. So that the program can handle a situation in which flow control is not honored, however, this check is made. If no room is available, the program goes to the label RXDOVF, which discards the character.

Lines 485 to 489 calculate the number of characters used in the input buffer in preparation for the flow control check later. Lines 490 and 491 store the received character in the input buffer. Lines 492 to 495 calculate the number of characters remaining empty in the input buffer. Lines 496 to 499 exit if RTS is already low or if more than 10 characters are empty in the input buffer. If 10 or fewer characters are empty in the input buffer and RTS is still high, lines 500 to 505 lower RTS and indicate this change by setting the INPUT_FLOW flag. Lines 506 to 514 exit the receive interrupt routine.

Lines 524 to 559 comprise the START_XMIT routine, which is called whenever output may need to be started by the sending of the first character from the buffer to the UART. Lines 524 to 526 save the registers this routine will use. Lines 527 and 532 exit the routine if the flag XMIT_ACT indicates that the output is already active. This check enables START_XMIT to be called whenever starting the output *may* be necessary, without the routine having to do extensive checking in each place it is called. Lines 533 and 534 exit the routine quickly if CTS is low, meaning that the remote device has asked for output to be held off for now. Lines 538 to 541 check the output buffer to see whether any characters are there to send. If not, the program exits quickly. If a character is in the output buffer, lines 546 to 551 extract the character and place it in the AL register. Lines 552 to 553 set XMIT_ACT to indicate that the output interrupt now can send all characters from the buffer (see "Understanding Output Operation" earlier in this chapter). Line 554 sets the flag XMIT_ACT to indicate that the output interrupt is active, and line 555 sends the character to the UART.

You need to keep some considerations in mind when using interrupts. On a fast computer, the UART may interrupt immediately when the character is written to the holding register. If lines 553 and 554 of this program were put in the reverse order (send character first and then set XMIT_ACT flag), the output interrupt occasionally could occur after the character was written to the UART but before the XMIT_ACT flag was set. If this character was the only character in the buffer, the transmit interrupt would mark XMIT_ACT clear and return with all output activity finished. Then the instruction to set the XMIT_ACT flag to 1 would execute. This step would leave the flag indicating that the output routine was active when it wasn't, and all output would halt for good! You need to do this type of thinking when you are writing interrupt routines in order to avoid routines that almost work but lock up erratically for no apparent reason.

Understanding the 16550 UART

So far, this chapter has described only the original 8250 or the improved 16450 UART. These UARTs become difficult or impossible to use at very high speeds, however, because they each have only a single input holding register. At a speed of 38,400 bps, a character arrives on the input serial line 3,840 times per second. Thus the computer must remove a character from the input register every 260 microseconds. If the removal takes as much as 265 microseconds even once, input data overrun occurs.

The 16550 UART has a 16-byte FIFO (first in, first out) input register that may be activated under software control. In this mode, the 16550 can hold up to 16 input characters rather than the 1 that a normal UART holds. Therefore, if the computer can read a character from the UART at an *average* of every 260 microseconds, an occasional delay of up to 4 milliseconds can occur without the loss of any input data. The diagram in figure 19.5 shows this FIFO structure.

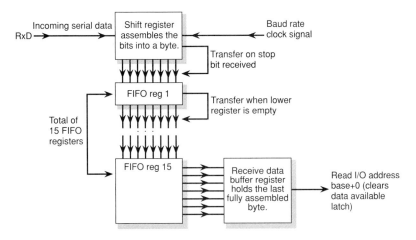

Fig. 19.5. 16550 UART input structure.

When the 16550 input FIFO is enabled, each character that is shifted in is put in FIFO Register 1. This character is passed in bucket-brigade fashion until the next FIFO Register is full. When the computer reads the Receive Data Buffer Register, each full FIFO Register shifts its character one down the line, and the last FIFO Register is placed in the Receive Data Buffer Register.

By forming this bucket brigade, the UART can hold up to 16 characters without the CPU reading a character from the UART before input data is lost. This capability becomes important when you realize that many high-speed disk controllers lock the CPU out for more than 500 microseconds, and some video or network cards do likewise. In such installations, using the disk or video while high-speed serial input is happening is impossible without losing data—unless you use the 16550 UART.

Using the FIFO Control Register (base+2)

To control the 16550 UART FIFO operation, a new register is added. This register is called the *FIFO Control Register* and is a write-only register at the I/O address base+2. This address does nothing in a normal 8250 or 16450 UART, so this addition is upwardly compatible. The format of the FIFO Control Register is as follows:

Bit	Description
7	Receiver trigger bit 1
6	Receiver trigger bit 0
5	Unused
4	Unused
3	DMA mode select
2	Transmit FIFO reset
1	Receive FIFO reset
0	FIFO enable

The 16550 UART resets with its FIFO disabled. In this mode, the UART's operation is identical to the 8250 or 16450 UART. Writing to the FIFO Control Register can turn FIFO mode either on or off.

If bit 0 is set to 1, the 16550 is changed to FIFO mode. If you write a 0 into this bit, the 16550 is changed back to the 8250-compatible modem.

Setting bit 1 to 1 clears all the input FIFO registers. If a character is partially received by the input shift register, that character is not affected. This bit self-clears after the reset occurs.

Setting bit 2 to 1 clears the output FIFO registers. This bit self-clears after the reset occurs.

Bit 3 controls the operation of two control pins on the UART. These pins are not used in an IBM standard serial port, so this bit has no effect. You should set it to 0.

Bits 6 and 7 set the input "trigger" threshold for a Receive Data Available Interrupt when the 16550 is operating in FIFO mode. You may choose any of the following four settings:

Bit 7	Bit 6	Threshold (in characters)
0	0	1
0	1	4
1	0	8
1	1	14

When this threshold is larger than 1, an interrupt is not generated until the number of characters indicated have been received. You can use this threshold along with some sophisticated programming techniques to reduce the number of interrupts that occur for input. For thresholds larger than 1, the 16550 also adds a time-out feature that starts a timer when one or more characters have been placed in the FIFO. If the threshold has not been reached, but four character times have elapsed with no received data, an alternate receive data available interrupt is generated.

How Other Registers Are Modified

When the 16550 is switched into FIFO mode, two other registers change, using previously unused bits. First, the Line Status Register bit 7 indicates whether an error exists in a character in the FIFO. You generally do not use this bit, but you need to know that it no longer is guaranteed to be a 0.

The major modifications take place in the Interrupt Identification Register. Bits 7 and 6 each become a 1 to indicate that the 16550 has switched to FIFO mode. These bits may be used to detect the presence of a 16550 UART. If you write to the FIFO Control Register on an 8250 or 16450 UART, nothing happens. Thus if you write a command to enable the FIFO mode, you then can check the high-order bits of the Interrupt Identification Register. If they remain 0, the UART is not a 16550. If they switch to 1, the serial port does have a 16550 UART.

Bit 3 of the Interrupt Identification Register is used in the 16550 FIFO mode to show the difference between a normal Receive Data Available Interrupt and one that occurs because fewer characters than the threshold were received and then timed out. If you use only the lower three bits of the Interrupt Identification Register, these two interrupts look the same. They also have the same priority (because they never can happen at the same time). If telling the difference between them is important to you, however, you can use bit 3. Bit 3 is set if the interrupt came from a received FIFO time-out.

Programming the 16550 UART

To modify the example program to use the FIFO feature of the 16550 UART, all you need to do is enable the FIFO buffers. To do this you must add code to write a 0x07 to the FIFO Control Register of the UART (base+2), which will enable the FIFO buffers and reset them.

After the FIFO is enabled, only the 16550 bits in the Interrupt Identification Register will become active. The program as shown already masks out those extra bits on line 385 so that the program will still properly interpret interrupts. Sending a 0x07 to the FIFO Control Register also sets the receiver interrupt threshold to one character so that the UART will interrupt as soon as it has any input data. This means that the program as written will properly handle input, even with the buffer enabled. And because lines 382 to 392 check for any remaining interrupts before they exit the interrupt, all characters will be removed from the input FIFO on a single interrupt.

The final consideration is that bit 7 of the Line Status Register (LSR) becomes active when the 16550 FIFOs are enabled. Line 384 of the program masks this bit so that its presence will not affect proper program operation. Therefore, you can fully support the 16550 with this program simply by adding the following lines after line 245.

```
ADD   DX,2      ;point to FIFO control register
OUT   DX,07h    ;enable 16550 if present
SUB   DX,2      ;restore port base address
```

The following program lists the code for the driver in Microsoft MASM assembler format. Refer to the sections under "Building a Serial Driver" earlier in this chapter for discussion of this program.

Program Listing 19.1

```
1                              NAME    IINT14
2                              PAGE    60,132
3                              TITLE   Replacement INT 14 serial Port Driver
4                      ;
5                      ;       Installs as a TSR and becomes an interrupt driven
6                      ;       replacement for the BIOS INT 14 routines.
7                      ;
8 0000                 CODE    SEGMENT PARA 'CODE'
9 0100                         ORG     100h
10                             ASSUME  CS:CODE,DS:CODE,ES:NOTHING
11                     ;
12 0100                MAIN    PROC    NEAR
13 0100   E9 0666 R            JMP     START               ;Begin Execution
14                     ;------------------------------------------------------------
15                     ;  Size Parameters
16                     ;
17 = 00C8              IN_SIZE         EQU     200         ;# bytes in input buffer
18 = 0064              OUT_SIZE        EQU     100         ;# bytes in output buffer
19                     ;
20                     ;------------------------------------------------------------
21                     ;  Data Area - interrupt buffers
22                     ;
23 0103   00C8[        INBUF1          DB      IN_SIZE DUP(0)   ;input buffer - COM1 (IRQ4)
24          00
25                  ]
26
27 01CB   00C8[        INBUF2          DB      IN_SIZE DUP(0)   ;input buffer - COM2 (IRQ3)
28          00
29                  ]
30
31 0293   0064[        OUTBUF1         DB      OUT_SIZE DUP(0)  ;output buffer - COM1 (IRQ4)
32          00
33                  ]
34
35 02F7   0064[        OUTBUF2         DB      OUT_SIZE DUP(0)  ;output buffer - COM2 (IRQ3)
36          00
37                  ]
38
```

```
39 035B  90                        EVEN                        ;assure on word boundary
40                         ;------------------------------------------------------------------
41                         ;  Following tables have one entry for each com port/IRQ
42                         ;  They are accessed by indexing by com port number.
43                         ;
44                         ;  Word tables are all first
45                         ;
46 035C  0103 R           IB_START    DW      INBUF1          ;inbuf start address
47 035E  01CB R                       DW      INBUF2
48                         ;
49 0360  01CB R           IB_END      DW      INBUF1+IN_SIZE  ;end of buffer + 1
50 0362  0293 R                       DW      INBUF2+IN_SIZE
51                         ;
52 0364  0103 R           IB_IPTR     DW      INBUF1          ;inbuf insertion pointer
53 0366  01CB R                       DW      INBUF2
54                         ;
55 0368  0103 R           IB_EPTR     DW      INBUF1          ;inbuf extraction pointer
56 036A  01CB R                       DW      INBUF2
57                         ;
58 036C  0293 R           OB_START    DW      OUTBUF1         ;outbuf start address
59 036E  02F7 R                       DW      OUTBUF2
60                         ;
61 0370  02F7 R           OB_END      DW      OUTBUF1+OUT_SIZE ;end of buffer + 1
62 0372  035B R                       DW      OUTBUF2+OUT_SIZE
63                         ;
64 0374  0293 R           OB_IPTR     DW      OUTBUF1         ;outbuf insertion pointer
65 0376  02F7 R                       DW      OUTBUF2
66                         ;
67 0378  0293 R           OB_EPTR     DW      OUTBUF1         ;outbuf extraction pointer
68 037A  02F7 R                       DW      OUTBUF2
69                         ;
70 037C  03F8             COMADR      DW      03F8h           ;COM 1 port address
71 037E  02F8                         DW      02F8h           ;COM 2 port address
72                         ;
73 0380  0030             IRQ_VEC     DW      030h            ;IRQ4 Vector offset
74 0382  002C                         DW      02Ch            ;IRQ3 Vector offset
75                         ;
76 0384  0546 R           IRQ_HAND    DW      GOTIRQ4         ;IRQ handler (COM1=IRQ4)
77 0386  054F R                       DW      GOTIRQ3         ; (COM2 = IRQ3)
78                         ;
79                         ;------------------------------------------------------------------
80                         ;    Byte tables with one entry/IRQ
81                         ;
82 0388  10               IRQ_MASK    DB      10h             ;IRQ4 bit mask
83 0389  08                           DB      08h             ;IRQ3 bit mask
84                         ;
85 038A  00 00            PORT_INIT   DB      0,0             ;>0 = initialized now
86                         ;
87 038C  00 00            XMIT_ACT    DB      0,0             ;>0 = xmitter active
88                         ;
89 038E  00 00            MODEM_SR    DB      0,0             ;modem status register save
90                         ;
91 0390  00 00            INPUT_FLOW  DB      0,0             ;>0 = RTS lowered
92                         ;
93                         ;------------------------------------------------------------------
94                         ;  Baud Rate Divisor table for setting port speed
95                         ;
96 0392  0417             BRG_DIVISOR DW      1047            ;110 baud
97 0394  0300                         DW      768             ;150 baud
98 0396  0180                         DW      384             ;300 baud
99 0398  00C0                         DW      192             ;600 baud
100 039A  0060                        DW      96              ;1200 baud
101 039C  0030                        DW      48              ;2400 baud
102 039E  0018                        DW      24              ;4800 baud
103 03A0  000C                        DW      12              ;9600 baud
104                        ;
```

```
105                                  PAGE
106          ;================================================================
107          ;    INT 14 service routine Entry Point
108          ;
109 03A2 FB          INT_14: STI                      ;re-enable interrupts
110 03A3 53                  PUSH    BX               ;save registers we will use
111 03A4 51                  PUSH    CX
112 03A5 52                  PUSH    DX
113 03A6 56                  PUSH    SI
114 03A7 57                  PUSH    DI
115 03A8 1E                  PUSH    DS
116 03A9 8C CB               MOV     BX,CS
117 03AB 8E DB               MOV     DS,BX            ;make DS: = CS: -> CODE
118 03AD 8B F2               MOV     SI,DX
119 03AF 03 F6               ADD     SI,SI            ;DX = 0 or 2
120 03B1 83 FE 02            CMP     SI,2             ;assure valid port address
121 03B4 76 09               JBE     PORTOK           ;it is
122 03B6 B4 80       PORTER: MOV     AH,80h           ;indicate error
123          ;
124          ;    Exit from INT 14 service call.  AH and AL = return values
125          ;
126 03B8 1F          PORTXT: POP     DS               ;restore user's registers
127 03B9 5F                  POP     DI
128 03BA 5E                  POP     SI
129 03BB 5A                  POP     DX
130 03BC 59                  POP     CX
131 03BD 5B                  POP     BX
132 03BE CF                  IRET                     ;return to caller
133          ;
134          ;    DX requested either COM1 or COM2 correctly
135          ;    AH = action code
136          ;
137 03BF 8B 94 037C R PORTOK: MOV    DX,[COMADR][SI]  ;DX = COM PORT Base Address
138 03C3 D1 EE               SHR     SI,1             ;SI = 0 or 1 port index
139 03C5 0A E4               OR      AH,AH            ;see if AH=0 (Initialize)
140 03C7 75 03               JNZ     POK1             ;no - check for other codes
141 03C9 E9 0468 R           JMP     DOINIT           ;yes
142          ;
143 03CC 80 BC 038A R 00 POK1: CMP   [PORT_INIT][SI],0 ;others must be to init'd port
144 03D1 74 E3               JZ      PORTER           ;not initialized, give error
145 03D3 FE CC               DEC     AH               ;see if AH=1 (send character)
146 03D5 74 10               JZ      DOSEND
147 03D7 FE CC               DEC     AH               ;see if AH=2 (read character)
148 03D9 74 33               JZ      DOREAD
149 03DB FE CC               DEC     AH               ;see if AH=3 (read status)
150 03DD 75 D7               JNZ     PORTER           ;other code is an error
151          ;
152          ;----------------------------------------------------------------
153          ;    Compose and return status for the AH=3 - Read Status call
154          ;
155 03DF E8 0505 R           CALL    LOAD_LINE_STATUS  ;set AH = "Line Status"
156 03E2 E8 0541 R           CALL    LOAD_MODEM_STATUS ;set AL = "Modem Status"
157 03E5 EB D1               JMP     PORTXT           ;return
158          ;
159          ;----------------------------------------------------------------
160          ;    Send character in AL to com port
161          ;
162 03E7 03 F6       DOSEND: ADD     SI,SI            ;make word address
163 03E9 8B 9C 0374 R DOSND1: MOV    BX,[OB_IPTR][SI] ;get output buffer pointer
164 03ED 8B CB               MOV     CX,BX            ;copy to CX
165 03EF 41                  INC     CX               ;bump pointer to next slot
166 03F0 39 8C 0370 R        CMP     [OB_END][SI],CX  ;is it past end?
167 03F4 75 04               JNZ     DOSND2           ;no
168 03F6 8B 8C 036C R        MOV     CX,[OB_START][SI] ;yes, "turn corner" to front
169 03FA 39 8C 0378 R DOSND2: CMP    [OB_EPTR][SI],CX ;is buffer full?
170 03FE 74 E9               JZ      DOSND1           ;yes - wait for spot to open up
171 0400 89 8C 0374 R        MOV     [OB_IPTR][SI],CX ;no, update insertion pointer
```

```
172 0404  88 07           MOV    [BX],AL              ;store our character in buffer
173 0406  E8 062A R       CALL   START_XMIT           ;start transmitter if 1st
174 0409  E8 0505 R       CALL   LOAD_LINE_STATUS     ;set AH = "Line Status"
175 040C  EB AA           JMP    PORTXT               ;return it
176                    ;
177                    ;--------------------------------------------------------------
178                    ;   Receive a character from com port to AL
179                    ;
180 040E  03 F6   DOREAD: ADD    SI,SI                ;make word address
181 0410  33 C9           XOR    CX,CX                ;long delay if waiting
182 0412  8B 9C 0368 R DORD1: MOV BX,[IB_EPTR][SI]    ;get input buffer pointer
183 0416  39 9C 0364 R    CMP    [IB_IPTR][SI],BX     ;is there a character available?
184 041A  75 06           JNZ    GOTCHR               ;yes, return it to caller
185 041C  E2 F4           LOOP   DORD1                ;no - delay a while waiting
186 041E  B4 80           MOV    AH,80h               ;timed out - return that status
187 0420  EB 96           JMP    PORTXT               ;and get out
188                    ;
189                    ;   Character is waiting, read it into AL
190                    ;
191 0422  8A 07   GOTCHR: MOV    AL,[BX]              ;yes, fetch the character
192 0424  43             INC    BX                   ;bump pointer to next slot
193 0425  39 9C 0360 R    CMP    [IB_END][SI],BX      ;is it past end?
194 0429  75 04           JNZ    DORD2                ;no
195 042B  8B 9C 035C R    MOV    BX,[IB_START][SI]    ;yes, "turn corner" to front
196 042F  89 9C 0368 R DORD2: MOV [IB_EPTR][SI],BX    ;update extraction ptr for next time
197 0433  8B 9C 0364 R    MOV    BX,[IB_IPTR][SI]
198 0437  2B 9C 0368 R    SUB    BX,[IB_EPTR][SI]     ;calculate # bytes now in buffer
199 043B  79 04           JNS    DORD3
200 043D  81 C3 00C8      ADD    BX,IN_SIZE           ;rounding corner, correct for it
201 0441  D1 EE   DORD3:  SHR    SI,1                 ;make byte index
202 0443  80 3E 0390 R 00 CMP   [INPUT_FLOW],0       ;have we lowered RTS?
203 0448  74 16           JZ     DORD4                ;no - all done
204 044A  83 FB 28        CMP    BX,40                ;are we less than 40 bytes used?
205 044D  73 11           JAE    DORD4                ;no - still halt input
206 044F  C6 06 0390 R 00 MOV   [INPUT_FLOW],0       ;yes, indicate we raised RTS again
207 0454  50             PUSH   AX                   ;save character we're reading
208 0455  83 C2 04        ADD    DX,4                 ;point to Modem Control Register
209 0458  EC             IN     AL,DX                ;read MCR
210 0459  0C 02           OR     AL,02h               ;raise RTS
211 045B  EE             OUT    DX,AL                ;write back to MCR in UART
212 045C  83 EA 04        SUB    DX,4                 ;restore base address
213 045F  58             POP    AX                   ;restore character we read
214 0460  D1 E6   DORD4:  SHL    SI,1                 ;restore word index
215 0462  E8 0505 R       CALL   LOAD_LINE_STATUS     ;set AH = "Line Status"
216 0465  E9 03B8 R       JMP    PORTXT               ;return character and status
217                    ;
218                    ;--------------------------------------------------------------
219                    ;   Initialize the selected COM port
220                    ;
221                    ;   SI = byte index, DX = port base address
222                    ;   AL = user's request parameters
223                    ;
224 0468  8A E0   DOINIT: MOV    AH,AL                ;save init params in AH
225 046A  8A D8           MOV    BL,AL                ;copy params to BL
226 046C  D0 C3           ROL    BL,1
227 046E  D0 C3           ROL    BL,1
228 0470  D0 C3           ROL    BL,1                 ;position baud rate bits to 2, 1, 0
229 0472  83 E3 07        AND    BX,0007h             ;isolate the baud rate param
230 0475  03 DB           ADD    BX,BX                ;multiply by 2 for word index
231 0477  8B 9F 0392 R    MOV    BX,[BRG_DIVISOR][BX] ;fetch baud rate divisor to BX
232 047B  83 C2 03        ADD    DX,3                 ;point DX to Line Control Register
233 047E  B0 80           MOV    AL,80h               ;Set DLAB=1 to
234 0480  EE             OUT    DX,AL                ;   allow access to BRG Divisor
235 0481  83 EA 03        SUB    DX,3                 ;point back to divisor register
236 0484  8A C3           MOV    AL,BL                ;low order divisor bits to AL
237 0486  EE             OUT    DX,AL                ;set divisor low order
238 0487  42             INC    DX                   ;point to divisor high order
239 0488  8A C7           MOV    AL,BH                ;high order divisor bits to AL
240 048A  EE             OUT    DX,AL                ;set divisor high order
241 048B  83 C2 02        ADD    DX,2                 ;point back to Line Control Reg
```

```
242 048E  8A C4            MOV     AL,AH               ;get rest of params in AL
243 0490  24 1F            AND     AL,01Fh             ;isolate parity, stop, and length
244 0492  EE               OUT     DX,AL               ;port is now programmed.
245 0493  83 EA 03         SUB     DX,3                ;put DX back to base address
246 0496  80 BC 038A R 00  CMP     [PORT_INIT][SI],0   ;is IRQ installed already?
247 049B  75 39            JNZ     INIT1               ;yes, don't do it again
248 049D  C6 84 038A R 01  MOV     [PORT_INIT][SI],1   ;no, indicate it is now
249                        ;
250                        ;   Install our interrupt handler for this COM port
251                        ;
252 04A2  06               PUSH    ES
253 04A3  33 DB            XOR     BX,BX               ;segment address = 0
254 04A5  8E C3            MOV     ES,BX               ;put in ES
255 04A7  03 F6            ADD     SI,SI               ;make word index
256 04A9  8B BC 0380 R     MOV     DI,[IRQ_VEC][SI]    ;ES:DI = IRQ address
257 04AD  FA               CLI                         ;halt interrupts during store
258 04AE  8B 9C 0384 R     MOV     BX,[IRQ_HAND][SI]   ;get IRQ offset
259 04B2  26: 89 1D        MOV     ES:[DI],BX          ;offset of vector
260 04B5  26: 8C 4D 02     MOV     ES:[DI][2],CS       ;CS:BX = our new handler
261 04B9  07               POP     ES                  ;restore interrupted ES:
262                        ;
263 04BA  42               INC     DX                  ;point to Interrupt Enable Register
264 04BB  B0 03            MOV     AL,03h              ;enable data avail and THRE ints only
265 04BD  EE               OUT     DX,AL               ;in UART
266 04BE  83 C2 03         ADD     DX,3                ;point to Modem Control Register
267 04C1  B0 0B            MOV     AL,0Bh              ;set OUT 2, RTS, and DTR high
268 04C3  EE               OUT     DX,AL               ;to connect interrupt to bus
269 04C4  83 EA 04         SUB     DX,4                ;point DX back to base address
270                        ;
271 04C7  D1 EE            SHR     SI,1                ;make byte index again
272 04C9  E4 21            IN      AL,21h              ;read 8259 enable register
273 04CB  8A A4 0388 R     MOV     AH,[IRQ_MASK][SI]   ;get IRQ bit for this port
274 04CF  80 F4 FF         XOR     AH,0FFh             ;make into a bit clear mask
275 04D2  22 C4            AND     AL,AH               ;clear this bit to enable interrupt
276 04D4  E6 21            OUT     21h,AL              ;put back in 8259
277                        ;
278 04D6  FA       INIT1:  CLI                         ;halt any interrupts
279 04D7  03 F6            ADD     SI,SI               ;make word index
280 04D9  8B 84 035C R     MOV     AX,[IB_START][SI]   ;get input buffer start pointer
281 04DD  89 84 0364 R     MOV     [IB_IPTR][SI],AX    ;reset buffer pointers
282 04E1  89 84 0368 R     MOV     [IB_EPTR][SI],AX    ;to empty buffer
283                        ;
284 04E5  8B 84 036C R     MOV     AX,[OB_START][SI]   ;get output buffer start pointer
285 04E9  89 84 0374 R     MOV     [OB_IPTR][SI],AX    ;reset buffer pointers
286 04ED  89 84 0378 R     MOV     [OB_EPTR][SI],AX    ;to empty buffer
287                        ;
288 04F1  D1 EE            SHR     SI,1                ;make byte pointer again
289 04F3  FB               STI                         ;allow interrupts again
290                        ;
291 04F4  E8 0505 R        CALL    LOAD_LINE_STATUS    ;set AH = "Line Status"
292 04F7  83 C2 06         ADD     DX,6                ;point to modem status register
293 04FA  EC               IN      AL,DX               ;read MSR from UART
294 04FB  83 EA 06         SUB     DX,6                ;restore base address
295 04FE  88 84 038E R     MOV     [MODEM_SR][SI],AL   ;init status register cell
296 0502  E9 03B8 R        JMP     PORTXT              ;return status - port initialized
297                        ;
298                        PAGE
299                        ;--------------------------------------------------------------
300                        ;   Subroutine to load LINE_STATUS into AH
301                        ;
302                        ;   On entry: SI = byte index for line, DX = port base address
303                        ;
304 0505           LOAD_LINE_STATUS:
305 0505  53               PUSH    BX                  ;save BX
306 0506  50               PUSH    AX                  ;and AX
307 0507  83 C2 05         ADD     DX,5                ;point to Line Status Register
308 050A  EC               IN      AL,DX               ;get UART Line Status
309 050B  83 EA 05         SUB     DX,5                ;point back to base address
310 050E  8A E0            MOV     AH,AL               ;put LSR value in AH
```

```
311 0510  80 E4 5E              AND    AH,5Eh                  ;clear T/O, THRE, and DA status
312 0513  03 F6                 ADD    SI,SI                   ;make word COM PORT index
313 0515  8B 9C 0364 R          MOV    BX,[IB_IPTR][SI]        ;get Input Insertion Pointer
314 0519  39 9C 0368 R          CMP    [IB_EPTR][SI],BX        ;is data available?
315 051D  74 03                 JZ     LLS1                    ;no - pointers the same
316 051F  80 CC 01              OR     AH,01h                  ;yes, set RDA status bit
317 0522  8B 9C 0374 R   LLS1:  MOV    BX,[OB_IPTR][SI]        ;get output insertion pointer
318 0526  43                    INC    BX
319 0527  39 9C 0370 R          CMP    [OB_END][SI],BX         ;is it at end?
320 052B  75 04                 JNZ    LLS2                    ;no
321 052D  8B 9C 036C R          MOV    BX,[OB_START][SI]       ;yes, "turn corner"
322 0531  39 9C 0378 R   LLS2:  CMP    [OB_EPTR][SI],BX        ;is output buffer full?
323 0535  74 03                 JZ     LLS3                    ;yes, "THR" is full
324 0537  80 CC 20              OR     AH,20h                  ;no, "THR" is empty
325 053A  D1 EE         LLS3:   SHR    SI,1                    ;return SI to byte index
326 053C  5B                    POP    BX                      ;original AL to BL
327 053D  8A C3                 MOV    AL,BL                   ;restore caller's AL
328 053F  5B                    POP    BX                      ;and caller's BX
329 0540  C3                    RET
330                         ;
331                         ;--------------------------------------------------------------
332                         ;    Subroutine to load MODEM_STATUS into AL
333                         ;
334                         ;    On entry: SI = byte index for line, DX = port base address
335                         ;
336 0541                   LOAD_MODEM_STATUS:
337 0541  8A 84 038E R          MOV    AL,[MODEM_SR][SI]       ;get UART Modem Status
338 0545  C3                    RET
339                         ;
340                             PAGE
341                         ;==============================================================
342                         ;    Interrupt Handler - Entered on Serial Port Interrupts
343                         ;
344                         ;       IRQ 4 interrupt entry
345                         ;
346 0546                   GOTIRQ4:
347 0546  50                    PUSH   AX                      ;save interrupted registers
348 0547  53                    PUSH   BX
349 0548  52                    PUSH   DX
350 0549  56                    PUSH   SI
351 054A  BE 0000               MOV    SI,0                    ;index for COM1
352 054D  EB 07                 JMP    SHORT IRQ_COM           ;enter common IRQ handler
353                         ;
354                         ;--------------------------------------------------------------
355                         ;       IRQ 3 interrupt entry
356                         ;
357 054F                   GOTIRQ3:
358 054F  50                    PUSH   AX                      ;save interrupted registers
359 0550  53                    PUSH   BX
360 0551  52                    PUSH   DX
361 0552  56                    PUSH   SI
362 0553  BE 0002               MOV    SI,2                    ;index for COM2
363                         ;
364                         ;--------------------------------------------------------------
365                         ;    Interrupt Handler Common Code
366                         ;
367                         ;    AX, BX & DX have been saved, SI = word index for port
368                         ;
369 0556                   IRQ_COM:
370 0556  1E                    PUSH   DS                      ;save DS
371 0557  8C C8                 MOV    AX,CS
372 0559  8E D8                 MOV    DS,AX                   ;set DS=CS=CODE:
373                         ;
374 055B  8B 94 037C R          MOV    DX,[COMADR][SI]         ;get port base address in DX
375                         ;
376 055F  B0 20                 MOV    AL,20h
377 0561  E6 20                 OUT    20h,AL                  ;Issue EOI to clear 8259
378                         ;
379                         ;    Now check (or re-check) UART for interrupt status
380                         ;
```

```
381 0563                         IRQ_XIT:
382 0563  83 C2 02                   ADD     DX,2                ;point to Interrupt I.D. Register
383 0566  EC                         IN      AL,DX               ;fetch IID
384 0567  83 EA 02                   SUB     DX,2                ;restore port base address
385 056A  24 07                      AND     AL,07h              ;isolate Type of int
386 056C  74 1B                      JZ      MODEM_STATUS        ;modem status interrupt
387 056E  A8 01                      TEST    AL,01h              ;is an interrupt pending?
388 0570  75 11                      JNZ     IRQXIT1             ;no -- all done, get out
389 0572  3C 04                      CMP     AL,04h              ;is it received data interrupt?
390 0574  74 5B                      JZ      RECV_DATA           ;yes
391 0576  3C 02                      CMP     AL,02h              ;is it Tx hold Register Empty?
392 0578  74 23                      JZ      TX_HOLD_EMPTY       ;yes
393                              ;
394                              ;   It's a Receiver Line Status Interrupt.  Shouldn't happen,
395                              ;   but clear it out anyway just in case
396                              ;
397 057A  83 C2 05                   ADD     DX,5                ;point to Line Status Register
398 057D  EC                         IN      AL,DX               ;read to clear interrupt
399 057E  83 EA 05                   SUB     DX,5                ;restore base address
400 0581  EB E0                      JMP     IRQ_XIT             ;see if another IRQ waiting
401                              ;
402 0583                         IRQXIT1:
403 0583  1F                         POP     DS                  ;retore interrupted registers
404 0584  5E                         POP     SI
405 0585  5A                         POP     DX
406 0586  5B                         POP     BX
407 0587  58                         POP     AX
408 0588  CF                         IRET                        ;return to interrupted program
409                              ;
410                              ;-------------------------------------------------------------
411                              ;   This was a Modem Status interrupt.  One of the signals
412                              ;   CD, CTS, DSR, or RI changed state.  Only one we are
413                              ;   handling here is CTS.  If it came high, we may need to
414                              ;   restart output if it was stopped by CTS low.
415                              ;
416                              ;   DX = port base address, SI = port word index
417                              ;
418 0589                         MODEM_STATUS:
419 0589  83 C2 06                   ADD     DX,6                ;point to Modem Status Register
420 058C  EC                         IN      AL,DX               ;fetch it
421 058D  83 EA 06                   SUB     DX,6                ;restore port base address
422 0590  D1 EE                      SHR     SI,1                ;make byte index
423 0592  88 84 038E R               MOV     [MODEM_SR][SI],AL   ;save status register
424 0596  D1 E6                      SHL     SI,1                ;make word index again
425 0598  E8 062A R                  CALL    START_XMIT          ;restart output if CTS cleared
426 059B  EB C6                      JMP     IRQ_XIT             ;done with this interrupt
427                              ;
428                              ;-------------------------------------------------------------
429                              ;   Got Transmit Hold Register Empty Interrupt.  Send
430                              ;   next character if there is one.
431                              ;
432                              ;   SI = word index for port, DX = port base address
433                              ;
434 059D                         TX_HOLD_EMPTY:
435 059D  D1 EE                      SHR     SI,1                ;make byte index
436 059F  F6 84 038E R 10            TEST    [MODEM_SR][SI],10h  ;Is CTS high?
437 05A4  74 22                      JZ      TX_HALT             ;no, stop output
438 05A6  D1 E6                      SHL     SI,1                ;yes, restore word index
439                              ;
440                              ;   It's ok to keep sending.  See if anything to send
441                              ;
442 05A8  8B 9C 0378 R               MOV     BX,[OB_EPTR][SI]    ;fetch extraction pointer
443 05AC  39 9C 0374 R               CMP     [OB_IPTR][SI],BX    ;anything in buffer to send?
444 05B0  74 14                      JZ      TX_DONE             ;no, all done
445                              ;
446                              ;   There is one to send, fetch it and put it in the UART
447                              ;
448 05B2  8A 07                      MOV     AL,[BX]             ;yes, fetch next character
449 05B4  43                         INC     BX                  ;bump extraction pointer
450 05B5  39 9C 0370 R               CMP     [OB_END][SI],BX     ;at end?
451 05B9  75 04                      JNZ     TXHE1               ;no
```

```
452 05BB  8B 9C 036C R            MOV     BX,[OB_START][SI]      ;yes, "turn corner"
453 05BF  89 9C 0378 R    TXHE1:  MOV     [OB_EPTR][SI],BX       ;save updated extraction pointer
454                       ;
455 05C3  EE                      OUT     DX,AL                  ;put char in transmitter of UART
456 05C4  EB 9D                   JMP     IRQ_XIT                ;check next interrupt
457                       ;
458                       ;    Indicate we quit sending output at this level
459                       ;
460 05C6                  TX_DONE:
461 05C6  D1 EE                   SHR     SI,1                   ;make byte index
462 05C8                  TX_HALT:
463 05C8  C6 84 038C R 00         MOV     [XMIT_ACT][SI],0       ;indicate output stopped
464 05CD  D1 E6                   SHL     SI,1                   ;make word index
465 05CF  EB 92                   JMP     IRQ_XIT                ;check next interrupt
466                       ;
467                       ;----------------------------------------------------------------
468                       ;  This is a Receive Data Available interrupt.  Read the character
469                       ;  from the UART and place it in the buffer.  If the buffer is
470                       ;  getting full (less than ten bytes free) then lower RTS to do
471                       ;  hardware handshake to prevent overrun.
472                       ;
473                       ;    DX = port base address, SI = port word index
474                       ;
475 05D1                  RECV_DATA:
476 05D1  EC                      IN      AL,DX                  ;read character from UART
477 05D2  8B 9C 0364 R            MOV     BX,[IB_IPTR][SI]       ;get insertion pointer
478 05D6  53                      PUSH    BX                     ;save 'store' pointer
479 05D7  43                      INC     BX                     ;bump insertion pointer
480 05D8  39 9C 0360 R            CMP     [IB_END][SI],BX        ;at end of buffer?
481 05DC  75 04                   JNZ     RECD1                  ;no
482 05DE  8B 9C 035C R            MOV     BX,[IB_START][SI]      ;yes, "turn corner"
483 05E2  39 9C 0368 R    RECD1:  CMP     [IB_EPTR][SI],BX       ;is buffer full?
484 05E6  74 3E                   JZ      RXDOVF                 ;yes, we were overrun
485 05E8  89 9C 0364 R            MOV     [IB_IPTR][SI],BX       ;update insertion pointer
486 05EC  2B 9C 0368 R            SUB     BX,[IB_EPTR][SI]       ;calculate number of free bytes
487 05F0  79 04                   JNS     RECD2                  ;BX = # used bytes
488 05F2  81 C3 00C8              ADD     BX,IN_SIZE             ;correct for backwards wrapping
489 05F6  8B D3           RECD2:  MOV     DX,BX                  ;save amount of used space in DX
490 05F8  5B                      POP     BX                     ;get store pointer
491 05F9  88 07                   MOV     [BX],AL                ;store the new character
492 05FB  B8 00C8                 MOV     AX,IN_SIZE             ;max # chars in output buffer
493 05FE  2B C2                   SUB     AX,DX                  ;# free characters to AX
494 0600  8B 94 037C R            MOV     DX,[COMADR][SI]        ;restore DX=base address
495 0604  D1 EE                   SHR     SI,1                   ;make byte index
496 0606  80 BC 0390 R 00         CMP     [INPUT_FLOW][SI],0     ;have we already lowered RTS?
497 060B  75 14                   JNZ     RECD3                  ;yes
498 060D  3D 000A                 CMP     AX,10                  ;are there at least 10 chars left?
499 0610  77 0F                   JA      RECD3                  ;yes, all is well
500 0612  83 C2 04                ADD     DX,4                   ;no, point to modem control register
501 0615  EC                      IN      AL,DX                  ;read MCR from UART
502 0616  24 FD                   AND     AL,0FDh                ;lower RTS signal
503 0618  EE                      OUT     DX,AL                  ;put back in MCR
504 0619  83 EA 04                SUB     DX,4                   ;restore base address to DX
505 061C  C6 84 0390 R 01         MOV     [INPUT_FLOW][SI],1     ;indicate we lowered RTS
506 0621  D1 E6           RECD3:  SHL     SI,1                   ;restore word index
507 0623  E9 0563 R               JMP     IRQ_XIT                ;process next interrupt
508                       ;
509                       ;  Input was overrun.  Apparently other end isn't listening to
510                       ;  our RTS indicator.  All we can do is ignore data until we
511                       ;  have space again.
512                       ;
513 0626  5B              RXDOVF: POP     BX                     ;even stack
514 0627  E9 0563 R               JMP     IRQ_XIT                ;check next interrupt
515                       ;
516                       ;----------------------------------------------------------------
517                       ;    START_XMIT
518                       ;
519                       ;    This routine will start output if it is not now active,
520                       ;    If CTS allows it, and if there is data waiting to be sent.
521                       ;
522                       ;    On entry, SI = word index of port, DX = port base address
523                       ;
```

```
524 062A                     START_XMIT:
525 062A 50                          PUSH    AX                      ;save AX & BX
526 062B 53                          PUSH    BX
527                          ;
528                          ;    See if output already running, or if CTS is holding it off
529                          ;
530 062C D1 EE                       SHR     SI,1                    ;make byte index
531 062E 80 BC 038C R 00             CMP     [XMIT_ACT][SI],0        ;is output running?
532 0633 75 2C                       JNZ     SXM1                    ;yes, all done
533 0635 F6 84 038E R 10             TEST    [MODEM_SR][SI],10h      ;does CTS allow output now?
534 063A 74 25                       JZ      SXM1                    ;no, all done
535                          ;
536                          ;    Ok to start output, is there anything to send?
537                          ;
538 063C D1 E6                       SHL     SI,1                    ;make work index again
539 063E 8B 9C 0378 R               MOV     BX,[OB_EPTR][SI]        ;get extraction pointer
540 0642 39 9C 0374 R               CMP     [OB_IPTR][SI],BX        ;anything waiting to go?
541 0646 74 1B                       JZ      SXM2                    ;no - all done
542                          ;
543                          ;    Something to send.  Fetch it and send it to UART while
544                          ;    indicating output is now started.
545                          ;
546 0648 8A 07                       MOV     AL,[BX]                 ;fetch char to send
547 064A 43                          INC     BX                      ;bump extraction pointer
548 064B 39 9C 0370 R               CMP     [OB_END][SI],BX         ;at end?
549 064F 75 04                       JNZ     SXM3                    ;no
550 0651 8B 9C 036C R               MOV     BX,[OB_START][SI]       ;yes, "turn corner"
551 0655 89 9C 0378 R        SXM3:   MOV     [OB_EPTR][SI],BX        ;update pointer
552 0659 D1 EE                       SHR     SI,1                    ;make byte pointer
553 065B C6 84 038C R 01             MOV     [XMIT_ACT][SI],1        ;indicate output is running
554 0660 EE                          OUT     DX,AL                   ;send character to UART Tx Hold Reg
555                          ;
556 0661 D1 E6               SXM1:   SHL     SI,1                    ;restore word index
557 0663 5B                  SXM2:   POP     BX                      ;restore registers
558 0664 58                          POP     AX
559 0665 C3                          RET                             ;exit
560                          ;
561                                  PAGE
562                          ;-----------------------------------------------------------------
563                          ;
564                          ;    Initial Program Load Entry Point
565                          ;
566                          ;    Connect to INT 14 and then exit as a TSR to remain resident
567                          ;
568 0666 BA 03A2 R          START:  MOV     DX,OFFSET INT_14        ;DS:DX = Handler Address
569 0669 B8 2514                     MOV     AX,2514h                ;replace INT 14 vector
570 066C CD 21                       INT     21h
571                          ;
572 066E BA 0688 R                   MOV     DX,OFFSET HDR_MSG
573 0671 B4 09                       MOV     AH,9
574 0673 CD 21                       INT     21h                     ;print installation message
575                          ;
576 0675 BA 0665 R                   MOV     DX,OFFSET START-1       ;point to end of resident code
577 0678 83 C2 0F                    ADD     DX,15                   ;round up to paragraph size
578 067B D1 EA                       SHR     DX,1
579 067D D1 EA                       SHR     DX,1
580 067F D1 EA                       SHR     DX,1
581 0681 D1 EA                       SHR     DX,1                    ;divide by 16 to get # paras
582 0683 B8 3100                     MOV     AX,3100h                ;TSR function, code=0
583 0686 CD 21                       INT     21h                     ;exit - we are installed
584                          ;
585 0688 0D 0A 2E 2E 2E 20   HDR_MSG DB      0Dh,0Ah,'... INT 14 Handler Installed',0Dh,0Ah,'$'
586      49 4E 54 20 31 34
587      20 48 61 6E 64 6C
588      65 72 20 49 6E 73
589      74 61 6C 6C 65 64
590      0D 0A 24
591                          ;
592 06A9               MAIN    ENDP
593 06A9               CODE    ENDS
594                            END     MAIN
```

Chapter Summary

In this chapter, you have learned the basics of programming an IBM-compatible serial port. Such programming is an advanced task, but isn't as difficult as is often made out to be. If you are a programmer who needs to write serial I/O applications, this information should be enough to enable you to build a working application.

This chapter has developed and fully explained a replacement BIOS INT 14 handler that is adequate for many serial applications. In the process, you have learned how to program the serial port and interrupts and how to handle flow control. You also have learned how the 16550 UART supports much higher speeds without the loss of input when the computer is doing other operations at the same time.

Part IV

Using Modems

20

Learning Modem Fundamentals

The task a modem performs is conceptually simple to understand: it converts computer serial signals to sound and back so that they can be sent over telephone circuits, which are designed for sound. The details, however, are more complicated, and you need to understand how modems operate in order to make an informed selection and install one correctly. If you have trouble with your modem and need to diagnose and fix problems, you certainly need to understand how a modem functions.

Chapter 2 introduced you to the modem—what it is, what it does, and the terminology associated with modems. This chapter explains in detail how today's basic modems work. (These modems operate at speeds up to 2400 bps.) You learn how a modem operates, how to install and set up a modem, and how to troubleshoot your setup if you have problems. Modems with advanced features, such as error-correcting options, as well as high-speed modems that can transmit data at speeds up to 14,400 bps are discussed in the next chapter.

Understanding the Modem's Task

Although modems can be used on specially conditioned leased telephone lines, they are used more commonly on the dial-up voice channels of a normal telephone. These telephone channels are designed to handle the human voice, which primarily uses a 3000Hz band of frequencies, and therefore are guaranteed to only carry frequencies between 300Hz and 3300Hz.

The telephone voice channel is a very low fidelity circuit, but it works fine for its intended purpose—transmitting the human voice. A modem using this channel is limited, however, because it can transmit serial data only with signals that fit this spectrum space, known as *bandwidth*. A second disadvantage is that a telephone channel specification allows distortion that may not affect your ability to understand a voice but can severely affect an electronic signal's wave form.

Although specialized modem experimentation began much earlier, electronic technology first became capable of making a useful telephone modem in the 1960s. At that time, Bell Telephone still had a monopoly on all telephone communications in the United States. When modems first were developed, Bell set the standards. The first modem widely available for dial-up operation was the Bell 103 modem, which was available only from Bell and connected directly to the telephone line via a data set adapter that Bell also provided.

The Bell 103 modem transmitted serial data at speeds of more than 300 bps. Because the primary purpose of such modems was to connect mechanical printing terminals that couldn't operate this fast, these modems were quite satisfactory. The normal speed at that time was 110 bps, or 100 words per minute—quite fast for a typewriter.

Over time, printer technology advanced. IBM introduced the Selectric typewriter, which was 30 percent faster than previous mechanical designs because of its print ball design. IBM also introduced the 136.5-bps speed to connect these printing terminals to computers. Later printers were even faster, and 150-bps printing terminals were considered high speed. Finally, thermal printing techniques allowed much faster printing terminals, and 300 bps became the common high speed. The Bell 103 modem handled all these tasks well.

Other companies also made modems, but they avoided connecting them directly to the telephone lines. Instead, they developed *acoustic coupling*, a technology in which the modem contained a speaker and a microphone. These modems had rubber suction cups into which the telephone handset was placed so that the sound could be coupled to the telephone as well as possible.

A few companies began to make direct connect modems compatible with the Bell 103, but these modems had to be connected to the telephone line through a Data Access Arrangement (DAA) coupler provided by Bell. Only modems connected in this way could automatically answer the telephone.

With the deregulation of the telephone industry, the FCC permitted anyone to build a direct interface to the telephone system into their modem. This situation continues today, and very few acoustically coupled modems remain. This process has been repeated in most countries around the world.

The Bell 103 Standard

The Bell 103 modem transmitted digital information as sound waves through a method known as *frequency shift keying*, or *FSK*. In FSK, the modem generates two frequencies, one frequency representing a binary 0 and the other a binary 1.

Modems compatible with Bell 103 use a 200Hz frequency shift to show the difference between 1 and 0. In order to operate *full-duplex* (both directions at the same time), the available audio band is divided into two portions. The calling and answering modems use two different sets of frequencies, leading to a distinction between *originate* and *answer* modems. Some early modems handled only the originate function in order to save parts and lower costs. Today, all modems can serve as the originate or answer modem.

A Bell 103-compatible modem originating a call uses the frequency 1070Hz to represent 0, and 1270Hz (a 200Hz frequency shift) to represent 1. The modem answering uses the frequency 2025Hz to represent 0, and 2225Hz (again a 200Hz shift) to represent 1. If you listen to the telephone while two Bell 103-compatible modems are connected, you can clearly hear this frequency shift when you press a key on the terminal.

This relatively wide shift (200Hz) made design of the receiving filters fairly simple, and the modem was quite reliable at speeds up to about 450 bps. Thus, a Bell 103 modem could easily handle the speeds then in common use—110 bps, 136.5 bps, 150 bps, and 300 bps. Today the Bell 103 standard is still used in what are known as 300-baud modems.

The circuitry for generating this type of modulation is quite simple, requiring only two tone generators that can be turned on and off according to the value of the serial input signal. The circuitry to receive this signal also is fairly simple; it consists of two filters, each tuned to one of the two frequencies. The output of the two filters is hooked to a comparing circuit; the filter outputting the strongest signal switches the output to a 1 or a 0.

The CCITT V.21 Standard

In Europe, the CCITT (International Telegraph and Telephone Consultative Committee) also was developing a standard for an FSK modem. In the early 1960s, the telephone systems of the United States and Europe had little interaction, so the CCITT V.21 standard (V.21 is the name of the standard) adopted for 300-baud and slower modems is different from the Bell 103 standard. A Bell 103 modem, therefore, cannot communicate with a CCITT V.21 modem even though they use the same modulation techniques and operate at the same speeds. Most basic modems today, however, include Bell 103 and CCITT V.21 300 bps capabilities and can communicate regardless of their country of origin.

CCITT V.21 standard modems use a 100Hz frequency shift and keep both carriers in the lower portion of the audio spectrum. As with the Bell 103 modems, the originate modem uses one set of frequencies—1080Hz for 0 and 1180Hz for 1—and the answer modem another—1750Hz for 0 and 1850Hz for 1.

The smaller shift (100Hz) and closer frequency spacing require better filters than the Bell 103 modem's 200Hz shift, and some early CCITT V.21 modems could reliably handle speeds up to 200 bps only. As mentioned previously, the printing terminal technology in use in the early 1960s (when the V.21 standard was first approved) didn't require speeds faster than 200 bps. As electronic technology improved, however, these modems could easily handle 300 bps; today a V.21 modem is assumed to be a 300-baud modem.

Echo Suppressors

On long distance lines, the modem signal transmission is accompanied by a noticeable time delay that causes echoes distracting to persons using the telephone lines. In order to remove these echoes, the telephone system adds equipment to the circuit known as an *echo suppressor*.

Although echo suppressors make talking on the telephone comfortable for people, they cause problems for modems because they filter out parts of the signal. In order for modems to operate correctly, the CCITT adopted a V.25 standard that enables modems to switch off any echo suppressors on the line. Telephone companies around the world have incorporated this standard into their echo suppression equipment.

With the V.25 standard, if a modem transmits a 2100Hz tone for more than 2.5 seconds but less than 4 seconds, any echo suppressors in the link turn themselves off. Although this standard is not required for calls within the same country or over a short distance, it generally is required for international calls.

This 2100Hz tone precedes any other modem handshake (or synchronization) sequence and is part of what you may hear as a two-tone start-up when a modem answers.

1200-bps Modems

As computer technology developed, the CRT display terminal created a demand for a faster modem. The FSK modulation technique is limited by the voice telephone bandwidth to a speed of about 2000 bps, and the closer this limit is pushed, the more precise the filters must be. Some modems (such as the Bell 202) were made that used FSK to create a half-duplex 1200-bps modem which could send data in only one direction at a time. This approach was economical, but users have demanded full-duplex, high-speed operation.

Because FSK modulation cannot produce a high-speed, full-duplex modem that operates on a dial-up voice telephone circuit, designers turned to another modulation method—*phase shift keying*, or *PSK* (PSK is discussed in more depth in the next section). This modulation method enabled the designers to play mathematical tricks and force more data speed into a 600-baud data channel. Because two 600-baud data channels can fit comfortably within the 3000Hz bandwidth of a voice telephone circuit, this method provided faster data speeds.

PSK provides these higher data rates by defining each signal element phase change as more than a single bit. Although mathematically PSK allows very high data rates, in actual practice, the number of discrete phase changes that can be detected is limited. In the early 1970s, however, electronic technology had advanced enough that designers were able to make reasonably priced stable circuits that could detect four different phase states on a 600-baud channel.

These four states represent four *dibits*, or two-bit data groups: binary 00, binary 01, binary 10, and binary 11. Because these four states could be represented within a single signal element, these modems could send data at 1200 bps on a 600-baud channel, marking the point at which baud and bps were no longer the same on a modem carrier.

Although Racal-Vadic was the first company to develop and market a working 1200-bps modem in 1972, the power of Bell Telephone to set standards still was strong enough that their Bell 212A modem design became the U.S. standard in 1976.

In 1980, the CCITT adopted the V.22 standard, which is quite similar to the Bell 212A standard. Today most basic 1200-bps modems support both the Bell 212A and the CCITT V.22 standards and can communicate regardless of the country of origin.

Phase Shift Keying

Phase shift keying appears complicated if not somewhat magical in its capability to get more data speed than signaling rate. PSK, however, really isn't very difficult to understand.

A pure tone generates an electronic pattern known as a *sine wave*, shown in figure 20.1. You may have seen this shape on the screen of an oscilloscope in a science fiction movie. This shape is constant regardless of frequency or tone, but with higher tones, more cycles occur in a single second; with lower tones, fewer cycles occur.

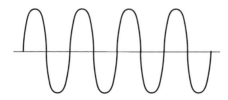

Fig. 20.1. *Electrical shape of a pure (single frequency) tone.*

Assuming the tone cycle begins when the wave form starts up from the center line, you can divide the wave into phases. Mathematically, these phases are marked in degrees, the same as angles within a circle. A single wave of 360 degrees, therefore, can be divided into four equal sections of 90 degrees each, as shown in figure 20.2.

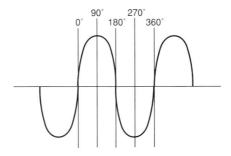

Fig. 20.2. The phase angles of a pure (single frequency) tone.

Phase shift keying then operates by switching the phase of the tone at the beginning of each signaling element. Figure 20.3 shows the wave form of dibit signaling transitions for each possible pair of bits. Notice that the phase of the wave form is shifted at the beginning of each signaling time element by omitting a portion of the cycle. The number of degrees omitted from the cycle is the phase shift angle. Signaling elements are equal time intervals. At 600 baud a signaling element is 1/600th of a second. By indicating two bits in each signaling element, a 1200 bps data rate is possible on a 600 baud channel.

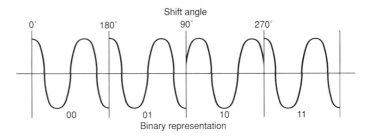

Fig. 20.3. Phase shift keying of a tone.

In theory, you can define as many different phase states as you want. The smaller the phase change, however, the more difficult it is to detect. A need exists, therefore, for more sophisticated filtering and sampling techniques.

The bandwidth of the channel as well as the noise or distortion that occurs are further limitations for phase shift keying. At higher frequencies and with a wider bandwidth, the signal can be divided more easily into more phases. Dividing a signal that fills the entire bandwidth of a voice telephone channel into 32 detectable phases is as far as present technology can go; this division is so close to the theoretical limits that it is very difficult to accomplish reliably.

Bell 212A/CCITT V.22 Operation

Bell 212A and CCITT V.22 modems use the same frequencies for originate and answer data channels. They also use the same general modulation methods, but V.22 modems also add a constant 1800Hz guard tone to the mix to prevent triggering telephone system control circuits that detect tones in the mid-range frequencies. These control circuits are rarely in use any more, but in some parts of the world, a 1200-bps modem call without this guard tone may be disconnected accidentally.

Because the modulation used by V.22 and Bell 212A modems is slightly different, they may not always be able to communicate. These two modem modulations, however, are much closer than the Bell 103 and CCITT V.21, and nearly all modems made to either standard accept the other as well. All modern basic modems incorporate circuitry to enable communication with either type of modem regardless of its country of origin.

Both types of modems use a 1200Hz frequency for the originate modem and a 2400Hz frequency for the answer modem. The signaling interval on each frequency is one 600th of a second, or 600 baud. Each signal interval is divided into four parts, and the phase shift from one interval to the next reflects a two-bit data combination, making the serial digital data rate 1200 bps.

In addition, the serial data is fed to a scrambler in order to ensure that the data patterns vary continuously. Phase shifts can be detected more reliably if the pattern varies constantly. This scrambling also spreads the power across the spectrum more evenly. After the carrier is modulated, therefore, you hear only a "whooshing" sound. You cannot hear the individual data stream as you can with an FSK modem.

When a 1200-bps modem answers the phone, it emits a series of tones before it connects and is ready to transfer data. These tones are present for several reasons. The following procedure explains the tones that are part of the 1200-bps modem connection.

1. When the modem first answers the telephone, it outputs an unscrambled binary 1 that sounds like a constant 2400Hz tone with a light buzz because the major frequencies lie between 2250Hz and 2550Hz. If the modem is a V.22 modem, an 1800Hz guard tone also is mixed with this tone. At the same time, the modem outputs the Data Set Ready (DSR) signal.

2. When the calling modem detects this tone (in about 200 milliseconds), it begins generating a constant binary 1 fed through the scrambler. This binary 1 is a whooshing tone with a pitch of about 1200Hz.

3. When the answering modem detects the calling modem's constant binary 1 (in about 250 milliseconds), the answering modem begins to modulate its 2400Hz tone with a scrambled binary 1 as well. After delaying 750ms to enable the calling modem to detect this change, the answering modem signals the assumed connection by raising the Carrier Detect (CD) and Clear To Send (CTS) signals.

4. When the calling modem detects that the answering modem has switched to a scrambled binary 1, it raises the Carrier Detect (CD) signal. The calling modem delays 750ms and then raises the Clear To Send (CTS) signal.

At this point, the modems are connected and may begin to send data normally. This connect sequence may have the V.25 echo suppressor defeat tone appended to the front of it as well, in which case a 2100Hz tone sounds for about three seconds before the sequence begins. An 1800Hz guard tone also may be appended if the modem is set to receive calls from certain countries that require this tone. If necessary, this guard tone follows the 2100Hz V.25 tone. As many as three distinct solid tones then may occur before the whooshing sound begins on a V.22 or Bell 212A connection.

If step 2 fails to occur, most modems drop back to check for a 300-bps connection. After a timeout period, the modem sends the 300-bps originate tone to try to initiate a 300-bps connection.

V.22bis 2400-bps Modems

By the early 1980s, CRT terminals and microcomputers, which could easily handle data at much higher speeds than 1200 bps, had become the most common terminal devices. In addition, microcomputer users were beginning to transfer data files with modems, increasing the demand for faster dial-up capabilities.

As electronic technology advanced, finer divisions of the PSK signal could be detected reliably. By 1985, filter technology was able to produce a modem that divided the PSK signal of the 1200-bps modems into eight portions instead of just four, and 2400-bps modems became available. This slight variation on the operation of the V.22 PSK modulation is called *quadrature amplitude modulation*, or QAM because amplitude modulation was added to the PSK modulation to make the phase states easier to detect.

Because of deregulation in the United States by this time as well as a greater interconnection of European and U.S. telephone systems, one standard was adopted for this new technology. All 2400-bps full-duplex modems use the CCITT V.22bis modulation standard.

The V.22bis standard is an enhancement of the V.22 CCITT standard 1200-bps design. (Standards use the Latin term *bis* and then *ter* to indicate significant enhancements in functionality.) V.22bis uses the same originate and answer frequencies (1200Hz and 2400Hz) and the same 600-baud signaling rate. The start-up sequence is modified slightly so that a V.22bis modem can detect whether it is being called by another V.22bis modem or a V.22 modem. This start-up sequence adjusts to 1200-bps V.22 operation if it cannot verify that the circuit or the modem can handle the smaller phase shifts of V.22bis.

The V.22bis answering sequence is a slight variation on the V.22 sequence described in the preceding section. Most V.22bis modems always send the V.25 echo suppressor override tone as well. The tones you hear when a 2400-bps modem answers the phone, therefore, are as follows:

1. The answering modem outputs the V.25 2100Hz tone for a period greater than 2.6 seconds and less than 4 seconds to cause the telephone company to turn off any echo suppressors.

2. The answering modem has a 75-millisecond silent period (+/– 20ms), after which it outputs an unscrambled 1200-bps binary 1 (like the V.22 modem) that sounds like a constant 2400Hz tone with a light "buzz" because the major frequencies lie between 2250Hz and 2550Hz. If the modem is a V.22 modem, an 1800Hz guard tone also is mixed with this tone. At the same time, the modem outputs the Data Set Ready (DSR) signal.

3. When the calling modem detects this tone (in about 200 milliseconds), it remains silent for another 450ms and then transmits the unscrambled double dibit pattern 00 and 11 at 1200 bps for 100 milliseconds. This pattern is the "I can do V.22bis—can you?"

announcement. After this 100ms transmission, a V.22bis modem then reverts to sending the constant binary 1 fed through the scrambler at 1200 bps, the same as a V.22 modem. This sounds like a whooshing tone with a pitch of about 1200Hz.

4. When the answering modem detects the 100-millisecond V.22bis pattern, it sends the same pattern for 100 milliseconds and then switches to a scrambled binary 1 at 2400 bps. After delaying 750ms to enable the calling modem to detect this change, the answering modem assumes it is connected and raises the Carrier Detect (CD) and Clear To Send (CTS) signals.

5. When the calling modem detects that the answering modem has switched to a 2400-bps scrambled binary 1, it raises the Carrier Detect (CD) signal. The calling modem then delays 600ms and switches to sending scrambled binary 1 at 2400 bps. After 200ms more delay, the calling modem assumes that the connection is complete and raises the Clear To Send (CTS) signal.

At this point, the modems are connected and may begin to send data normally. If the answering modem fails to detect the 100-millisecond special V.22bis sequence, it falls back to the V.22 1200-bps connection detailed previously. As with the V.22 sequence, if the answering modem fails to stimulate step 3 in V.22 or V.22bis form, it falls back and issues a 300-bps modem answer tone to try for a connection at that speed.

The V.22bis standard does not incorporate a method for one end of the connection to verify that the other end has fallen back to 1200bps. If the telephone line has exactly the correct distortion properties, one modem may decide that the connection is a V.22bis 2400-bps connection while the other may interpret it to be a V.22 1200-bps connection. The result is a continuous stream of garbage characters from both modems, and the only corrective action is to disconnect and redial the call. If this problem occurs repeatedly, you should investigate your telephone connection to be sure a local problem is not causing the distortion.

Because a V.22bis modem needs more exact wave form reproduction than slower modems to operate correctly, it requires a connection that is more closely equalized for frequency distortion. As a result, active *equalization*— in which the modems adjust an equalizer filter at the beginning of the connection to "tune" the line for the best frequency response—was introduced with the V.22bis modem.

Telephone lines rarely maintain the same electrical characteristics for many minutes or hours, however, so the concept of *retraining* also was introduced with the V.22bis modem. V.22bis modems constantly monitor themselves to determine how many errors or near errors are occurring. When one modem determines that the line quality has become unacceptable, it initiates a retrain sequence: the modem stops transmitting data and lowers CTS to indicate that it is retraining the line, signaling the remote modem to enter retrain with it. Together the two modems reequalize the line for the best overall response.

A V.22bis retrain sequence generally lasts about 1 second, after which CTS is restored and the flow of data through the modems resumes. Normally, the modem not initiating the retrain sequence receives a few garbage characters when the retrain occurs.

Selecting a Basic Modem

When you are selecting a basic 300-, 1200-, or 2400-bps modem, your first decision is whether to purchase an internal or external modem. Internal modems make a tidy installation because they fit inside the case of your computer. External modems, however, have signal indicator lights that can be very helpful in diagnosing incorrect operation. You can use an internal modem only in the type of computer for which it is made, but you can connect an external modem to any computer that has an RS-232 port. (RS-232 ports are explained in Chapter 2.)

Heat is the most common cause of electronic failure. No matter what modem you select, you should be sure it doesn't overheat. With an internal modem, you may have to leave a slot open to provide the modem a sufficient air flow. In extreme cases, you may have to add an additional cooling fan. With an external modem, you should leave adequate space around the modem to ensure proper ventilation. If you can, test the modems you are considering before you buy. If all other features are nearly equal and you detect a noticeable difference in how warm a particular modem runs, choose the cooler modem.

Almost all basic modems are of sufficient quality for everyday use. Integrated circuit technology has made basic modems reliable and widely available at very low cost. Price may well be the most important determining factor in your purchase. As with all electronic equipment, check the warranty and service arrangements, but if the retailer has a good return policy in the first few weeks, you should be fine. Modems tend to break in the first few weeks or not at all (unless you abuse them).

Installing and Setting Up a Modem

When installing an external modem, make sure that you have the correct cable. (You learned about cable signals in Chapter 2.) Make sure that you have all the wires your modem requires and that they are connected properly. Improper cables are a frequent cause of difficulty in modem installations.

Next, check any necessary switch or jumper settings. Your terminal software should indicate what settings are required, but if it doesn't, make sure that Carrier Detect and Data Terminal Ready are set to normal operation. Many modems are shipped from the factory with these signals set forced (always present regardless of true conditions). If your modem has Dip Switches, these signals are often on switches 1 and 6. Make sure these switches are set off and not on. Incorrect settings for these two signals are responsible for many installation problems.

Some modems have no switches or jumpers; configuration settings are kept in a special memory—*NOVRAM* or *nonvolatile RAM*—that is not lost when power is turned off. For these modems, you must use extended AT commands to set the software switches held in this memory. The standard command AT&C1&D2&W usually sets Carrier Detect and Data Terminal Ready to normal operation.

If you are installing an internal modem, make sure you set it up for the proper com port. The modem must not be set to the same address as any other serial port you have installed. Make sure that you know what port the modem is set to because you have to tell your software how to locate the port.

Many modems have two jacks for your telephone cord to plug into. On some modems, you can plug the cord from the wall jack into either plug, but on many modems, you must plug it into the proper jack, or the modem does not function properly. Double-check the labels on the jacks (or in the manual) to determine which jack the cord from the wall should plug into. Many modem installation problems result from the simple mistake of plugging the telephone cord from the wall into the wrong jack on the modem. Both jacks look the same, and the cord plugs into either one. They often are quite different electrically, however, and the modem doesn't work if the correct one is not used.

When you install an external modem, make sure that it gets adequate ventilation. The amount of ventilation required varies from modem to modem, but you should avoid covering the modem and blocking the air flow. You also should not locate an external modem near a television set, fan motor, or other source of electrical interference that can impede proper operation.

Testing and Troubleshooting Your Installation

One reason serial communications are considered difficult is that when something goes wrong, often one symptom only appears—nothing works at all. Because total failure is such a common symptom, the process can be frustrating and seem very difficult. You can successfully examine your setup and diagnose and repair problems in your communications installation, however, if you proceed slowly in a logical sequence.

Chapter 23 presents a detailed modem troubleshooting procedure, but in the next sections you examine the most common causes of difficulty with a modem setup. The following particular problems are addressed:

- The modem doesn't respond.
- The modem dials but doesn't connect.
- The modem connects but receives garbage characters.

The Modem Doesn't Respond

When your modem doesn't appear to respond at all, first make sure that all the cables are connected and that your modem is turned on. This advice may sound condescending, but even after 26 years of working with computer communications, I find myself wasting time because I have forgotten to turn the modem on. Every serial communications installation has several cables, and the power to an external modem usually is separate from the rest of the system. You can save a great deal of time and trouble by making this simple check of your connections first.

In the same spirit, make sure that you have the telephone cord to the wall plugged into the correct jack on the modem. If the modem has two jacks, one jack is for the cord to the wall and the other is for a telephone you can use when the modem is off. If you have reversed these connections, the modem may not work.

If you have absolutely no response to typing at the modem from your terminal program, type *ATE1Q0* to be sure that your modem is set to echo your input. If you still have no response, make sure that you told the software the correct com port to which the modem is connected (or set to emulate if you are using an internal modem). If you give the software incorrect information about the modem's electrical location, the modem will not operate properly. Again, this check is simple, but important.

If you see an echo from your modem when you type, try typing the letters *at* and pressing Enter. The modem should respond with OK or 0. If it doesn't respond with either of these codes, type *ATQ0* to make sure that your modem is set to give responses. If you still have no response, check the speed setting on your terminal program. If you set your terminal program's port speed faster than the modem's top speed, your modem cannot understand your commands or respond. If you set your terminal program's port speed to 2400 bps, for example, a 1200-bps modem cannot respond. Make sure that your port speed is set correctly.

If your terminal program's port speed is set correctly and the modem still doesn't respond, check for a jumper or switch setting with a label such as "Disable AT command set." Make sure this switch is set to enable AT command recognition; otherwise, your modem will not respond to commands. The purpose of this switch is to turn your smart modem into a dumb modem for special applications. For normal modem operation, however, you want the smart modem you paid for.

If you have made all the checks described in this section and your modem still doesn't respond properly, try to determine which part of your equipment is causing the problem by checking each part of the circuit you can see in a logical sequence to find out how far data gets before it fails.

If you have an external modem, look at the lights on the modem. When you activate your terminal software, the TR light should light up. When you deactivate your software, this light normally goes off. With most terminal programs, if you tell your terminal program to disconnect the phone (using the Alt-H command on many terminal programs), the TR light should flash off and back on. If the TR light passes this test, the terminal program has correctly located the modem's com port. You also have proved that the DTR signal wire in the cable between your modem and the serial port is correct.

If the TR light does not pass this test, check your serial port or cable for a problem. By far the most common problems in this area are a loose or incorrect cable or an incorrect com port address configured in the terminal program.

When the TR light operates correctly, observe the SD light as you type a few keys. The SD light should flash with each letter you type, indicating that code is being transmitted to the modem. If you do not see such a flash, check the cable wiring carefully. If the cable is wired correctly and the SD light still does not flash properly, make sure that you have correctly told your terminal program which interrupt the serial port (or internal modem) is set to use.

Next, observe the RD light. If it does not flash along with the SD light, enter the command *ATE1* and press Enter. This command tells your modem to echo all command characters you type to it. After you have entered this command, the RD light should flash along with the SD light. If you still don't see the characters you type printed back on your screen, check the cable to make sure that all the wires are good. If they are, check the interrupt setting on your com port. If you have chosen COM1, IRQ4 should be selected. If you have chosen COM2, IRQ3 should be selected.

If the interrupt is correct, look for other I/O cards that may be using the same interrupt. The IBM PC bus cannot share the same interrupt for more than a single I/O card. If you have a hard drive, network card, or tape unit that uses the same interrupt, you have to switch the com port to another setting. When troubleshooting, unplugging other I/O cards to see whether the problem goes away can be helpful. If removing other cards solves the problem, you almost certainly have a conflict in the I/O address or interrupt. The solution is to change the configuration of your com port.

The Modem Dials but Doesn't Connect

If the modem dials another modem but doesn't connect properly, look at the connection to the phone jack on the wall. Make sure that you don't have any other modems plugged into the same telephone circuit. Check especially for such problems as a laptop computer modem plugged in from another room. Two modems plugged into the same phone circuit often cause problems even if one is turned off.

If your modem's speaker isn't already on, turn it on by entering the ATM1 command. Then listen closely to the speaker when you dial. You first should hear a dial tone followed by the dialing. If you don't hear a dial tone, make sure that no other telephone is off the hook on this circuit. Also make sure that you have plugged the modem into the wall jack correctly and that you are using the proper telephone cord plug on your modem. Finally, verify that the telephone cord is plugged into the proper jack on your modem.

If you continue to hear a dial tone after dialing starts, you probably have selected the wrong dialing method. To select tone dialing, you use the string ATDT; to select pulse dialing, use the string ATDP in your terminal program Init strings.

If you hear crackling sounds while the modem is dialing, you probably have a loose wire somewhere in the phone circuit. Check the cable from your modem to the wall as well as all other telephones connected to this circuit. A bad cord anywhere on the circuit can cause trouble for a modem.

The Modem Connects but Gets Garbage

Modem noise is one of the most difficult problems to deal with. One of the reasons is that several conditions can look like noise which are not noise at all. If you have the improper speed or parity set, your screen can look like it is noisy. Usually in this case you never see anything that looks correct because of the incorrect settings. True line noise occurs in bursts. You see sections of correct data followed by bursts of incorrect data.

Another condition that can appear to be noise is caused by an incorrect terminal emulation setting (see Chapter 16). In this case, character sequences intended to control your display are not interpreted properly and display directly on your screen. They appear to be sequences of numbers or punctuation and can be mistaken for noise. Rule out these incorrect setups before you conclude that you truly have a noise problem.

What seems like line noise can have many causes. If you notice such noise on several different calls, you probably can correct much of it on your end. First check for loose cables as indicated in the preceding section; a loose cable is one of the most frequent causes for this type of problem. Also make sure that the phone cord you are using to connect the modem to the wall jack is in good condition. If possible, exchange it with another one to be certain.

Make sure that the modem isn't located near a color TV, fan motor, or other source of electrical noise. Electrical noise can easily enter a modem through the air and cause what seems to be line noise problems. The solution in this case is to reposition the modem.

Finally, check all other telephones that are hooked to the same line. Unplug them one by one to determine whether one is causing the problem. Sometimes telephones that work fine can cause distortions of the modem's signal, inducing apparent noise.

Chapter Summary

Basic modems can provide reliable low-cost connections at speeds up to 2400 bps. In this chapter, you learned how these modems work as well as how to install and operate them. You also learned some basic modem troubleshooting techniques. If you take care with the installation of these modems, they can give you many years of trouble-free service.

Today modems also are available with many advanced features. The next chapter examines such features as error correction, data compression, and modem speeds higher than 2400 bps.

21

Understanding Advanced Modems

I n Chapter 20 you learned about basic modems capable of speeds up to 2400 bps. Technology never stands still, however, and modems now are available with such advanced features as error correction and data compression. Although 2400 bps has been called high speed, today's high-speed modems are capable of raw transmission speeds up to 14,400 bps. Some of these modems cost much more than the simpler basic modems, but their prices are falling rapidly.

In this chapter, you learn what these advanced features do so that you can determine the features you may need. You learn about such capabilities as MNP, V.42, and V.42bis error correction and data compression. You also learn how to evaluate the added speed claims from data compression options and what impact these features have on installing and using advanced modems.

Understanding High-Speed Modems

One of the most exciting developments in modems is the very high speeds that now are possible. When you can transfer data at speeds up to 1800 characters per second, entire applications that are not reasonable at lower speeds become feasible.

To get a feeling for the difference such speeds make, consider the time involved to transfer a 100K computer data file that contains more than 15,000 typewritten words. To determine the length of time required to transfer this data file under ideal conditions, divide the number of characters in the file by the number of characters per second the modem can transfer, as in the following formula:

$$\text{Transfer Time (secs)} = \frac{\text{\# of characters in file}}{\text{modem speed (in chars/sec)}}$$

If your serial format settings are any of the most commonly used (8 data bits and no parity, or seven data bits with parity), a total of ten bits are in a character including the start and stop bits. The number of characters per second (*cps*) therefore is the serial bps rate divided by 10.

At 2400 bps (the top speed of the basic modems discussed in the preceding chapter), files transfer at a maximum of 240 cps. If this maximum rate is obtained, transferring the 100K file takes more than 7 minutes. At 9600 bps (the top speed of a V.32 modem), this same file requires about 1 minute 45 seconds for the transfer. At the top speed of a V.32bis modem (14,400 bps), this file takes just slightly more than 1 minute to transfer.

Similarly, a 1M file can be sent in 10 minutes or less with these high-speed modems. At 2400 bps, a 1M file requires more than 1 hour to send. You can easily see that this speed increase enables new applications for PC telecommunications that before simply were not feasible.

Understanding Proprietary High-Speed Modems

When V.32 full-duplex 9600-bps modems were still very expensive, manufacturers tried several ways to meet the growing demand for faster modem speeds at a reasonable price. Most high-speed modems produced during this period were half-duplex 9600- or 14,400-bps modems that slowed down considerably when used to send data in both directions at the same time. The protocols used by these modems also were proprietary to a specific manufacturer; that is, a modem made by one company could not communicate at high speed with a modem from another company. Most of these modems disappeared quickly after V.32 modems became available at a reasonable price.

Two technologies from this period remain important enough to consider, however: the U.S. Robotics HST (high speed technology) protocol and the Telebit PEP (packetized ensemble protocol). The HST protocol remains interesting primarily because it achieved very high market penetration and provided one-way speeds up to 14,400 bps. Special file-transfer protocols (for example, YMODEM-g and streaming ZMODEM) enabled HST modems to transfer files about 40 percent faster than V.32 9600-bps modems. In addition, the cost of the HST technology was significantly lower than V.32 technology until quite recently.

Because of its large installed base, the HST protocol likely will remain common for a year or two. But the availability of reasonably priced standard V.32bis modems that operate full-duplex at 14,400 bps means that HST eventually will be replaced with nonproprietary standard modems. This process most likely will take awhile because of the market position HST modems have achieved.

The Telebit PEP modulation method, however, may have a longer life. Unlike the other proprietary protocols, PEP is not a variation on the phase shift keying techniques discussed in Chapter 20. Instead, PEP modulation divides the 3000Hz telephone bandwidth into hundreds of very narrow channels only a few Hertz wide. The modems then constantly evaluate each channel and use only those channels that are providing good transmission. This technique makes the PEP extremely error resistant on very noisy telephone lines.

The PEP has a very long delay between sending and receiving data, which can be very annoying for a person trying to type through a PEP connection. This delay is not a problem, however, if the primary function is transferring data files. Because of its superior noise immunity, the PEP is likely to remain available for niche applications in which error resistance is essential.

In general, proprietary high-speed modems can communicate at high speed only with another modem of the same brand. With the exception of the PEP, the only reason to buy a proprietary high-speed modem is cost. The cost of V.32 and V.32bis modems is falling rapidly, however, and the compromise of paying less to get a modem that cannot communicate at high speed with most other modems is becoming less desirable.

Some proprietary high-speed modem manufacturers are making a last effort to obtain market share by selling modems at extremely low prices. Although one or more of these modems may attain a brief period of usefulness, the price of standard high-speed technology will continue to fall, so you should consider such a purchase very carefully. If the cost justifies using the proprietary modem for a short period of time (until the cost of standard modems drops further), then such a purchase still may make sense.

Understanding V.32 Modems

A V.32 modem is a 9600-bps modem built to comply with the CCITT V.32 standard. Any V.32 9600-bps modem can connect with and transfer data at 9600 bps with any other V.32 modem regardless of who manufactured it. This compatibility across brands is the purpose of industry standards and what sets V.32 modems apart from the proprietary high-speed modems discussed in the previous section. In addition to cross-brand compatibility, a V.32 modem allows full-speed 9600 bps transmission of serial data in both directions at the same time (*full duplex*), while the proprietary modems we looked at earlier can send data only at high speed in a single direction at a time. This section examines how a V.32 modem accomplishes this seemingly impossible feat.

In Chapter 20, you learned that 1200- and 2400-bps modems can send digital data at a faster speed than their signaling rate by using phase shift modulation. In order to produce full-duplex operation, these modems establish two different channels by dividing the available telephone bandwidth in half. This bandwidth (3000Hz in the range 300Hz to 3300Hz) is the limiting factor in modem speeds.

The development of filter technology made dividing and detecting the tone cycle possible. The amount of reliable division possible when two channels are active in this small frequency range, however, is restricted. As the limits are approached, the two channels begin to interfere with each other. The 2400-bps V.22bis modems generally represent the limit of this approach.

If a single tone is placed at 1800Hz (the exact center of the telephone frequency range) and data is sent in only one direction at a time, however, then high speeds become possible by slightly extending the technology for V.22bis 2400-bps modems. Now, instead of trying to cram the signal into a 600-baud channel (the maximum possible when the channel is divided in half), the entire bandwidth of the voice channel is available. When the other channel is not present, therefore, with good filters, a 2400-baud single channel that fits within the voice telephone band can be obtained.

When the same modulation techniques are applied to this 1800Hz tone as to the V.22bis 600-baud channel (covered in Chapter 20), a data transfer rate of 2400 baud times 4 phase states—9600 bps—becomes possible. Although these techniques appear productive, this scheme almost certainly results in a higher than desirable error rate because it is pushing the limits of the available bandwidth as well as the filter technology.

In order to reduce this higher error rate, the data to be sent can be encoded mathematically so that it contains extra bits. If these bits are constructed properly, the receiving end has enough redundant information to reconstruct the initial message if one bit error occurs in a group of bits. Line noise therefore does not cause data errors until two bits fail in the same character.

One such method of encoding data as it is modulated is known as *trellis coded modulation*, which can tolerate twice the noise power as quadrature amplitude modulation without errors. It also is much less susceptible to *impulse noise*, a quite common form of noise on a dial-up telephone circuit. In fact, with trellis coded modulation, better noise resistance can be achieved at the limits of the channel bandwidth than at 2400 bps with V.22bis modulation.

Although very high data transfer rates in one direction are possible, many users require a full-duplex modem with these high transfer rates in both directions at the same time. In the mid-1980s, engineers devised a way to overcome the problem of the limited bandwidth in the telephone channel.

When a modem's receiver listens to the telephone channel, it hears the combination of the tones it is sending and receiving. Because the modem knows exactly what tones it is sending, finding a way to subtract those tones from the combined signal would enable the receiver to determine exactly what tones the other end is sending even if they are on the same frequency. This process would enable both modems to use the same frequency at the same time without compromising the data.

The reality of a voice telephone circuit, however, does not make this process easy. Part of the signal sent by the telephone is reflected back so that it appears to come from the other modem. In order to determine accurately what portion of the combined signal the receiver is picking up comes from the other modem, complex filtering and time-shifting techniques are required. In the 1980's, digital signal-processing microcomputers made producing circuitry capable of handling these conditions technically possible.

Digital signal processors (DSP) are special microcomputers that use the active processing power of a computer to do extremely sophisticated filtering. Digital signal processing enables very accurate filters to be constructed with far fewer parts than the previous passive filter circuitry required. But a far more important implication of digital signal processing is that it allows realistic circuitry to implement very sophisticated mathematical filtering models that previously could not be built or were not stable enough to use outside of a laboratory.

The sounds a V.32 modem makes when connecting are quite complicated. In fact, a V.32 connection takes several seconds to complete. The following description explains what takes place when a V.32 modem connects:

1. The answering modem outputs the V.25 2100Hz tone to disable any echo suppressors on the line. Unlike the basic modems discussed in the last chapter, however, you hear a "clicking" on this tone, indicating that the phase of the 2100Hz tone is reversed every 450ms to inform the telephone network to disable any echo cancellation equipment along with the echo suppressors. V.32 modems do their own echo cancellation.

2. The calling V.32 modem does not wait for the end of the 2100Hz tone. As soon as the calling modem detects the tone, it outputs a constant 1800Hz tone. By sending this tone early (during the V.25 tone), the calling modem announces to the answering modem that it also is a V.32 modem.

3. When the answering modem completes the V.25 2100Hz tone (a period of 3.3 seconds +/– 0.7 seconds), it tries to connect as a V.32 modem by sending a combined 3000Hz and 600hz tone. About every 30 milliseconds, the answering modem reverses the phase of this tone combination.

4. When the calling modem detects this phase reversal, it reverses the phase of its 1800Hz tone to signal the answering modem that it knows the modems are timing the echos.

5. When the answering modem detects the calling modem reversing the phase of its 1800Hz tone, the answering modem immediately reverses the phase of its two-tone signal again. This "handshake" of phase reversal of the two signals enables the two modems to time the propagation delay of the telephone circuit accurately. *(Propagation delay* is the time it takes for the sound to travel from one modem to the other and back.) This knowledge is essential for the modem's signal processor to be able to account for and exactly cancel out the echoes of its own tones later.

6. After this stage is complete, the modems then do a "flip-flop" exchange of full-spectrum modulated signals in order to train their adaptive line equalizers. During this process, each modem adjusts its equalizing filters while the other modem sends a signal for a period of 600 milliseconds to 3 seconds. In this sequence, the answering modem transmits first and then goes silent.

7. The originating modem responds with a similar signal but then leaves its signal on while the answering modem responds one more time. During this final exchange, the modems negotiate the line quality; they both now are using the full-audio spectrum simultaneously. At this point, they determine whether they can reliably attain 9600-bps transmission or whether they must fall back to 4800 bps.

8. After this negotiation is complete, the modems switch to sending scrambled binary 1s for 50 milliseconds. They are then ready to transmit data normally.

   During step 3, if the answering modem doesn't hear the 1800Hz tone from the calling modem, it tries for about three seconds to connect as a V.22bis modem (see Chapter 20 for details of this sequence). If it doesn't stimulate a V.22 or V.22bis response, it repeats the attempt to connect as a V.32 modem in case the calling modem failed to detect the first attempt. If this second attempt fails, the answering modem falls back to attempt a connection at 300 bps.

The CCITT V.32 standard uses these techniques to produce a full-duplex 9600-bps modem. Such modems have been available for quite awhile, but originally they cost as much as $10,000 because of the complex electronic tasks they perform. As integrated circuit technology has progressed, these modems have become available for well under $1,000, and the price is falling rapidly.

Understanding V.32bis Modems

In 1990, electronic technology advanced again, and digital filtering techniques improved enough at a reasonable price that even faster modems were possible. A modified version of the CCITT V.32 standard called V.32bis was produced; this version divided the phase states again, creating 50 percent more discrete elements. This increase enabled the base data rate to advance to 14,400 bps full-duplex.

A V.32bis modem is capable of connecting to any V.32 modem at 9600 bps full-duplex or to any V.32bis modem at 14,400 bps. In addition, all modern V.32 and V.32bis modems can connect to 300-, 1200-, and 2400-bps modems. Today's V.32bis modems provide very reliable high-speed connections, and their price also is falling rapidly.

Understanding the Limits of Future Speeds

In the near future, a V.32ter standard may become available that will allow speeds up to 19,200 bps. This speed is nearly the theoretical limit of modem speeds on the available bandwidth of voice telephone lines. Any additional speed increases will have to occur from new data compression techniques or through a theoretical breakthrough that creates a totally new approach to modem technology.

Understanding Error-Free Modems

The modulation protocols discussed in this chapter and Chapter 20 (Bell 103, Bell 212A, V.21, V.22, V.22bis, V.32, and V.32bis) specify the modulation scheme a modem uses to transfer data at a certain speed. They also specify the connection procedure so that modems can negotiate the highest speed they share in common.

Modems also use other protocols that can be added on top of any of the speed protocols. The two most commonly used error-correcting protocols are MNP (Microcom Networking Protocol) and V.42. These protocols provide error-free transmission and guarantee that the data sent to one modem emerges from the other modem unchanged because of line noise or distortion.

You should not confuse error-correcting protocols with the basic modem speed protocols. Any error-correcting protocol may be added to any modem speed protocol; therefore, any combination of 1200-, 2400-, or 9600-bps modems and MNP or V.42 is possible. When comparing modems, you should evaluate any error-correcting protocol a modem provides separately from its speed protocol.

MNP Error-Correcting Protocols

In the 1980's, the Microcom modem company developed a set of modem-to-modem protocols called the Microcom Networking Protocol (MNP). Microcom initially licensed these protocols for a fee to other modem producers. Because of the cost, however, few manufacturers adopted the protocols. In an effort to make these protocols industry standards, Microcom eventually released many of them into the public domain. Because these protocols enhanced the operation of the higher speed modems, other modem manufacturers then began to use several of the MNP protocols (known as levels).

The MNP protocols take advantage of the fact that modern modems have a computer inside them. They add a layer of software inside the modem that provides error-free transfer by error-checking the data transmission between the modems. If any data transmission has an error, the receiving modem requests a retransmission of the bad information and sends only good information from the modem's serial port. Because this process takes place inside the modems themselves, it is transparent to the modem user. In many ways, this operation is similar to the file-transfer protocols examined in Chapter 3, except that it occurs inside the modem, not in the terminal program software.

As the error-correcting portion of the MNP protocols developed, they were assigned different levels. MNP levels 1, 2, and 3 are the basic error-correcting protocols.

MNP level 1 often is referred to as Block mode. It is similar in concept to the XMODEM file-transfer protocol discussed in Chapter 3. In fact, MNP level 1 originally was intended for use by software outside of the modem. In MNP 1, the data is placed in packets with a numbered header and a mathematical CRC (cyclic redundancy check) verification. As each block of data is sent, the receiving modem calculates the CRC of the data received, verifying that it is correct. If the CRC matches, the receiving modem sends an acknowledgment, and the sending modem sends the next packet of data. If the packet has an error, the receiving modem indicates the error, and the sending modem sends the packet again.

Because the sending modem waits between each block of data for an acknowledgment or retransmission request, this protocol is a half-duplex protocol. The verification process reduces the overall speed of the modem to about 70 percent of its original value. As a result, MNP 1 rarely is used today, and many modems no longer support it.

MNP level 2 often is called Stream mode. It still uses the modem's asynchronous channel to package data in the same way as MNP 1, but it provides overlap by sending several packets ahead and then waiting for the acknowledgment or retransmission request. This type of operation is called *sliding windows*, and it makes MNP 2 a full-duplex protocol. Because it is full-duplex, MNP 2 reduces the modem speed much less than MNP 1. MNP 2 provides about 84 percent of the original modem speed while assuring error-free transmission. Thus, a 2400-bps modem using only MNP 2 can send data at about 202 cps.

Most modems today support MNP 2 internally even though they rarely use it. Its presence does enable software on one end of the connection to simulate an MNP modem with a basic modem that doesn't have MNP internally. This type of software was popular when MNP modems were very expensive, but the use of such software has decreased dramatically.

MNP level 3 incorporates level 2 but is more efficient. MNP 3 makes use of the fact that V.22, V.22bis, and V.32 modems are inherently synchronous modems even though they may be used in an asynchronous mode. With MNP 3, the sending modem strips off the start and stop bits from the asynchronous data sent to the modem and instead packages "raw bits" in larger data packets that are delimited by special header information.

As Chapter 1 explained, the start and stop bits are 2 of the 10 bits in a character. Removing these 2 bits results in a 20 percent speed increase that recovers the overhead of the extra characters in the packet headers and CRC. Thus, MNP 3 actually speeds up a modem to 108 percent of its original speed so that a 2400-bps modem can deliver data at speeds up to 254 cps.

MNP level 4 carries the improvements of MNP 3 even further, taking two separate steps to achieve greater speeds. First, MNP 4 streamlines the packet header information to eliminate some of the *protocol overhead* (extra data sent by the protocol that is not part of the user's data). Second, it sends larger packets than MNP 3, making the added protocol characters a smaller percentage of the total characters sent and thus slowing transmission less.

The use of large packet sizes to reduce protocol overhead has one drawback—larger packets take longer to resend when an error occurs. To minimize this disadvantage, MNP 4 adjusts the packet size according to the error rate it is correcting currently. MNP 4 theoretically can improve speeds by as much as 50 percent in some cases, but on average it provides a speed about 20 percent faster than the original modem speed. A 2400-bps modem with MNP 4 therefore can transmit data at average speeds of more than 270 cps, and a 9600-bps modem with MNP 4 can transmit data at speeds of more than 1100 cps.

The MNP protocol begins operation after the modem carriers connect normally. After it makes the carrier connection, an MNP modem sends a link request negotiation packet indicating that it is capable of MNP operation. This packet indicates the highest of the first three MNP levels (1, 2, or 3) the modem can support. It also carries information about any other MNP protocols the modem supports.

If the other modem can do MNP, it responds with a negotiation packet of its own, and the modems determine which level to use. If they select MNP 3, the modems then check whether they both support MNP 4. If they do, the MNP 4 protocol is used for error correction.

If an MNP modem connects with a basic (non-MNP) modem, this negotiation packet appears as normal data output from the non-MNP modem. The MNP negotiation packets begin with the ASCII SYN character (0x16) and can be filtered out by software. If they are not filtered, you see them on the screen of your terminal program as data.

Because an MNP modem can connect with a non-MNP modem, the negotiation sequence stops if a proper answer packet is not received within about four seconds. The modem then drops back to non-MNP operation unless you have programmed it to accept only MNP connections, in which case it disconnects.

The CCITT V.42 Standard

Even though MNP level 4 has been widely implemented, it has had some significant performance problems. When the CCITT decided to produce an error-correcting modem protocol standard, it devised a different protocol known as V.42. As a response to the large installed base of MNP modems that existed at the time the V.42 standard was written, MNP levels 2, 3, and 4 are included in V.42 as an alternate protocol so that V.42 modems can establish error-free connections with MNP modems.

V.42 has its own error-correcting protocol called LAPM (Link Access Procedure for Modems), which performs better than MNP. LAPM, like MNP, copes with phone line impairments by automatically retransmitting data corrupted during transmission, thus assuring that only error-free data passes through the modems. The LAPM protocol provides higher average speeds, however, as well as much smoother operation. The smoothness of the operation is very important when you are using a modem interactively to type to a remote computer and wait for a response. This effect is discussed in detail later in this chapter under the section "Understanding Modem Buffering."

As with MNP, the V.42 protocol begins after the modem carriers connect normally. The originating V.42 modem sends out a stream of XON characters of alternating parity (0x11, 0x91) separated by 8 to 16 mark idle bits (1 bits) for about 750 milliseconds. During this time, the originating modem listens for a response from the answering modem. As with the MNP negotiation sequence, these characters appear as output at the serial port of a receiving modem that does not support V.42.

If no response to the V.42 negotiation is received, the V.42 answering modem sends MNP link request packets and attempts an MNP connection as detailed previously (unless you have programmed it not to do so). If these requests receive no response, the modem falls back to a normal connection without error correction (unless you have programmed it not to).

When you are setting up a V.42 modem, you usually have the option of programming any sequence of operations you want. You can select only V.42 or normal, only MNP or normal, only V.42 or MNP, only V.42, or only MNP operation. If you have restricted the modem to certain modes of operation, it disconnects when it fails to establish a connection in one of the allowable modes.

Understanding Data Compression Modems

After protocols were in place between modems, engineers realized that they could make modems appear faster than they actually were through data compression. You may be familiar with data compression because of its use in archive programs such as ARC or ZIP. If a modem can compress the data it is given as input and then decompress the data before outputting it, then fewer bits need to be sent between the modems. This technique makes the modems appear faster because they can send uncompressed data at a higher rate.

MNP Data Compression Protocols

The first data compression protocol to appear was from Microcom and is called MNP level 5. MNP 5 compresses data by detecting redundant units of data and encoding them in shorter units of fewer bits. Examples include letters used more frequently than others (such as *e* and *s*) and repeated groups of letters. On ASCII text data, MNP 5 can provide a maximum increase of up to 100 percent in throughput. Many 2400-bps modem manufacturers therefore advertise their modems as capable of 4800 bps if they support MNP 5. This claim, however, is true only if you are transferring uncompressed ASCII text files written in English, or database files with long strings of repeated characters such as spaces.

The actual performance of MNP 5 depends on the type of data being sent. Text files provide the highest increase in speed, but program files cannot be compressed as much, resulting in a smaller increase. On precompressed data (files already compressed with ARC, PKZIP, or other compression software), MNP 5 actually can expand the data, resulting in a decrease in performance. For this reason, MNP 5 often is disabled on BBSs because most files being transferred actually transfer faster without MNP 5 because they already have been compressed with ZIP, ARC, ARJ, and so on.

Microcom also developed MNP 7, which provides better compression than MNP 5. Whereas MNP 5 has a maximum theoretical compression speed gain of 200 percent, MNP 7 can obtain almost 300 percent improvement in speed with the proper data. MNP 7, however, requires a much more powerful computer inside the modem and usually is inferior to data compression protocols devised more recently. MNP 7, therefore, is not likely to proliferate beyond those modems that currently have it.

The CCITT V.42bis Standard

By the time the CCITT developed a standard for modem data compression, it had the benefit of observing how MNP 5 and other data compression schemes had worked in the real world. As a result, the CCITT V.42bis data compression standard employs an algorithm that requires little computer power but theoretically can compress data enough to increase the effective speed by almost 400 percent.

A second advantage of V.42bis is that it compresses data only when it can gain speed from doing so. Each block of data is analyzed, and if compression will increase the speed of transferring that data, compression is enabled; otherwise, the data is not compressed. Because files on BBSs or commercial data services already are compressed (using ARC, PKZIP, and similar programs), speeds with MNP 5 actually can decrease throughput on this type of data. V.42bis, however, does not result in lower speeds because compression is added only when a benefit is realized.

As mentioned earlier, the V.42 error-correcting protocol includes MNP as an alternative for compatibility. Because V.42bis data compression can be used only if the modem also supports the V.42 error correction protocol, however, V.42bis compression is not compatible with MNP 5 compression. Therefore, a V.42bis modem that does not also support MNP 5 compression still can establish an MNP 4 error-free connection with MNP-capable modems (because V.42bis includes V.42, which includes MNP 2, 3, and 4 as alternate protocols).

Understanding Modem Buffering

Buffering is the use of memory to hold one or more blocks of data for processing. In a transmission stream this processing may be protocol related, a change of speeds, or both. A memory buffer can accept input data at one speed and send the data out at a different speed. A memory buffer also can accept input data in one format and send it out in another format.

When MNP or V.42/V.42bis protocols are added to a modem, the modem must be buffered. The sending modem must analyze the data you want to send and turn it into packets. The receiving modem then unpacks the data into its internal buffers and sends it to the serial port.

In addition, if you want to take advantage of the increased speeds that data compression offers, you have to feed the data to the modem at a faster rate than the modem's carrier speed. Because data compression is variable, the best throughput is obtained by force-feeding the modem so that it can "throttle" the flow of data to the optimum speed. Buffers also may be needed with error-correcting protocols, which may experience delays when they are hiding a high telephone line error rate.

In order to exploit the advantages of error-correcting and data-compression protocols, you must set the computer's serial port rate higher than the modem rate and use hardware (RTS/CTS) flow control (explained in Chapter 3). The modem's buffers can accept data at a higher rate than the modem can transmit data and can use the CTS flow control to prevent the computer from sending more data than the buffers can hold.

Figure 21.1 illustrates this buffering process. After a block of data is in the modem's input buffer, the computer in the modem then can compress the data and place the compressed data in error-correcting protocol packets. These packets then are sent between the modems and received in the other modem's memory buffers. The receiving modem can expand the data to its original size and send it at the faster port speed to the other computer's serial port. This results in the fastest possible data speed between the computer ports.

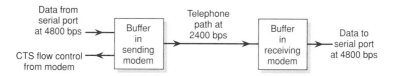

Fig. 21.1. *The buffering process.*

Because compression gains vary based on the actual data being transmitted, some data blocks will compress more than others and therefore will transmit faster. The buffering process enables the modems to send data as fast as possible at all times and gain the maximum advantage from the data compression.

If transmission errors occur on the telephone line, the modems have to resend some of the error-correcting protocol packets. This is possible because the data is in the memory buffers in the modem. Resending packets does take time and slows down the overall transmission rate a small amount. The buffers enable this variability of speed to be handled because the hardware flow control stops data input to the modem when the buffers are full and restarts input when more room is available in the modem buffers. A high telephone line error rate therefore will appear to the modem user as a slower overall transmission speed. The data will be received error-free and just a bit slower because of the buffer to buffer retransmission.

Flow Control Methods

The best method of flow control is *hardware flow control*, which uses the RTS and CTS signals discussed in Chapter 3, because you legally can send any data characters in this mode.

An alternate method of flow control, called *software flow control*, uses two data characters (XOFF and XON) to indicate when the computer should stop or start sending data. With this method, the modem sends an XOFF (0x13) character to the serial port when it wants the input data to stop. The modem sends an XON (0x11) character to indicate when it is ready for the computer to resume sending data.

This flow control method is called software flow control because your software has to monitor the data in order to control the flow. When you use this method of flow control, you also have to decide whether the modem should pass these characters to the other end as well as act on them. You usually do not need to use this method, but in some cases you may, so buffered modems enable you to select either type of flow control.

Because software flow control uses data characters as control, you cannot send binary data or use file-transfer protocols such as XMODEM with this type of flow control. The modem and receiving software cannot differentiate between XOFF intended as a true data character and XOFF intended as flow control. You therefore should use hardware flow control if possible.

Effects of Modem Buffering

The buffering required by error correction and data compression protocols also affect the person who is typing interactively. A perceptible delay, which may be unsettling, occurs between pressing a key and seeing an echo. The V.42 protocol is markedly superior to the MNP protocol in smoothing out these delays, but they cannot be avoided completely.

In addition, because the modem buffers a substantial amount of data, the screen the user is viewing is not synchronized with the data the computer is sending. This lack of synchronization can cause confusion if the user performs an action resulting in a rapid change of output; this change is delayed while the data already in the buffers is sent. On systems with hot-key responses or other highly interactive operations, you must be aware that the modem removes the tight one-to-one feeling you have when these protocols are not in use.

Selecting an Advanced Modem

Selecting a modem, like most other purchases, is primarily a trade-off between cost and benefits. The primary problem you face in selecting an advanced modem is understanding enough about the benefits to determine what they are worth in your application.

Today two price points exist that are based on a modem's fundamental speed. These points occur at 2400 bps and at 9600 bps. To decide which type of modem is best for your application, consider the amount of data you need to transfer and the amount of time transferring that data will take. In general, if all else is equal, a 9600-bps takes one-fourth the time a 2400-bps modem takes to transfer the same information. If a task takes one hour with a 2400-bps modem, for example, the same task takes about 15 minutes with a 9600-bps modem.

You must decide if this difference in time (including any long-distance charges affected by this time) is worth the difference in price. In general, if you transfer large quantities of data over long-distance telephone lines on a regular basis, a 9600-bps modem can pay for itself quickly by reducing your long-distance charges.

If you decide you want a 9600-bps modem, you then must decide which type. At this point in the evolution of modem technology versus price, buying a proprietary protocol high-speed modem is probably not a good idea unless you are buying all modems for all ends of an enclosed application. In this case, you still may have a small price advantage in purchasing a proprietary high-speed modem, especially if you view the purchase as covering a period of one or two years with no recovery value required from the modems at the end.

If compatibility with other modems is a priority, you should purchase a modem that operates with standard protocols. A V.32 modem can connect and communicate with any other V.32 modem at 9600 bps. A V.32bis modem communicates with any other V.32bis modem at 14,400 bps and any other V.32 modem at 9600 bps. The decision here again is one of price versus the extra 50 percent speed increase.

If you are planning to use the modem heavily in the BBS community, you may want to investigate the U.S. Robotics HST Dual Standard modem, which accepts the HST protocol as well as V.32 and V.32bis protocols. This modem is somewhat more expensive, but the benefit of having the HST protocol—for the time it remains important in the BBS community—may be worth the price.

When you are examining error correction and data compression features in a modem, you must decide between MNP and V.42 modems. Note that MNP and V.42 modems are available with or without the associated MNP 5 and V.42bis data compression options.

I recommend that you consider only a modem with the V.42bis protocol—which always includes V.42 and MNP 4 as a subset—instead of a modem with only the MNP 5 protocol—which includes all other MNP levels as a subset—unless the price difference is large. MNP 5 compression adds very little in most cases, and BBSs generally turn it off in order to obtain maximum speed with compressed files. The V.42bis compression always is better on any transfer except with precompressed files, and in such cases, it does not slow down the data transmission. In addition, the V.42 error-free protocol results in much smoother operation when you are typing than does MNP.

Understanding Software Considerations

If you use any modem with advanced V.42 or MNP capability, extra setup considerations exist for the modem and for your software. Most importantly, you must enable flow control in your terminal program and in the modem. You normally should use hardware flow control (RTS/CTS) if possible because it imposes no restrictions on the type of data you can send through the modem.

If you are using data compression, you should set the port speed of your software higher than the baud rate of the modem; otherwise, you do not receive any benefit from the compression feature. If you are using MNP 5 compression, set the port speed to the modem at least double the carrier speed of the modem. For V.42bis compression, set the port speed at least four times the carrier speed of the modem for best results.

When you set the port speed higher than the carrier speed on a modem, you also have to program the modem using AT commands to lock its port speed at the higher rate. Figuring out the proper commands to use to set the modem can be a chore because few of these commands are standardized.

The one quasi-standard method in use is from Microcom, the company that developed MNP. For modems using this method, you lock the port with the AT\J0 command and set proper RTS/CTS flow control with the AT\Q3\G0 command.

For other modems, you have to figure out what commands to use if your software doesn't indicate them. Hayes and U.S. Robotics modems illustrate how different these commands can be. On the Hayes Ultra modem, the command ATS36=7 locks the port, and the command AT&K3 selects the proper RTS/CTS flow control. On U.S. Robotics modems, AT&B1 locks the port and AT&H1&R2 sets the proper RTS/CTS flow control. You can learn these settings only by carefully examining your modem's manual. This area of serial communications is developing so quickly that no standards exist yet.

Several other considerations arise because of the more complex nature of these modems. Because the expanded features create many added AT commands to control these features, the initialization strings for these advanced modems can become quite lengthy. Some programs may not have enough room in their init strings to hold all of the settings required. In this case, you have to select some of those settings to be stored in the NOVRAM through the use of the &W command and ensure that an ATZ is part of the start-up sequence for your software. Finally, some terminal programs have an *autobaud* feature that changes the port speed to match the connect message speed. If you lock the modem port speed, you have to disable this feature for proper operation.

Understanding Hardware Considerations

With high-speed advanced modems, you may find that you need to set your modem and computer port speed at 19,200 bps or even 38,400 bps to obtain maximum speed and throughput from the modem. These speeds are very fast and actually exceed the ratings of the early IBM PC computer serial ports. Very few computers can handle sustained data flows at these speeds with the normal serial port hardware.

The IBM PC computer was designed to handle a maximum serial port speed of 9600 bps. Although the serial port could be set to faster speeds, the overall design of the machine was done with the assumption that 9600 bps would be as fast as it would ever be used in a general way. Even though new computers are much faster, this consideration still affects what you can do when you use your serial ports.

At a speed of 19,200 bps, your computer is sending or receiving a character every 520 microseconds (that's 0.00052 seconds or 1/1920th of a second). At 38,400 bps this is halved to 260 microseconds or 0.00026 seconds. This isn't much time. The design of the standard IBM serial port is such that it can hold only a single character while it is receiving the next. Therefore, if you are receiving data and the computer cannot read a character out of the serial port 3,840 times every second—never delaying longer than 0.00026 seconds in the process—then an incoming character still will be in the holding register of the UART when the next character arrives. The UART will have no place to put the new character, so the UART will store the new character over the old one the computer hasn't read yet. This is known as data overrun and results in the loss of a character.

New computers such as the 80386 or 80486 are very fast, so why is this a problem? Because the I/O bus is set up so that some peripherals (most notably hard disks and video displays) can lock the CPU out for up to 700 microseconds. This means that you have no chance to avoid data overrun if the disk or video is allowed to run while you are receiving data. This is life in the fast lane of communications.

Not many video adapters have this problem any more. This problem mostly shows on portables or laptops that have LCD or Gas Plasma video displays, which are much slower. Most high-speed disk drive controllers, however, cause this problem. When you examine the specifications of the IBM PC

I/O bus, you wonder how these high-speed disks can be made to work at all—they require everything this bus has to work. This means they usually do lock out the CPU during disk activity.

How can you tell if you have such a problem? The simplest way is to attempt a high-speed download using a streaming protocol such as YMODEM-g. If you get errors repeatedly, you probably are suffering from data overrun. Many terminal programs use RTS/CTS flow control when they write to the disk in an attempt to avoid this problem, but it isn't always successful. Besides, you are getting somewhat slower speed this way because the data flow is being stopped and started. If you use a multitasker (such as DESQview, DoubleDos, and so on) or run anything in the background while doing communications, you almost certainly will suffer data overrun with high-speed modems.

What is the solution? Two solutions really are available—one good and one cheap. The cheap solution is to reduce the interface speed of your new modem to 9600 bps. This solves the problem because you are running within the original IBM design at this speed and is cheap because you don't have to buy anything new to reduce the speed. This solution is unsatisfying, however, because you bought this modem to go faster than that. The second way is to substitute a 16550 UART for the 8250 or 16450 now present in your serial port. Because the 16550 UART can buffer up to 16 input characters before it overflows, the problem will disappear and you can get full use of your high-speed modem.

A 16550 solves the data overrun problem because it has 16 places to hold input characters instead of only one. This means that as long as the CPU empties the UART before 16 characters are backed up in it, no data will be lost. So even at 38,400 the computer can be delayed 16×0.00026 seconds or up to 4 milliseconds without data loss. This is five or six times longer than the longest delays I/O devices cause, so you no longer have a problem.

Installing the 16550 in your serial port is quite easy if your current UART is in a socket. You simply unplug your old UART and plug in the 16550 instead. The old UART can be found by looking for a large 40-pin chip with the numbers 8250, 82C50, 16450, or 16C50 written on it. Remove the old chip carefully, using an IC puller (available at Radio Shack) or a small screwdriver to pry from each end in rotation until the chip comes out. You can make the new chip go into the socket more easily by placing it on a desk and pressing down on each side to align and center the pins. Ensure that all 40 pins go into the socket when you plug in the new chip—they are all there for a reason and they all count.

What do you do when you find the UART in your present serial port is soldered? First kick the dog, then curse—your job just got a lot harder. If you aren't handy with a soldering iron, you either need to find someone who is, or buy a serial card that has its UART in a socket.

If you are handy with a soldering iron, you can take the old UART out by carefully cutting off all of its pins. Throw the now useless body of the chip away and unsolder each pin stub from the card one at a time. Use Solder wick to clear the holes in the card. Now, invest in a 40-pin socket (about 80 cents at Radio Shack). Install the socket into the empty holes so that you can plug in your 16550 UART. You can just solder in your 16550, but using a socket means you don't have to go through this again if anything ever happens to your UART. You also can move your 16550 to a new serial port if you ever change your computer setup.

Most new terminal software is "16550 aware" and you won't need to do anything to make use of your 16550. If you have older software that isn't aware of how to use a 16550, you can obtain the program 16550.EXE from many BBS systems. This program should be run before you load your terminal software to turn on the 16550 buffers.

In addition to the UART, you must take extra care with cables at these high speeds. At 38,400 bps you cannot use a cable longer than 12 feet and still have a reliable connection. You also must be sure that your cables are in good shape, as any damage to wires or connections in the cable affect these high-speed signals immediately. Problems with high-speed serial communications often are solved with a new serial cable. Low capacitance cables are a plus at these speeds, but not strictly necessary. Be sure the cable is shielded (all new cables must be per FCC regulations). Remember that a cable which works fine at 2400 bps may not work at all at 38,400 bps.

Chapter Summary

In this chapter you learned about the various types of advanced error-free and high-speed modems. You learned the difference between a standard V.32 or V.32bis modem and proprietary high-speed modems. You also learned about how the internal error-correcting protocols MNP and V.42 provide error-free transmission to modems of all speeds. And you learned how the data compression protocols MNP 5 and V.42bis can make modems transmit data at speeds faster than their actual carrier speeds.

Today's advanced modems provide a great deal of performance for the money. They introduce new considerations, however, that you should not overlook if you want good results. The primary considerations are that high speeds of data transmission can require special serial port hardware to avoid data loss, and that the modems introduce a "choppy" feel to the modem link which can be troublesome to you as a user.

When you are comparing these modems, remember that the following three types of protocols for speed, error correction, and data compression can occur in any combination:

Protocol	Meaning
Bell 103	300 bps
V.21	300 bps
Bell 212A	1200 bps
V.22	1200 bps
V.22bis	2400 bps
V.32	9600 bps
V.32bis	14,400 bps
HST	Asymmetrical 14,400 bps
PEP	Half-duplex 16,000 bps
MNP 1	Error correction
MNP 2	Error correction
MNP 3	Error correction
MNP 4	Error correction
V.42	Error correction
MNP 5	Data compression
V.42bis	Data compression

Be careful not to confuse a 2400-bps modem with V.42/V.42bis (advertised as "up to 9600 bps") with a true V.32 or V.32bis modem. Sorting out these numbers is a large part of the task of comparing modems.

Remember that error-correcting and data-compressing modems offer nothing unless they connect with another modem that supports the same protocol. When they connect with a basic modem of the same speed, they become basic modems themselves.

If you choose and install an advanced modem carefully, it can give you many years of reliable use. After you experience data transfers at speeds over 1000 cps, you never will want to go back to the basic modems again.

22

Modem
Troubleshooting
Guide

When you install a modem and it doesn't work, finding out what is wrong can be very difficult. You cannot easily determine whether the product is defective or whether you just have a cable or simple setting wrong. The only symptom may be that the modem doesn't work. Attacking the problem at random with no real understanding of what is happening almost never leads to a satisfactory solution.

Modems rarely are truly broken. Usually an incorrect setting or cable connection causes the malfunction. With today's advanced modems and software, however, you may have to deal with literally hundreds of settings; when one is wrong, knowing where to look can be quite difficult.

This chapter presents an organized approach to most modem problems in a sequence of steps (a sort of flow chart). If you follow the numbered steps slowly, and carefully perform each indicated step, you usually can discover what is wrong with your installation. If you have a particularly difficult problem, you may need to understand some of the reasons behind the steps to discover what is wrong. Chapters 2, 20, and 21 explain how the hardware serial ports and your modem work.

Understanding Modem Firmware

Modern modems contain powerful computers and large software programs to control their operation. These programs are called *firmware* because they are firmly stored in read-only memory chips (ROMs) inside the modem. Today's advanced modems contain complex programs that are subject to the same software bugs as any computer software. To correct these bugs, modem manufacturers update the modem software from time to time.

Modem firmware problems most often show themselves as intermittent modem lockups that require the modem to be reset to continue operation. Because today's modems contain more firmware than hardware, firmware bugs can be the cause of nearly any modem malfunction. Modems are used much more frequently for dialing out than for automatic answering, so most modems are tested less in auto-answer mode. Firmware bugs therefore tend to occur more frequently in auto-answer mode than in dial-out mode, especially early in a modem's design life.

If you have persistent and unexplainable problems with a modem, your modem's firmware may have bugs. Call the modem manufacturer to find out whether your modem has the latest version of the firmware. Most modem makers will update your modem firmware for a small fee (and sometimes for no charge). If your modem has a firmware bug, you can repair it only by upgrading the ROM chips in the modem.

Using This Troubleshooting Guide

This chapter assumes that you have a terminal program (such as PROCOMM PLUS, Telix, or Qmodem) and know how to alter the communications parameter settings (as explained in Chapters 1 and 3). The steps make use of the status lights that appear on external modems to assist in diagnosing the problem. Internal modems have no lights; therefore, narrowing down the cause of certain problems can be a bit more difficult. If you have an internal modem, you may need to open your computer and check the modem itself in some steps, so you should be prepared with the appropriate screwdrivers or other necessary equipment.

To use the sequence of steps, find the starting point that most closely indicates the problem you are having. The steps are grouped under four problem headings: "The Modem Doesn't Respond," "The Modem Dials but Doesn't Connect," "The Modem Connects but Gets Garbage," and "The Modem Fails When Doing File Transfers." If your problem is not listed, then start at the beginning and perform the tests indicated under "The Modem Doesn't Respond." Somewhere you should find a problem that you can fix, and often that will be the only problem.

For this guide to help you successfully find and fix your modem problems, you must follow all the steps in order, even if you don't understand why they are important. If you skip a step, the guide may not lead you to the problem, and it may even indicate a problem that you don't have.

Testing and Troubleshooting Your Installation

One reason serial communications are considered so difficult is that when something is wrong, you usually see only one symptom—nothing works. Because total failure is such a common symptom, the process can be very frustrating. Examining your setup and diagnosing and repairing problems in your communications installation really is not that difficult, however, if you proceed slowly in a logical sequence.

The Modem Doesn't Respond

Begin here if your modem appears not to respond at all when you use it. In other words, when you load your terminal program and type to the modem, nothing happens.

Step 1

Check that the modem is plugged in and turned on.

If it is not, plug it in and turn it on. Your modem should be ready now, so check again to see if you still have any problems.

If the modem is plugged in and turned on, go to Step 2.

Step 2

External Modem: Check all cable connections to make sure they are connected tightly. Wiggle each connection while pressing it into its socket.

Internal Modem: Check that the modem card is fully seated in its socket inside the computer. Wiggle the modem card while pressing it into its socket. If it is plugged into the PC expansion bus, make sure the card top is level so that the connections are solid.

After performing these steps, retest your installation to see whether it works. If it still does not work, go to Step 3.

Step 3

Check that the telephone cord from the wall is plugged into the correct jack on the modem. Most modems have two or more jacks for telephone cords, but only one jack is the correct one for the cord that connects to the wall. The correct jack usually is marked Line or TelCo.

If your telephone cord is in the wrong jack, move it to the proper jack and retest your system to see if your modem now works.

If your telephone cord is already in the proper jack, proceed to Step 4.

Step 4

Set your terminal program speed to 300 bps. Then type the following strings, pressing Enter after each string:

    ATZ
    ATE1Q0

If you do not see any response, proceed to Step 5.

If you see a response of OK or 0 (zero) on your screen after typing the last string, then your modem and serial port are properly connected to your computer and operating. You may rule out the modem, serial port, or data wire connections in the cable as a source of your problems.

Step 5

Verify that you have set your terminal program to the same com port to which your modem is connected. For external modems, verify the switch settings of your serial port as COM1 or COM2. For internal modems, these switch settings are on the modem itself. Consult the documentation of the serial port adapter or internal modem to make sure that you have the modem connected to (or configured as) the com port you think you do. Then check the terminal program setup screen to make sure that the terminal program is also set to the same com port.

If you find an error in the com port setting or configuration, correct it and retest your installation to see if your modem now operates.

If you find no error, proceed to Step 6.

Step 6

Verify that you are using the proper type of cable between your modem and serial port. If you have questions about the cable type, refer to Chapter 2 for details on cable construction.

If you are certain you have the proper cable type, check your modem's documentation for a jumper or Dip switch labeled Disable AT Command Set. The label on your modem may be slightly different (Disable Command Recognition, for example), but it should indicate in some way that it enables the modem to recognize AT commands or not. Make sure this switch is set to enable the modem to recognize the AT command set.

If you find such a switch and change it, retest your modem installation to make sure it now works properly.

If your modem has no such switch or it was already set properly, go to Step 7 if you have an external modem or to Step 12 if you have an internal modem.

Step 7

If you have an external modem, identify the indicator light that shows Data Terminal Ready, usually labeled TR. If your modem does not have a TR indicator light, proceed to Step 8. When you activate your terminal program, the TR light should light up, and it should go off when you deactivate the program. With most terminal programs, if you tell your terminal program to disconnect the phone (using the Alt-H command on many terminal programs), the TR light should flash off and back on.

If the TR light passes this test, you have verified that the DTR wire in your modem cable is correct, that you are properly addressing your serial port, and that your modem is properly programmed to process DTR. Go to Step 10.

If the TR light fails this test, it may do so in one of two ways. If the TR light does not go off, go to Step 8. If the TR light never lights, go to Step 9.

Step 8

Check for a Dip switch or jumper on your modem labeled DTR Normal/ Ignore. Make sure it is set to the normal (sometimes called *active*) position.

If you find such a switch and change it, repeat the Step 7 test. If you find no such switch or it is already set properly, type the following string and press Enter:

    AT&F&C1&D2&W

If your modem has NOVRAM and software switches, this string enables the DTR and Carrier detect signals to operate properly. Repeat the Step 7 test. If the test still fails, go back to Step 5.

Step 9

If the TR light does not come on, check the cable. The DTR wire that connects to pin 20 of the modem likely is not present or wired incorrectly. If you verify that the wire is indeed present and correct, then your serial port or modem may be malfunctioning. By far the most common problems here are a loose or incorrect cable, or an incorrect com port address configured in the terminal program (which you should have already verified before reaching this step). You will have to correct the wiring problem.

Step 10

Locate the Send Data indicator light on your modem, usually labeled SD. Closely observe the SD light as you type a few keys. It should flash with each letter you type, indicating that key data is being transmitted to the modem.

If you see a flash as you press each key, proceed to Step 11.

If you do not see such a flash, check the cable wiring for pins 2 and 3 of the modem (TxD and RxD). If the cable is wired correctly and the SD light still does not flash properly, make sure that you have correctly told your terminal program which interrupt the serial port is set to use. For COM1, the serial port should be set to IRQ4; for COM2, the serial port should be set to IRQ3.

Verify that the switch settings on your serial port are correct. If they are, you may have an IRQ conflict with another I/O card in your computer. Such problems commonly occur with network adapter cards (which frequently use IRQ3) and some tape backup units (which also can use IRQ3). If you have such a conflict, you must change the serial port to the other address (COM1 to COM2, or COM2 to COM1) or change the IRQ used by the conflicting card to correct the problem.

Step 11

Locate the Receive Data indicator light on your modem, usually labeled RD. Press several keys and observe the RD light when the SD light flashes. (Step 8 verified that the SD light flashed with each keypress; if the SD light doesn't flash, you have skipped a step.) If the RD indicator light also does not flash, type the following string and press Enter:

ATE1

After this command, the RD and SD lights should flash; each key you press should echo back from the modem, printing on-screen.

If you see the characters echo, type *ATZ*; the modem should respond with OK. If it does, your modem, cable, and serial port are installed correctly.

If the RD light flashes but the characters do not print on-screen as you type, check the cable pins 2 and 3 to the modem to make sure that the wire is not broken. If the cable is good, then you most likely have a conflict or your serial port's IRQ is set incorrectly. For COM1, the serial port should be set to IRQ4; for COM2, the serial port should be set to IRQ3.

Verify that the switch settings on your serial port are correct. If they are, you may have an IRQ conflict with another I/O card in your computer. Such problems commonly occur with network adapter cards (which frequently use IRQ3) and some tape backup units (which also can use IRQ3). If you have such a conflict, you must change the serial port to the other address (COM1 to COM2, or COM2 to COM1) or change the IRQ used by the conflicting card (usually changeable by a jumper or switch on the card).

Step 12

If you arrive at this step, then you most likely have a conflict or an incorrect setting of your modem's IRQ. For COM1, the modem should be set to IRQ4; for COM2, the modem should be set to IRQ3. Some modems configure the com port number and the IRQ number separately. Verify that any IRQ switch settings on modem match the com port you have selected.

If the settings are correct, you may have an IRQ conflict with another I/O card in your computer. Such problems commonly occur with network adapter cards (which frequently use IRQ3) and some tape backup units (which also can use IRQ3). If you have such a conflict, you must change the modem to the other address (COM1 to COM2, or COM2 to COM1) or change the IRQ used by the conflicting card. You can verify the presence of a conflict by removing as many I/O cards as possible. If your modem begins to work, then you have an addressing or IRQ conflict.

The Modem Dials but Doesn't Connect

Begin here if your modem appears to dial another modem correctly but doesn't connect properly.

Step 1

If you are using the dialing function of your terminal program, you must be sure that your modem is programmed to return the proper result codes your terminal program requires to detect a connection. Many terminal programs will do nothing at all on a dial function, even if the modem connects correctly, if the modem does not return the expected result codes.

To check this, you should place your terminal program in terminal mode and directly enter the dialing string manually. If you are dialing the telephone number 699-8222, for example, you should type the following command and press Enter to tone dial this number:

    ATDT699-8222

If you must use pulse dialing, you should type the following command and press Enter:

    ATDP699-8222

When you manually dial, you see the result codes the mode returns. This may indicate that you need to reprogram the result code format for your terminal program to dial correctly. Result code programming varies with the brand of modem, but the ATV1X4 command often sets the codes to the correct condition for most terminal programs. See your modem's manual and Appendix C for more information on programming the result codes of your modem.

If your modem is returning proper result codes, then you should proceed to Step 2.

Step 2

Look closely at the connection to the phone jack on the wall. Make sure that you do not have any other modems plugged into the same telephone circuit. Check especially for such things as a laptop computer modem plugged in from another room. Two modems plugged into the same phone circuit often cause problems even if one is turned off. If any other phones or modems are plugged into the phone line, disconnect them and retest.

If the modem now connects properly, put the other devices back one at a time to identify the problem.

If the modem still does not connect properly, go to Step 3.

Step 3

Type the following string and press Enter:

    ATM1

After you enter this string, you should hear a dial tone before the modem begins to dial.

If you don't hear a dial tone, proceed to Step 4.

If you hear a dial tone followed by dialing, proceed to Step 7.

Step 4

Make sure no other telephone is off the hook on this circuit. Also check that you really have the modem telephone cord plugged into the wall correctly. Check that the telephone cord from the wall is plugged into the correct jack on the modem. Most modems have two or more jacks for telephone cords, but only one is the correct jack for the cord that connects to the wall. The correct jack is usually marked Line or TelCo.

If these checks lead you to a problem that you can correct, you can stop here.

If you still don't hear a dial tone, go to Step 5.

Step 5

Check whether your modem has a feature to *blind dial*, or dial without waiting for a dial tone. If it does, make sure that it is turned off. If you are using a modem in an office setting, make sure that you have indicated in your dialing string any prefix dialing required to obtain an outside line (dialing 9 before the phone number, for example).

> **NOTE**
>
> For most Hayes compatible modems, blind dialing is controlled by the ATXn command. ATX1 causes the modem to wait only a fixed amount of time and then dial regardless of whether a dial tone is present. ATX2 causes the modem to wait until it actually detects a dial tone to begin dialing. See Appendix C for more details on this command.

If these checks lead you to a problem that you can correct, you can stop here.

If you still have no dial tone, proceed to Step 6.

Step 6

Plug a regular telephone into the wall socket to make sure the phone line is working.

If it is not, you need to repair the telephone circuit.

If a regular telephone works properly in the wall jack, proceed to Step 7.

Step 7

If you continue to hear a dial tone after the dialing starts, you probably have selected the wrong type of dialing. To select tone dialing, use the string ATDT to dial. To select the pulse dialing method, use the string ATDP.

If you have selected the wrong dialing type, correct and retest to make sure that your modem is now working properly.

If you hear a dial tone, then dialing, and then ringing, the dialing is working correctly. If you then hear the other modem answer and send tones but your modem hangs up before the connection finishes, you should check the call timeout in both your modem and your terminal program. The modem call timer is set with the ATS7=nn command where nn is the number of seconds the modem allows from the time it begins dialing until a connection must occur. To set the timeout to 45 seconds, type the command *ATS7=45* and press Enter. (See Appendix C for details on this command.) For high-speed modems using V.32 or V.32bis modulation, or for overseas telephone calls, you often have to set the S7 value higher than the factory default to obtain proper operation.

If dialing type or modem timeout are not the problem, proceed to Step 8.

Step 8

If you hear crackling sounds as the modem dials (before the other modem answers, when it should be quiet except for the ring), you probably have a loose wire in your phone circuit. Check the cable from your modem to the wall as well as all other telephones connected to this circuit (You should have checked these in Step 2).

The Modem Connects but Gets Garbage

Begin here if your modem connects to another modem but the characters you type do not appear properly on the other end. This portion of the troubleshooting flow chart checks for noise problems and improper settings.

Step 1

Verify that you have the baud rate (or bps rate) set the same on both ends of the connection. If you do not have the speeds set the same, you will either see nothing at all or garbage characters. Garbage from the wrong baud rate setting usually displays many lowercase letter x and tilde ($\sim$) characters.

If you are certain that the baud rate is set correctly and you still have garbage characters showing, you should determine if this occurs on all (or most) calls you make, or only when you call certain telephone numbers.

If you see this line noise or garbage on all (or most) calls, go to Step 2.

If you see it only on some calls, go to Step 3.

Step 2

What appears to be intermittent line noise can have many causes. If such noise appears on all (or most) calls, you probably can correct it on your end. First check for loose cables, one of the most frequent causes of noise problems. Also check that the phone cord you are using to connect the modem to the wall jack is in good condition. If possible, trade it with another one to be certain.

Make sure that the modem isn't located near a color TV, fan motor, or other source of electrical noise. This type of noise can easily enter a modem through the air and cause what appears to be line noise problems. The solution in this case is to move the modem to a different location.

Finally, check all other telephones that are hooked to the same line. Unplug them one by one to determine whether any of them are causing the problem. Sometimes telephones that work fine by themselves can cause distortions of the modem's signal, inducing apparent noise when the modem is used.

Step 3

Do the errors occur only with every other letter of the alphabet? For example, do A, C, E, and so on print correctly, whereas B, D, F, and so on always show as garbage (or vice versa)?

If the answer is no, go to Step 4 if your modem has V.42 or MNP. Go to Step 5 if your modem does not have V.42 or MNP.

If the answer is yes, your terminal program's parity setting is incorrect for the computer you are calling. Change the parity setting to the correct one for this system. Your characters should be displaying properly now.

Step 4

Do characters disappear in groups or batches? In other words, does the output to your screen look fine when you type slowly, but when you type faster or ask for a larger display, only the first few lines look correct and then characters are missing?

If the answer is yes, you have a problem with flow control. Check that your terminal program is set for hardware (RTS/CTS) flow control. Also check that your modem is set for bi-directional hardware flow control. This setting varies according to the modem brand; consult your modem's manual to determine the correct setting. The following are the proper AT commands for setting flow control on some popular modems:

U.S. Robotics HST and HST/DS:

> AT&R2&H1&W

Hayes V-Series:

> AT&K5&W

MicroCom modems:

> AT\Q3&W

Set your modem and terminal program for proper RTS/CTS flow control. Your characters should now display properly.

If the answer to this question is no, go to Step 5 if the garbage occurs on a normal (not reliable MNP or V.42) connect.

If the garbage occurs on a reliable (MNP or V.42) connect, go to Step 6.

Step 5

If the garbage occurs only on certain telephone connections, your modem may have modulation incompatibilities with the modem on that particular computer. Such incompatibility is a rare problem today, but does still occur between some combinations of different modem brands. The other possibilities are that the computer you are calling is connected to a line that has call waiting, and other incoming calls are generating the noise, or that the other computer is connected to a telephone line with a poor connection.

Step 6

V.42 and MNP connections cannot have line noise because the modems remove line noise. If your modem is not damaged, then you have some terminal program setting that is incompatible with the computer you are calling. The most common setting that causes this problem is incorrect terminal emulation. Refer to Chapter 16 for an explanation of terminal emulation and possible emulations settings.

The Modem Fails When Doing File Transfers

Use this section if your modem appears to work properly but fails when you attempt a file transfer.

Step 1

If the failure occurs on all protocols, go to Step 2.

If the failure occurs only on some protocols such as XMODEM or YMODEM but not on other protocols such as KERMIT, your modem is set to intercept the XON/XOFF characters instead of passing them through. This modem flow control problem primarily occurs on V.42 or MNP modems. Check your modem's manual to set XON/XOFF to pass through the modem (that is, XON/XOFF flow control in the modem itself should be off). The exact procedure varies, depending on the manufacturer.

Step 2

If these failures do not occur at lower speeds, go to Step 3.

If the failures occur at several different speeds, your modem or terminal program (or both) are not set for RTS/CTS flow control. Check the terminal program to make sure that it is set for hardware (RTS/CTS) flow control. Also make sure that the modem is programmed to use RTS/CTS bi-directional flow control. Finally, make sure that the cable to your modem has the wires to pins 4 and 5 of the modem properly connected. All these portions of RTS/CTS flow control must be set correctly for proper operation.

Step 3

ailures on file transfers at high speeds only (19,200 bps or 38,400 bps) indicate that you need a 16550 UART in your serial port. You need to upgrade your serial port to make it reliable at these high speeds.

Using Your Modem's Self-Tests

Many Hayes-compatible modems have several self-testing capabilities built into them. These tests are useful when you want to make sure that the modem does not have an internal failure. This section explains how to use these capabilities and analyze the results of the tests. Before learning how to use these tests, however, you need to understand the modem components that each test is checking.

Your modem has two major components. One deals with the analog signals (tones) that are sent to and received from the telephone line. The other component is the digital computer that is part of your modem and that deals with the data. This component has memory to hold your commands and a control program (called *firmware*) that is stored permanently in the modem's read-only memory (ROM). The overall structure of these components when two modems are connected over a telephone line as well as the direction of data flow for each piece are illustrated in figure 22.1.

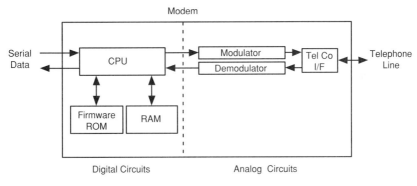

Fig. 22.1. *The digital and analog components of a modem.*

The Modem Memory Self-Test

Your modem's computer can fail. The most difficult form of computer failure to detect is the failure of one or more memory locations. For this reason, most modems have a memory self-test. This test varies with different brands of modems, so you should read your modem manual carefully to learn how to run the test on your modem. On Hayes modems (and many Hayes-compatibles), the test is performed by typing the string *ATI2* and pressing Enter. If the modem responds with OK, then its memory has passed the test. If it responds with ERROR, then the memory has failed the test and the modem needs to be repaired.

Modem Data Loopback Tests

The remaining modem self-tests are called *loopback tests* because they provide a switch at a particular point on the circuit shown in figure 22.2. This switch bridges the sending and receiving sides, enabling the modem to send out signals it has generated and have those signals returned via the *loopback*. Because the modem generated the signals, it knows what the return signals should look like, and it can verify whether the process succeeded or failed.

Modem loopback tests can occur at three points with dial-up telephone modems. Two of these points are inside a single modem and do not require the modem to be connected to a telephone line. Figure 22.2 shows these loopback points.

NOTE

Many of the newer full-featured modems with V.42 and MNP capabilities require you to turn off these features before you can run these local tests. Read your modem's manual carefully to determine whether your modem requires any special setup. Many modems that require you to turn off such features simply respond with an ERROR response if you try to run the tests without turning those features off, giving the impression that your modem does not support the self-tests when in fact it does.

The Local Digital Loopback Test

The first point at which a loopback test can be performed is at the interface between the digital and analog portions of the modem. This test is known as a *local digital loopback test*.

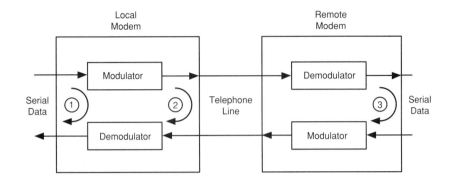

① − Local Digital Loopback
② − Local Analog Loopback
③ − Remote Digital Loopback

Fig. 22.2. *Location of modem loopback test points.*

When this test is requested, the modem ties its digital transmitter directly to its digital receiver, bypassing all of its analog circuitry. In effect, this action routes the modem's serial port received data directly to the modem's transmit data. This test verifies the modem's digital circuitry, the data cable between the modem and the computer, and the serial port.

To perform this test with a Hayes-compatible modem, type the string *AT&T3* and press Enter. On some modems, this command is sent by typing *ATS16=4* and pressing Enter.

After the local digital loopback command is sent to the modem, you can type on your terminal's keyboard to verify that the keys you press display correctly on-screen. If you see any errors, check your cable and serial port carefully. If they are correct, then the modem has an internal problem in its digital circuitry.

To end the local digital loopback test, press the plus key (+) three times and wait two seconds for the modem to respond with OK. Then type *AT&T0* and press Enter to end the test.

The Local Analog Loopback Test

After you verify that your modem, serial port, and cable are working properly (using the local digital loopback test), you then can check the modem's analog modulator and demodulator circuits by using the *local analog loopback test*. This test connects the modem's modulator output directly to the modem's demodulator input. The modem then internally generates a data pattern and sends it around the loop, verifying the received data pattern and indicating any errors that occur.

Some modems only have a manual analog loopback test, whereas others have a manual and an automatic test. The manual test works in the same way as the local digital loopback test described in the preceding section. You must input data manually and verify that the modem displays the same as you type. With an automatic test, the modem generates the data patterns and compares the received data.

To initiate a manual local analog loopback test, type the string *AT&T1* and press Enter. You then can test the entire modem circuitry (digital and analog) by pressing keys on your keyboard and verifying that the same characters print on-screen. To end the manual local analog loopback test, press the plus key (+) three times and wait two seconds for the modem to respond with OK. Then type *AT&T0* and press Enter to end the test.

To initiate an automatic local analog loopback test, type the string *AT&T8* and press Enter. The modem tests itself by sending and verifying a data pattern through its entire analog circuitry. Allow this test to run for a few minutes and then press the plus key (+) three times and wait two seconds for the modem to respond with OK. Then type *AT&T0* and press Enter to end the test. The response is a three-digit number giving the number of errors the test detected. If this number is 000, no errors occurred.

The Remote Digital Loopback Test

The remote digital loopback test tests the two modems and the telephone line between them. In this test, the *remote* modem (the modem at the other end of the telephone line) loops back all data sent to it at the modem's serial port.

NOTE

Both modems must be Hayes-compatible and support remote digital loopback testing in order to perform this test.

Because this test involves two modems and a telephone line, it is more complicated than the local loopback tests. The first step is to set up the remote modem. The operator of the remote modem must place that modem in auto-answer mode by typing the string *ATS0=1* and pressing Enter. You then must use your terminal program to dial the remote modem using the speed and modulation method you want to test (see Chapters 20 and 21 for an explanation of modulation methods). When the modems connect, each operator must regain control of the local modem by pressing the plus key (+) three times and waiting for the modem to respond with the OK response code.

Now that the connection is established and each operator can enter modem commands, the remote operator types the command *AT&T4* and presses Enter, enabling your modem to request a remote digital loopback test.

You now have two options (if your modem supports both). You can run a manual remote loopback test by typing the string *AT&T6* and pressing Enter, and then pressing keys on your keyboard to verify that the same letters print on-screen. Each key you press is modulated onto the telephone line, transmitted to the remote modem, demodulated back to serial data, looped back to the remote modulator, put back on the phone line, demodulated by your modem, and transmitted to your screen.

Alternatively, you can run an automatic remote loopback test by typing *AT&T7* and pressing Enter. In this test, your modem generates data patterns and sends them around the entire circuit. Your modem compares the received data patterns with the ones it transmitted and reports how many errors occurred.

To end a manual or an automatic remote digital loopback test, press the plus key (+) three times and wait two seconds for the modem to respond with OK. Then type *AT&T0* and press Enter to end the loopback test. If you are running an automatic test, the modem responds with a three-digit number indicating the number of errors it detected.

An automatic remote digital loopback test is an inexpensive way to test the quality of a telephone line and modem connection. If you let the test run for many minutes, the error count can give you an idea of how effective the connection will be for normal use.

Chapter Summary

Modem and serial data problems can be very difficult to find and fix if you don't take a methodical approach. If you simply begin to change settings at random, hoping to solve the problem, you may make the situation worse.

This chapter has given you a series of steps to follow to diagnose and repair problems with your modem, serial port, and data cables. If you combine this approach with the knowledge you gained from Chapters 1, 2, and 3 of how serial communications work and the knowledge of modems you gained from Chapters 20 and 21, you will be able to solve most modem communications problems you encounter.

Glossary of Terms

Abort. End the currently running task on a computer.

A/B Switch. An external switch box that is used to switch a serial port among two or more external devices.

Address. 1. Number or text string that uniquely identifies your electronic mailbox to a store-and-forward electronic mail system. 2. Location of memory in a computer. See also *DNS*.

Analog Signal. A signal that varies continuously. Analog signals have the properties of frequency (which is heard as tone or pitch) and amplitude (which is heard as volume). Telephone systems are designed to carry analog signals, and modems are used to convert the digital signals used by a computer to the analog signals required by a telephone circuit. See also *Digital Signal*.

ANSI. 1. Acronym for American National Standards Institute, a standards organization. 2. The X3.64 ANSI standard, which indicates how screen position and color may be sent to a remote terminal. See also *Emulation*.

Answer Mode. See *Originate Mode*.

ASCII. Acronym for American Standard Code for Information Interchange. This standard seven-bit coding method assigns the numbers 0 to 127 to the upper- and lowercase letters of the alphabet, numbers, punctuation, and 32 control codes. Systems that add extra characters (such as graphics), using the numbers 128 to 255, often are referred to as "extended ASCII," but only the lower 128 characters truly are standardized. Because computers internally understand only numbers, this coding method enables computers to manipulate and store text.

Asynchronous. Serial data transmission in which the time between each pair of characters sent may vary arbitrarily. To enable this type of transmission, each character is framed with a start bit and a stop bit, which resynchronizes transmission for each character sent.

AT Command Set. The industry standard method of controlling a modem from a computer. The name derives from the fact that each command begins with the letters AT. The AT command set also is known as the Hayes command set.

Auto Answer. An option provided by modems and software to enable unattended answering of incoming calls. The modem detects the ring signal, answers the telephone, and indicates to the computer that another modem has called, enabling the software to handle the connection automatically.

Auto Dial. An option provided by software (and internally by some modems) to enable a telephone number to be dialed automatically in response to commands from the computer.

Auto Line Feed. A parameter in many terminal programs that automatically adds a line feed (LF) locally when a carriage return (CR) is received.

Backspace. Move the cursor one space to the left (backwards) on the computer screen or printer paper.

Baud Rate. The number of signal elements transmitted in one second. Baud rate is always equivalent to bits per second on a digital circuit, but on the analog telephone circuit, baud rate is often lower than the bits per second rate. The reason is that advanced modulation techniques represent more than a single bit with a single signal element.

BBS. See *Bulletin Board System*.

Bulletin Board System (BBS). A computer set up to answer computer telephone calls and enable callers to access information remotely. The strict definition of a computer bulletin board system centers on the message system, but today a wide variety of remote access functions such as file transfer, remote access databases, on-line games, and so on usually are included in the definition.

Bell Compatible. A modem compatible with the Bell 103 or Bell 212A standards. Such a modem can communicate at 300 bps (Bell 103) or 1200 bps (Bell 212A) with any other modem that also is Bell compatible.

Binary. 1. A numbering system that uses only the numbers 0 and 1. 2. Computer data that includes data bytes not representing printable text, such as a program file.

Bit. The smallest unit of data storage. Bit is a contraction of the term *binary digit* and may have a value only of 1 or 0.

Bits per Second (bps). A measurement of serial transmission speed. It is the number of bits transmitted in one second. See also *Baud Rate*.

Block. A group of data bytes transmitted as a single unit. Typically, data blocks have headers that identify and sequence the blocks and a trailer that contains a calculated value verifying the integrity of the data. This trailer is usually a CRC or a checksum. See also *Checksum* and *CRC*.

Break. A special asynchronous serial data signal in which a single start bit lasts longer than a normal character length. A break can be distinguished as unique from any possible data character and thus often is used for expedited control during serial communications. The name derives from the fact that in the early days of Teletype transmissions, this signal was generated by a switch that would break the connection for a specified period of time.

Buffer. Memory used as temporary storage for data in transit. Buffers are used to absorb speed differences and also to enable the data within the buffer to be processed (assembled into a data block).

Byte. A group of eight bits that represents one data storage element. For text data, one character requires one byte of storage.

Capture. The process of storing the text that is being displayed on your computer screen. You may capture incoming data into a buffer in memory or into a disk file.

Carriage Return. The ASCII character that represents the Enter key. This character is used in serial transmission to indicate the end of a line of text. A carriage return is a request to position the cursor at the beginning of the display line.

Carrier. The tone used by a modem to "carry" information. This tone is modulated by the modem to encode the serial data into an analog signal for transmission over the telephone lines.

Carrier Detect (CD). The RS-232 interface signal that indicates the presence or absence of a received carrier.

Cathode Ray Tube (CRT). The CRT is the display screen on your computer.

CCITT. The International Telegraph and Telephone Consultative Committee. A prominent international standards group that adopts and publishes standards on many areas of communications.

CD. See *Carrier Detect*.

Central processing unit (CPU). This term refers to the active portion of your computer.

Character. A single letter, number, punctuation mark, or control code. A character requires one byte of storage in a computer but may require fewer than eight bits to represent. (ASCII characters, for example, require only seven bits for full character representation.)

Character Length. The number of data bits used to represent a character. Seven data bits are sufficient to represent ASCII text, but eight data bits are required to represent binary data such as program files.

Characters per Second (cps). A measurement of transmission speed. It is the number of data characters transmitted in one second.

Checksum. A mathematical process that adds all characters in a data block to arrive at the total. This value is used to check the data block for transmission errors by the receiver. The checksum process is much easier to implement than the CRC process (which serves the same function), but the checksum method does not detect errors as well. See also *Block*.

Clear To Send (CTS). The RS-232 interface signal indicating that the computer is "clear" to send data to the modem. This signal is used for hardware flow control. See also *Request To Send*.

Communications Parameters. The settings used by software to indicate the speed, number of data bits, parity, and number of stop bits to be used in serial communications. See also *Default* and *Parameter*.

Compatible. Capable of operation with the other elements of a system. This term is used to indicate that one brand of product may be used interchangeably with another. (Any modem that is V.22bis compatible, for example, is able to communicate with any other modem that is V.22bis compatible at 2400 bps.)

Computer Mediated Communication (CMC). Computer message conferencing. Traditionally this term is used to refer to topic-based conferences such as Cosys, Picospan, Caucus, and so on. CMC often is used, however, to refer to all on-line message systems other than private electronic mail.

Conference. 1. An interactive real-time conversation system, sometimes called an on-line chat system. 2. A message system that enables discussions among group members. See also *SIG*.

Configuration. 1. A specific arrangement of hardware and software. 2. The specific setting of parameters in a program.

Connect. The moment when two modems complete the negotiation process and are ready to transmit data. This readiness often is indicated by the message CONNECT, which is sent by the modems.

Connect Time. The amount of time spent connected to another computer. This period is a common charging unit on commercial services, beginning when the modems connect and ending when you log off the system. The connect time generally is expressed in hours, minutes, and seconds.

Console. The operating or control terminal for an on-line system. The terminal sometimes is called the "local console" or "operator's console."

Control Character. A nonprinting character that you generate by holding down the Ctrl key on your keyboard while pressing a letter key. Documentation of control characters refers to them as either Ctrl-X or ^X. Control characters are used to issue commands to many on-line systems.

Control Key. The special key on your keyboard marked Ctrl. You use this key to generate control characters.

Control Sequence. A special multiple-character sequence that controls the operation of the remote terminal. These sequences begin with a control character and perform such operations as cursor position, reverse video, and color selection.

CPU. See *Central Processing Unit*.

CRC. Cyclic redundancy check. This term refers to a mathematical calculation performed on a data block to detect any transmission errors. The CRC method is more effective than the checksum procedure at detecting errors. See also *Block*.

CRT. See *Cathode Ray Tube*.

CTS. See *Clear To Send*.

Cursor. The small (usually blinking) line or block on your screen that indicates where the next character will be displayed.

Data. Information of any type.

Database. A collection of data that has been organized so that you can search for specific information and retrieve it easily.

Data Bits. The number of bits in a serial data element that represents the data character being sent.

Data Set Ready (DSR). The RS-232 interface signal indicating to the computer that the modem is powered on and ready to make a connection.

Data Terminal Ready (DTR). The RS-232 interface signal indicating that the computer is ready for the modem to make a connection. If this signal is removed by the computer during a modem connection, the modems usually will hang up and clear the circuit.

DB-9 Connector. The D-shaped nine-pin connector used on some IBM-compatible serial ports.

DB-25 Connector. The D-shaped 25-pin connector used on most RS-232 connections on modems and on many computer serial ports.

Data Communications Equipment (DCE). This term refers to the configuration of the RS-232 connector found on a modem. See also *DTE*.

Data Terminal Equipment (DTE). This term defines the pin configuration of the RS-232 interface found on your computer. See also *DCE*.

DCE. See *Data Communications Equipment*.

Default. A setting, instruction, or parameter that is used by a program if you do not supply a specific value. See also *Communications Parameters* and *Parameter*.

Delimiter. A character that separates multiple parameters in a text string.

Demodulate. The process of recovering the digital information from the carrier. See also *Modulate* and *Modem*.

Dial Up. The process of calling one computer with another, using modems and the telephone system. A dial-up system is a computer that you can access through your telephone system and a modem.

Digital Signal. A signal that has only specific discrete levels representing binary data. See also *Analog Signal*.

Directory. A listing of files or messages available in a specific area of a BBS or on-line service.

Disconnect. Terminate a connection with another modem or on-line service.

DNS. See *Domain Name System.*

Domain. A subportion of the hierarchical naming used in the domain name system. A domain is usually a separate administrative unit but does not have to be. See also *Address* and *DNS.*

Domain Name System (DNS). This system of electronic mail addressing is used by Internet and many commercial mail services. See also *Address* and *Domain.*

Doors. Special programs that are run remotely. These programs are used to extend BBS functions beyond those the bulletin board software itself supports. Almost any function may be presented as a door. To the caller, a door appears to be part of the BBS.

Download. The process of receiving a file from a remote computer.

DSR. See *Data Set Ready.*

DTE. See *Data Terminal Equipment.*

DTMF. Acronym for dual-tone modulated frequency, which is the term used by the telephone company to indicate the tones used to dial a number.

DTR. See *Data Terminal Ready.*

Duplex. Indicates that a communications channel is capable of handling data in both directions at the same time. Half duplex indicates that a channel can communicate only in one direction at a time. To contrast with the term half duplex, the term full duplex often is used to indicate that a circuit is capable of duplex operation.

EBCDIC. See *Extended Binary Coded Decimal Interchange Code.*

Echo. To repeat. If your terminal program is set to half duplex, the program usually echoes any keyboard input directly to the display. In a full-duplex system, your keyboard characters travel to the host computer, which echoes them back to your display.

EchoMail. A term that originated in the FidoNet store-and-forward network but is used elsewhere now to describe a method of creating a distributed message conference. Such a message conference connects several separate BBS message systems and enables national or international distributed message conferences. The effect is to create a single message area in which questions asked on one system may be answered by users of another system as though they were all on a single system.

Electronic Mail. The exchange of messages on-line. Also called E-mail. Traditionally, electronic mail refers to messages that may be read by only the sender or the recipient (and possibly the system administrator). Public message areas sometimes are called electronic mail but usually are called conferences or bulletin boards.

Emulation. The act of imitating something else. Generally used when a computer is programmed to operate the same as a specific model of data terminal in a process called terminal emulation.

Enter. Type text in response to a prompt.

Equalization. A process used by modems to compensate for telephone line frequency irregularities. Equalization involves altering the frequency response of the circuit.

Error-Free Modem. A modem that uses either the MNP or V.42 protocol to guarantee that all data is transmitted without errors.

Escape. A key used by many terminal programs to return to command mode. The ASCII escape code (0x1B) also is used in a data stream to introduce a control sequence that modifies the display operation.

Even Parity. The condition in which a parity bit is added to the data and is set so that the character data field always contains an even number of 1 bits when the parity bit is included. See also *Odd Parity*.

Extended Binary Coded Decimal Interchange Code (EBCDIC). IBM developed this eight-bit code. It serves the same purpose as ASCII and today is seen rarely outside the IBM mainframe environment.

External Modem. A self-contained modem that requires an RS-232 connection to a computer. An external modem generally can be used with any computer if the proper cable is available.

FidoNet. An international dial-up store-and-forward network that connects bulletin board system message areas. It also enables the sending of electronic mail among all systems.

File. A collection of data treated as a unit.

File Name. An identifier given to a file so that the computer can store and locate it.

File Transfer Protocol. See *Protocol*.

Filter. 1. Electronic circuitry that removes unwanted frequencies from a telephone line. 2. Software that removes unwanted characters from a data stream. The filter is usually a function of a terminal program.

Flow Control. The method that enables the computer and modem to regulate the flow of data between themselves. Software flow control uses the ASCII characters XOFF (0x13) and XON (0x11) to start and stop the data flow. Hardware flow control uses the RS-232 signals RTS and CTS to start and stop the data flow. Hardware flow control can be used with binary data because no confusion arises between the data and the control signals. See also *XON/XOFF*.

Framing Bits. The bits that "frame" the serial character in the data stream. See also *Start Bit* and *Stop Bit*.

Full Duplex. See *Duplex*.

Gateway. Software or hardware that connects two separate computer systems or message systems. A gateway is responsible for any data format conversions required to match two systems that normally are not compatible. A gateway may operate in real time or as a store-and-forward system.

Half Duplex. See *Duplex*.

Handshaking. Any process of signal exchange that is used to synchronize operations at two ends of a serial link. Handshaking may be done through hardware signals (such as RTS/CTS flow control) or through software signals (such as file transfer protocols).

Hardware. The physical components of a computer system.

Hayes Command Set. See *AT Command Set*.

Hayes Compatible. A modem that handles the AT command set in a manner similar enough to a Hayes modem that the modem may be used with most software.

Hertz (Hz). A measurement unit of frequency. One Hertz is one cycle per second.

Hexadecimal (hex). A base 16 numbering system that uses the letters A through F in addition to the digits 0 through 9 to represent numbers from 0 to 15 in a single digit. Because each hexadecimal digit represents 4 binary bits, hexadecimal is a convenient numbering system to use when discussing computers. A byte is 2 hexadecimal digits (8 bits) in length.

Host Software. Software that enables a computer to receive calls through a modem and accept commands from a remote user. Bulletin board system software (BBS) is host software, as are remote control software and terminal program host mode software.

Host System. A computer that enables one or more simultaneous callers to connect to it and execute functions remotely.

Initialization String. A group of modem commands used at the start-up of a communications program. These commands are sent to the modem to program it to the operating state required by the software.

Interface. The point of connection between two components of a computer system.

Internal Modem. A modem installed entirely within a computer. An internal modem takes all power and signals from within the computer and can be used only by the particular brand or model of computer for which the modem was designed.

Internet. An interconnected group of computers that makes possible the exchange of electronic mail and files. The computers are connected by TCP/IP protocol links and also support direct logon from each other if authorized. The Internet computers are networked in smaller groups that are connected to form the Internet. The Internet forms the backbone of the domain name system (DNS) electronic mail system.

Jumper. An electrical connection between two terminals on a printed circuit board. Jumpers rather than switches sometimes are used in modems for configuration of functions that rarely are changed.

Kermit. A file transfer protocol that was designed for maximum portability by Columbia University.

Kilobyte (K). A unit of memory capacity equal to 1,024 bytes.

KiloHertz (KHz). A unit of frequency measure equal to 1,000 cycles per second.

Line Feed. The ASCII character (0x0A), which indicates that the cursor should move to the next line.

Line Noise. Electrical signals induced into a telephone line by electrical machinery (such as motors), lightning, or improper cabling or equipment malfunction. These signals disrupt the normal operation of modems and can cause scrambled data unless you are using error-free modems or an error-checking transfer protocol.

Login. The act of establishing a software connection with a remote system. Usually a user identifier and possibly a password are required to complete the logon process and be connected fully and ready to use the remote system.

Logoff. The act of terminating the connection with the remote software by executing a specific logoff command.

Logon. See *Login*.

Macro. Character strings stored by your software and sent in response to a single key press. Macros are used to ease typing and to reduce errors on frequent operations, such as logging on.

Mark. An electrical state that represents a binary 1.

Mark Parity. The condition in which a parity bit is added to the data bits in a serial character and always is set to 1 regardless of the data.

Megabyte (M). A unit of storage equal to 1,048,576 bytes.

MegaHertz (MHz). A unit of frequency measure equal to 1,000,000 cycles per second.

Message. An electronic note left by one person for another and stored on a computer. Messages often are arranged in public conferences to allow discussion groups.

Microcom Networking Protocol (MNP). A protocol that provides modems with the capability to send data with no errors. Levels through MNP 4 are error-correcting protocols. MNP 5 is a data-compression protocol that can increase effective modem speed with some data.

MNP. See *Microcom Networking Protocol*.

Modem. MOdulator/DEModulator. The device is used to translate the digital serial signals from your computer serial port into analog signals that can be sent on a telephone line (modulation). A modem also translates the received analog signals back to digital signals for your computer to read (demodulation). See also *Demodulate* and *Modulate*.

Modulate. The act of altering the carrier tone to encode information. See also *Demodulate* and *Modem*.

Multiuser System. A remote system (BBS or commercial service) that enables more than one user to be connected and use the system at the same time.

Network. 1. A group of computers that have agreed upon a protocol enabling them to exchange messages left on one computer for a user on another computer. 2. A group of computers connected with special hardware and software so that they can share files and other resources such as printers. See also *Store and Forward*.

No Parity. The condition in which a serial character has only data bits and no added parity bit.

Nonvolatile Random-Access Memory (NRAM). This memory retains information even when the power is turned off. NRAM is used in some modems to hold settings that do not change when the modem is powered off and back on.

NOVRAM. See *NRAM*.

NRAM. See *Nonvolatile Random-Access Memory*.

Null Modem. A special cable that enables you to connect two DTE (terminal) devices directly by fooling each into thinking it is connected to a modem. The trick involves cross-connecting the wiring in a special way.

Odd Parity. The condition in which a parity bit is added to the data and is set so that the data character or field always contains an odd number of 1 bits when the parity bit is included. See also *Even Parity*.

Off-line. The state in which a computer is not connected to another computer.

On-line. The state in which a computer is connected to another computer.

Originate Mode. In a full-duplex modem, the mode that uses the "originating" tone set. (A full-duplex modem has two sets of tones, one for each direction. The answer mode uses the other set of tones, the "answer" tone set.)

Packet. See *Block*.

Packet-Switched Network. A data communications service that transmits data from one computer to another in the form of packets. With this method, the data for several computers can be sent at the same time on the same lines, reducing the cost. Commercial packet-switched networks such as Telenet and Tymnet sell data-connection services on their networks.

Parallel. A connection in which more than one data bit is transmitted at the same time.

Parameter. Any of a program's settings that you can change to affect the program's operation. See also *Default* and *Communications Parameters*.

Parity Bit. A bit that is added to the serial data. This bit is not part of the data but is set to ensure that a given number of 1 bits are contained in each serial character. Parity is a primitive form of error detection that has proven ineffective but still is used in some older systems.

Parity Checking. The act of verifying that the proper number of 1 bits (either even or odd) is contained in a received serial character.

Polling. Calling another system to see whether any messages or files are waiting. This process usually is automated.

Port. A point of connection between the computer and an external device such as a modem or a printer. A port is seen as a specific type and configuration of connector.

Profile. Information about your computer and terminal configuration that is stored by a remote computer. The profile enables the computer to remember between calls your preferential settings of certain parameters on the remote system.

Protocol. A mutually agreed-upon set of rules used in data transmission. File transfer protocols are the most visible occurrence.

Request To Send (RTS). The RS-232 interface signal used by the computer to tell the modem that it is allowed to send data to the computer. This signal is used in hardware flow control. See also *Clear To Send*.

RS-232. The Electronic Industries Association (EIA) standard that defines the method of electrical interconnection for serial communications ports.

RTS. See *Request To Send*.

Rubout. The ASCII character (0x7F), which is used on some systems to indicate the Backspace character. On other systems, this character is used as a padding character and is ignored. The character is named rubout because on old paper-tape systems it occurred when every hole was punched. The character thus could be used to "rub out" mistakes when a tape was being punched manually, because it could be punched over any other character.

Script. A set of stored commands that instructs a terminal program to perform automatically a predefined function (such as log on, read all messages, log off, and so on).

Serial Communications. The sequential transmission of bits over a single circuit.

Serial Port. The interface between a computer and a serial communications device.

SIG. See *Special Interest Group*.

Signal. Any data transmitted as variances in electrical impulses or tone frequencies.

Sign Off. See *Logoff*.

Sign On. See *Logon*.

Software. Programs that control the operation of hardware.

Space. An electrical state that represents a binary 0.

Space Parity. The condition in which a parity bit is added to serial data but always is set to 0.

Special Interest Group (SIG). This area of a BBS or commercial service usually combines message boards with file areas, all of which deal with a specific subject of mutual interest to the service's users. See also *Conference*.

Start Bit. The transition from 1 to 0 (mark to space), which indicates that a character transmission is beginning in asynchronous serial communications.

Stop Bit. The final bit of an asynchronous serial character, which is always a 1 (mark) to "frame" the end of the character electrically.

Store and Forward. A system that stores messages addressed to another computer and forwards them in batches later.

Synchronous. A format of serial communications in which bits are sent continuously. Because each clock interval is a new bit at all times, start or stop bits are not needed, and only data bits are sent. Synchronous transmission requires advanced software drivers but provides at least 20 percent faster throughput than asynchronous transmission. Because of its complexity, however, synchronous transmission is used primarily on dedicated links between computers.

SYSOP. Abbreviation for SYStem OPerator. This person is responsible for the operation and administration of a bulletin board system or subsection of a commercial on-line service.

T1 Line. A special leased telephone circuit capable of data transmission at speeds of 1.544 megabits per second.

TCP/IP. See *Transmission Control Protocol/Internet Protocol*.

Telecommunications. Data communications over telephone lines, using computers, terminals, and modems.

Telecommuting. Using telecommunications to avoid having to go to the office on a part-time or full-time basis.

Telecomputing. Using a computer to communicate with another computer over telephone lines. Generally, telecomputing includes at least some remote program operation.

Teletype (TTY). This terminal configuration is the most primitive configuration possible.

Terminal. A device that has a keyboard, display (or printing) device, and a serial port. A terminal is capable of transmitting key presses to the remote computer and displaying received data.

Terminal Emulation. See *Emulation*.

Terminal Mode. A mode of operation in which software on your PC emulates the operation of a terminal.

Transmission Control Protocol/Internet Protocol (TCP/IP). This set of synchronous protocols is used to connect the backbone computers in the Internet. TCP/IP was developed in the early 1970s by the Defense Advanced Research Projects Association (DARPA). It is a packet-switched protocol.

TTY. See *Teletype*.

UART. See *Universal Asynchronous Receiver/Transmitter*.

Universal Asynchronous Receiver/Transmitter (UART). The UART is the integrated circuit that translates the characters from the computer to and from the serial port.

Upload. The act of sending a data file to a remote computer.

V.22. CCITT modem protocol for 1200-bps modems.

V.22bis. CCITT modem protocol for 2400-bps modems.

V.32. CCITT modem protocol for 9600-bps modems.

V.32bis. A CCITT extension to V.32 protocol that enables modems to achieve full-duplex speeds of 14,400 bps.

V.42. A CCITT error-correcting protocol that enables modems to provide error-free data transmission. V.42 incorporates MNP levels 2, 3, and 4 as a subset and may be used on any speed modem.

V.42bis. A CCITT data-compression protocol that increases transmission speed as much as four times with some data. This advanced compression protocol always improves transmission speed slightly, as contrasted with MNP 5, which may reduce effective speed with some data patterns. Can be used on any speed modem.

Video display terminal (VDT). See *CRT*.

Wild Card. A selection mask formed by mixing text with the special masking characters question mark (?) and asterisk (*). This mask matches any character in which the special characters are present and thus can be used to select a group of names from a list based on selection criteria.

Word. 1. The largest group of data that the computer treats as a single entity. For an IBM PC or compatible, a word is 16 bits or 2 bytes. 2. The length of an asynchronous serial data character including start, stop, and parity bits. For PC communications, a serial data word is usually 10 bits.

Word Length. See *Character Length*.

X.25. A packet-switching protocol that commonly is used by public data networks. It also is used by private data networks. See also *Packet-Switched Network*.

XMODEM. The original error-correcting file transfer protocol. Xmodem breaks files into data blocks and sends them one at a time, waiting for verification or resend requests between each block.

XON/XOFF. Software flow control. The characters XON (0x13) and XOFF (0x11) are used to signal that data should stop and start. See also *Flow Control*.

YMODEM. A file transfer protocol that is similar to XMODEM except that it uses larger blocks. YMODEM also is capable of sending several files in a single transfer and keeping exact sizes and dates intact.

ZMODEM. An advanced file transfer protocol that adapts to the conditions of the circuit. It also has an auto-start feature on the receiving end, which makes ZMODEM much more convenient to use than most protocols.

B

ASCII and Extended IBM Codes

This appendix presents the ASCII, Extended IBM Graphic, and the Extended-IBM Function Key codes. In the tables, a ^ represents the Control (Ctrl) key. For example, ^C represents Ctrl-C.

ASCII Codes

The codes for the American Standard Code for Information Interchange (ASCII) are presented in the following table:

Decimal	Hex	Octal	Binary	Graphic Character	ASCII Meaning
0	0	0	00000000		^@ NUL (null)
1	1	1	00000001	☺	^A SOH (start-of-header)
2	2	2	00000010	●	^B STX (start-of-transmission)
3	3	3	00000011	♥	^C ETX (end-of-transmission)
4	4	4	00000100	♦	^D EOT (end-of-text)
5	5	5	00000101	♣	^E ENQ (enquiry)
6	6	6	00000110	♠	^F ACK (acknowledge)
7	7	7	00000111	·	^G BEL (bell)
8	8	10	00001000	■	^H BS (backspace)
9	9	11	00001001	○	^I HT (horizontal tab)
10	A	12	00001010	◙	^J LF (line feed - also ^Enter)
11	B	13	00001011	♂	^K VT (vertical tab)
12	C	14	00001100	♀	^L FF (form feed)
13	D	15	00001101	♪	^M CR (carriage return)
14	E	16	00001110	♫	^N SO

Decimal	Hex	Octal	Binary	Graphic Character	ASCII Meaning
15	F	17	00001111	☼	^O SI
16	10	20	00010000	►	^P DLE
17	11	21	00010001	◄	^Q DC1
18	12	22	00010010	‼	^R DC2
19	13	23	00010011	‖	^S DC3
20	14	24	00010100	¶	^T DC4
21	15	25	00010101	§	^U NAK
22	16	26	00010110	▬	^V SYN
23	17	27	00010111	↕	^W ETB
24	18	30	00011000	↑	^X CAN (cancel)
25	19	31	00011001	↓	^Y EM
26	1A	32	00011010	→	^Z SUB (also end-of-file)
27	1B	33	00011011	←	^[ESC (Escape)
28	1C	34	00011100	∟	^\ FS (field separator)
29	1D	35	00011101	↔	^] GS
30	1E	36	00011110	▲	^^ RS (record separator)
31	1F	37	00011111	▼	^_ US
32	20	40	00100000		Space
33	21	41	00100001	!	!
34	22	42	00100010	"	"
35	23	43	00100011	#	#
36	24	44	00100100	$	$
37	25	45	00100101	%	%
38	26	46	00100110	&	&
39	27	47	00100111	'	'
40	28	50	00101000	(	(
41	29	51	00101001	)	)
42	2A	52	00101010	*	*
43	2B	53	00101011	+	+
44	2C	54	00101100	,	,
45	2D	55	00101101	-	-
46	2E	56	00101110	.	.
47	2F	57	00101111	/	/
48	30	60	00110000	0	0
49	31	61	00110001	1	1
50	32	62	00110010	2	2
51	33	63	00110011	3	3
52	34	64	00110100	4	4
53	35	65	00110101	5	5
54	36	66	00110110	6	6
55	37	67	00110111	7	7
56	38	70	00111000	8	8
57	39	71	00111001	9	9
58	3A	72	00111010	:	:
59	3B	73	00111011	;	;
60	3C	74	00111100	<	<
61	3D	75	00111101	=	=
62	3E	76	00111110	>	>
63	3F	77	00111111	?	?
64	40	100	01000000	@	@
65	41	101	01000001	A	A
66	42	102	01000010	B	B

Decimal	Hex	Octal	Binary	Graphic Character	ASCII Meaning
67	43	103	01000011	C	C
68	44	104	01000100	D	D
69	45	105	01000101	E	E
70	46	106	01000110	F	F
71	47	107	01000111	G	G
72	48	110	01001000	H	H
73	49	111	01001001	I	I
74	4A	112	01001010	J	J
75	4B	113	01001011	K	K
76	4C	114	01001100	L	L
77	4D	115	01001101	M	M
78	4E	116	01001110	N	N
79	4F	117	01001111	O	O
80	50	120	01010000	P	P
81	51	121	01010001	Q	Q
82	52	122	01010010	R	R
83	53	123	01010011	S	S
84	54	124	01010100	T	T
85	55	125	01010101	U	U
86	56	126	01010110	V	V
87	57	127	01010111	W	W
88	58	130	01011000	X	X
89	59	131	01011001	Y	Y
90	5A	132	01011010	Z	Z
91	5B	133	01011011	[	[
92	5C	134	01011100	\	\
93	5D	135	01011101	]	]
94	5E	136	01011110	^	^
95	5F	137	01011111	_	_
96	60	140	01100000	`	`
97	61	141	01100001	a	a
98	62	142	01100010	b	b
99	63	143	01100011	c	c
100	64	144	01100100	d	d
101	65	145	01100101	e	e
102	66	146	01100110	f	f
103	67	147	01100111	g	g
104	68	150	01101000	h	h
105	69	151	01101001	i	i
106	6A	152	01101010	j	j
107	6B	153	01101011	k	k
108	6C	154	01101100	l	l
109	6D	155	01101101	m	m
110	6E	156	01101110	n	n
111	6F	157	01101111	o	o
112	70	160	01110000	p	p
113	71	161	01110001	q	q
114	72	162	01110010	r	r
115	73	163	01110011	s	s
116	74	164	01110100	t	t
117	75	165	01110101	u	u
118	76	166	01110110	v	v

Decimal	Hex	Octal	Binary	Graphic Character	
119	77	167	01110111	w	w
120	78	170	01111000	x	x
121	79	171	01111001	y	y
122	7A	172	01111010	z	z
123	7B	173	01111011	{	{
124	7C	174	01111100	|	|
125	7D	175	01111101	}	}
126	7E	176	01111110	~	~
127	7F	177	01111111	Δ	Del

The Extended IBM Character Graphics Display Codes are presented in the following table:

Decimal	Hex	Octal	Binary	Graphic Character
128	80	200	10000000	Ç
129	81	201	10000001	ü
130	82	202	10000010	é
131	83	203	10000011	â
132	84	204	10000100	ä
133	85	205	10000101	à
134	86	206	10000110	å
135	87	207	10000111	ç
136	88	210	10001000	ê
137	89	211	10001001	ë
138	8A	212	10001010	è
139	8B	213	10001011	ï
140	8C	214	10001100	î
141	8D	215	10001101	ì
142	8E	216	10001110	Ä
143	8F	217	10001111	Å
144	90	220	10010000	É
145	91	221	10010001	æ
146	92	222	10010010	Æ
147	93	223	10010011	ô
148	94	224	10010100	ö
149	95	225	10010101	ò
150	96	226	10010110	û
151	97	227	10010111	ù
152	98	230	10011000	ÿ
153	99	231	10011001	Ö
154	9A	232	10011010	Ü
155	9B	233	10011011	¢
156	9C	234	10011100	£
157	9D	235	10011101	¥
158	9E	236	10011110	₧
159	9F	237	10011111	ƒ
160	A0	240	10100000	á
161	A1	241	10100001	í

Decimal	Hex	Octal	Binary	Graphic Character
162	A2	242	10100010	ó
163	A3	243	10100011	ú
164	A4	244	10100100	ñ
165	A5	245	10100101	Ñ
166	A6	246	10100110	ª
167	A7	247	10100111	º
168	A8	250	10101000	¿
169	A9	251	10101001	⌐
170	AA	252	10101010	¬
171	AB	253	10101011	½
172	AC	254	10101100	¼
173	AD	255	10101101	¡
174	AE	256	10101110	«
175	AF	257	10101111	»
176	B0	260	10110000	░
177	B1	261	10110001	▓
178	B2	262	10110010	▓
179	B3	263	10110011	│
180	**B4**	264	10110100	┤
181	**B5**	265	10110101	╡
182	**B6**	266	10110110	╢
183	**B7**	267	10110111	╖
184	**B8**	270	10111000	╕
185	**B9**	271	10111001	╣
186	**BA**	272	10111010	║
187	**BB**	273	10111011	╗
188	**BC**	274	10111100	╝
189	**BD**	275	10111101	╜
190	**BE**	276	10111110	╛
191	**BF**	277	10111111	┐
192	C0	300	11000000	└
193	C1	301	11000001	┴
194	C2	302	11000010	┬
195	C3	303	11000011	├
196	C4	304	11000100	─
197	C5	305	11000101	┼
198	C6	306	11000110	╞
199	C7	307	11000111	╟
200	C8	310	11001000	╚
201	C9	311	11001001	╔
202	CA	312	11001010	╩
203	CB	313	11001011	╦
204	CC	314	11001100	╠
205	CD	315	11001101	═
206	CE	316	11001110	╬
207	CF	317	11001111	╧
208	D0	320	11010000	╨
209	D1	321	11010001	╤
210	D2	322	11010010	╥
211	D3	323	11010011	╙
212	D4	324	11010100	╘

Decimal	Hex	Octal	Binary	Graphic Character
213	D5	325	11010101	╞
214	D6	326	11010110	╥
215	D7	327	11010111	╫
216	D8	330	11011000	╪
217	D9	331	11011001	╝
218	DA	332	11011010	╭
219	DB	333	11011011	█
220	DC	334	11011100	▄
221	DD	335	11011101	▌
222	DE	336	11011110	▐
223	DF	337	11011111	▀
224	E0	340	11100000	∝
225	E1	341	11100001	β
226	E2	342	11100010	Γ
227	E3	343	11100011	π
228	E4	344	11100100	Σ
229	E5	345	11100101	σ
230	E6	346	11100110	μ
231	E7	347	11100111	τ
232	E8	350	11101000	Φ
233	E9	351	11101001	θ
234	EA	352	11101010	Ω
235	EB	353	11101011	δ
236	EC	354	11101100	∞
237	ED	355	11101101	ϕ
238	EE	356	11101110	∈
239	EF	357	11101111	∩
240	F0	360	11110000	≡
241	F1	361	11110001	±
242	F2	362	11110010	≥
243	F3	363	11110011	≤
244	F4	364	11110100	⌠
245	F5	365	11110101	⌡
246	F6	366	11110110	÷
247	F7	367	11110111	≈
248	F8	370	11111000	°
249	F9	371	11111001	·
250	FA	372	11111010	·
251	FB	373	11111011	√
252	FC	374	11111100	ⁿ
253	FD	375	11111101	²
254	FE	376	11111110	■
255	FF	377	11111111	

Extended IBM Keyboard Codes

Certain keys cannot be represented by the standard ASCII codes. To represent the codes, a two-character sequence is used. The first character is always an ASCII NUL (0). The second character and its translation are listed in the following table. Some codes expand to multikeystroke characters. If an asterisk (*) appears in the column Enhanced Only, the sequence is available only on the IBM Enhanced Keyboards (101- and 102-key keyboards).

Enhanced Only	Decimal Meaning	Hex	Octal	Binary	Extended Code
*	1	01	001	00000001	Alt-Esc
	3	03	003	00000011	Null (null character)
*	14	0E	016	00001110	Alt-Backspace
	15	0F	017	00001111	Shift-Tab (back-tab)
	16	10	020	00010000	Alt-Q
	17	11	021	00010001	Alt-W
	18	12	022	00010010	Alt-E
	19	13	023	00010011	Alt-R
	20	14	024	00010100	Alt-T
	21	15	025	00010101	Alt-Y
	22	16	026	00010110	Alt-U
	23	17	027	00010111	Alt-I
	24	18	030	00011000	Alt-O
	25	19	031	00011001	Alt-P
*	26	1A	032	00011010	Alt-[
*	27	1B	033	00011011	Alt-]
*	28	1C	034	00011100	Alt-Enter
	30	1E	036	00011110	Alt-A
	31	1F	037	00011111	Alt-S
	32	20	040	00100000	Alt-D
	33	21	041	00100001	Alt-F
	34	22	042	00100010	Alt-G
	35	23	043	00100011	Alt-H
	36	24	044	00100100	Alt-J
	37	25	045	00100101	Alt-K
	38	26	046	00100110	Alt-L
*	39	27	047	00100111	Alt-;
*	40	28	050	00101000	Alt-'

Enhanced Only	Decimal Meaning	Hex	Octal	Binary	Extended Code
*	41	29	051	00101001	Alt-'
*	43	2B	053	00101011	Alt-\
	44	2C	054	00101100	Alt-Z
	45	2D	055	00101101	Alt-X
	46	2E	056	00101110	Alt-C
	47	2F	057	00101111	Alt-V
	48	30	060	00110000	Alt-B
	49	31	061	00110001	Alt-N
	50	32	062	00110010	Alt-M
*	51	33	063	00110011	Alt-,
*	52	34	064	00110100	Alt-.
*	53	35	065	00110101	Alt-/
*	55	37	067	00110111	Alt-* (keypad)
	57	39	071	00111001	Alt-space bar
	59	3B	073	00111011	F1
	60	3C	074	00111100	F2
	61	3D	075	00111101	F3
	62	3E	076	00111110	F4
	63	3F	077	00111111	F5
	64	40	100	01000000	F6
	65	41	101	01000001	F7
	66	42	102	01000010	F8
	67	43	103	01000011	F9
	68	44	104	01000100	F10
	71	47	107	01000111	Home
	72	48	110	01001000	↑
	73	49	111	01001001	PgUp
	74	4A	112	01001010	Alt-– (keypad)
	75	4B	113	01001011	←
	76	4C	114	01001100	Shift-5 (keypad)
	77	4D	115	01001101	«
	78	4E	116	01001110	Alt-+ (keypad)
	79	4F	117	01001111	End
*	80	50	120	01010000	»
*	81	51	121	01010001	PgDn
*	82	52	122	01010010	Ins (Insert)
	83	53	123	01010011	Del (Delete)
	84	54	124	01010100	Shift-F1
	85	55	125	01010101	Shift-F2
	86	56	126	01010110	Shift-F3
	87	57	127	01010111	Shift-F4

Enhanced Only	Decimal Meaning	Hex	Octal	Binary	Extended Code
	88	58	130	01011000	Shift-F5
	89	59	131	01011001	Shift-F6
	90	5A	132	01011010	Shift-F7
	91	5B	133	01011011	Shift-F8
	92	5C	134	01011100	Shift-F9
	93	5D	135	01011101	Shift-F10
	94	5E	136	01011110	Ctrl-F1
	95	5F	137	01011111	Ctrl-F2
	96	60	140	01100000	Ctrl-F3
	97	61	141	01100001	Ctrl-F4
	98	62	142	01100010	Ctrl-F5
	99	63	143	01100011	Ctrl-F6
	100	64	144	01100100	Ctrl-F7
	101	65	145	01100101	Ctrl-F8
	102	66	146	01100110	Ctrl-F9
	103	67	147	01100111	Ctrl-F10
	104	68	150	01101000	Alt-F1
	105	69	151	01101001	Alt-F2
	106	6A	152	01101010	Alt-F3
	107	6B	153	01101011	Alt-F4
	108	6C	154	01101100	Alt-F5
	109	6D	155	01101101	Alt-F6
	110	6E	156	01101110	Alt-F7
	111	6F	157	01101111	Alt-F8
	112	70	160	01110000	Alt-F9
	113	71	161	01110001	Alt-F10
	114	72	162	01110010	Ctrl-PrtSc
	115	73	163	01110011	Ctrl-←
	116	74	164	01110100	Ctrl-→
	117	75	165	01110101	Ctrl-End
	118	76	166	01110110	Ctrl-PgDn
	119	77	167	01110111	Ctrl-Home
	120	78	170	01111000	Alt-1 (keyboard)
	121	79	171	01111001	Alt-2 (keyboard)
	122	7A	172	01111010	Alt-3 (keyboard)
	123	7B	173	01111011	Alt-4 (keyboard)
	124	7C	174	01111100	Alt-5 (keyboard)
	125	7D	175	01111101	Alt-6 (keyboard)
	126	7E	176	01111110	Alt-7 (keyboard)
	127	7F	177	01111111	Alt-8 (keyboard)
	128	80	200	10000000	Alt-9 (keyboard)

Enhanced Only	Decimal Meaning	Hex	Octal	Binary	Extended Code
	129	81	201	10000001	Alt-0 (keyboard)
	130	82	202	10000010	Alt--(keyboard)
	131	83	203	10000011	Alt-= (keyboard)
	132	84	204	10000100	Ctrl-PgUp
*	133	85	205	10000101	F11
*	134	86	206	10000110	F12
*	135	87	207	10000111	Shift-F11
*	136	88	210	10001000	Shift-F12
*	137	89	211	10001001	Ctrl-F11
*	138	8A	212	10001010	Ctrl-F12
*	139	8B	213	10001011	Alt-F11
*	140	8C	214	10001100	Alt-F12
	141	8D	215	10001101	Ctrl-↑/8 (keypad)
	142	8E	216	10001110	Ctrl--(keypad)
	143	8F	217	10001111	Ctrl-5 (keypad)
	144	90	220	10010000	Ctrl-+ (keypad)
	145	91	221	10010001	Ctrl-↓/2 (keypad)
	146	92	222	10010010	Ctrl-Ins/0 (keypad)
	147	93	223	10010011	Ctrl-Del/. (keypad)
	148	94	224	10010100	Ctrl-Tab
*	149	95	225	10010101	Ctrl-/ (keypad)
*	150	96	226	10010110	Ctrl-* (keypad)
*	151	97	227	10010111	Alt-Home
*	152	98	230	10011000	Alt-↓
*	153	99	231	10011001	Alt-Page Up
*	155	9B	233	10011011	Alt-←
*	157	9D	235	10011101	Alt-→
*	159	9F	237	10011111	Alt-End
*	160	A0	240	10100000	Alt-↑
*	161	A1	241	10100001	Alt-Page Down
*	162	A2	242	10100010	Alt-Insert
*	163	A3	243	10100011	Alt-Delete
*	164	A4	244	10100100	Alt-/ (keypad)
*	165	A5	245	10100101	Alt-Tab
*	166	A6	256	10100110	Alt-Enter (keypad)

Enhanced Keyboard Function Key Codes

The following extended codes are available only with the Enhanced Keyboards (101/102-key keyboards); the codes are available for key reassignment only under DOS 4.0 and later. The keys include the six-key editing pad and the four-key cursor-control pad. To reassign these keys, you must include the DEVICE = ANSI.SYS /X command in CONFIG.SYS or issue the enable extended function codes escape sequence (Esc[1q). All extended codes are prefixed by 224 decimal (E0 hex).

Decimal Meaning	Hex	Octal	Binary	Extended Code
71	47	107	01000111	Home
72	48	110	01001000	↑
73	49	111	01001001	Page Up
75	4B	113	01001011	←
77	4D	115	01001101	→
79	4F	117	01001111	End
80	50	120	01010000	↓
81	51	121	01010001	Page Down
82	52	122	01010010	Insert
83	53	123	01010011	Delete
115	73	163	01110011	Ctrl-←
116	74	164	01110100	Ctrl-→
117	75	165	01110101	Ctrl-End
118	76	166	01110110	Ctrl-Page Down
119	77	167	01110111	Ctrl-Home
132	84	204	10000100	Ctrl-Page Up
141	8D	215	10001101	Ctrl-↑
145	91	221	10010001	Ctrl-↓
146	92	222	10010010	Ctrl-Insert
147	93	223	10010011	Ctrl-Delete

C

The AT
Command Set

The Hayes AT command set consists of a basic set of commands and extended set of commands. All Hayes compatible modems implement the basic command set. The extended command set, however, is not as standard and the AT commands used to control advanced modem features such as MNP and V.42 have very little standardization. This appendix discusses the basic command set as well as those extended AT commands that are nearly always implemented in newer modems. The following table shows the basic AT command set.

Command	Function
AT	Prefixes all other commands. By itself results in an OK response indicating the modem is accepting commands.
A/	Reissues the last command given
A	Manually answers an incoming call
C0	Disables transmit carrier
C1	Enables transmit carrier
DP	Dials using pulse method (number follows)
DT	Dials using DTMF tone method (number follows)
E0	Disables echo of command input
E1	Enables echo of command input

Command	Function
F0	Sets half duplex mode
F1	Sets full duplex mode
H0	Places phone on hook (hang up)
H1	Takes phone off hook (pick up receiver)
I	Returns modem product code
I1	Returns internal ROM checksum
I2	Performs modem internal memory test
L1	Sets speaker to low volume
L2	Sets speaker to medium volume
L3	Sets speaker to high volume
M0	Turns speaker off
M1	Speaker on when dialing, off when carrier detected
M2	Speaker always off
O	Returns on-line if connected
Q0	Returns modem command result codes
Q1	Does not return modem command result codes
Sn=x	Sets S register to value. These registers control several internal timing and modem features
Sn?	Displays current value of an S register
V0	Returns numeric result codes
V1	Returns text result codes
Xn	Sets result code completeness level to n. Higher values enable call progress monitoring, baud rate detection, and other extended result codes
Z	Resets the modem internally, and restores last stored values for all parameters
+++	Enters command mode if now in data mode and this string is preceded and followed by a pause

AT (Attention)

All Hayes AT commands are sent to the modem prefixed by the attention sequence AT. This informs the modem that what follows is a command it should interpret and take action on. The command buffer in a Hayes modem can hold command strings up to a maximum of 40 characters in length. Some other brands of modems allow more than 40 characters in a command, but if you limit all AT commands to 40 characters or less, any modem which is Hayes compatible will correctly interpret the command. The shorthand command A/ will again perform the last AT command the modem received and is the only command that does not begin with AT and end with an Enter key press.

The letters AT must be both uppercase or both lowercase. This is because the modem uses these letters to determine the baud rate to which your computer has set the serial port. In this way the modem can recognize the commands at any port speed the modem supports. Some modems accept only an uppercase AT.

A (Answer Telephone)

The A command causes the modem to take the telephone line off hook and generate a carrier tone.

B (Mode Select)

The B command selects either CCITT V.22 or Bell 212A modulation for 1200 bps. A number follows the letter B to indicate which modulation method you want the modem to use. ATB0 selects CCITT V.22 modulation, and ATB1 selects Bell 212A modulation.

C (Carrier Set)

The C command enables most Hayes compatible modems to turn the carrier signal on or off. A number follows the letter C to indicate whether on or off is desired. ATC0 turns the carrier off, ATC1 turns the carrier on. This is a special purpose command; for normal use you will always want ATC1.

D (Dial)

This command is the "workhorse" of the Hayes compatible modem. It tells the modem to dial a specified telephone number and which dialing type (Tone or Pulse) to use. The dialing command is a string consisting of the telephone number and the following modifier characters:

P	Use Pulse Dialing
T	Use Tone Dialing
R	Reverse the tones and use the Answer tone set after dialing
W	Causes the modem to delay the number of seconds indicated in the S7 register for a dial tone. *Note:* Not all modems support this command.
,	A comma causes a brief delay before dialing the next digit. The length of this delay is set in the S8 register.
;	If the command ends in a semicolon, the modem returns to the command state after dialing.

Examples:

To tone dial the telephone number 699-8222, use the following command:

ATDT6998222

To pulse dial the telephone number 699-8222 from an office where you must first dial a 9 to get an outside line, use the following command:

ATDP9,,6998222

Some modems add extra capability to the dial command by supporting other control characters. The characters included here are the basic ones.

E (Echo)

The echo command tells the modem whether or not you want it to echo each command character you type back to your screen. A number follows the letter E to indicate what you want. ATE1 requests the modem to echo command letters, and ATE0 tells the modem not to echo the command letters. The modem interprets and executes all commands identically regardless of this setting.

F (Duplex)

This command is used by some Hayes modems to select full- or half-duplex operation of the carrier at 300 bps. ATF1 selects full-duplex (normal) operation, and ATF0 selects half-duplex operation. This is a special purpose command; you normally will use ATF1.

H (Hook)

This command indicates whether you want the modem to have the telephone line *On hook* or *Off hook*. ATH1 causes the modem to take the telephone line off hook (pick up the telephone), and ATH0 causes the modem to place the telephone line on hook (hang up the phone).

I (Information)

This command asks the modem to send product identification information to your terminal screen as follows:

ATI or ATI0	Returns the product's 3-digit identification code
ATI1	Returns a 3-digit checksum of the modem's ROM firmware
ATI2	Causes the modem to test its ROM firmware checksum and return either an OK or an ERROR. ERROR indicates hardware failure.

L (Loudness)

This command selects the modem's speaker volume. Some modems have a manual volume control, and for them this command does nothing. Other modems have no volume knob on the speaker and this command sets the speaker volume as follows:

ATL0   Lowest volume

ATL1   Low volume

ATL2   Medium volume

ATL3   High volume

M (Mute)

This command determines when the modem has its speaker on, and when the speaker is muted, as follows:

ATM0   Speaker always off

ATM1   Speaker is turned on when the modem is dialing or answering the telephone, but turns off when a carrier signal is detected

ATM2   Speaker is always on

ATM3   Speaker is turned on after the last digit is dialed and stays on until carrier is detected. *Note:* Not all Hayes-compatible modems support ATM3.

O (On-line)

The command ATO places the modem back into data mode after an escape sequence (+ + +) enters the command mode.

Q (Quiet)

This command tells the modem whether or not you want it to report the results of action with result code messages (for example, CONNECT, OK, and so on). ATQ0 causes the modem to operate without reporting any result codes. ATQ1 causes the modem to report result codes for each operation.

S (Set Registers)

Hayes compatible modems have a series of registers called *S registers*, which this command can either set or interrogate. Each S register has a number which is part of this command. To demonstrate the command, the following example uses register number 1.

ATS1=5   Sets register 1 to a value of 5

ATS1?   Causes the modem to return the value currently in register 1

At the end of this appendix is a chart of the basic Hayes S registers. Advanced modems have many S registers (as many as 100 for some modems). Each modem uses its S registers for different purposes, however, and you have to read the documentation of your modem to see what S registers it has and what they do.

V (Verbose)

This command enables you to select one of two types of result codes from the modem. ATV1 selects the verbose text codes like CONNECT, OK, and so on. ATV0 selects a numeric equivalent for each code. Text codes typically are used in manual mode because they are easy for people to read and understand. Numeric codes are used by many software programs because they are shorter and therefore more immune to noise.

X (eXtended Result Codes)

This command selects the number of different result codes the modem will return. The exact meaning of this command varies with different modem brands, but the higher the number after the X command the more status a modem will return. Basic Hayes modems use ATX1, ATX2, ATX3, and ATX4 to define four levels of result-code reporting. On most modems, X values higher than 1 also enable dial tone detection when dialing. X1 waits a specified time and then dials regardless of the presence or absence of a dial tone. This is called *blind dialing*. X2 causes the modem to wait until it detects a dial tone before dialing.

Z (Reset)

The command ATZ causes the modem to perform a complete reset and restores all settings to their default values. Some modems also take the time to do a full modem diagnostic when this command is issued.

+++ (Escape)

When your modem is in data mode and you want to issue a modem command, you must ask the modem to switch from data to command mode. You issue the data string +++ to cause this to happen. In order to avoid switching to command mode in case true data has this string in it, you must wait at least 1 second before typing this data, and again after typing it before the modem will switch to command mode. The one second actually is the number of seconds you program into register S12. The escape character (+) also can be changed by programming a new value into register S2. After you escape to command mode and issue the modem commands you want, you return to the data mode with the ATO command.

S Registers

The Hayes modem supports the following standard S registers. Most compatibles also support these registers. S registers above these basic ones, however, vary in function with each brand of modem.

Register	Description	Range	Default
S0	Select ring to answer on	0-255 Rings	0
S1	Ring Count (read only)	0-255 Rings	—
S2	Escape Sequence Character	0-127 ASCII	43
S3	Carriage Return Character	0-127 ASCII	13
S4	Line Feed Character	0-127 ASCII	10
S5	Backspace Character	0-127 ASCII	8
S6	Wait before blind dialing	2-255 sec	2
S7	Wait time for carrier detect	1-255 sec	30
S8	Duration of delay for comma	0-255 sec	2
S9	Carrier Detect response time	1-255 1/10 sec	6
S10	Delay carrier loss to hang up	1-255 1/10 sec	14
S11	Duration/Spacing of dialing tones	50-255 msec	95
S12	Escape Sequence Guard Time	0-255 1/50 sec	50

Result Codes

The basic Hayes modems report the following result codes:

Numeric	Verbal
0	OK
1	CONNECT
2	RING
3	NO CARRIER

Numeric	Verbal
4	ERROR
5	CONNECT 1200
6	NO DIALTONE
7	BUSY
8	NO ANSWER
10	CONNECT 2400

Extended AT commands

With the introduction of the Hayes 2400 modem, Hayes no longer used dip switches to configure modem settings. Most modem manufacturers today have followed along, at least partially. In order to program these extended modem features, extended commands were required. A basic set of extended commands exist which most modems support. Much less standardization exists among extended commands, however, so you should check your modem manual carefully to see what commands it supports.

&C (Carrier Detect)

This command sets the modem to have the RS-232 signal Carrier Detect either always be true, or to track the presence of a carrier as follows:

AT&C0 Forces carrier detect ON all the time

AT&C1 Makes the modem's carrier detect signal reflect the presence or absence of a carrier signal

&D (Data Terminal Ready)

This command sets the modem to either honor or ignore the RS-232 Data Terminal Ready signal as follows:

AT&D0 Modem ignores DTR signal

AT&D1	Modem returns to command state when DTR changes from ON to OFF
AT&D2	Modem goes on-hook (hangs up) and will not answer the telephone when DTR changes from ON to OFF. The modem automatically resumes answering the phone when DTR returns to the ON state.
AT&D3	Modem hangs up and performs a reset when DTR changes from ON to OFF

&F (Factory Settings)

The AT&F command returns all modem NOVRAM settings to their factory default settings.

&G (Guard Tones)

When a modem is used (or receiving calls from) outside the United States, it should answer with guard tones (see Chapter 20 and 21 for an explanation of guard tones). This command sets the modem to handle guard tones as follows:

AT&G0	No guard tones are sent
AT&G1	Modem sends a 550Hz guard tone
AT&G2	Modem sends an 1800 Hz guard tone

&T (Test)

Most modems that have extended command sets support several self-test modes. See Chapter 22 for a description of how to use these self-test modes. The commands testing are as follows:

AT&T0	End any testing, return modem to normal mode
AT&T1	Initiate local analog loopback test

AT&T3	Initiate local digital loopback test
AT&T4	Allow this modem to grant a request from another modem for a remote digital loopback test
AT&T5	Ignore any request from another modem to perform a remote digital loopback test
AT&T6	Initiate manual remote digital loopback test
AT&T7	Initiate remote digital loopback test and perform automatic self test sequence
AT&T8	Initiate local analog loopback test and perform automatic self test sequence

&W (Write Changes to NOVRAM)

This command stores any changes to the extended parameters into the modem's nonvolatile RAM (NOVRAM). This means the modem now will reset to the new settings with either an ATZ command or when the power is turned off and on. Some modems enable multiple profiles to be stored and add a number to the AT&W command to indicate which profile should be stored.

Cable Diagrams

T he theory behind the wiring of the cables shown here is explained in Chapter 2. This appendix provides a quick reference to the most commonly required serial cables.

Standard 25-Pin IBM PC Modem Cable

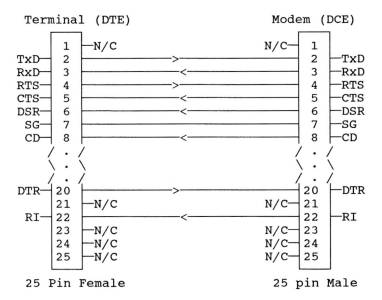

Standard 99-Pin IBM PC Modem Cable

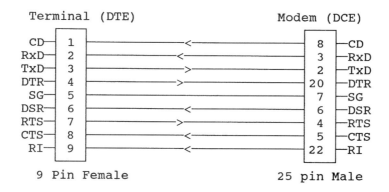

25-Pin RS-232 Full Handshake Null Modem Cable

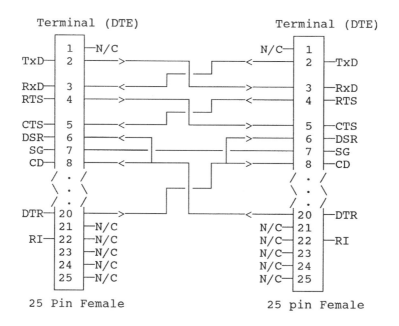

E

Using PROCOMM Lite

Chapter 3 introduced you to the fundamentals of PC communications software. PROCOMM Lite, a PC communications program from the publishers of the popular PROCOMM PLUS communications package, is included free with this book. Using this program, you can connect your computer to electronic bulletin boards and on-line services, and upload and download files by using the program's built-in file-transfer protocols. This appendix helps you get started using PROCOMM Lite by describing the commands in the program and explaining their primary functions.

Installing PROCOMM Lite

PROCOMM Lite consists of a single program file, PCLITE.EXE, stored on the disk found in the back of this book. The program can be used from either a floppy disk or from a hard disk.

If you want to use PROCOMM Lite from a floppy disk, you first should make a backup copy. You can use the DISKCOPY command, for example, to make an exact duplicate of the disk, or you can use the COPY command to copy PCLITE.EXE to another disk. Place the original disk in a safe place and use the copy.

To use PROCOMM Lite from a hard disk, first use the MKDIR (or MD) command to create a directory for the program. You may want to create a directory named PCLITE, for example. To create a directory named PCLITE, type the following command at the DOS command line and press Enter:

MD \PCLITE

After you create the directory, copy the PROCOMM Lite program file to the new directory. To copy the program from a floppy disk in drive A to the PCLITE directory on drive C, for example, type the following command at the DOS command line and press Enter:

COPY A:PCLITE.EXE C:\PCLITE

Change to the directory that contains the PROCOMM Lite program and you are ready to get started.

Getting Started

PROCOMM Lite works with any PC-, AT-, or PS/2-style keyboard or equivalent. The program uses practically every key on the keyboard, so you need to be familiar with your keyboard. This documentation uses the labels found on an IBM Enhanced keyboard to describe keys you should press.

Begin the PROCOMM Lite program by entering the following command at the DOS prompt:

PCLITE

You also can start PROCOMM Lite using one of the following special start-up commands:

PCLITE /C

Use this command only if your computer "freezes-up" whenever you run PROCOMM Lite. This command starts the program without initializing the computer's serial port or the modem. Use the Alt-P (Line/Port Setup) command to select a different serial port. Exit from the program and then try starting PROCOMM Lite again, without /C.

PCLITE /K

This command disables keyboard speedup. Some older PC compatibles may not work properly with the normal keyboard speedup feature enabled.

> PCLITE /B
>
> Use this command if you are running PROCOMM Lite on a computer with a monochrome or LCD screen.
>
> PCLITE /F*script*
>
> This start-up command causes PROCOMM Lite to run a script immediately upon beginning the program. Refer to the discussion on script language that appears in the section "Using the PROCOMM Lite Script Language" at the end of this appendix.

After you start the program, PROCOMM Lite displays its logo screen. This screen displays the message PRESS ANY KEY TO ENTER TERMINAL MODE. When you press Enter, the screen shown in figure E.1 appears.

This screen is called the Terminal Mode screen. You need to become familiar with the Terminal Mode screen because it is the program's main screen. The Terminal Mode screen displays text transmitted by the computer with which you are communicating.

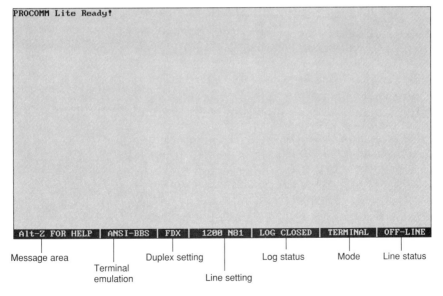

Fig. E.1. The PROCOMM Lite Terminal Mode screen.

PROCOMM Lite should greet you at the upper left of the Terminal Mode screen with the message PROCOMM Lite Ready! and place a blinking underscore, the cursor, below the P in PROCOMM Lite.

If your screen does not display this message, PROCOMM Lite has not been able to correctly initialize your modem. Ensure that your modem is turned on and determine to which serial port the modem is connected. By default, PROCOMM Lite assumes that your modem is connected to serial port COM1. If your modem is attached to a different port, refer to the section "Using the Alt-P (Line Port Setup) Command" for help in changing PROCOMM Lite's port setting.

At the bottom of the screen, the Terminal Mode screen's status line is divided into the following seven sections:

- **Message Area.** Contains instructions for displaying the PROCOMM Lite help system, but occasionally displays other status messages. This area also shows the name of any script file you may be running.

- **Terminal Emulation.** Terminals are special computers with the sole purpose of connecting to a larger computer for data entry and retrieval. When PROCOMM Lite emulates a type of terminal, the program impersonates the terminal type in order to communicate with a large computer that expects only terminals to be connected. Figure E.1 indicates that the type of terminal emulated by PROCOMM Lite is ANSI-BBS.

- **Duplex Setting.** Indicates whether PROCOMM Lite is set to full duplex (FDX) or half duplex (HDX). You can toggle this setting by pressing Alt-E.

- **Line Settings.** Indicates transmission speed setting as well as current parity, data bits, and stop bits settings. For example, 1200 N81 means that PROCOMM Lite is set to communicate at a transmission rate of 1200 bps, with no parity bit, 8 data bits, and 1 stop bit. You can change this setting by pressing Alt-P and choosing another set of line settings.

- **Log Status.** Indicates whether you have instructed PROCOMM Lite to save the current session to a log file on disk. By default, PROCOMM Lite does not save to a disk file (or capture) the information that scrolls across your screen. The system therefore usually displays LOG CLOSED in the status line, as shown in figure E.1. If you turn on the Log File feature, PROCOMM Lite changes the message to LOG OPEN.

- **Mode.** Tells you that you are at the Terminal Mode screen.

- **Line Status.** Indicates the line status—whether you are connected to another computer. When you start PROCOMM Lite, the message displays `OFF-LINE`, meaning that your PC is not communicating with another computer. When you connect to another computer, this status line message changes to `ON-LINE`.

Although the largest portion of the Terminal Mode screen is blank, this screen comes alive after you connect to another computer. The first 24 lines are where all the action occurs.

Using the Command Menu

The PROCOMM Lite Command Menu is an excellent learning tool and memory aid. If you are new to PROCOMM Lite, you can routinely display the Command menu before executing a command. To display the PROCOMM Lite Command Menu, press Alt-Z (see fig. E.2). If you forget a command, you can display this screen to remind you. After you become familiar with the commands, you can execute them without first displaying the Command Menu.

To execute a command listed on this menu, you press the key or keystroke combination indicated. The first command in the menu, for example, is the Dialing Directory command, and Alt-D is listed to its right. This entry tells you that you must press Alt-D to display the PROCOMM Lite dialing directory.

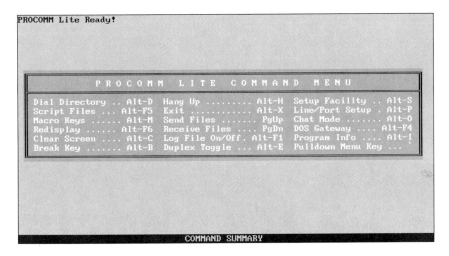

Fig. E.2. The PROCOMM Lite Command Menu.

The descriptions of the Command Menu commands are divided into the following two sections to help you find the commands you need:

- **Set Up and other functions.** Relates to configuring and operating PROCOMM Lite. These commands include the following:

Alt-S	Setup Facility
Alt-P	Line/Port Setup
Alt-F4	DOS Gateway
Alt-I	Program Information
	Pulldown Menu Key

- **Communications.** Used in connection with a communications session. These commands include the following:

Alt-D	Dialing Directory
Alt-H	Hang Up
Alt-X	Exit
Alt-F5	Script Files
Alt-F6	Redisplay
Alt-C	Clear Screen
Alt-B	Break Key
PgUp	Send Files
PgDn	Receive Files
Alt-F1	Log File On/Off
Alt-E	Duplex Toggle
Alt-O	Chat Mode
Alt-M	Macro Keys

Because most of the commands listed in the Command menu are Alt-key combinations, this group of commands is referred to as the Alt-key commands. Only PgUp (Send Files), PgDn (Receive Files), and the Pulldown Menu Key (`) do not use the Alt key.

When you are instructed to press one of these Alt-key commands, the command name is included in parentheses, as in Alt-C (Clear Screen). This method can help you learn to identify the keystroke with the command.

Using the Menu Line

In addition to the commands listed in the Command menu, you can invoke most PROCOMM Lite features by selecting options from a pull-down menu.

PROCOMM Lite doesn't automatically display a menu at the top of the Terminal Mode screen. To access the top-line menu system, you must be at the Terminal Mode screen. Press the Pulldown Menu Key or click either mouse button. The Pulldown Menu Key is the ` on the same key as the tilde (~). PROCOMM Lite displays the menu line at the top of the screen with the first pull-down menu extended (see fig. E.3).

You can reassign the menu-line key to a different key by using the Setup utility. If you forget which key is the menu-line key, you can refer to the Command Menu, which lists the menu-line key as the last command. You do not have to display the Command Menu, however, before you can display the menu line.

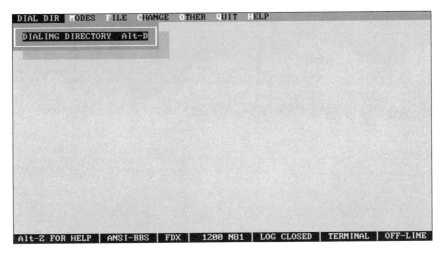

Fig. E.3. The PROCOMM Lite pull-down menu.

If you press the right-arrow key or move the mouse to the right, PROCOMM Lite pulls down the second menu, the Modes menu, which contains the options Redisplay and Chat. The options contained on the seven pull-down menus are as follows:

Menu	Option(s)	Equivalent command
DIAL DIR	Dialing Directory	Alt-D
MODES	Redisplay	Alt-F6
	Chat	Alt-O
FILE	Download	PgDn
	Upload	PgUp
	Run Script	Alt-F5
	Toggle Log	Alt-F1
CHANGE	Setup	Alt-S
	Line Settings	Alt-P
	Macro Keys	Alt-M
	Echo (Duplex)	Alt-E
OTHER	DOS Gateway	Alt-F4
	Hangup	Alt-H
	Send Break	Alt-B
	Clear Screen	Alt-C
	Information	Alt-F
QUIT	Exit to DOS	Alt-X
HELP	Display Help Menu	Alt-Z

While the menu line is displayed, you execute a PROCOMM Lite command by selecting the appropriate option from the pull-down menu. This menu operates in a manner familiar to most PC users. If you are using the keyboard to access a command, you can press Alt and the first letter of the menu name to select a pull-down menu. After you select a menu name, a pull-down menu displays with one or more options.

To select an option on the pull-down menu, press the first letter of an option or use the up- or down-arrow key to select an option. Press Enter to invoke the option. The first-letter technique usually is the quickest method, but use the method most comfortable for you.

If you are using a mouse, you can point the mouse cursor to a menu name by moving the mouse to the left or right. After you pull down a menu, you can make a selection from the menu by moving the mouse up or down to highlight the option you want. Click the left or right mouse button to select the option.

When you extend a pull-down menu, notice that to the right of the menu option is the equivalent Alt-key command. You cannot use this Alt-key command, however, while the menu line is displayed.

After you select a menu option, PROCOMM Lite executes the command and removes the menu line from the screen. To remove the menu from the screen without executing an option, press Esc or click the right mouse button.

As a general rule, the Alt-key commands are the fastest way to execute PROCOMM Lite features. After you commit the commonly used Alt-key commands to memory, you may decide to forgo use of the menu line. If you find that using top-line menus is more user-friendly than memorizing keyboard commands, don't hesitate to use the PROCOMM Lite menu line.

Using Setup and Other PROCOMM Lite Commands

The following Command Menu commands deal with the setup and operation of PROCOMM Lite off-line. These are the Alt-S (Setup), Alt-P (Ports), Alt-F4 (DOS Gateway) and ` (Pulldown menu) commands.

Using the Alt-S (Setup Facility) Command

To customize PROCOMM Lite for your computer, use the Alt-S command to access the setup facility. The screen shown in figure E.4 appears.

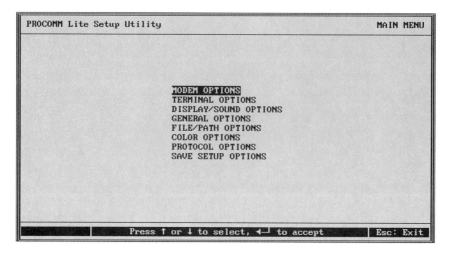

Fig. E.4. *The PROCOMM Lite Setup Utility main menu.*

With the Setup Utility menu, you can choose to set up the following options for your modem and computer:

- **Modem Options.** Displays the Modem Options menu with four choices. General Options enables you to set up modem options, including the time to wait for a connection, the time between calls, setting autobaud detect, DTR, and CD settings. Modem Commands enables you to set up the initialization command for your modem, the dialing command (such as ATDT), the dialing command suffix, and hangup command. The Result Messages setup screen enables you to inform PROCOMM Lite of the connect messages generated by your modem. The Port Assignments setup screen enables you to choose the interrupt and base port address for your COM port.

- **Terminal Options.** Enables you to choose the duplex setting, flow control, line wrap, screen scroll, CR translation, BS translation, break length, enquiry (ENQ) setting, and strip 8th bit.

- **Display/Sound Options.** Enables you to choose sound and visual options for the screen.

- **General Options.** Enables you to choose the CD high setting, keyboard speed, pause character, transmit pacing, pull-down menus, pulldown menu key, DTR setting, chat mode setting, and mouse sensitivity.

- **File/Path Options.** Enables you to choose the default file names for your log files, as well as a default download directory.

- **Color Options.** Enables you to choose color settings for your terminal and for display windows.

- **Protocol Options.** Enables you to choose your protocol timings, the action to be taken for aborted downloads or CD loss, and ASCII transfer options.

When you set any of these parameters, choose the Save Setup Options setting on the menu to record your selections to disk. Then press Esc to return to the Terminal Mode screen.

Using the Alt-P (Line/Port Setup) Command

The Line/Port settings command enables you to choose the COM port assignment, your baud rate, parity, data bits, and stop bits. A menu is displayed from which you can choose these options. Use Alt-S after choosing to record your selections to disk. Press Esc to exit this screen.

Using the Alt-F4 (DOS Gateway) Command

You may exit temporarily to the DOS prompt while in PROCOMM Lite (even if you are on-line) by pressing Alt-F4. You may want to do this, for example, to look for a file to upload. To get back into PROCOMM Lite, enter *exit* at the DOS prompt.

Using the ` (Pulldown Menu Key) Command

To cause the pull-down menu to appear, you can press the ` key. This key is usually at the upper left of the keyboard, on the same key as the ~ (tilde). You then can access the menu as described in the earlier section "Using the Menu Line." Press Esc to remove the menu from the Terminal Mode screen.

Using PROCOMM Lite Communications Commands

The following commands are those used during a PROCOMM Lite communications session. They include the Alt-D (Dialing Directory), Alt-H (Hang Up), Alt-X (Exit), Alt-F5 (Script Files), Alt-F6 (Redisplay), Alt-C (Clear Screen), Alt-B (Break Key), PgUp (Send Files), PgDn (Receive Files), Alt-F1 (Log File On/Off), Alt-E (Duplex Toggle), Alt-O (Chat Mode), and Alt-M (Macro Keys).

Using the Alt-D (Dialing Directory) Command

The first step toward communicating with another computer—whether next door, across town, or across the country—usually is to dial the telephone number. PROCOMM Lite enables you to store commonly used phone numbers in its dialing directory. With this directory, you can easily access phone numbers from a list on-screen rather than having to remember them or look them up.

Phone dialing, however, is not the only purpose of the Dialing Directory screen. This screen helps you take care of several other chores necessary for successful PC communications. Through the Dialing Directory screen, you establish your computer's communications settings, such as transmission speed, parity, data bits, and so on.

This section describes how to use the Dialing Directory screen to control the process of dialing and connecting to remote computers. You learn how to build dialing directories and how to add, edit, and delete directory entries and dialing codes. This section then shows you how to dial another computer by using a single dialing directory entry.

To display the Dialing Directory screen, press Alt-D (Dialing Directory) from the Terminal Mode screen. As shown in figure E.5, this directory initially contains no entries.

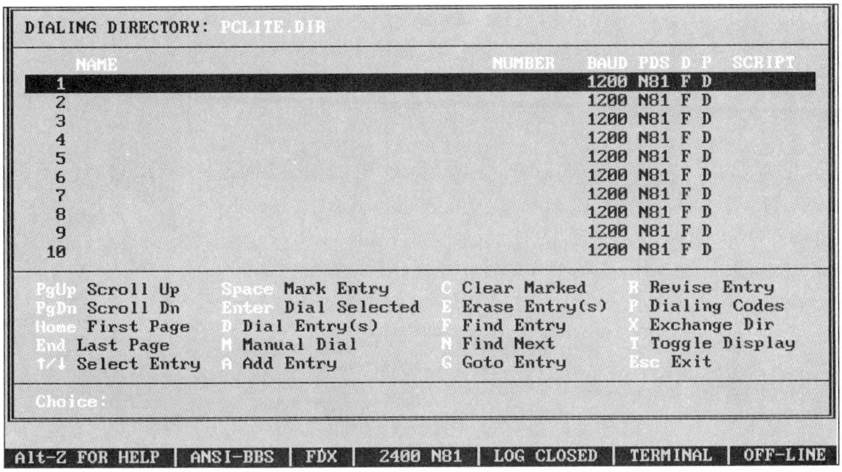

Fig. E.5. The Dialing Directory screen.

Note that the Dialing Directory screen is divided horizontally into five sections. For convenience and clarity, you can refer to these sections, from top to bottom, as follows: the name line, the entry section, the command section, the choice line, and the status line.

The first section—the name line—displays the name of the current dialing directory. The default PROCOMM Lite directory file is named PCLITE.DIR.

The main portion of the Dialing Directory screen is the entry section. This part of the screen contains the list of phone numbers that you want PROCOMM Lite to dial. PROCOMM Lite normally displays only 10 entries at a time, but you can scroll this portion of the screen to display any of the other entries in the directory. Each directory can contain up to 200 entries. In addition to the telephone number, each entry in the directory stores other information about the setting such as transmission speed, line settings (parity, data bits, stop bits), duplex mode, COM port, and more.

Below the entry section of the screen is the command section, which lists the keyboard commands available while dialing. You can divide these commands into four groups according to their functions: editing commands, entry-selection commands, dialing commands, and global commands (see table E.1).

Table E.1
PROCOMM Lite Dialing Directory Commands

Key	Name	Function
Editing Commands		
R	Revise Entry	Revises highlighted entry
A	Add Entry	Adds a new entry at first open spot in directory
E	Erase Entry(s)	Erases highlighted entry
P	Dialing Codes	Displays Dialing Codes window
Entry-Selection Commands		
PgUp	Scroll Up	Scrolls entry section up 10 rows
PgDn	Scroll Dn	Scrolls entry section down 10 rows
Home	First Page	Moves highlight to row 1
End	Last Page	Moves highlight to row 191
↑	Select Entry	Moves highlight up one row
↓	Select Entry	Moves highlight down one row
F or /	Find Entry	Searches for entry by name or phone number
N	Find Next	Searches for next matching entry
G	Goto Entry	Selects an entry by entry number
Dialing Commands		
Enter	Dial Selected	Dials highlighted entry or dialing queue
D	Dial Entry(s)	Dials one or more entries by entry number
M	Manual Dial	Dials the number you type manually
Space bar	Mark Entry	Adds entry to dialing queue
C	Clear Marked	Clears all dialing queue marks
Global Commands		
X	Exchange Dir	Loads a different directory
T	Toggle Display	Displays other communications settings
Esc	Exit	Returns to Terminal Mode screen

Beneath the command section is the choice line, which contains only the word `Choice`, followed by a blinking cursor. This line is a prompt asking you to enter a command. PROCOMM Lite never displays anything else in this section. The last section of the Dialing Directory screen is the status line, which is the same line that appears when PROCOMM Lite is in Terminal Mode.

Working with Dialing Directory Entries

Although the primary purpose of a dialing directory is to dial the telephone, the dialing directory is more than just a speed dialer. When you use a directory entry to dial a computer, PROCOMM Lite automatically changes your computer's communications settings to those specified in the entry (if necessary), instructs the modem to dial the phone, and runs any script file specified in the entry.

The next several sections discuss how to add entries to the Dialing Directory screen, including how to specify communications settings and a script file; how to use the entry-selection commands to move to particular entries; and how to modify and delete directory entries.

Adding an Entry

To add an entry to the directory currently displayed on the Dialing Directory screen, press A (Add Entry). PROCOMM Lite displays a pop-up Revise Entry dialog box (see fig. E.6).

When PROCOMM Lite first displays the Revise Entry dialog box, the Name field is highlighted. In this field, enter the name of the computer system you want to call. Type up to 24 characters (letters, numbers, symbols, and spaces), and press Enter. The cursor-movement and editing keys listed in table E.2 are available for you to use while entering or editing a name in this field. In fact, you can use these keys in any line in the Revise Entry dialog box and in input fields throughout PROCOMM Lite.

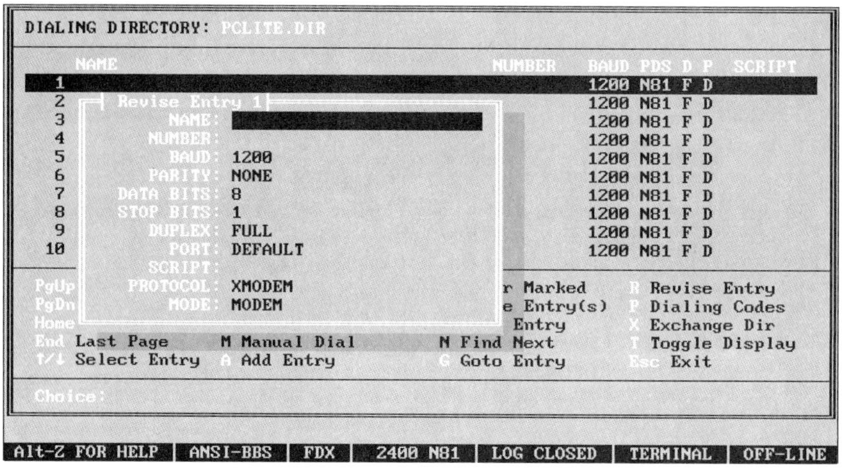

```
┌──────────────────────────────────────────────────────────────────┐
│ DIALING DIRECTORY: PCLITE.DIR                                      │
│     NAME                                    NUMBER  BAUD PDS D P  SCRIPT │
│  1                                                  1200 N81 F D        │
│  2   ┌─ Revise Entry 1 ──────────────┐              1200 N81 F D        │
│  3   │    NAME:                      │              1200 N81 F D        │
│  4   │  NUMBER:                      │              1200 N81 F D        │
│  5   │    BAUD: 1200                 │              1200 N81 F D        │
│  6   │  PARITY: NONE                 │              1200 N81 F D        │
│  7   │DATA BITS: 8                   │              1200 N81 F D        │
│  8   │STOP BITS: 1                   │              1200 N81 F D        │
│  9   │  DUPLEX: FULL                 │              1200 N81 F D        │
│ 10   │    PORT: DEFAULT              │              1200 N81 F D        │
│      │  SCRIPT:                      │                                  │
│ PgUp │PROTOCOL: XMODEM               │ r Marked    R  Revise Entry      │
│ PgDn │    MODE: MODEM                │ e Entry(s)  P  Dialing Codes     │
│ Home └──────────────────────────────┘  Entry      X  Exchange Dir      │
│ End  Last Page     M Manual Dial     N Find Next   T  Toggle Display    │
│ ↑/↓  Select Entry  A Add Entry       G Goto Entry  Esc Exit            │
│                                                                    │
│ Choice:                                                            │
│                                                                    │
├────────────────────────────────────────────────────────────────────┤
│ Alt-Z FOR HELP │ ANSI-BBS │ FDX │ 2400 N81 │ LOG CLOSED │ TERMINAL │ OFF-LINE │
└──────────────────────────────────────────────────────────────────┘
```

Fig. E.6. *The Dialing Directory Revise Entry dialog box.*

Table E.2
Dialing Directory Cursor-Movement and Editing Keys

Key	Function
←	Moves cursor left one character
→	Moves cursor right one character
Home	Moves cursor to left end of line
End	Moves cursor to right end of line
Ins	Toggles Insert/Overtype modes
Del	Deletes character at cursor
Backspace	Deletes character to left of cursor
Ctrl-Backspace	Deletes entire line
Ctrl-End	Deletes characters from cursor to right end

When you press Enter from the Name field, PROCOMM Lite moves the highlight to the Number field. Type the telephone number of the system you want to call and press Enter.

After you type the telephone number and press Enter, PROCOMM Lite moves the highlight to the Baud field. This field already contains the default transmission speed, 1200. After you highlight this field, PROCOMM Lite displays another dialog box that lists the available transmission speeds. If you don't need to change the setting, just press Enter. Otherwise, press the up- or down-arrow key to highlight the appropriate speed and press Enter. PROCOMM Lite removes the list of transmission speeds from the screen and moves the highlight to the next field in the Revise Entry dialog box, Parity.

You complete the remaining fields in the Revise Entry dialog box in a similar manner. For each field, the default setting is filled in; after you highlight the field, another dialog box appears from which you can choose another setting. You press Enter to accept the default or highlight another choice in the dialog box and press Enter.

After you make all the selections concerning an entry, PROCOMM Lite displays a dialog box asking three questions in sequence. The first prompt asks the following:

```
CLEAR LAST DATE AND TOTAL? (Y/N)
```

This question has no significance when you are adding a new directory entry, so just press Enter to accept the default value of No. PROCOMM Lite displays the next prompt:

```
ACCEPT THIS ENTRY? (Y/N)
```

Press Y to accept all the settings you have entered. This action adds the entry to the directory currently loaded in memory but does not add the entry to the directory file on the disk. If you answer No by pressing N, PROCOMM Lite moves the highlight back to the first line in the Revise Entry dialog box without erasing any entries you have made.

After you accept the entry, PROCOMM Lite displays one final prompt:

```
SAVE ENTRY TO DISK? (Y/N)
```

Press Y, and PROCOMM Lite saves the entry to the current directory file on disk (the default file is PCLITE.DIR) and returns to the Dialing Directory screen.

You can, of course, answer No to the last question in the Revise Entry dialog box. PROCOMM Lite then returns to the Dialing Directory screen without saving the new entry to disk. The entry still is listed on-screen but is available for the current session only. When you quit PROCOMM Lite, this temporary entry is erased from memory. Using this feature, you can create a temporary directory entry for a computer system with which you expect to connect only once.

Selecting an Entry

The simplest way to select a dialing directory entry is with the up- and down-arrow and PgUp and PgDn keys. When you want to move to the end of the directory, press End (Last Page). Regardless of which entry is current when you press End, PROCOMM Lite displays entries 191 through 200, with entry 191 selected. To move directly to the top of the directory, press Home (First Page). PROCOMM Lite displays entries 1 through 10 and selects entry 1.

PROCOMM Lite provides two methods that enable you to go directly to an entry without scrolling. If you know the entry number of the entry you want to select, press G (Goto Entry) from the Dialing Directory screen. PROCOMM Lite then displays a dialog box containing the prompt ENTRY TO GOTO. At this prompt, type the number of the entry you want to select and press Enter. PROCOMM Lite moves the highlight directly to the specified entry.

More often than not, however, you don't know the number of the entry. PROCOMM Lite thus provides a quick and easy way to search for the entry by name or telephone number. From the Dialing Directory screen, press F to select Find Entry. At the SEARCH FOR prompt, type the search criterion—any portion of the entry's name or the entry's telephone number. You can use upper- or lowercase letters. Press Enter, and PROCOMM Lite searches the directory for the first match—an entry in which the name or phone number contains the characters specified in the search criterion.

When searching, the program starts at the currently selected entry and searches entries numerically toward the end of the directory. If PROCOMM Lite does not find a match by the time it reaches the end of the directory, the program starts at entry number 1 and searches down toward the current entry. If no match is found, PROCOMM Lite displays the message TARGET NOT FOUND and then removes the search window from the screen.

Of course, several entries in your directory may contain a given string of characters. The first entry found may not be the one you want. PROCOMM Lite, however, provides another command that enables you to continue the search, using the same criterion, until you locate the appropriate entry. This command is N (Find Next).

Before you can use the N (Find Next) command, you must have used the F (Find Entry) command to establish the search criterion. If the F (Find Entry) command does not select the entry you want, press N (Find Next). PROCOMM Lite continues the search, using the same criterion and searching from the current entry to the bottom of the directory and then from the top of the directory to the current entry. PROCOMM Lite moves the highlight to the next matching entry, if one exists. If PROCOMM Lite finds no more matches, it beeps but displays no message and does not move the highlight.

Modifying an Entry

The procedure for modifying an entry in the Dialing Directory screen is nearly identical to the procedure for adding a new entry. Highlight the entry to modify, and then press R (Revise Entry). PROCOMM Lite displays the Revise Entry dialog box with the values from the selected entry already filled in.

To change the entry's name, use the cursor-movement and editing keys listed in table E.2, or just type a new name over the old one. To leave a setting as it is and move to the next setting, press Enter.

The only item that requires additional thought when you are modifying an entry is the first question listed on the second Revise Entry dialog box:

        CLEAR LAST DATE AND TOTAL? (Y/N)

As you use a particular entry to connect to another computer, PROCOMM Lite maintains a simple call history. The program records the date of the last successful connection made with the entry and keeps a tally of the number of connections. This Revise Entry prompt is asking whether you want to set the call-history date back to the default value of 00/00/00 and set the tally to 0. The default answer is No. If you press Enter, PROCOMM Lite does not clear the date or tally. When you want to start running a new tally, press Y in response to the question. If you are modifying the entry so that it dials a different remote computer, for example, you probably will want to abandon the old LAST DATE value and start the tally again at 0.

Deleting an Entry

Occasionally you may want to purge your directory of entries you no longer use. To delete a single directory entry, select the entry and press E (Erase Entry). PROCOMM Lite displays the prompt

        ERASE ENTRY NUMBER n? (Y/N)

where *n* is the selected entry number. Press Y, and PROCOMM Lite deletes the entry from the directory. PROCOMM Lite does not, however, move other entries up to fill the empty space.

If you want to remove two or more entries from the directory, you do not have to delete each one separately. Select the first entry you want to delete and press the space bar. PROCOMM Lite marks the entry by placing a triangular-shaped pointer on the Dialing Directory screen just to the left of the entry number. Repeat this procedure for each entry you want to delete. Finally, press E (Erase Entry). PROCOMM Lite displays the prompt

```
ERASE n MARKED ENTRIES? (Y/N)
```

where *n* is the number of entries you have marked. Press Y, and PROCOMM Lite deletes the marked entries. As is the case when you delete only one entry, PROCOMM Lite does not move other entries into the lines vacated by the deleted entries.

Working with Dialing Codes

As you build your PROCOMM Lite dialing directory, you may find that you are entering a certain string of numbers over and over again. If your long-distance carrier requires that you dial a special sequence of numbers before dialing the regular telephone number, for example, you must include that access sequence with every directory entry's telephone number. With PROCOMM Lite's special shorthand dialing codes, you can incorporate a string of numbers by reference without typing the entire string. You can establish up to 10 dialing codes in each dialing directory.

Adding or Modifying a Dialing Code

Press P (Dialing Codes) from the Dialing Directory screen to display the Dialing Codes dialog box, as shown in figure E.7. This box contains 10 lines, each labeled with a letter from A through J. Each line can hold one dialing code. The cursor blinks near the lower left corner of the window, to the right of a small triangular pointer. To enter a new dialing code, press the up- or down-arrow key to move the cursor to one of the 10 lines, and press R (Revise).

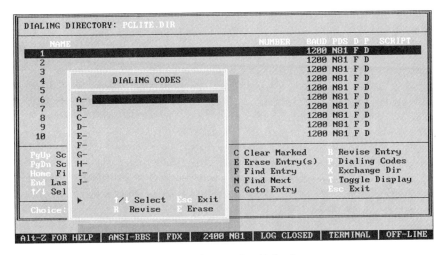

Fig. E.7. *The Dialing Directory Dialing Codes dialog box.*

In the selected dialing code line, type the number you want to establish as a dialing code. (Keep in mind that while the cursor is in a dialing code line, the cursor-movement and editing keys listed in table E.2 are available and have the same effect they do when you are editing in the Revise Entry dialog box.) Each code can contain up to 24 characters, which can be any displayable character but must be recognizable by your modem in order to be of any value. (Consult your modem's manual for available dialing codes.) Most typically, you type numbers, hyphens, and parentheses in dialing codes. The numbers instruct your modem to dial the telephone, and the hyphens and parentheses make the phone number easier for you to read but have no effect on the dialing.

Sometimes you may want the modem to pause during the dialing procedure. If you want the dialing code to dial a long-distance telephone carrier, for example, you may need the modem to pause after dialing the long-distance access code. If you have a Hayes-compatible modem, you can add commas to the dialing code to tell the modem to pause. (If you don't have a Hayes-compatible modem, check your modem manual for a pause code.) Each comma pauses the modem for two seconds. If you want your modem to pause the dialing for six seconds—between the dialing of a phone number and a five-digit extension, for example—you add three commas. When you use this code in a dialing directory telephone number, the modem dials the telephone number 1112222, pauses for six seconds, and then dials the extension 33333.

Many office telephone systems require that you dial 9 or some other digit to get an outside line. You then must wait for a second dial tone before you continue to dial. For this reason, most Hayes-compatible modems recognize the letter W as a command to pause and wait for a second dial tone before continuing to dial the telephone number. (Check your modem documentation to make sure that your modem recognizes this command.) For example, you can create the following dialing code:

9W111-2222,,,33333

When used in a directory entry telephone number, this code tells the modem to dial 9 and then wait for a second dial tone. As soon as the modem detects the dial tone, the modem continues dialing the rest of the number, pauses for six seconds, and dials the five-digit extension.

After you add the dialing codes you want, press Esc (Exit) to return to the Dialing Directory screen.

If you later decide that you need to modify a dialing code, just press P (Dialing Codes) to redisplay the Dialing Codes dialog box. Press the up- or down-arrow key to select the dialing code that you want to modify, and use the cursor-movement and editing keys to make your changes. Press Enter to accept the changes and Esc (Exit) to return to the Dialing Directory screen.

Using a Dialing Code

You can use a dialing code in any entry in the dialing directory. Typically, you will use dialing codes in many entries, or you probably would not have gone to the trouble of creating the dialing code in the first place.

Suppose that you decide to create a directory entry to call BCSNET (a Boston Computer Society bulletin board), which requires the long-distance access code shown in dialing code A. Rather than retype the access code, you can use dialing code A in your directory entry. From the Dialing Directory screen, press A (Add Entry) to add dialing information to the next empty entry. Create the entry as usual, except that when you type the telephone number, type the letter *a* before you type the area code. When you use this entry to dial BCSNET, PROCOMM Lite first sends the number stored in dialing code A to the modem and then sends the telephone number.

Erasing a Dialing Code

Erasing a dialing code is similar to erasing a dialing directory entry. First, from the Dialing Directory screen, press P (Dialing Codes) to display the Dialing Codes dialog box. Press the up- or down-arrow key to select the code you want to erase. Finally, press E (Erase). PROCOMM Lite erases the dialing code without asking for any confirmation. Press Esc (Exit) to return to the Dialing Directory screen.

Dialing a Number

The ultimate purpose of each entry in a PROCOMM Lite dialing directory is to set up your computer for communication with another computer and then to instruct your modem to dial that computer's telephone number. This section describes several ways to use the Dialing Directory screen, including two ways to dial a number by using a single directory entry and two ways to create a dialing *queue*—a series of entries for PROCOMM Lite to use in sequence. In this section, you also review the Manual Dial feature.

When you use one of the methods described in this documentation to initiate a call, PROCOMM Lite first sets the transmission speed, line settings (parity, data bits, and stop bits), duplex setting, error-checking protocol, terminal emulation, and other settings according to the values stored in the entry's fields. PROCOMM Lite then instructs the modem to dial the telephone number stored in the entry's Number field.

Using a Directory Entry

After you add an entry to your PROCOMM Lite dialing directory, you can use that entry when you want to connect to a remote computer. You can use one of the following two methods:

- When you can remember the number of the entry, press Alt-D to display the Dialing Directory screen and then press D (Dial Entry). PROCOMM Lite displays a dialog box with the prompt Entry(s) to dial, with the cursor blinking to the right of this prompt.

  Type the entry number and press Enter. PROCOMM Lite moves the highlight to the entry, adjusts the transmission speed and line settings, and instructs the modem to begin dialing the number. When the program successfully connects to the other computer, the Terminal Mode screen appears. The next time you display the Dialing Directory screen during the same PROCOMM Lite session, the entry that was selected when the connection was made is still selected.

- Select the dialing directory entry and then press Enter. PROCOMM Lite adjusts the transmission speed and line settings and instructs the modem to begin dialing the number. Even if PROCOMM Lite is unsuccessful in connecting to the other computer, if you return to the Terminal Mode screen and then later redisplay the Dialing Directory screen, this entry is still selected. You can use the entry again by pressing Enter.

Monitoring Call Progress

PROCOMM Lite informs you of a call's progress in the PROCOMM Lite Dialing box, which appears after you specify the entry to dial. The Dialing box helps you visually monitor the status of the call. In this window, PROCOMM Lite lists the name and telephone number of the system your modem is dialing and keeps you informed about several other parameters.

If your modem detects no dial tone when it begins to dial the telephone, the modem stops dialing and sends the code NO DIALTONE to PROCOMM Lite. PROCOMM Lite in turn displays this message in the Last Call field of the Dialing box. Similarly, the call may proceed to the point at which the modem detects a busy signal. The modem sends the message BUSY, which PROCOMM Lite then displays in the Last Call field.

When the modem sends either of these messages, it also stops dialing the number and hangs up the phone line (often referred to in modem documentation as going on-hook). PROCOMM Lite then displays the message PAUSING... in the Dialing field.

While pausing the dialing procedure, PROCOMM Lite counts the number of seconds that have elapsed since the last attempted call, displaying the count in the Elapsed Time field. PROCOMM Lite waits the number of seconds indicated in the Pause Between Calls field, increments the Pass Number field by 1, and sends an instruction to the modem to dial the number again. The Pause Between Calls setting is necessary to enable the modem to get ready to redial. The length of time PROCOMM Lite pauses between calls is set by default to four seconds. You can increase or decrease this time by using the setup utility.

The Pause Between Calls field is set at 4, so the modem waits four seconds, as indicated in the Elapsed Time field, and dials again. PROCOMM Lite continues this procedure until it successfully connects to the other modem or until you stop the dialing by pressing Esc (Abort).

The last two fields in the Dialing box keep track of the time of day when the dialing procedure started and the time that dialing for the most recent call started.

Even if your modem does not recognize a busy signal, PROCOMM Lite has another feature that can help you. After sending the instruction to the modem to dial the telephone number, PROCOMM Lite waits the length of time listed in the Wait for Connection field of the Dialing box. If the modem does not complete a connection in that length of time—because the line is busy or because the other end does not answer—PROCOMM Lite automatically puts the phone line back on-hook and pauses the dialing. After the time span indicated in the Pause Between Calls field, PROCOMM Lite tries the call again. The program repeatedly attempts to connect to the modem on the other end of the line until the connection is successful or until you press Esc (Abort)—whichever comes first.

When your modem successfully connects to the modem of the computer with which you want to communicate, your modem sends a message to PROCOMM Lite such as CONNECT 1200. Whatever the message, PROCOMM Lite displays it in flashing characters in the Last Call field of the Dialing box. In case you don't notice this visual connect signal, PROCOMM Lite also sounds several beeps on your computer to alert you. This feature enables you to leave the room while you wait for your modem to get a clear line and connect to the remote computer. Just don't forget to be listening for the beeps.

Using the Manual Dial Feature

Occasionally, you may decide to call a computer that you expect to never call again. Rather than go to the effort of creating a dialing directory entry that you will just erase later, you can use PROCOMM Lite's Manual Dial feature.

To dial another computer without creating an entry, press M (Manual Dial) from the Dialing Directory screen. PROCOMM Lite displays a dialog box in the center of the screen with the prompt Number to dial.

At the prompt, type the phone number you want the modem to dial. Include any digits necessary to get an outside line. You also can use commas (to pause dialing) and dialing codes. When you finish typing the number, double-check it to make sure that you typed it correctly, and then press Enter.

PROCOMM Lite replaces the Number to Dial dialog box with the Dialing dialog box. Instead of displaying an entry name in the Dialing field, PROCOMM Lite displays the message MANUAL DIAL and begins dialing the number. Because PROCOMM Lite has no entry from which it can take transmission speed, line settings, and other communications settings, the program uses whatever settings are current, as indicated in the Dialing Directory screen status line. The current settings are determined by the last dialing directory entry used. If no entry has been used during the current PROCOMM Lite session, the program uses the default communications settings.

Except for the message in the Dialing field, the Dialing box operates in the same way as when you are using a directory entry to dial a remote computer.

If you use the M (Manual Dial) feature again during the same PROCOMM Lite session, the telephone number you typed still appears in the Number to Dial dialog box. Press Enter to use the same number, or edit the number if you want to call a different computer.

Managing Your Dialing Directory

A PROCOMM Lite dialing directory is a small database—a collection of information arranged in fields (columns) and records (rows). In PROCOMM Lite, each dialing directory entry is a record in the database, and each column (Name, Number, Baud, and so on) is a field. Up to this point, this appendix has concentrated on how to create, edit, delete, and use directory entries—one at a time, or in groups (as a dialing queue). But as is true with any database program, PROCOMM Lite enables you to perform certain functions on the entire database—the entire dialing directory. The remainder of this section describes how to use these directory-wide functions to manage your dialing directory database.

You first learn how to change the display in the Dialing Directory screen so that it shows a set of fields different from what you normally see. You then learn how to print other dialing directories.

Toggling the Display

You probably have noticed by now that you enter more information in the Revise Entry dialog box about a particular entry than is displayed in the Dialing Directory screen. Each entry's name, telephone number, transmission rate, parity, data bits, stop bits, duplex setting, port, and script normally are shown in the Dialing Directory screen. The Revise Entry box, however, includes a setting for a default file-transfer protocol and a terminal-emulation setting. Neither of these settings shows up in the Dialing Directory screen. You also may recall that PROCOMM Lite keeps a call history on each directory entry—the date of the last connection and the total number of connections made with the entry. This call-history information normally does not display in the Dialing Directory screen. PROCOMM Lite does provide a way, however, for you to see all this information that usually is hidden. This procedure is called *toggling* the display. To display other settings for an entry, press T (Toggle Display). This causes PROCOMM Lite to display other terminal settings.

Building and Using Other Dialing Directories

Although PROCOMM Lite limits you to 200 entries per dialing directory, the program does not limit the number of directories you can build. In effect, then, you can manage as many directory entries as you want. PROCOMM Lite can, however, work with only one directory file at a time.

When you want to create another dialing directory in addition to the default PCLITE directory, display the Dialing Directory screen and press X (Exchange Dir). PROCOMM Lite displays a dialog box with the prompt DIRECTORY TO LOAD. Type a valid DOS file name, but do not type a period or an extension. Press Enter. PROCOMM Lite creates a file on disk with the specified name and the extension DIR, removes the current directory from the screen, and displays the new directory with blank entries. You then can add, modify, delete, and use entries in this new directory, just as you did in the PCLITE dialing directory.

To switch among your existing directories, display the Dialing Directory screen and press X (Exchange Dir). In the box that appears, type the file name of the directory you want to access, omitting the period and extension, and press Enter. PROCOMM Lite loads the specified dialing directory. You can use an unlimited number of dialing directories in this manner, each with the capacity to hold up to 200 entries.

Using the Alt-H (Hang Up) Command

To disconnect a call, press the Alt-H key while in Terminal Mode. This "hangs up" the phone and disconnects you from a communications session. Care must be taken not to disconnect from your session without first logging off of the computer you are "talking" to—particularly if you are being charged for connect time. Hanging up does not always end the connect charge accumulation.

Using the Alt-X (Exit) Command

Before you turn off PROCOMM Lite, you should sign off or disconnect from any active communications session. When the status line at the bottom of the screen indicates that PROCOMM Lite is OFF-LINE, you can quit the program. From the Terminal Mode screen, press Alt-X (Exit). Near the top of the screen, PROCOMM Lite opens a small window with the following prompt:

    EXIT TO DOS? (Y/N)

Yes is the default response to this prompt. You can affirm this response by pressing Enter or the letter Y. PROCOMM Lite exits and returns you to the operating system. Press N to answer No and to continue with your PROCOMM Lite session.

If you forget to disconnect from an active communications session before exiting PROCOMM Lite, the program reminds you with the following prompt:

    HANG-UP LINE? (Y/N)

Normally, you answer Yes to this query. If you press N to choose No, however, PROCOMM Lite quits and returns to DOS, but does not drop the line. This feature enables you to exit from PROCOMM Lite without disconnecting the communications session—you can run another DOS program without losing the line. When you restart PROCOMM Lite later, your connection still will be intact.

Using the Alt-F5 (Script Files) Command

You can create scripts using the PROCOMM Lite script language, and run them in PROCOMM Lite. These files have the extension ASP. When you press the Alt-F5 command, you are prompted to enter the name of the script file to run. Type the name and press the Enter key, and the script begins. See the later section "Using the PROCOMM Lite Script Language" for more information on the script language.

Using the Alt-F6 (Redisplay) Command

PROCOMM Lite continually saves the information that scrolls across your screen. The data is saved to a portion of memory called the redisplay buffer. This buffer is a temporary storage area that is emptied every time you exit PROCOMM Lite and return to DOS. The redisplay buffer usually can hold up to 63,000 characters. The default size is 20,000 characters. You can set the buffer size in the Setup Utility Display/Sound Options screen. When the redisplay buffer fills, PROCOMM Lite discards the oldest data first to make room for the new data on your screen.

To see the data stored in the redisplay buffer, press Alt-F6 (Redisplay). PROCOMM Lite displays the last 22 lines that scrolled off the Terminal Mode (or Chat mode) screen.

Obviously, you cannot view all 20,000 characters stored in the redisplay buffer at once. Because many lines may contain only a few characters, the redisplay buffer often holds many screens that scrolled off the top of the Terminal Mode screen. Several commands can help you move quickly around the buffer. The last line of the Redisplay Buffer screen contains a list of available keyboard commands. These commands are described in table E.3

Table E.3
Redisplay Buffer Screen Keyboard Commands

Command	Name	Effect
PgUp	Pages	Scrolls screen up one page/screen (22 lines)
PgDn	Pages	Scrolls screen down one page/screen (22 lines)
↑	Scrolls	Scrolls screen up one line
↓	Scrolls	Scrolls screen down one line
Home	Top	Jumps screen to top of buffer
End	Bottom	Jumps screen to bottom of buffer
F	Find	Finds specific character string
W	Write	Writes buffer information to file or printer
Esc	Exit	Returns to Terminal Mode screen

You sometimes may be looking for a particular word or phrase that you recall seeing on a previous screen. Before you begin searching for a particular character string, press Home (Top) to move the screen to the top of the redisplay buffer. Then press F (Find) to display the following prompt in the bottom line of the Redisplay Buffer screen:

    String to find:

Type the string of characters you want to find. You can include any displayable characters as well as spaces. You also can use upper- or lowercase letters. After you type the string, press Enter. PROCOMM Lite searches through the buffer, looking for the string in any portion of a word. If you search for the string *com*, for example, PROCOMM Lite finds the following words:

    becoming
    Welcome
    compiled
    PROCOMM Lite
    COMPOSE
    CompuServe

When PROCOMM Lite finds the text you requested, it displays the line that contains the matching word or phrase in the top line of the screen.

To repeat a Find, press F and then press Enter. After you press F, you see that the text of your last Find is still active. After you press Enter, PROCOMM Lite locates the next match to the text you requested. If no matches to the string are found, PROCOMM Lite displays the following message:

    String not found

Occasionally you may want to keep a copy of the redisplay buffer. The buffer may contain a list of telephone numbers, for example, that you want to keep. To save the redisplay buffer to disk, press W (Write). PROCOMM Lite displays a dialog box with the following prompt:

    Please enter filename:

Type a valid DOS file name (*phone#.txt*, for example) and press Enter. If the buffer contains more than a screen full of information, you are prompted with the following message:

    Write Entire buffer or Mark a block to write (E or M)

If you press E, PROCOMM Lite places the contents of the buffer into the file. If you press M, the program prompts you to mark the text that you want to save to the file. Use the up- and down-arrow keys to place the cursor at the beginning of the text that you want to save and press Enter. Next, place the cursor at the end of the text you want to save and press Enter again. PROCOMM Lite saves this block of text to the file.

You can print the contents of the redisplay buffer. Press W to choose the Write command and type *prn* or *lpt1* as the file name. DOS knows that PRN specifies the default printer port and LPT1 specifies the first parallel port. Be sure to turn on your printer, load it with paper, and connect it to the parallel port. Press Enter to send the contents of the buffer to your printer. If your printer is connected to the second parallel port on your computer, type *lpt2*.

To return to the Terminal Mode screen (or to the Chat mode screen if you were in Chat mode when you pressed Alt-F6), press Esc (Exit).

Using the Alt-C (Clear Screen) Command

The Alt-C command is used to clear the Terminal Mode screen. If you have confidential information on the screen, for example, pressing Alt-C will erase the information on the screen.

Using the Alt-B (Break Key) Command

A *break* is a signal sent to the computer to get its attention. If you press Alt-B (Break Key), PROCOMM Lite temporarily interrupts the data transmission with a space condition on the transmit line for at least the length of a data word. The break length can be set on the Terminal Options screen.

Using the PgUp (Send Files) Command

File transfer is a two-way street. PROCOMM Lite enables you to send and receive computer files to and from the remote computer. The steps you take to send a file are similar to but not exactly the same as the steps necessary to receive a file. This section describes how to send one or more files to a remote computer.

In PROCOMM Lite, the phrases *send a file*, *upload a file*, and *transmit a file* mean the same thing. They refer to the act of transferring a computer file from your computer to a remote computer. Sometimes new users are confused by the term upload, thinking that it refers to uploading a file to your computer—which would be receiving a file. When used in PROCOMM Lite, however, the phrase *upload a file* means to send a file to another computer. This usage is easiest to understand in the context of sending files to a host computer, such as a bulletin board. When you send a file to the bulletin board, you are uploading the file to the host. Receiving a file from the host is called *downloading* a file.

Before you can send a file by using PROCOMM Lite, you must be connected to the remote computer, and you must be at the Terminal Mode screen. If the remote computer is a PC that is not operating a bulletin board, you must inform the remote computer's operator that you are about to upload a file. For file transfer to occur, the operator must begin the appropriate download procedure soon after you execute your upload procedure.

When you are connected to a host system (an on-line service, such as CompuServe, or a PC bulletin board, for example), you inform the host program that you intend to upload a file by selecting an appropriate menu option or command. (The exact command depends on the host program with which you are communicating.) You usually have to type the name of the file you are going to upload to the host and indicate a file-transfer protocol. (File-transfer protocols are explained in the paragraphs that follow.) When the host instructs you to begin your transfer, you begin the upload procedure.

To upload a file, press PgUp (Send Files). PROCOMM Lite displays the Upload menu, which contains a list of upload protocols available for you to use when you upload a file. A *protocol* is a mutually agreed-on set of rules. A *file-transfer protocol* is an agreed-on set of rules that control the flow of data between the two computers.

When you press PgUp (Send Files) to display the Upload menu, PROCOMM Lite lists the current file-transfer protocol in the bottom line of the menu:

```
(or press ENTER for current_protocol)
```

where *current_protocol* represents the name of the current protocol. XMODEM is the current protocol when you start PROCOMM Lite.

The file-transfer list includes the following protocols:

XMODEM
YMODEM
YMODEM-G
1K-XMODEM
1K-XMODEM-G
ASCII

After you select an upload (file-transfer) protocol, PROCOMM Lite removes the Upload menu from the screen and displays a dialog box with the following prompt:

```
Please enter filename:
```

Type the name of the file you want to transmit to the remote computer and press Enter. Include the full DOS path if the file is not in the current DOS directory. This documentation refers to this file as the transfer file.

After you start the upload procedure and the remote computer begins its download procedure, PROCOMM Lite displays the following message at the bottom of the screen:

```
File transfer in progress... Press ESC to abort
```

PROCOMM Lite also displays a window on the right side of the screen that displays the progress of the upload. The window includes the following information:

- **Protocol.** Indicates the file-transfer protocol PROCOMM Lite is using

- **File Name.** Shows the name of the transfer file

- **File Size.** Lists the size of the transfer file in bytes

- **Block Check.** Refers to the method the file-transfer protocol uses to detect errors in the data

- **Total Blocks.** Tells how much data you must send for a successful transfer of the file

- **Transfer Time.** Indicates an approximate amount of time that the file transfer should take using the chosen upload protocol

- **Transmitted.** Indicates as a percentage the portion of the file that has been transmitted successfully to the remote computer

- **Byte Count.** Denotes the number of bytes of data that have been transmitted

- **Block Count.** Shows the number of blocks of data that have been transmitted successfully. The block count is equal to the byte count divided by the number of bytes in a block.

- **Corrections.** Shows the number of errors in transmission that were corrected. When XMODEM detects an error, the protocol requests that the block be re-sent. If XMODEM detects an error in the same blocks 15 times, the protocol aborts the transfer. Otherwise, if the block arrives without an error, XMODEM acknowledges that it has received the block correctly, and the next block is sent. The occurrence of many errors usually means that the modems are not communicating clearly—probably as a result of poor telephone transmission.

- **CPS.** Shows the average number of characters per second being transmitted.

When the file-transfer protocol detects an error, the protocol sends a message to your computer indicating the type of error. The most recent message generated by the file-transfer protocol is displayed in the Last Message field.

When the upload is finished, you are returned to the Terminal Mode Screen.

Using the PgDn (Receive Files) Command

PROCOMM Lite enables you to send and receive computer files to and from a remote computer. The steps you take to receive a file are similar to, but not exactly the same as, the steps necessary to send a file. This section describes how to receive a file.

In PROCOMM Lite, the phrases *receive a file* and *download a file* mean the same thing. They refer to the act of receiving a computer file sent to your computer by a remote computer. The term *download* is easiest to understand when used in the context of obtaining a file from a host computer, such as a bulletin board. Receiving a file from the host is called downloading a file from the host. Conversely, when you send a file to the bulletin board, you are uploading a file to the host.

Before you can receive a file from a remote computer, you must be at the Terminal Mode screen. If the remote computer is a PC not operating a bulletin board or running PROCOMM Lite Host mode, you must inform the remote computer's operator that you are about to start downloading a file. The operator should begin the upload procedure first, and then you should immediately begin your download procedure. Most protocols are receiver-driven; that is, the sender will not actually begin uploading data until it receives a signal to start from the receiver. The only exception is the ASCII file-transfer protocol. The receiver should begin the download procedure before the sender begins uploading. Otherwise, the first few characters of the file will be lost.

When you are connected to a host system, you usually inform the host program that you intend to download a file by selecting an appropriate menu option or command. You usually have to type the name of the file you are going to download from the host and select a file-transfer protocol. When the host instructs you to begin your transfer procedure, you begin the download procedure.

To download a file, press PgDn (Receive Files). PROCOMM Lite displays the Download menu, which contains a list of download protocols available for you to use when you download a file. PROCOMM Lite also indicates at the top edge of the Download menu the free space available on your working disk.

XMODEM is the current protocol when you start PROCOMM Lite. After you call a remote system by using the dialing directory, the dialing directory entry establishes the current protocol. Otherwise, each time you select a file-transfer protocol during an upload or download operation, the se-lected protocol becomes the current protocol. When you want to use the current file-transfer protocol, press Enter. To use a different protocol, type its letter (listed in the Download menu) and press Enter.

When you select a download (file-transfer) protocol, PROCOMM Lite removes the Download menu from the screen and displays a dialog box prompting you to enter a file name. Type the name of the file you want to download from the remote computer (or any name you want the file to have on your system) and press Enter.

Often you type the file name of the transfer file on-screen just before you press PgDn (Receive Files). For example, when you are downloading a file from a host computer, such as a bulletin board system, the host requests you to enter the file name and file-transfer protocol and then instructs you to begin your download. When you press PgDn (Receive Files), PROCOMM Lite prompts you to enter a file name. After you enter a file name,

PROCOMM Lite searches the download directory for a file with the same name. If a file already exists with the name you specified, PROCOMM Lite displays the following message:

```
File already exists. Overwrite it? (Y/N)
```

Press Y to answer Yes if you want to continue with the download procedure and overwrite the existing file. Press N to answer No if you don't want to replace the existing file with the new file. Unless you chose the ASCII protocol, you are prompted again with the following message:

```
Please enter filename:
```

At this point, you may type a different name for the file to be named on your system so that the downloaded file does not overwrite the existing file. If you decide that you don't want the transfer file, simply press Esc to abort the download procedure. If you chose ASCII and you get the prompt File already exists. Overwrite it? (Y/N), answer No to cause incoming data to be appended to the end of the existing file.

After you start the download procedure and the remote computer begins its upload procedure, PROCOMM Lite displays the following message at the bottom of the screen:

```
File transfer in progress... Press ESC to abort
```

PROCOMM Lite also displays a window at the right side of the screen.

As soon as your computer receives the entire file, your computer beeps, and the word COMPLETED flashes in place of the word PROTOCOL in the top line of the Progress window. After several seconds, PROCOMM Lite removes the Progress window from the menu and returns to the normal Terminal Mode screen.

Using the Alt-F1 (Log File On/Off) Command

PROCOMM Lite provides a Log File feature that enables you to capture part or all of a session directly to disk without using the Redisplay command. To capture to a disk information that scrolls off the Terminal Mode screen, press Alt-F1 (Log File On/Off). PROCOMM Lite opens a dialog box containing the following prompt:

```
Enter log filename, or CR for default:
```

Specify a disk file to receive the captured data by typing a valid DOS file name and pressing Enter. (If you don't specify a DOS directory, PROCOMM Lite places the log file in the current directory.) If you press Enter without typing a file name, PROCOMM Lite uses the default log file name established in the File/Path Options screen of the setup utility. When you install PROCOMM Lite, the default log file name is PCLITE.LOG.

After you press Enter at the log file name prompt, PROCOMM Lite begins capturing to the specified disk all file characters that subsequently are displayed to your Terminal Mode screen. If the log file you are using already contains data, PROCOMM Lite adds the new data to the end of the file. To remind you that PROCOMM Lite is capturing the session to disk, the program displays the message LOG OPEN in the status line.

The Log File feature does not capture information already on-screen when you press Alt-F1 (Log File On/Off).

Using the Log File feature is much like turning on your VCR to record a ball game or movie for viewing later. Even if you miss something the first time through, you have the chance to look at it again as often as you like.

Even while you're using the Log File feature, you can skip information that you don't want to save to the log file. Suppose that you have been reading your electronic mail and saving a copy to disk, but you are about to read a particularly sensitive piece of correspondence that you don't want to save.

To end the Log File feature, press Alt-F1 again. This closes the log file. During Chat mode, you cannot activate the Log File feature. You can turn on the feature in the Terminal Mode screen before entering Chat mode, however, and PROCOMM Lite continues to save incoming and outgoing information to disk while messages are sent back and forth to another computer. You must return to the Terminal Mode screen before you can turn off the Log File feature.

Using the Alt-E (Duplex Toggle) Command

Duplex is one of those computer terms that through continual usage has acquired a meaning that is a bit different from its original definition. When used most accurately, *duplex* refers to the capability of a modem to send and receive information at the same time. A modem that is capable of sending and receiving data simultaneously is referred to as a *full-duplex* modem. *Half-duplex* modems, however, operate like a one-lane bridge. When receiving data, a half-duplex modem cannot send data. When sending data, the modem cannot receive data. The modem continually switches from send mode to receive mode and from receive mode to send mode in order to communicate with another modem—much as a CB radio does.

When your computer is connected to a host computer (such as an on-line service or a bulletin board) through a full-duplex modem, the computer at the other end typically echoes back to your screen any character you type on your keyboard. This echo provides a crude but effective way to confirm that the other computer received the character you typed. Consequently, when the host computer echoes characters to your screen, PROCOMM Lite refers to the mode as *Full-duplex mode*.

Sometimes the computer on the other end does not echo characters to your screen. PROCOMM Lite refers to this mode as *Half-duplex mode*. When your computer is connected to another computer that is using Half-duplex mode, PROCOMM Lite must provide the echo in order for you to see what you are typing.

The status line at the bottom of the Terminal Mode screen shows the current duplex mode setting. The characters FDX in the status line indicate that PROCOMM Lite is in Full-duplex mode. The characters HDX in the status line indicate that PROCOMM Lite is in Half-duplex mode.

Just as other communications settings must match at both ends of the PC communications line, the duplex mode of your computer's modem must match the mode of the computer on the other end. Otherwise, one of two problems occurs: double characters or no characters.

Double characters occur when the computer on the other end is echoing the characters you send back to your screen, but you have set PROCOMM Lite to Half-duplex mode. You may type *Hello*, for example, but see HHeelllloo on-screen. To solve this problem, press Alt-E (Duplex Toggle) to switch PROCOMM Lite to Full-duplex mode.

No characters occur when the other computer is not echoing characters you transmit back to your screen, and you have set PROCOMM Lite to Full-duplex mode. You may type *Hello*, for example, but not see any characters on-screen. Again, press Alt-E (Duplex Toggle) to switch your modem to Half-duplex mode.

Using the Alt-O (Chat Mode) Command

When you carry on a two-way conversation with the operator of another computer, you may have difficulty separating what you are typing from what the person at the other end is typing. If you both type at the same time, the letters all run together on your screen, resulting in unintelligible alphabet soup. To help you solve this problem, PROCOMM Lite provides Chat mode.

To start Chat mode, press Alt-O (Chat Mode) from the Terminal Mode screen. PROCOMM Lite clears the screen and displays the Chat mode screen. The Chat mode screen is split into two sections. The top section is labeled REMOTE, and the bottom section is labeled LOCAL.

The Remote section of the Chat mode screen contains information received from the computer at the other end of the line (the remote computer). You see only the remote user's side of the conversation in this portion of the screen, which can display up to 18 lines of information at a time.

Incoming text starts at the bottom of the Remote section and, as more text comes in, scrolls up the screen. When the section fills and another line is received, the information at the top scrolls off the screen. Your side of the conversation appears in the Local section of the Chat mode screen.

In addition to the convenience of being able to distinguish easily between your input and the remote user's input, you will note two other advantages of using Chat mode rather than typing at the normal Terminal Mode screen.

First, while you are using Chat mode, you do not have to worry about whether PROCOMM Lite is in Full-duplex or Half-duplex mode. The CR Translation setting doesn't matter either. PROCOMM Lite always displays exactly what you type.

Second, if you are in Chat mode, PROCOMM Lite does not send a line of text until you press Enter. Before you press Enter, you can read and edit the line without the remote user seeing it. This feature eliminates the embarrassment of knowing that someone is watching as you make typing errors. You can review each line before sending it to the remote computer. You can change this so that PROCOMM Lite sends each character as you type it on-screen by using the Setup Utility General Options screen.

To edit a line in the Local section of the Chat mode screen (before you press Enter), use the Backspace key to erase characters. Each time you press Backspace, PROCOMM Lite erases the character to the left of the cursor. After you erase the incorrect characters, type the correct information and press Enter to send the line.

To exit Chat mode, press Esc. PROCOMM Lite scrolls all characters off the top of the screen and displays a blank Terminal Mode screen.

You may at first be tempted to use Chat mode all the time, but you quickly will find that few of the keyboard commands available from the Terminal Mode screen have any effect from the Chat mode screen. Only Alt-F6 (Redisplay) and the keyboard macros (Alt-0 to Alt-9) operate in Chat mode. You probably will want to use the Chat mode screen when you are conversing directly with someone who is typing messages to you on the other end of the line. Several on-line services, for example, offer a capability to chat with other users who are connected to the system. These on-line service features are sometimes called CB simulators, referring to the types of communications popular on CB radios.

Working with Macro Keys (Alt-M)

Many aspects of PC communications become routine; you type certain words or phrases frequently. For example, you log on to an on-line computer service by typing your identification number and then a password; each time you log on to a bulletin board, you type your name and a password. Typing these items is not hard work, but it does take time and holds a potential for typing errors. If your ID number is 2974,ARQ, for example, you may type 2947,AQR by mistake. To help you save time and keystrokes and to reduce typographical errors, PROCOMM Lite enables you to assign any string of up to 50 characters (including spaces) to an Alt-key combination referred to as a keyboard macro key.

A *macro key* defines a string of characters that you type in, such as a log-on code, a command to a remote computer, and so on. By using macro keys, for example, you can condense a text entry that you would have to type. Then, at the time when you are prompted to enter information, you simply type the macro key associated with the prompt and PROCOMM Lite types the response to the prompt on-screen for you.

Creating Macro Keys

The first step to creating a macro key is to display the Macro Keys screen. Press Alt-M (Macro Keys) from the Terminal Mode screen, and PROCOMM Lite displays the screen shown in figure E.8. Using this screen, you can create up to 10 macro keys. The left side of the screen contains a list of 10 Alt-key combinations from Alt-1 to Alt-0, the macro key names. Next to the key name is a blank entry area. You create a macro key command by typing characters in this macro key entry area.

To add a macro key command to the Macro Keys screen, use the up- or down-arrow key to move the highlighted bar up or down in the list. Position the highlight on a blank macro key entry and press R (Revise). The cursor then jumps to the contents field. In the macro key (contents) entry area, type up to 50 characters to comprise your macro key. You can use the cursor-movement and editing keys listed in table E.4 while typing characters in the macro key entry area.

After you enter the contents of the macro key, press Enter. You now can use the up- and down-arrow keys to go to another key to revise. To save the macro keys you entered, press S. To exit and return to the Terminal Mode screen, press Esc.

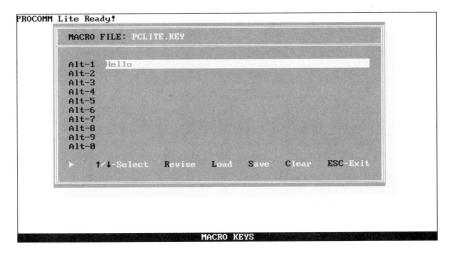

Fig. E.8. *The PROCOMM Lite Macro Keys Screen.*

Table E.4
Cursor-Movement and Editing Keys on the Macro Keys Screen

Key	Function
←	Moves cursor one space to the left
→	Moves cursor one space to the right
Home	Moves cursor to left end of macro key entry area
End	Moves cursor one space to right of last character in macro key entry area
Insert	Toggles Insert/Overtype modes
Delete	Deletes character at cursor
Backspace	Deletes character to left of cursor
Tab	Deletes entire macro key entry
Ctrl-End	Deletes characters from cursor to right end
Ctrl-Backspace	Deletes all characters in contents area

When you later execute a macro key, letters and numbers translate into their corresponding keystrokes. Suppose that you type *BR549* in the contents entry area on the Macro Keys screen. Later, when you execute the macro key, PROCOMM Lite types the same characters: BR549.

A number of special macro key codes are referred to as control codes because the first character in each code, the caret (^), represents the Ctrl key on the keyboard. When you execute a macro key that contains the characters ^ C, for example, PROCOMM Lite does not type the caret (^) and then the letter C. Instead, PROCOMM Lite transmits to the remote computer the same code that is sent if you press Ctrl-C, a keystroke combination that may have a special meaning to a particular on-line service or bulletin board. Many on-line services and bulletin boards use Ctrl-key combinations, such as Ctrl-C, Ctrl-X, Ctrl-S, and Ctrl-Q, to enable you to control the flow of information across your screen or to cancel an operation in midstream.

In general, you can create any Ctrl-key combination by typing in the macro key entry area the caret (^) followed by a letter (A through Z, upper- or lowercase) or one of the following characters:

[ ] \ ^ _

Several of the macro key control codes translate into keystrokes that you normally would not expect and that do not involve the Ctrl key. Suppose that you type the characters $^\wedge M$ in a macro key entry. When you execute the macro key, PROCOMM Lite sends a carriage return to the remote computer for the $^\wedge$ M. In other words, when you want the macro key to "press" Enter, use the code $^\wedge$ M. See table E.5 for other macro key control codes.

Table E.5
Special Macro Key Control Codes

Code	Executed keystroke
$^\wedge$ M	Enter
$^\wedge$ H	Backspace
$^\wedge$ I	Horizontal tab (Tab key)
$^\wedge$ J	Line feed
$^\wedge$ K	Vertical tab
$^\wedge$ [	Esc

You may want to create a macro key to enter your CompuServe ID number, for example. Suppose that your ID is 12345,6789. To create the ID macro key, press Alt-M (Macro Keys) to display the Macro Keys screen, move the highlighted bar to an empty entry, and press R (Revise). Type *12345,6789* $^\wedge$ *M* to the right of Alt-1 and press Enter. The macro key ends with $^\wedge$ M so that the macro key sends the carriage return symbol. Press Enter after you enter the contents of the macro key, and then press S to save the macro key file. Press Esc to save the key and return to the Terminal Mode screen. Now, whenever you want to type your CompuServe ID number, you simply press Alt-1, and it appears on the Terminal screen.

The pause character has a special meaning in a PROCOMM Lite macro key entry. The pause character is one of the PROCOMM Lite settings you can change on the Setup Utility General Options screen. The default setting for the pause character is the tilde ($\sim$). When you use the pause character in a macro key entry, PROCOMM Lite translates the character into a half-second pause during execution of the macro key.

Don't confuse the pause character ($\sim$) with the comma (,) that is used to pause dialing by a Hayes-compatible modem.

Saving Macro Keys to a Macro Key File

After you type a macro key entry and press Enter, PROCOMM Lite stores the macro key in memory (RAM). The macro key is not automatically saved permanently to a disk file. Unless you take steps to save the macro key to disk, you can use the macro key for the current PROCOMM Lite session only.

To save the current macro key entry or entries as they appear on the Macro Keys screen, press S (Save). PROCOMM Lite saves the entries to the file indicated at the top of the Macro Keys screen. The default macro file name for PROCOMM Lite is PCLITE.KEY. Each time you start the program, PROCOMM Lite assumes that your macro keys are stored in the file PCLITE.KEY.

To create and save the macro keys under a different file name, first press L (Load) at the Macro Keys screen. PROCOMM Lite prompts you to enter a name for the macro key file. If you enter a new name not currently on disk, a new, blank Macro Key screen appears. When you place information on this screen and press S (Save), PROCOMM Lite saves a new macro file to disk under the new name. This structure enables you to create more than 10 macro keys. No more than 10 macro keys fit on one Macro Keys screen, but by saving macro keys in separate files, you can create an unlimited number of macro keys.

You may want to use one group of 10 macro keys only when connected to the CompuServe on-line service, for example. Perhaps each of these macro keys takes you directly to a different service of CompuServe. Because these macro keys are of no use for any other purpose, they need to be accessible only while you are using CompuServe. To create this set of CompuServe macro keys, first press Alt-M to display the Macro Keys screen. Then press L (Load) to load a new screen, name the file CSERVE, and add the 10 macro keys, replacing any macro keys that are there. Press S (Save) to save your CompuServe macro keys to the new disk file CSERVE.KEY. Your CompuServe macro keys are still current—available for use when you connect to CompuServe. The next time you start the program, however, PROCOMM Lite again reads the set of macro keys found in the file PCLITE.KEY.

Modifying Macro Keys

The procedure for modifying a macro key you have created is almost the same as the procedure for creating a macro key. Press Alt-M (Macro Keys) to display the Macro Keys screen. Use the cursor-movement keys listed in table E.4 to move the highlighted bar to the macro key entry you want to modify. Press R (Revise), and PROCOMM Lite moves the blinking cursor to the left end of the highlighted macro key entry area. You then can use any of the editing keys listed in table E.4 to modify the macro key entry. After you make the needed changes, press Enter. PROCOMM Lite saves the change to memory. Don't forget to save the modified version of the macro key entry to the disk file.

Loading a Macro Key File

Each time you start PROCOMM Lite, it reads only the macro key file named PCLITE.KEY. When you want to use a different set of macro keys, you must load into memory the macro key file that contains the set of macro keys you need. You can use only those macro key commands that are associated with the current macro file in memory, and you can have only one macro file in memory at one time.

To load a macro key file, press Alt-M (Macro Keys) to display the Macro Keys screen; then press L (Load). PROCOMM Lite displays a small window that contains the prompt `File to load`. Type the name of the macro key file you want PROCOMM Lite to load—but don't type the extension—and press Enter. If you don't remember the name of the file you want to load, press Enter at the `File to load` prompt, and PROCOMM Lite displays a list of the files currently on disk. Use the up- and down-arrow keys or the mouse to highlight the macro file you want to load. Then press Enter or click the mouse button. PROCOMM Lite loads the file's set of macro keys and immediately displays the keys in the Macro Keys screen.

When you finish using the auxiliary set of macro keys, you may want to return to the set contained in the file PCLITE.KEY. To access the PCLITE.KEY file, press Alt-M (Macro keys) to display the Macro Keys screen, press L (Load), type *pclite*, and press Enter. PROCOMM Lite reloads the macro keys from the disk file PCLITE.KEY and displays the keys in the Macro Keys screen.

Executing Macro Keys

After you create a macro key, using it is simple. To execute a macro key, press the Alt-key combination listed to the left of the macro key on the Macro Keys screen—the key combination you assigned when you typed the macro key at that position.

Using the PROCOMM Lite Script Language

This last section of the appendix describes the programmable capabilities of the PROCOMM Lite script language.

Understanding PROCOMM Lite Scripts

PROCOMM Lite script language programs are referred to as *scripts*. A PROCOMM Lite script file consists of a series of lines of ASCII characters, with each line containing one command from the PROCOMM Lite script language. PROCOMM Lite essentially executes commands one-by-one from top to bottom. Any text or other characters that appear to the right of a semicolon (;)—except when within a quoted string—are ignored when PROCOMM Lite executes the script. Use this text to incorporate comments (often called *internal documentation*) into the script for future reference. The DOS file name of each PROCOMM Lite script must end in the extension ASP.

Use PROCOMM Lite scripts to control and automate the operation of PROCOMM Lite. The scripts further simplify the use of this already easy-to-use communications program. You can create several scripts executed through the dialing directory, for example, each of which logs on to a different computer system.

Creating a PROCOMM Lite Script

You can use any text editor that can create ASCII files to write a script or to modify an existing script. The file-name extension of each script file must be ASP. You cannot include more than one PROCOMM Lite script language command on one line.

You can type PROCOMM Lite script language commands in upper- or lowercase letters, but the commands must be spelled out completely.

Running a Script

PROCOMM Lite provides several options for executing, or running, a script. You can run a PROCOMM Lite script from the DOS prompt at the PROCOMM Lite start-up, from the Dialing Directory screen, or from the Terminal Mode screen. After PROCOMM Lite begins to execute the script, the program displays the name of the script in the left section of the Terminal Mode screen status line.

Running a Script from the DOS Prompt

PROCOMM Lite provides a way to run a script from DOS. This method causes a script to run immediately when you start PROCOMM Lite. To run a PROCOMM Lite script from DOS, add /f and the script name to the PROCOMM Lite start-up command. Change to the DOS directory that contains the script you want to run and type the following command:

> pclite /fscriptname

In this command, replace scriptname with the file name of the script. You do not have to type the ASP file-name extension. Do not leave a space after /f. When you press Enter, PROCOMM Lite starts and immediately plays the script without stopping at the logo screen.

To run from DOS a script that is named MYFILE.ASP, for example, type the following command and press Enter:

> pclite /fmyfile

PROCOMM Lite loads and the script is executed.

To start PROCOMM Lite from DOS, the script must be in the same DOS directory as the PROCOMM Lite program files.

Running a Script from the Dialing Directory

If you create a script and want PROCOMM Lite to run it after the program connects to a particular computer, type the name of the script in the SCRIPT line of the Dialing Directory's Revise Entry window. PROCOMM Lite runs the script every time you connect to the remote computer.

Running a Script from Terminal Mode

The third way to execute a PROCOMM Lite script is available from the Terminal Mode screen. When you press Alt-F5 (Script Files), PROCOMM Lite displays a dialog box with the following message:

```
SCRIPT SELECTION (Enter for .ASP file list)
```

The box also displays the message:

```
Please enter filename:
```

After pressing Alt-F5, you have two alternatives for indicating which script you want to run:

- Type the name of the script file and press Enter. PROCOMM Lite assumes that the script has the file-name extension ASP; you do not need to type it. PROCOMM Lite executes the script.

- If you are not sure how to spell the name of the script file, press Enter to see a list of all the script files in the working directory (all files with the ASP file name extension). PROCOMM Lite displays up to 15 script file names in a tall, thin window. While this second window is displayed, you can use the up- and down-arrow, PgUp, and PgDn keys to highlight any PROCOMM Lite script name. Highlight the name of the script that you want to run and press Enter. PROCOMM Lite runs the script.

You can run your script named MYFILE, for example, by pressing Alt-F5 (Script Files), typing *myfile*, and pressing Enter.

Terminating the Execution of a Script

From time to time, you may decide that you don't want to run a script after all. To terminate a script in midstream, press Esc. PROCOMM Lite displays a dialog box with the following prompt:

```
EXIT SCRIPT (Y/N)
```

If you answer No, the script resumes; if you answer Yes, the script ends.

Understanding the PROCOMM Lite Script Language Commands

PROCOMM Lite provides seven script commands:

- **ALARM.** Sounds alarm for the time limit specified in the setup.

```
alarm   ; inform user logon is complete
```

This command causes the alarm to sound for the number of seconds specified in the Display/Sound Options screen of the PROCOMM Lite Setup Utility (the default is 5 seconds).

- **CLEAR.** Clears the screen.

- **KEYGET.** Waits for the user to press a key.

- **MESSAGE.** Displays a string on a local screen. The string must be enclosed in quotation marks.

- **PAUSE.** Halts execution for a specified number of seconds.

```
pause 3  ; wait three seconds
```

This command causes the program to wait three seconds after the call is made before executing the next script command.

- **TRANSMIT.** Sends a specified string to a remote computer. The string must be enclosed in quotation marks.

- **WAITFOR.** Halts execution until a specified string is received or 30 seconds have elapsed. An example of this is found in the CSERVE.ASP file.

```
waitfor "User ID:"    ;wait for "User ID:" to come in

transmit "12345,1234" ;send your ID
```

This command instructs the program to wait until the User ID: prompt is sent from the remote computer before transmitting the ID. This combination of commands is important in automating procedures that depend on prompts from other computers before information can be transmitted to the remote computer.

The following sample script uses five of these commands. This script demonstrates how you can use a script to log on to an on-line information service—CompuServe in this example. The script also goes directly to the DATASTORM tech support forum on CompuServe.

```
;*****************************************************
;*                                                   *
;* A sample script file for logging onto CompuServe  *
;*                                                   *
;*****************************************************
;
CLEAR                               ; Clear the screen
MESSAGE "Logging onto COMPUSERVE..."  ; display message
TRANSMIT "^C"                        ; Transmit CTRL-C
WAITFOR "User ID:"                   ; Wait for "User ID:"
TRANSMIT "12345,1234"                ; Send your ID
TRANSMIT "^M"                        ; Send carriage return
WAITFOR "Password:"                  ; Wait for "Password:"
TRANSMIT "my password"               ; Send your password
TRANSMIT "^M"                        ; Send carriage return
WAITFOR "!"                          ; Wait for prompt
TRANSMIT "g dstorm^M"                ; Go to DATASTORM support
ALARM                                ; Sound alarm
MESSAGE "Log on complete!"           ; Display message locally
```

To create this script, use a text editor that produces ASCII text files. Insert your CompuServe user ID in place of 12345,1234 and insert your password in place of *my password*. Give the script a file name that has the extension ASP. To run the script, add the script's file name to the SCRIPT line of the Dialing Directory entry that you use to call CompuServe.

Summary

This brief documentation for PROCOMM Lite has introduced you to the commands available in the program. You have learned how to use PROCOMM Lite's menus, dialing directory, setup facility, macros, file-transfer capability, and scripts. With PROCOMM Lite and this book, you are well on your way to mastering the intriguing world of PC communications.

Index

C

N

Q-R

S

W

X-Z

Enhance Your Personal Computer System
With Hardware And Networking Titles From Que!

Upgrading and Repairing PCs

Scott Mueller

This book is the ultimate resource for personal computer upgrade, maintenance, and troubleshooting information! It provides solutions to common PC problems and purchasing descisions and includes a glossary of terms, ASCII code charts, and expert recommendations.

IBM Computers & Compatibles

$29.95 USA

0-88022-395-2, 724 pp., 7 3/8 x 9 1/4

Hard Disk Quick Reference

Que Development Group

Through DOS 4.01

$8.95 USA

0-88022-443-6, 160 pp., 4 3/4 x 8

Introduction To Personal Computers, 2nd Edition

Katherine Murray

IBM, Macintosh, & Apple

$19.95 USA

0-88022-758-3, 400 pp., 7 3/8 Xx9 1/4

Networking Personal Computers, 3rd Edition

Michael Durr & Mark Gibbs

IBM & Macintosh

$24.95 USA

0-88022-417-7, 400 pp., 7 3/8 x 9 1/4

Que's Computer Buyer's Guide, 1992 Edition

Que Development Group

IBM & Macintosh

$14.95 USA

0-88022-759-1, 250 pp., 8 x 10

Que's Guide to Data Recovery

Scott Mueller

IBM & Compatibles

$29.95 USA

0-88022-541-6, 500 pp., 7 3/8 x 9 1/4

Que's PS/1 Book

Katherine Murray

Covers Microsoft Works & Prodigy

$22.95 USA

0-88022-690-0, 450 pp., 7 3/8 x 9 1/4

Using Novell NetWare

Bill Lawrence

Version 3.1

$29.95 USA

0-88022-466-5, 728 pp., 7 3/8 x 9 1/4

Using Your Hard Disk

Robert Ainsbury

DOS 3.X & DOS 4

$29.95 USA

0-88022-583-1, 656 pp., 7 3/8 x 9 1/4

To Order, Call:
(800) 428-5331 OR (317) 573-2500

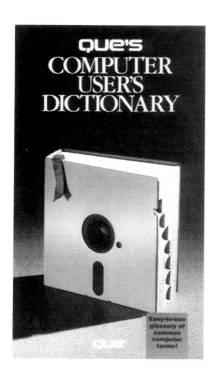

Find It Fast With Que's Quick References!

Que's Quick References are the compact, easy-to-use guides to essential application information. Written for all users, Quick References include vital command information under easy-to-find alphabetical listings. Quick References are a must for anyone who needs command information fast!

1-2-3 for DOS Release 2.3 Quick Reference
Release 2.3
$9.95 USA
0-88022-725-7, 160 pp., 4 3/4 x 8

1-2-3 Release 3.1 Quick Reference
Releases 3 & 3.1
$8.95 USA
0-88022-656-0, 160 pp., 4 3/4 x 8

Allways Quick Reference
Version 1.0
$8.95 USA
0-88022-605-6, 160 pp., 4 3/4 x 8

AutoCAD Quick Reference, 2nd Edition
Releases 10 & 11
$8.95 USA
0-88022-622-6, 160 pp., 4 3/4 x 8

Batch File and Macros Quick Reference
Through DOS 5
$9.95 USA
0-88022-699-4, 160 pp., 4 3/4 x 8

CorelDRAW! Quick Reference
Through Version 2
$8.95 USA
0-88022-597-1, 160 pp., 4 3/4 x 8

dBASE IV Quick Reference
Version 1
$8.95 USA
0-88022-371-5, 160 pp., 4 3/4 x 8

Excel for Windows Quick Reference
Excel 3 for Windows
$9.95 USA
0-88022-722-2, 160 pp., 4 3/4 x 8

Fastback Quick Reference
Version 2.1
$8.95 USA
0-88022-650-1, 160 pp., 4 3/4 x 8

Hard Disk Quick Reference
Through DOS 4.01
$8.95 USA
0-88022-443-6, 160 pp., 4 3/4 x 8

Harvard Graphics Quick Reference
Version 2.3
$8.95 USA
0-88022-538-6, 160 pp., 4 3/4 x 8

Laplink Quick Reference
Laplink III
$9.95 USA
0-88022-702-8, 160 pp., 4 3/4 x 8

Microsoft Word Quick Reference
Through Version 5.5
$9.95 USA
0-88022-720-6, 160 pp., 4 3/4 x 8

Microsoft Works Quick Reference
Through IBM Version 2.0
$9.95 USA
0-88022-694-3, 160 pp., 4 3/4 x 8

MS-DOS 5 Quick Reference
Version 5
$9.95 USA
0-88022-646-3, 160 pp., 4 3/4 x 8

MS-DOS Quick Reference
Through Version 3.3
$8.95 USA
0-88022-369-3, 160 pp., 4 3/4 x 8

Norton Utilities Quick Reference
Norton Utilities 5 & Norton Commander 3
$8.95 USA
0-88022-508-4, 160 pp., 4 3/4 x 8

PC Tools 7 Quick Reference
Through Version 7
$9.95 USA
0-88022-829-6, 160 pp., 4 3/4 x 8

Q&A 4 Quick Reference
Versions 2, 3, & 4
$9.95 USA
0-88022-828-8, 160 pp., 4 3/4 x 8

Quattro Pro Quick Reference
Through Version 3
$8.95 USA
0-88022-692-7, 160 pp., 4 3/4 x 8

Quicken Quick Reference
IBM Through Version 4
$8.95 USA
0-88022-598-X, 160 pp., 4 3/4 x 8

UNIX Programmer's Quick Reference
AT&T System V, Release 3
$8.95 USA
0-88022-535-1, 160 pp., 4 3/4 x 8

UNIX Shell Commands Quick Reference
AT&T System V, Releases 3 & 4
$8.95 USA
0-88022-572-6, 160 pp., 4 3/4 x 8

Windows 3 Quick Reference
Version 3
$8.95 USA
0-88022-631-5, 160 pp., 4 3/4 x 8

WordPerfect 5.1 Quick Reference
WordPerfect 5.1
$8.95 USA
0-88022-576-9, 160 pp., 4 3/4 x 8

WordPerfect Quick Reference
WordPerfect 5
$8.95 USA
0-88022-370-7, 160 pp., 4 3/4 x 8

To Order, Call:

If your computer uses
3 1/2-inch disks . . .

While many personal computers use 5 1/4-inch disks to store information, newer computers are switching to 3 1/2-inch disks for information storage. If your computer uses 3 1/2-inch disks, you can return this form to Que to obtain a 3 1/2-inch disk to use with this book. Simply fill out the remainder of this form, and mail to:

Book Disk Exchange
Que Corporation
11711 N. College
Carmel, IN 46032

We will send you, free of charge, the 3 1/2-inch version of the software.

Book Title _____

Name _____   Phone _____

Company _____   Title _____

Address _____

City _____   St _____ ZIP _____